T0122592

Springer Theses

Recognizing Outstanding Ph.D. Research

Aims and Scope

The series "Springer Theses" brings together a selection of the very best Ph.D. theses from around the world and across the physical sciences. Nominated and endorsed by two recognized specialists, each published volume has been selected for its scientific excellence and the high impact of its contents for the pertinent field of research. For greater accessibility to non-specialists, the published versions include an extended introduction, as well as a foreword by the student's supervisor explaining the special relevance of the work for the field. As a whole, the series will provide a valuable resource both for newcomers to the research fields described, and for other scientists seeking detailed background information on special questions. Finally, it provides an accredited documentation of the valuable contributions made by today's younger generation of scientists.

Theses are accepted into the series by invited nomination only and must fulfill all of the following criteria

- They must be written in good English.
- The topic should fall within the confines of Chemistry, Physics, Earth Sciences, Engineering and related interdisciplinary fields such as Materials, Nanoscience, Chemical Engineering, Complex Systems and Biophysics.
- The work reported in the thesis must represent a significant scientific advance.
- If the thesis includes previously published material, permission to reproduce this must be gained from the respective copyright holder.
- They must have been examined and passed during the 12 months prior to nomination.
- Each thesis should include a foreword by the supervisor outlining the significance of its content.
- The theses should have a clearly defined structure including an introduction accessible to scientists not expert in that particular field.

More information about this series at http://www.springer.com/series/8790

Gonzalo Manzano Paule

Thermodynamics and Synchronization in Open Quantum Systems

Doctoral Thesis accepted by
the Complutense University of Madrid, Madrid, Spain

Author
Dr. Gonzalo Manzano Paule
Scuola Normale Superiore
Pisa, Italy

and

International Center for Theoretical Physics
Trieste, Italy

Supervisors
Prof. Juan M. R. Parrondo
Universidad Complutense de Madrid
Madrid, Spain

Prof. Roberta Zambrini
Universitat de les Illes Balears—CSIC
Palma de Mallorca, Spain

ISSN 2190-5053 ISSN 2190-5061 (electronic)
Springer Theses
ISBN 978-3-030-06757-1 ISBN 978-3-319-93964-3 (eBook)
https://doi.org/10.1007/978-3-319-93964-3

Printed on acid-free paper

This Springer imprint is published by the registered company Springer International Publishing AG part of Springer Nature
The registered company address is: Gewerbestrasse 11, 6330 Cham, Switzerland

Supervisors' Foreword

Even a century after its birth, we are still challenged by fundamental questions posed by quantum mechanics and learning how to use its striking features for practical applications. The field of quantum information explores the computational power of quantum systems, which can greatly outperform classical computers; entanglement can be used to secure communications; the high sensitivity of quantum superpositions to external parameters is being used to design new sensors. Quantum computation, quantum communication, quantum sensing, and metrology are examples of the new technologies that make use of genuine quantum effects, like coherence and entanglement.

In recent years, a new field has been added to these attempts to exploiting quantum effects: *quantum thermodynamics*. Furthermore, the progress toward systems composed of several elementary units with increasing complexity requires the study of emergent phenomena, such as *quantum synchronization*. These timely topics are treated in the thesis of Gonzalo Manzano within the common theoretical framework of open quantum systems. Manzano's original contributions lead to a better understanding of a number of basic phenomena: synchronization, decoherence, thermalization, and irreversibility in open systems, whose control is crucial to implement the aforementioned quantum technologies. The thesis also includes the design and analysis of quantum thermal machines that can achieve higher efficiencies than classical engines and refrigerators.

As known, isolated systems are a useful idealization to start with, but only a more realistic approach considering interactions with the environment can explain most physical phenomena, taking into account the effects of exchanges of energy, matter, and information. In the classical regime, the interaction with the environment is necessary for basic tasks that require friction, such as walking, as well as to relax toward equilibrium, like when ice melts in our drink, and plays a prominent role in emergent phenomena and dissipative structures, ranging from synchronization of heart pacemaker cells to cyclones. Therefore, the study of open systems is a well-established topic intersecting with most research fields including, but not limited to, thermodynamics.

Still, it is in the quantum regime where the study of open systems becomes even more relevant and powerful, since open quantum systems play a fundamental role in establishing the conditions for the emergence of a classical world out of fundamental quantum laws valid in the microscopic regime. Superposition of states is at the heart of quantum physics, and the process of decoherence provides a mechanism for the transition to classical mixed states, with prominent experimental demonstrations in atomic systems known from the 1990s.

This thesis explores the behavior of open quantum systems in a variety of contexts. Its first part is a very valuable, self-contained, and exhaustive introduction (Part I) that covers the main aspects of (a) basic quantum theory, including several quantities to assess the correlation between quantum systems, (b) the theory of quantum open systems, including the formalism of quantum maps and stochastic trajectories, and (c) the more recent framework of stochastic thermodynamics and quantum thermodynamics.

The rest of the Gonzalo Manzano's thesis is devoted to his original contributions to these fields. Part II focuses on synchronization of quantum oscillators. Among complex phenomena, mutual synchronization is a paradigmatic one, reported in physical, biological, chemical, and social contexts, allowing for the adjustments of the rhythms of different systems. A natural question addressed in recent years is about the persistence of this phenomenon in the quantum regime, as well as its connection with quantum correlations. The thesis contains the first study in which mutual synchronization is actually found to witness the presence of quantum discord and entanglement, in a fundamental model of coupled oscillators (Chap. 4). Furthermore, it addresses the intriguing possibility that synchronization not only persists in the presence of quantum noise, but is also induced by dissipation into the environment. Quantum synchronization is discussed in bosonic models allowing for more complex forms of interaction with the environment, acting not only independently and identically on different system components, but also collectively or locally (Chaps. 4 and 5). Spontaneous synchronization can arise either during a pre-thermalization transient or also in the stationary state, when more than two detuned oscillators are considered. The last case is studied in connection with persistent decoherence-free subspaces (Chap. 5).

Can mesoscopic systems like quantum complex networks synchronize? In Chap. 6, it is shown that bosonic networks not only can display mutual synchronization induced by dissipation, but also have the possibility to be tuned locally (only at one node) to make the whole network synchronous, and therefore also strongly quantum correlated, or to select synchronous clusters. Furthermore, the conditions to synchronize and entangle two nodes through a network are also established. These results provide a comprehensive description of quantum synchronization in the framework of bosonic networks.

Part III is devoted to quantum fluctuation theorems (QFTs) that characterize the fluctuations of work, heat, and other quantities related to entropy production, along arbitrary nonequilibrium processes. Chapter 7 introduces a QFT for quantum maps and operations, a generic formalism that describes the evolution of open systems. The novelty of this QFT is that it is independent of the details of the environment and can be applied when the physical mechanism behind a phenomenon is not fully understood, such as decoherence and the collapse of the wave function. Another general QFT, in this case for bipartite systems, is derived in Chap. 8. This theorem helps to clarify some aspects of the previous QFT for maps and also allows one to split the entropy production along nonequilibrium processes into two terms that obey respective "second laws": the adiabatic entropy production, which quantifies the entropy production due to nonequilibrium constraints, such as temperature gradients, and the nonadiabatic entropy production, which accounts for the local irreversible relaxation due to driving. All those results are illustrated in a number of relevant physical examples.

The last part of the thesis (Part IV) focuses on quantum thermal machines, that is, quantum systems that are in contact with thermal baths at different temperatures and are able to perform different tasks, like converting heat into work (motors) or pumping heat from cold to hot reservoirs (refrigerators). Over the last decade, a vast literature has developed on these machines, which is reviewed in the introduction, whereas Part IV of the thesis contains the original contributions to the field. Chapter 11 analyzes how the efficiency of these machines depends on the dimension of the Hilbert space. Another interesting aspect is the consideration of nonequilibrium reservoirs. In quantum mechanics, one can modify the state of an equilibrium thermal bath, by squeezing or adding coherences, to obtain a nonequilibrium reservoir. Then, we can imagine a thermal machine working with those reservoirs. In fact, there are already experimental realizations of thermal motors working with squeezed baths. Chapter 10 analyzes in detail several thermodynamic cycles between squeezed thermal baths using QFTs and shows that one can have motors and refrigerators that greatly outperform cycles with equilibrium reservoirs.

Summarizing, here you will find a number of relevant and original contributions that help to better understand the collective and thermodynamic properties of open quantum systems. Gonzalo Manzano has also included an exhaustive and self-contained introduction that makes the thesis an excellent resource for learning more about all these new developments, which are crucial for understanding both the fundamental aspects of quantum mechanics and the possibilities and limitations of quantum technologies.

Madrid, Spain
Palma de Mallorca, Spain
April 2018

Prof. Juan M. R. Parrondo
Prof. Roberta Zambrini

Abstract

Dissipation effects have profound consequences in the behavior and properties of quantum systems. The unavoidable interaction with the surrounding environment, with whom systems continuously exchange information, energy, angular momentum, or matter, is ultimately responsible for decoherence phenomena and the emergence of classical behavior. However, there exists a wide intermediate regime in which the interplay between dissipative and quantum effects gives rise to a plethora of rich and striking phenomena that has only started to be understood. In addition, the recent breakthrough techniques in controlling and manipulating quantum systems in the laboratory have made this phenomenology accessible in experiments and potentially applicable. In this thesis, we aim to explore from a theoretical point of view some of the connections between dissipative and quantum effects regarding two main aspects: the thermodynamical behavior of quantum systems and the relation between dynamical and quantum correlations shared between them.

Quantum correlations are one of the most surprising characteristics of nature, attracting a long-standing interest from the formulation of quantum theory. The understanding of the mechanisms creating, preserving, or destroying quantum correlations becomes of major importance when exploring the quantum-to-classical boundary, while being essential to designing schemes in which decoherence phenomena can be avoided in practical applications. An important type of dynamical correlations with a more classical flavor is synchronization phenomena, which have been studied in a broad range of physical, chemical, and biological systems. Synchronization may arise as a spontaneous cooperative behavior of different oscillatory units that, when coupled, adapt their rhythms to a common frequency. This *mutual synchronization* phenomenon has been recently considered in the quantum regime, mostly from a classical point of view, while genuine quantum traits of synchronization are now starting to be investigated.

A first main objective of this thesis is to determine the possible connections between mutual synchronization and quantum correlations, as measured by entanglement or quantum discord. In order to investigate this connection, we use the machinery of open quantum systems theory. More precisely, we consider many-body systems consisting of interacting quantum harmonic oscillators coupled to the environment. The environment will be modelled in two main different ways, which will be compared. In the first case, all the units in the many-body system feel the same dissipation modelled as a common heat bath. In the second case, each unit is assumed to feel an independent dissipation modelled by separate thermal baths. We start with the simplest case of two quantum harmonic oscillators in Chap. 4 which allows us to identify the basic mechanisms leading to transient synchronization and its relation with the slow decay of quantum correlations. We find that both phenomena are produced due to the presence of collective dissipation. We then consider the case of three oscillators in Chap. 5, in which a richer phenomenology appears while still allowing an analytical treatment in several cases of interest. Finally, we scale the system up to complex harmonic networks in Chap. 6, where a broader class of local/global dissipation can be addressed, and our previous findings let us engineer the normal modes of the network. We can then obtain synchronization and protection of quantum correlations in the whole network or in a selected cluster, by simply tuning one or few parameters, such as one frequency or certain coupling strengths. The importance of the results presented in this part of the thesis relies on the fact that they show for the first time that synchronization is related to genuine quantum features and that it may emerge, even in linear systems, due to the presence of dissipation.

The remaining parts of the thesis are dedicated to explore the thermodynamic features of open quantum systems. In particular, we explore the quantum versions of *fluctuation theorems*. These theorems are universal relations which introduce constraints in the statistics followed by quantities such as work, or entropy, defined as stochastic fluctuating variables in processes occurring arbitrarily far from equilibrium. They can be understood as a refined version of the second law of thermodynamics for small systems dominated by fluctuations where the laws of thermodynamics are only expected to be fulfilled on average.

Work fluctuation theorems have been extensively investigated in the quantum regime under an inclusive Hamiltonian approach. Also, fluctuation theorems for the exchange of heat and particles in transient and steady-state regimes have been established, as well as entropy production fluctuation theorems. Other approaches considered specific open-system dynamics, including unital measurements, quantum trajectories, or Lindblad master equations. However, the different attempts to generalize those results to general completely positive and trace-preserving (CPTP) maps are limited by the presence of an efficacy (correction) term. Furthermore, the characterization of entropy production in situations going beyond the assumption of ideal equilibrium reservoirs constitutes an open challenge.

The second main objective we pursue in this thesis is the development of fluctuation theorems valid for quantum CPTP maps, together with the interpretation of the quantities fulfilling them. This theoretical development may then be applied to gain insight into the characterization of entropy production in general quantum evolution and the thermodynamic description of specific configurations. We define thermodynamic protocols generating trajectories by means of quantum measurements and the occurrence of the quantum operations which compose the CPTP maps. The probabilities of such trajectories then must be compared with those of their time-reversed twins, defined in a suitable way. The application to specific situations will require as well an adequate modelling of the dynamical evolution. In Chap. 7, we develop a general fluctuation theorem for a large class of quantum CPTP maps. The theorem is based on the properties of the invariant states of the dynamics. We discuss the meaning of the quantity fulfilling the theorem in many situations of physical interest as different versions of the entropy production. This interpretation is then clarified in Chap. 8, where we characterize entropy production from first principles and explore the conditions under which it splits into adiabatic and nonadiabatic contributions, each of them fulfilling an independent fluctuation theorem. In Chap. 9, we illustrate our findings with some particular models of interest in quantum thermodynamics and discuss their implications.

Thermodynamic theory was developed from the analysis of real heat engines, such as the steam engine along the nineteenth century. Those macroscopic engines have quantum analogues, whose analysis constitutes an important branch of quantum thermodynamics. A quantum thermal machine is intended as a small quantum system operating between different thermal reservoirs (or more general reservoirs) and possibly subjected to external driving. The machine performs a thermodynamic task such as work extraction, refrigeration, heat pumping, or information erasure. Quantum thermal machines provide simple setups in which quantum thermodynamics can be studied at the fundamental level, but also tested experimentally.

Clarifying the impact of quantumness in the operation and properties of the machines represents a major challenge. Quantum effects may be incorporated, e.g., by means of nonequilibrium reservoirs. There have been different works in the literature pointing that nonequilibrium quantum reservoirs may be used to increase both power and efficiency. Nevertheless, a solid understanding of this enhancement and their optimization has remained elusive, as it requires a precise formulation of the second law of thermodynamics in such nonequilibrium situations. Furthermore, the sole fact that energy levels are discretized may also introduce limitations when trying to improve the performance of machines by means of increasing the number of levels. Indeed, the scaling properties of small thermal machines have not been yet established.

A final general objective of this thesis is to provide insight into the role played by quantumness in the performance and operation of quantum thermal machines. We perform a thermodynamic analysis of the quantum Otto cycle for a single bosonic mode in the presence of a nonequilibrium squeezed thermal reservoir. Equipped with the findings about entropy production in quantum processes and the generalized formulation of the second law previously developed, we will perform an entropic analysis of this setup in Chap. 10. We identify nonequilibrium features introduced by the squeezed thermal reservoir in the operation of the engine, optimize it, and discuss its many striking consequences such as the appearance of multitask regimes in which the heat engine may extract work and refrigerate a cold reservoir at the same time. Finally, we study the performance of multi-level autonomous thermal machines in terms of the number of levels in Chap. 11. We first identify the primitive operation of autonomous machines and then characterize the different elements determining their performance. This allows us to compare different ways of scaling the system by adding extra levels. Fundamental limitations to improve the performance of the machine then naturally arise, leading to a novel statement about the third law of thermodynamics in terms of the Hilbert space dimension of the machine.

Acknowledgements

This work would not have been possible without the invaluable help of many people, who directly or indirectly contributed to push forward this thesis. It is my intention to dedicate here some words of gratitude to all of them.

In the first place, I want to acknowledge the advice and dedication of my supervisors, Juan M. R. Parrondo and Roberta Zambrini. At this point in time, I can say that I have learned quite much from them during these years. They have generously shared with me their wide knowledge of quantum theory and nonequilibrium thermodynamics and taught me how to make difficult problems understandable in rather simpler terms and how to solve them by making use of a broad range of techniques. I also learned from them to immerse myself in the literature to catch the state of the art of some topic and to communicate my research results in worthy scientific English. They have always helped me when I needed in both scientific and administrative areas, and given to me the opportunity to interact with other distinguished physicists of different institutions.

Next, it is a pleasure for me to acknowledge the helpful comments and advices I received from all the people of my group at Universidad Complutense de Madrid, the *Group of Statistical Mechanics.*[1] I am especially grateful to Luis Dinis for their vigorous support, including their recommendations when I was collaborating with him in teaching *Laboratorio de Física para Biólogos*, to Léo Granger for his inspiration and the very interesting conversations about thermodynamics we frequently maintained during his postdoc in Madrid, and to Jordan M. Horowitz for his precious help and splendid work during our collaboration. I would like also to show my gratitude to the whole GISC group[2] for giving me the opportunity to participate in their annual workshops.

Thanks to the members of the *Institute For Cross-Disciplinary Physics and Complex Systems* (IFISC) in Mallorca, who welcomed me during my many visits and provided for me all the comforts to carry out my work there. At IFISC, I have

[1] Web site: http://seneca.fis.ucm.es.

[2] Grupo de Física Interdisciplinar y Sistemas Complejos.

had the opportunity to interact with many people working on different fields, attend to outstanding seminars on a variety of topics, and to present some of the results of my research. It has been a pleasure to collaborate with Fernando Galve during these years, who shared interesting perspectives with me and always offered his crucial help. Thanks also to Gian-Luca Giorgi for his help with master equations and the rest of my collaborators in the synchronization part of this thesis, and Pere Colet and Emilio Hernández-García, for their respective contributions.

During my two stays in the *H. H. Wills Physics Laboratory* at the University of Bristol, Prof. Sandu Popescu took care of me and integrated me as a full member of his group. I am very grateful to him for his kind hospitality, the very interesting conversations he is capable of provoking, and his inspirational suggestions that opened my mind in a number of ways. I also acknowledge the rest of the people of the *Quantum Information and Foundations Group* for hosting me. In particular, I would like to thank my collaborators Ralph Silva and Paul Skrzypczyk for sharing their excellent work and interesting viewpoints, and Tony Short for helping me with the crazy stuff of the stay certificate. Thanks also to Nicolas Brunner from Université de Genève for his contribution to our work on autonomous multi-level quantum thermal machines.

I must also recall the importance of the professors who awake my interest in physics when I still was an undergraduate student at the Universidad Complutense de Madrid trying to understand something. I have been actually lucky, because one of them has been indeed my supervisor during my Ph.D. research, Prof. Juan M. R. Parrondo. I also acknowledge the insightful lessons on classical mechanics of Prof. Enrique Maciá Barber, from which I could catch his passion for understanding the underlying "melody of the universe." Thanks also to Alvaro de la Cruz-Dombriz for his instructive teaching of space–time structure and cosmology matters, to Prof. Ricardo Brito, who introduced me for the first time to out-of-equilibrium processes, and to the exemplary teacher Prof. Joaquín Retamosa, who is resting in peace now. Furthermore, during my master's degree in fundamental physics I had the opportunity to meet other great scientists whose teaching I really appreciated. In particular, I want to thank Isabel Gonzalo Fonrodona and David Gómez-Ullate Otzeida.

The research presented in this thesis has been developed with the financial support of the Spanish MINECO (FPI grant No. BES-2012-054025) and has been partially supported by the COST Action MP1209 "thermodynamics in the quantum regime."

Finalmente quisiera dar las gracias a toda la gente que me ha apoyado en el plano personal y animado durante estos años. Estoy seguro de que sin este entorno vital y social no hubiera llevado esta tesis a cabo. Es por tanto que se podrí-a decir que su contribución ha sido la más importante e imprescindible. Gracias a mi madre Esther por su caniño y cuidado incondicionales. A mis abuelos Concesa y Victoriano, a Honorio, a mi tía Mariángeles y al resto de mi familia por su ayuda y su consejo. Mi último agradecimiento va dedicado a todos mis amigos y amigas con los cuales he podido compartir momentos importantes y forjar unos lazos sólidos a través del tiempo. Como saben, son y han sido siempre un gran apoyo para mi.

Contents

Acronyms

CB	Common bath
COP	Coefficient of performance
CP	Completely positive
CPTP	Completely positive and trace-preserving
DFS	Decoherence-free subspace/subsystem
ETH	Eigenstate thermalization hypothesis
FT	Fluctuation theorem
GGE	Generalized Gibbs ensemble
LB	Local bath
LOCC	Local operations and classical communication
NS	Noiseless subsystem
NEMS	Nanoelectromechanical structures
NMR	Nuclear magnetic resonance
NSD	No sudden death
TMP	Two-measurement protocol
POVM	Positive-operator-valued measure
PPT	Positive partial transpose (criterion)
QND	Quantum nondemolition (measurement)
RWA	Rotating wave approximation
SB	Separate baths
SD	Sudden death
SDR	Sudden death and revivals
SME	Stochastic master equation
SSE	Stochastic Schrödinger equation

Fundamental Constants

ħ Planck constant ($6.62607004 \times 10^{-34}$ m^2 kg s^{-1})
k_B Boltzmann constant ($1.38064852 \times 10^{-23}$ m^2 kg s^{-2} K^{-1})

Part I
Introduction to Open Quantum Systems and Quantum Thermodynamics

Chapter 1
Basic Concepts

Any realistic quantum system cannot be completely isolated. In general, it is unavoidably coupled to a larger environment and thus, even if this interaction is weak, needs to be regarded as an open system, just like happens when one considers small classical particles. The environment, which is frequently (but not only) regarded as a thermal *reservoir* or *bath*, influences the quantum system under consideration in a non-negligible way, which must be taken into account when describing its dynamical evolution and properties. System and environment are continuously sharing information, which is manifested in the building up of correlations between them. This information is no longer available in general, as it involves a huge number of uncontrollable degrees of freedom. Indeed, obtaining a complete microscopic description of the whole ensemble involved in the problem is both intractable and generally not needed from a practical point of view, but a rather simpler probabilistic approach is highly desirable. The theory of open quantum systems provides such an effective description, allowing the treatment of complex systems by means of a small number of relevant variables. The irrelevant degrees of freedom are instead described only approximately, which results in the appearance of dissipative and stochastic terms in the final form of the effective equations of motion, a characteristic feature of an irreversible evolution [1, 2].

Open quantum systems theory has been widely studied and applied by many scientists from different communities in the last half-century. Nowadays it constitutes an everyday tool in modern quantum optics, atomic physics, condensed matter, chemical physics, quantum information science or the novel field of quantum thermodynamics. A more rigorous treatment of open quantum systems from a mathematical point of view complements this heterogeneity and provides consistency to the theory (see e.g. [3] and references therein).

The study of open quantum systems is also of special importance for fundamental questions about the quantum description of nature. One example is quantum measurement theory, as long as any measurement requires a description in terms of the

G. Manzano Paule, *Thermodynamics and Synchronization in Open Quantum Systems*, Springer Theses, https://doi.org/10.1007/978-3-319-93964-3_1

interaction of the system to be measured with an apparatus (a second quantum system), which records the result and leads to fundamental back-action on the former. Another example comes from the fact that the interaction of a quantum system with its environment leads to the well-known phenomenon of *decoherence*, through which superpositions of quantum states are irreversibly lost producing the emergence of classical behavior [4]. It is thus clear that the detailed study of open quantum systems constitutes a key point if one wants to benefit from quantum phenomena in practical applications, as become patent in modern quantum computation, quantum metrology or quantum cryptography [5, 6].

This chapter aims to provide an introduction to some of the most important concepts employed when dealing with open quantum systems. In particular, we review and illustrate the essential concepts and methods which are going to be used in this thesis, while skipping rigorous demonstrations and referring to more specific books or reviews on this topic. We organized the chapter as follows. In Sect. 1.1 we start by reviewing the necessary elements of quantum theory needed for the description of open quantum systems, the dynamical evolution of closed (completely isolated) quantum systems, and its relation with the open system dynamics experienced by one of its constituents in the case of many-body systems. Next, in Sect. 1.2, we focus on the case of qubits and harmonic oscillators, as they are two prototypical systems where the basic elements of the theory can be well illustrated. In Sect. 1.3 we review the basics of quantum measurement theory, its general mathematical formulation in terms of operations and effects, and introduce the most important classes of measurements. Finally, in Sect. 1.4, we define classical and quantum correlations, introducing different quantifiers such as entanglement, mutual information, and discord, discussing their main physical interpretations.

1.1 Quantum Mechanics

In the general framework of quantum mechanics, each state of an isolated quantum system can be represented by a normalized state vector $|\psi\rangle$ in an associated Hilbert vectorial space $\mathcal{H}$. Any measurable quantity on this system is represented by an hermitian (or self-adjoint) operator, $\hat{\mathcal{O}} = \hat{\mathcal{O}}^\dagger$, in the same space, whose eigenvalues represent possible results (or outcomes) of a quantum measurement, as we will see in more detail in Sect. 1.3. Quantum theory is intrinsically random and the *pure* state $|\psi\rangle$ contains all the information one can know about the probability of obtaining different outcomes for all different observables of the system. To illustrate this point let us decompose the operator $\hat{\mathcal{O}} = \sum_n o_n |o_n\rangle\langle o_n|$ where $\{|o_n\rangle\}$ is the set of eigenvectors (or eigenstates) of $\hat{\mathcal{O}}$ providing a basis of $\mathcal{H}$, and o_n its corresponding (non-degenerate) eigenvalues. The probability of obtaining the result o_n in a measurement of the observable $\hat{\mathcal{O}}$ is the scalar product $0 \leqslant |\langle\psi|o_n\rangle|^2 \leqslant 1$, as given by Born rule [7]. Moreover the mean value of some observable in the state $|\psi\rangle$ is given by the quantum mechanical expectation value $\langle\hat{\mathcal{O}}\rangle = \langle\psi|\hat{\mathcal{O}}|\psi\rangle$, representing the mean of different results when the observable is measured, weighted with their different probabilities to occur. In the same manner the variance of $\hat{\mathcal{O}}$ on $|\psi\rangle$ reads $\sigma^2(\hat{\mathcal{O}}) = \langle\hat{\mathcal{O}}^2\rangle - \langle\hat{\mathcal{O}}\rangle^2$. The latter is zero if and only if the state $|\psi\rangle$ is an eigenstate of the operator, that is, when $\hat{\mathcal{O}}|\psi\rangle = \alpha|\psi\rangle$, being α a real number (then $|\psi\rangle = |o_n\rangle$ and $\alpha = o_n$ for some n). At difference from classical physics, $\sigma(\hat{\mathcal{O}})$ cannot be simultaneously zero for all observables $\hat{\mathcal{O}}$, as the Heisenberg uncertainty principle asserts [8]. Indeed for any quantum system, non-commuting observables such as position and momentum, $[\hat{x}, \hat{p}] \equiv \hat{x}\hat{p} - \hat{p}\hat{x} = i\hbar\mathbb{1}$ being $\mathbb{1}$ the identity operator, cannot share any common eigenstate. A general form of the Heisenberg uncertainty principle for arbitrary observables $\hat{\mathcal{O}}$ and $\hat{\mathcal{O}}'$ is the Robertson uncertainty relation [9] Robertson uncertainty relation.

$$\sigma(\hat{\mathcal{O}})\sigma(\hat{\mathcal{O}}') \geqslant \frac{1}{2}|\langle\psi|[\hat{\mathcal{O}}, \hat{\mathcal{O}}']|\psi\rangle|, \tag{1.1}$$

which unravels the connection between the commutativity of observables and the complementarity of their uncertainties [10].

1.1.1 The Density Operator

When considering open quantum systems we need to incorporate in the description new sources of randomness other than the intrinsic uncertainty of quantum states, coming e.g. from our lack of knowledge about the specific state of the environment, the preparation procedure, or the correlations built up in the interaction between system and surroundings. In this case we represent the state of our open system by a density operator (or density matrix) ρ,[1] firstly introduced by von Neumann [11] and Landau [12] in 1927. It characterizes our state of knowledge about the system and represents the quantum analogue to the phase-space probability distribution of classical statistical mechanics. The use of the density operator allows us to work with statistical mixtures of state vectors:

$$\rho = \sum_k p_k|\psi_k\rangle\langle\psi_k|, \qquad \text{with } k = 1, 2, \ldots, N, \tag{1.2}$$

where p_k are the probabilities $\big(0 \leqslant p_k \leqslant 1,\ \sum_k p_k = 1\big)$ of being our microscopic system in each of the N pure states $|\psi_k\rangle$, and the operators $|\psi_k\rangle\langle\psi_k|$ are *projectors* onto the state $|\psi_k\rangle$. The density operator is self-adjoint ($\rho = \rho^\dagger$), positive-semidefinite ($\rho \geqslant 0$), and has unit trace ($\mathrm{Tr}[\rho] = 1$).

In principle any *mixed* state ρ can be decomposed into a mixture of pure states in an infinite number of ways,[2] but there is only one in which the states $|\psi_k\rangle$ in

---

[1] We omit the hat symbol ^ used to distinguish between operators and scalars for the density operator.

[2] All of them related by a unitary transformation. Furthermore the same density operator can be also decomposed in a mixture of *mixed* states in an infinite number of ways.

the decomposition are mutually orthogonal between them, i. e. for which they verify $\langle\psi_k|\psi_l\rangle = \delta_{k,l}$. This is guaranteed by the spectral theory of density operators, as long as ρ has only a countable set of strictly positive eigenvalues [1]. In such case, we may call Eq. (1.2) the spectral decomposition of ρ, the probabilities p_k and the states $|\psi_k\rangle$ being respectively the eigenvalues and eigenstates of the ρ, and N the dimension of the Hilbert space $\mathcal{H}$ (which may be infinite). Given a density operator, ρ, the most likely pure state the system is in, is given by the eigenstate $|\psi_k\rangle$ corresponding to the largest eigenvalue p_k [13].

It is worth mentioning that the density operator ρ is sufficient to describe all the possible results of measurements on the system for any observable. Indeed the expression for the mean value introduced earlier, can be now rewritten for the case of a mixed state as

$$\langle\hat{\mathcal{O}}\rangle = \mathrm{Tr}[\hat{\mathcal{O}}\rho] = \sum_k p_k \langle\psi_k|\hat{\mathcal{O}}|\psi_k\rangle, \tag{1.3}$$

In a similar way we may use the trace to rewrite the expression for the variance $\sigma^2(\hat{\mathcal{O}})$ in terms of ρ, or the general uncertainty relation in Eq. (1.1).

Another important property of the density operator is that it always verify $\mathrm{Tr}[\rho^2] \leqslant \mathrm{Tr}[\rho] = 1$, where the equality is only reached in the case of a pure state $\rho = |\psi\rangle\langle\psi|$, when the information about the state of the microscopic system is complete. On the opposite side, the maximally mixed state reads $\rho = \mathbb{1}/N$, where N again denotes the dimension of the system Hilbert space $\mathcal{H}$. This corresponds to the case in which all the possible physical pure states of the microscopic system are equally probable. Hence we may define the quantity $\mathcal{P}(\rho) \equiv \mathrm{Tr}[\rho^2]$, called the *purity* of a state, in order to quantify its degree of mixedness. This quantity takes values in the range $1/N^2 \leqslant \mathcal{P}(\rho) \leqslant 1$, with the upper bound reached for pure states and the lower bound reached for maximally mixed states.

It is however important to distinguish a mixture of pure states, as given in Eq. (1.2), from a *superposition* of the form

$$|\psi\rangle = \sum_k c_k|\psi_k\rangle, \qquad \text{with } k = 1, 2, \ldots, N, \tag{1.4}$$

where $c_k = \langle\psi_k|\psi\rangle$ are a set of complex numbers such that $\sum_k |c_k|^2 = 1$. The existence of such states, as motivated by the superposition principle, lies at the heart of quantum theory. The differences between mixture and superposition states are fundamental. While the former simply describes our lack of knowledge in the specific pure state the system is in, the latter corresponds to a single pure state. Hence we can no longer interpret the system being in different states $|\psi_k\rangle$ with certain probability, but we have to really consider that the system is in all those states at once. Let us assume the set of states $\{|\psi_k\rangle\}$ to form a basis of the Hilbert space of the system with dimension N, and compare the density operator $\rho = |\psi\rangle\langle\psi|$ for the superposition state (1.4) with the one of the mixed state in Eq. (1.2). The state (1.2) has only diagonal elements (using the basis $\{|\psi_k\rangle\}$) given by the probabilities

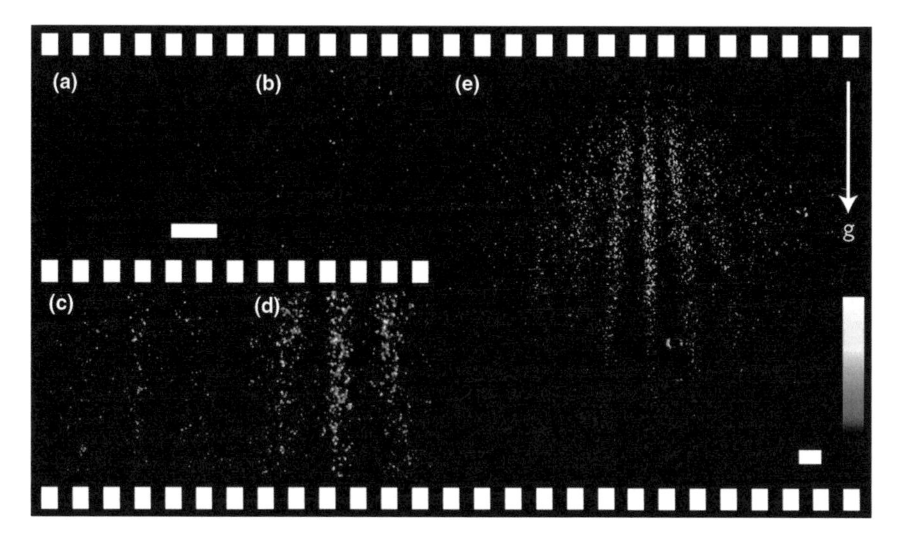

Fig. 1.1 Interference pattern for stochastically arriving single PcH_2 molecules in a modern double-slit experiment. The images correspond to selected frames from a false-color movie recorded with an EMCCD camera. Obtained from Ref. [15]

p_k, while the state (1.4) gets diagonal elements $\rho_{kk} = \langle\psi_k|\rho|\psi_k\rangle = |c_k|^2$, but also off-diagonal ones $\rho_{kl} = \langle\psi_k|\rho|\psi_l\rangle = c_k^* c_l$ for $k \neq l$. Off-diagonal terms are called *coherences* between the states $\{|\psi_k\rangle\}$, and are responsible of the interference effects due to the wave-particle complementarity of quantized matter, just as in Young's famous double-slit experiment [14]. Modern *which-path* experiments within different setups have considerably evolved from the 90s [2], being nowadays able to test some of the most famous thought-experiments formulated from the very beginning of the quantum theory for larger and larger systems (electrons, atoms, molecules), testing in the laboratory the connections between complementarity and decoherence (see Fig. 1.1).

1.1.2 Liouville–von Neumann Equation

The time evolution of a (non-relativistic) isolated quantum system in terms of its density operator, ρ, is given by the Liouville–von Neumann equation

$$i\hbar\frac{d}{dt}\rho(t) = [\hat{H}(t), \rho(t)], \tag{1.5}$$

being $\hat{H}(t)$ the Hamilton operator representing the energy of the system. Notice that we have included the possibility of time-dependent Hamilton operators, allowing for the description of external driving. The Liouville–von Neumann equation describes the reversible evolution of the quantum system and when ρ is a pure state it is equivalent to the usual Scrödinger equation, first introduced in 1926 [16]. The formal solution of Eq. (1.5), given the initial state $\rho(t_0)$ at time t_0, reads $\rho(t) = \hat{U}(t, t_0)\rho(t_0)\hat{U}(t, t_0)^{\dagger}$, with

$$\hat{U}(t, t_0) \equiv \hat{T}_{+} \exp\left(-\frac{i}{\hbar}\int_{t_0}^{t} ds H(s)\right), \tag{1.6}$$

the unitary evolution operator, $\hat{U}\hat{U}^{\dagger} = \hat{U}^{\dagger}\hat{U} = \mathbb{1}$, fulfilling

$$i\hbar\frac{d}{dt}\hat{U}(t, t_0) = \hat{H}(t)\hat{U}(t, t_0), \tag{1.7}$$

and with initial condition $\hat{U}(t_0, t_0) = \mathbb{1}$. It fulfills the chain rule $\hat{U}(t, t_0) = \hat{U}(t, t_1)$ $\hat{U}(t_1, t_0)$ for $t \leqslant t_1 \leqslant t_0$. Due to the fact that the Hamilton operator may not commute with itself at different times, we introduced in the integral above the time-ordering operator, $\hat{T}_{+}$, implying that in general the unitary evolution operator can be only calculated from an infinite series in the form

$$\begin{aligned}\hat{U}(t, t_0) = \mathbb{1} + \sum_{n=1}^{\infty}\left(\frac{-i}{\hbar}\right)^{n}\int_{t_0}^{t} ds_n \hat{H}(s_n)\int_{t_0}^{s_n} ds_{n-1}\hat{H}(s_{n-1}) \ldots \\ \ldots \int_{t_0}^{s_3} ds_2 \hat{H}(s_2)\int_{t_0}^{s_2} ds_1 \hat{H}(s_1),\end{aligned} \tag{1.8}$$

where time ordering implies $t > s_n > s_{n-1} > \cdots > s_2 > s_1$, an expression known as the Dyson series. When the Hamilton operator in Eq. (1.6) is independent of time the unitary evolution operator reduces to

$$\hat{U}(t, t_0) = \hat{U}(t - t_0) = \exp\left(-\frac{i}{\hbar}\hat{H}(t - t_0)\right), \tag{1.9}$$

and then $\hat{U}^{\dagger}(t - t_0) = \hat{U}(t_0 - t)$, corresponding to the evolution operator when time is reversed.

1.1.3 Heisenberg and Interaction Pictures

The above Eq. (1.5) gives us the evolution of the density operator $\rho(t)$ in the *Schrödinger picture*. An equivalent formulation, the so-called *Heisenberg picture*, is obtained by assuming the state of the system fixed and letting the observables evolve in time. Then the equation of motion for an arbitrary observable $\hat{\mathbb{O}}(t)$, can be written as

$$\frac{d}{dt}\hat{\mathcal{O}}(t) = \frac{i}{\hbar}[\hat{H}(t), \hat{\mathcal{O}}(t)] + \left(\frac{\partial \hat{\mathcal{O}}}{\partial t}\right)_H, \tag{1.10}$$

whose solution is given by $\hat{\mathcal{O}}(t) = \hat{U}^\dagger(t, t_0)\ \hat{\mathcal{O}}\ \hat{U}(t, t_0)$, $\hat{\mathcal{O}}$ being the initial (Schrödinger picture) observable and $\hat{U}$ given by Eq. (1.6). Here we denote $\left(\frac{\partial \hat{\mathcal{O}}}{\partial t}\right)_H = \hat{U}^\dagger(t, t_0)\left(\frac{\partial \hat{\mathcal{O}}}{\partial t}\right)\hat{U}(t, t_0)$. It's straightforward to check that both pictures produce identical expectation values for all observables.

A third frame, the *interaction picture*, can be also introduced by splitting the Hamiltonian into time-independent and time-dependent parts, which we denote as $\hat{H} = \hat{H}_0 + \hat{V}(t)$. Typically $\hat{H}_0$ is easy to deal with, and represents the Hamilton operator of two or more non-interacting systems, while $\hat{V}(t)$ usually represents a time-dependent interaction term. In this case we split the evolution operator into a product of two unitary operators

$$\hat{U}(t, t_0) = \hat{U}_0(t - t_0) \times \hat{U}_I(t, t_0), \tag{1.11}$$

where $\hat{U}_0(t - t_0) \equiv \exp\left(-\frac{i}{\hbar}\hat{H}_0(t - t_0)\right)$ is generated by the time-independent part of the Hamiltonian, and $\hat{U}_I(t, t_0)$ is given by Eq. (1.6) replacing $\hat{H}(t)$ by $\hat{U}_0^\dagger \hat{V}(t)\hat{U}_0$. Hence the operator $\hat{U}_0$ governs the evolution of observables, while the density operator evolves accordingly with $\hat{U}_I$. By redefining the density operator and observables, we have the following time-evolution equations:

$$\begin{aligned} \rho_I(t) &= \hat{U}_I\ \rho(0)\ \hat{U}_I^\dagger, \quad &\text{with} \quad \rho_I(t) &\equiv \hat{U}_0^\dagger\ \rho(t)\ \hat{U}_0, \\ \hat{\mathcal{O}}_I(t) &= \hat{U}_0^\dagger\ \hat{\mathcal{O}}\ \hat{U}_0, \quad &\text{with} \quad \hat{\mathcal{O}}_I(t) &\equiv \hat{U}_I\ \hat{\mathcal{O}}(t)\ \hat{U}_I^\dagger, \end{aligned} \tag{1.12}$$

where we call $\rho_I(t)$ and $\hat{\mathcal{O}}_I(t)$ the *interaction frame* density operator and observables respectively. The *interaction picture* has proven very useful in deriving and solving the dynamics for open quantum systems, as we will see in the next sections. It allows to split the effects of the interaction between a system and its surroundings from the (isolated) free-evolution, simplifying considerably the mathematical treatment.

1.1.4 The Microreversibility Principle

The microreversibility principle is a crucial symmetry of time evolution in isolated quantum systems. It relates the unitary evolution operator of a non-autonomous quantum system, as introduced in Eq. (1.6), with the one describing the time-reversed evolution [17, 18]. Let us assume a quantum system evolving from time $t = 0$ to time τ under the action of some Hamiltonian $\hat{H}(\lambda(t))$, whose time-dependence arises from external manipulation through a control parameter $\lambda(t)$. Consider that this parameter vary in time according to some prescribed protocol $\Lambda = \{\lambda(t) \text{ for } 0 \leqslant t \leqslant \tau\}$. The unitary time evolution operator for the system, $\hat{U}(t,0)[\Lambda]$, obeys

$$i\hbar\frac{d}{dt}\hat{U}(t,0)[\Lambda] = \hat{H}(\lambda(t))\hat{U}(t,0)[\Lambda], \tag{1.13}$$

in the interval $t \in [0,\tau]$ where the protocol Λ is defined.

Now we compare this evolution with the one generated by the *time-reversed* protocol $\tilde{\Lambda} = \{\tilde{\lambda}(t) \text{ for } 0 \leqslant t \leqslant \tau\}$, where $\tilde{\lambda}(t) = \lambda(\tau - t)$, i.e. the control parameter takes on exactly the inverse sequence of values. The corresponding time-evolution operator $\hat{U}(t,0)[\tilde{\Lambda}]$ generated by the Hamiltonian $\hat{H}(\tilde{\lambda}(t))$ now reads:

$$i\hbar\frac{d}{dt}\hat{U}(t,0)[\tilde{\Lambda}] = \hat{H}(\tilde{\lambda}(t))\hat{U}(t,0)[\tilde{\Lambda}] \tag{1.14}$$

where again $t \in [0,\tau]$. The microreversibility principle ensures the following relation between *forward* and *backward* evolutions [18]:

$$\hat{U}^{\dagger}(\tau,t)[\Lambda] = \hat{\Theta}^{\dagger}\, \hat{U}(\tau - t, 0)[\tilde{\Lambda}]\, \hat{\Theta}, \tag{1.15}$$

where $\hat{\Theta}$ is the anti-unitary time-reversal operator in quantum mechanics, $\Theta\Theta^{\dagger} = \Theta^{\dagger}\Theta = \mathbb{1}$ and $\hat{\Theta}(a|\psi\rangle + b|\phi\rangle) = a^{*}\hat{\Theta}|\psi\rangle + b^{*}\hat{\Theta}|\phi\rangle$.[3] It is responsible of sign inversion of odd variables under time-reversal such as linear and angular momenta, spin or magnetic field, while leaving even variables, such as position, unaltered [19]. The microreversibility principle in Eq. (1.15) is always fulfilled provided the Hamilton operator is invariant under time-reversal, $\hat{\Theta}^{\dagger}\hat{H}(\lambda(t))\hat{\Theta} = \hat{H}(\lambda(t))$ (for a proof see [18]). Otherwise the Hamiltonian governing the time-reversed evolution can be set as

$$\hat{H}_R(\tilde{\lambda}(t)) \equiv \hat{\Theta}\hat{H}(\tilde{\lambda}(t))\hat{\Theta}^{\dagger}, \tag{1.16}$$

in Eq. (1.14) [instead of $\hat{H}(\tilde{\lambda}(t))$]. The latter implies the change in sign of the odd variables appearing in $\hat{H}$, such as external magnetic fields [17]. We provide a proof of this claim in Appendix.

The microreversibility principle relates the evolution from some arbitrary initial state $\rho(0)$ to $\rho(\tau) = \hat{U}(t,0)[\Lambda]\ \rho(0)\ \hat{U}^{\dagger}(t,0)[\Lambda]$, to the evolution from the time-reversed final state $\tilde{\rho}(0) = \hat{\Theta}\ \rho(\tau)\ \hat{\Theta}^{\dagger}$ to the time reversed initial state $\tilde{\rho}(\tau) = \hat{\Theta}\ \rho(0)\ \hat{\Theta}^{\dagger}$ as:

$$\tilde{\rho}(\tau) = \hat{U}(\tau,0)[\tilde{\Lambda}]\ \tilde{\rho}(0)\ \hat{U}^{\dagger}(\tau,0)[\tilde{\Lambda}], \tag{1.17}$$

as is illustrated in Fig. 1.2. It is worth noticing that the notion of time-reversal here corresponds to an *operational* point of view, as it is defined via the time-reversed protocol for the external drive controlling the parameter $\lambda(t)$. We finally stress that

---

[3]This antilinearity property is what differentiates anti-unitary from unitary operators. Unitary operators fulfills linearity, while anti-unitary ones fulfills anti-linearity, and for both of them $\hat{\Theta}\hat{\Theta}^{\dagger} = \hat{\Theta}^{\dagger}\hat{\Theta} = \mathbb{1}$.

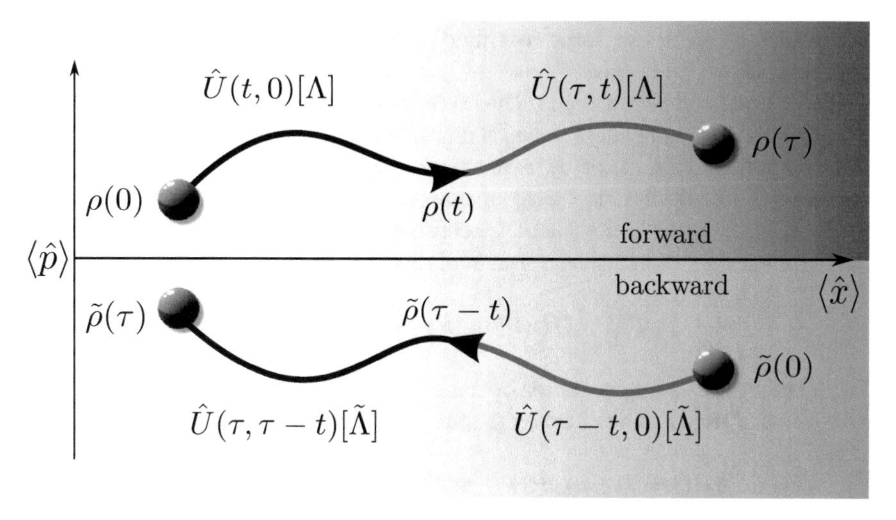

Fig. 1.2 Illustration of the microreversibility principle. The unitary time-evolution operators of the *forward* evolution (top) and the *backward* one (bottom) are related by the time-reversal operator $\hat{\Theta}$. The two unitary operators appearing in Eq. (1.15) are marked in blue. In the axis we put the mean values of position and momenta in order to emphasize the effect of time-reversal in odd variables

the microreversibility principle is a crucial symmetry property in deriving the so-called *fluctuation theorems* for quantum evolutions which will be the subject of Part III of this thesis.

1.1.5 Composite Quantum Systems

Let us now consider a quantum system composed by two different interacting subsystems, A and B, with associated Hilbert spaces $\mathcal{H}_A$ and $\mathcal{H}_B$ respectively. The subsystems may correspond to different physical systems (particles, atoms, molecules, ...) or also to different degrees of freedom of the same entity. In any case, the compound system AB has an associated Hilbert space given by the tensor product of the subsystems Hilbert spaces $\mathcal{H}_{AB} = \mathcal{H}_A \otimes \mathcal{H}_B$. This larger Hilbert space has dimension equal to the product of the dimensions of $\mathcal{H}_A$ and $\mathcal{H}_B$, and any arbitrary observable of the compound system takes the form

$$\hat{\mathcal{O}} = \sum_k \hat{\mathcal{O}}_A^{(k)} \otimes \hat{\mathcal{O}}_B^{(k)}, \tag{1.18}$$

where $\hat{\mathcal{O}}_A^{(k)}$ acts on $\mathcal{H}_A$ and $\hat{\mathcal{O}}_B^{(k)}$ acts on $\mathcal{H}_B$ for all k. Local observables corresponding to subsystem A alone are then expressed by $\hat{\mathcal{O}}_A \otimes \mathbb{1}_B$. Similarly for local observables of subsystem B, we have $\mathbb{1}_A \otimes \hat{\mathcal{O}}_B$. As can be immediately noticed, the local observables of the above form constitute only a very small portion of the set of possible observables in AB.

Imagine our subsystems A and B have not interacted before some arbitrary time t_0, hence being completely uncorrelated at this moment. In this case the density operator of the compound system can be expressed as the product state $\rho_{AB}(t_0) = \rho_A(t_0) \otimes \rho_B(t_0)$, $\rho_A(t_0)$ being the local state of subsystem A at that time (and similarly for B). Then let the two systems interact accordingly to some global Hamilton operator $\hat{H}(t)$, in such a way that at a later instant of time t, the state of the compound system is

$$\rho_{AB}(t) = \hat{U}(t, t_0)\rho_A(t_0) \otimes \rho_B(t_0)\, \hat{U}^\dagger(t, t_0), \tag{1.19}$$

$\hat{U}(t, t_0)$ being the unitary evolution operator given by Eq. (1.6). The local state of the subsystems at time t is obtained by partial tracing over the complementary Hilbert space:

$$\rho_A(t) = \mathrm{Tr}_B[\rho_{AB}(t)] \quad \text{and} \quad \rho_B(t) = \mathrm{Tr}_A[\rho_{AB}(t)], \tag{1.20}$$

where $\mathrm{Tr}_{A(B)}[\cdot] = \sum_i \langle \psi_i^{A(B)} | \cdot | \psi_i^{A(B)} \rangle$, with the set $\{|\psi_i^{A(B)}\rangle\}$ a orthonormal basis of the Hilbert space $\mathcal{H}_{A(B)}$. The use of the partial trace operation is justified as it can be proven to be the unique operation which provides the correct description of local observables for subsystems of a composite system [5]. The states appearing in Eq. (1.20) are called the *reduced states* (or local states) of subsystems A and B, which retain only the local information determining the statistics of measurements of local observables. Indeed, we cannot express any more the global state as a tensor product of the reduced counterparts, $\rho_{AB}(t) \neq \rho_A(t) \otimes \rho_B(t)$, since the state $\rho_{AB}(t)$ contains in general much more information than $\rho_A(t)$ and $\rho_B(t)$. More precisely, this happens whenever the unitary evolution cannot be expressed as a tensor product of local unitary evolutions in each subsystem $\hat{U}(t, t_0) = \hat{U}_A(t, t_0) \otimes \hat{U}_B(t, t_0)$, or if $\hat{U}$ is the complete SWAP operation exchanging the states of the subsystems, in the case of equal dimensions [20]. We then say that the state $\rho_{AB}(t)$ is correlated, meaning that the two subsystems have exchanged information during the interaction.

In particular, if the global state of the system at time t is pure, $\rho_{AB}(t) = |\Psi\rangle\langle\Psi|$, the reduced states $\rho_A(t)$ and $\rho_B(t)$ have the same eigenvalues. This follows from the Schmidt decomposition theorem, which asserts that there always exists a unique decomposition of $|\Psi\rangle$ reading

$$|\Psi\rangle = \sum_k \alpha_k |\psi_A^{(k)}\rangle \otimes |\psi_B^{(k)}\rangle, \tag{1.21}$$

where $\{|\psi_A^{(k)}\rangle\}$ and $|\psi_B^{(k)}\rangle$ are respectively orthonormal basis of $\mathcal{H}_A$ and $\mathcal{H}_B$, and the complex amplitudes α_k (where $\sum_k |\alpha_k|^2 = 1$) are called Schmidt coefficients. We say that $|\Psi\rangle$ is *entangled* if it cannot be expressed as a product $|\Psi\rangle = |\psi_A\rangle \otimes |\psi_B\rangle$ of some states of the subsystems, that is, if it has more than one non-zero Schmidt coefficients. Furthermore we say that $|\Psi\rangle$ is a *maximally entangled* state if all non-zero Schmidt coefficients are equal [1].

Consider for instance the case of two spin-$\frac{1}{2}$ particles (or more generally *two-level systems* or *qubits*), with basis states $|0\rangle_i$ and $|1\rangle_i$ for $i = A, B$. The maximally entangled states in this case, commonly called *Bell states*, read as follows:

$$|\Phi_\pm\rangle \equiv \frac{1}{\sqrt{2}} \left(|0\rangle_A |0\rangle_B \pm |1\rangle_A |1\rangle_B\right), \tag{1.22}$$

$$|\Psi_\pm\rangle \equiv \frac{1}{\sqrt{2}} \left(|0\rangle_A |1\rangle_B \pm |1\rangle_A |0\rangle_B\right), \tag{1.23}$$

to be compared with an uncorrelated state of the type $|\Theta\rangle = |\phi\rangle_A \otimes |\psi\rangle_B$. In contrast to the separable state $|\Theta\rangle$, in the states (1.22) and (1.23) the particles A and B do not have a definite pure state vector characterizing its quantum state. Their corresponding reduced density matrices read $\rho_A = \rho_B = \mathbb{1}/2$, i.e. the maximally mixed states in the two-dimensional Hilbert spaces $\mathcal{H}_A = \mathcal{H}_B = \mathbb{C}^2$. Hence, from a local point of view, the two particles are with equal probabilities in either $|0\rangle$ or $|1\rangle$. Imagine now that we perform a measurement on particle A, obtaining that it is in the state $|0\rangle_A$. Hence the particle A 'collapses' to this state, and the state of particle B should also immediately 'collapse' to the correspondent state $|0\rangle_B$ for Eq. (1.22) or $|1\rangle_B$ for Eq. (1.23), even if the two particles are arbitrarily far away from each other. This effect represents the so-called 'spooky action' at the heart of the EPR 'paradox' [21]. Any posterior measurement on B is then no longer random, but it can be predicted with probability 1, in sharp contrast with what would happen if we do not measure A or even if we simply don't know the result of this measurement. Nevertheless, for an observer in solitary confinement who has only access to particle B, there is no way to determine if particle A has been previously measured or not (the observer always obtain the same statistics). In any case, the local measurement results in subsystems A and B will always be strongly correlated (see Fig. 1.3). This reveals a *non-local* character of quantum physics, which is encoded in the global entangled quantum state of AB, non accessible from a local point of view, but making the subsystems statistically dependent of each others, in a way that any local and deterministic (hidden variable) theory can never reproduce [22].

We will turn on the concept of entanglement in Sect. 1.4 where we introduce and discuss different quantifiers of entanglement for pure and mixed states, together with their role as an indicator of the quantumness of correlations.

1.1.6 Quantum Entropies

Here we introduce two important notions in order to characterize the information contained in a quantum state given by a density operator ρ, namely the *von Neumann entropy* and the *relative entropy*. The von Neumann entropy is of major importance in quantum statistical mechanics and quantum thermodynamics, and we will reefer to it in general as simply the entropy of a state. It is defined as the functional introduced

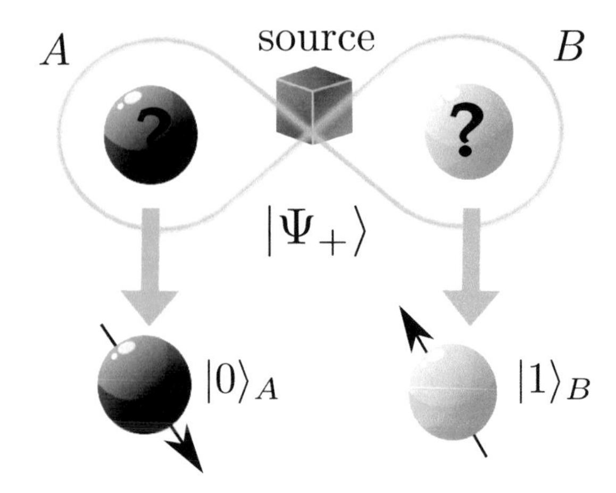

Fig. 1.3 Scheme of the composite quantum system discussed in the text. A source emits an entangled pair of spin-$\frac{1}{2}$ particles in the Bell state $|\Psi_+\rangle$ of Eq. (1.23). The results of local measurements (indicated by the solid arrows) produce correlated results which lead to complete certainty in the prediction of results in one subsystem when the result of the measurement in the other subsystem is known

by von Neumann in 1927:

$$S(\rho) \equiv -\mathrm{Tr}[\rho \ln \rho] = -\sum_k p_k \ln p_k, \tag{1.24}$$

where in the second equality we have used the spectral decomposition of ρ, as given in Eq. (1.2), and the convention $0 \ln 0 = 0$ is adopted. Von Neumann entropy measures the amount of uncertainty (or the lack of information) about the specific (pure) state the system is in. Indeed it is equivalent to the *Shannon entropy*, $H(\{p_k\})$, of the distribution $\{p_k\}$ for the ρ eigenstates, that is, the average value of the *surprise*, $\ln(1/p_k)$, of finding the quantum system in state $|\psi_k\rangle$ when measuring it. We immediately notice that, as long as the state ρ can be written as a mixture of pure states in several ways, we may obtain different values for the Shannon entropies of the corresponding (different) distributions, e.g. $\rho = \sum_k p'_k |\psi'_k\rangle\langle\psi'_k|$, with $\langle\psi'_k|\psi'_l\rangle \neq \delta_{k,l}$. In this sense, the von Neumann entropy corresponds to the minimum of all those uncertainties, that is, it describes our uncertainty about the state when the measurement process allows for perfect distinction between the pure states of the mixture ρ.

The von Neumann entropy is non-negative for all density operators, $S(\rho) \geqslant 0$, and invariant under unitary transformations, $S(\hat{U}\rho\hat{U}^\dagger) = S(\rho)$. It vanishes only in the case of a pure state, for which one has complete knowledge of the system state. Furthermore, it is bounded from above by $S(\rho) \leqslant \ln N$, N being the dimension of the system Hilbert space, where the equality is reached for the maximally mixed state. This implies that the von Neumann entropy is also a measure of the 'mixedness' of the sate ρ, alternative to the previously introduced *purity*. Another important property of $S(\rho)$ is that it constitutes a concave functional of ρ, i. e. for any two positive numbers $\lambda_1 + \lambda_2 = 1$ we have:

$$S(\lambda_1\rho_1 + \lambda_2\rho_2) \geqslant \lambda_1 S(\rho_1) + \lambda_2 S(\rho_2), \tag{1.25}$$

the equality holding only when $\rho_1 = \rho_2$. In addition, for composite systems, it fulfills the subadditivity condition:

$$S(\rho_{AB}) \leqslant S(\rho_A) + S(\rho_B), \tag{1.26}$$

where $\rho_A = \text{Tr}_B[\rho_{AB}]$ and $\rho_B = \text{Tr}_A[\rho_{AB}]$ are the reduced density operators for the subsystems A and B respectively. Here the equality holds only when $\rho_{AB} = \rho_A \otimes \rho_B$, corroborating that beyond product states, ρ_{AB} contains more information than the sum of the informations we can extract from the subsystems, due to the correlations between them. Furthermore if the global state ρ_{AB} is pure, we have that $S(\rho_A) = S(\rho_B) \geqslant 0$ (see also Sect. 1.4).

We also introduce the *relative entropy* of a density operator ρ to another density operator σ, as:

$$D(\rho||\sigma) \equiv \text{Tr}[\rho(\ln\rho - \ln\sigma)] = -S(\rho) - \text{Tr}[\rho\ln\sigma] \tag{1.27}$$

which constitutes a measure of the distinguishability between the two quantum states. More specifically, it corresponds to the extra amount of information required to encode ρ in the eigenstates of σ. In the definition (1.27) we adopt the convention $-s\ln 0 = \infty\ \forall s > 0$, which leads to $D(\rho||\sigma) = \infty$ when the support of ρ intersects with the kernel of σ. In this case ρ cannot be encoded in the eigenstates of σ, and the states are perfectly distinguishable. If we introduce the spectral decompositions of ρ and σ as:

$$\rho = \sum_{k=1}^{N} p_k|\psi_k\rangle\langle\psi_k| \quad \text{and} \quad \sigma = \sum_{k=1}^{N} q_k|\phi_k\rangle\langle\phi_k|.$$

Then the relative entropy in Eq. (1.27) can be rewritten as:

$$D(\rho||\sigma) = \sum_{k=1}^{N} (p_k \ln p_k - r_k \ln q_k)\,, \tag{1.28}$$

with $r_k = \sum_j p_j|\langle\phi_k|\psi_j\rangle|^2$, which reduces to the Kullback–Leibler divergence [23] between the distributions $\{p_k\}$ and $\{q_k\}$ when $\langle\phi_k|\psi_j\rangle = \delta_{k,j}$, i.e. when the two density operators commute, $[\rho, \sigma] = 0$. The Kullback–Leibler divergence between two distributions is a logarithmic measure of the probability of incorrectly guessing via hypothesis testing the distribution $\{p_k\}$ to be the source of a large sequence of data being truly generated by the distribution $\{q_k\}$ [24]. This argument applies as well to the quantum case, in which the probability of guessing the state ρ after performing n measurements on σ (for n large) is $e^{-nD(\rho||\sigma)}$ (for a more detailed discussion see [25] and references therein).

The relative entropy is an asymmetric distance, which is always non-negative, $D(\rho||\sigma) \geqslant 0$, and zero if and only if $\rho = \sigma$, as follows from Klein's inequality [5]. As the von Neumann entropy, it is invariant under unitary transformations, $D(\rho||\sigma) = D(U\rho U^{\dagger}||U\sigma U^{\dagger})$, and fulfills joint convexity:

$$D(\lambda_1\rho_1 + \lambda_2\rho_2||\sigma) \leqslant \lambda_1 D(\rho_1||\sigma) + \lambda_2 D(\rho_2||\sigma), \tag{1.29}$$

$$D(\rho||\lambda_1\sigma_1 + \lambda_2\sigma_2) \leqslant \lambda_1 D(\rho||\sigma_1) + \lambda_2 D(\rho||\sigma_2), \tag{1.30}$$

where again $\lambda_i \geqslant 0$ $(i = 1, 2)$ and $\lambda_1 + \lambda_2 = 1$. It is also worth stressing that partial tracing cannot increase the relative entropy

$$D(\rho||\sigma) \geqslant D(\text{Tr}_P[\rho]||\text{Tr}_P[\sigma]), \tag{1.31}$$

where P denotes the part of the system which is traced over, as ignoring part of the information about a system can never help to better distinguish between its states.

The relative entropy is an essential tool to characterize correlations in composite systems and also to quantify irreversibility in dynamical evolution, as we will discuss in more detail in Sect. 1.4 and in Chap. 3. Let us also point that most of the measurements of entanglement can be derived from relative entropy [25].

1.1.7 Distance Measures

Distance measures are introduced to quantify how close are two quantum states. We have already introduced the relative entropy, $D(\rho||\sigma)$ in Eq. (1.27), but it lacks some desirable properties one would expect from a proper distance, such as symmetry in the arguments. There is indeed a variety of distance measures which have been introduced and found convenient in different contexts related to quantum information [5]. Here we introduce two important and widely used measures of the distance between two quantum states: the *trace distance* and the *fidelity*, both of them obtained throughout generalization of concepts in classical probability theory, and playing an important role in the description of open quantum systems.

The trace distance between two generic quantum states ρ and σ is defined as

$$T(\rho,\sigma) = \frac{1}{2}\text{Tr}[|\rho - \sigma|] = \frac{1}{2}\text{Tr}\left[\sqrt{(\rho-\sigma)^2}\right] = \frac{1}{2}\sum_k |\lambda_k|, \tag{1.32}$$

where for any operator $\hat{A}$, we define $|\hat{A}| = \sqrt{\hat{A}^{\dagger}\hat{A}}$, and $\{\lambda_k\}$ are the set of (not necessarily positive) eigenvalues of $\rho - \sigma$. If ρ and σ commute, the trace distance reduces to the classical Kolmogorov distance between probability distributions, $T(p_i, q_i) = \sum_i |p_i - q_i|/2$, $\{p_i\}$ and $\{q_i\}$ being the eigenvalues of ρ and σ respectively. An alternative way of writing the trace distance is

$$T(\rho,\sigma) = \max_{\hat{\Pi}} \text{Tr}[\hat{\Pi}(\rho - \sigma)], \tag{1.33}$$

where the maximum runs over all projectors $\hat{\Pi}$. This expression can be indeed extended to all positive operators $0 \leqslant \hat{\Pi} \leqslant \mathbb{1}$ (see Sect. 1.3 below), leading to interpret the trace distance as the maximum difference in probabilities when an arbitrary measurement is performed [5].

The trace distance constitutes a proper distance as $T(\rho, \sigma) = 0$ if and only if $\rho = \sigma$, it is symmetric in its arguments, and fulfills the triangle inequality

$$T(\rho, \chi) \leqslant T(\rho, \sigma) + T(\sigma, \chi), \tag{1.34}$$

which establish that the trace distance is a metric [5]. Moreover we have $0 \leqslant T(\rho, \sigma) \leqslant 1$, where $T(\rho, \sigma) = 1$ if and only if ρ and σ have orthogonal supports, corresponding to the maximum distance between the two states. Some other important properties of the trace distance are the following. First, it is preserved under unitary transformations, $T(\rho, \sigma) = T(\hat{U}\rho\hat{U}^\dagger, \hat{U}\sigma\hat{U}^\dagger)$. Furthermore, it is *strong convex*, a more general property than *joint convexity* implying the later, which reads

$$T\left(\sum_i \rho_i, \sum_i q_i\sigma_i\right) \leqslant T(p_i, q_i) + \sum_i p_i T(\rho_i, \sigma_i), \tag{1.35}$$

for probabilities p_i and q_i with $\sum_i p_i = \sum_i q_i = 1$ and density operators ρ_i and σ_i. Finally, as the relative entropy, it never increases under partial tracing

$$T(\rho, \sigma) \geqslant T(\mathrm{Tr}_P[\rho]||\mathrm{Tr}_P[\sigma]), \tag{1.36}$$

P being the part of the system which is averaged over.

We now introduce a second distance between quantum states, the *fidelity*, measuring the closeness of two quantum states:

$$F(\rho, \sigma) = \mathrm{Tr}[\sqrt{\rho^{\frac{1}{2}}\sigma\rho^{\frac{1}{2}}}]. \tag{1.37}$$

The fidelity, unlike the trace distance, does not constitute a metric on density operators. It is bounded by $0 \leqslant F(\rho, \sigma) \leqslant 1$, where $F(\rho, \sigma) = 1$ if and only if $\rho = \sigma$. Furthermore, as the trace distance, fidelity is also symmetric in its arguments and is invariant under unitary transformations. It is related to the trace distance by

$$1 - F(\rho, \sigma) \leqslant T(\rho, \sigma) \leqslant \sqrt{1 - F^2(\rho, \sigma)}, \tag{1.38}$$

where the upper bound is reached when ρ and σ are pure states. The fidelity behaves qualitatively as the contrary of the trace distance, that is, it increases when the two states are less distinguishable and decreases when they become more distinguishable. As a consequence, the fidelity fulfills an *strong concavity* property in analogy to the strong convexity property of the trace distance, c.f. Eq. (1.35).

Two simple examples in which a closed form of the fidelity can be obtained are the case of commuting density operators, and for σ being a pure state. In the first case the fidelity becomes

$$F(\rho, \sigma) = \sum_i \sqrt{p_i q_i}, \tag{1.39}$$

where again $\{p_i\}$ and $\{q_i\}$ are the eigenvalues of ρ and σ. In this case the fidelity reduces to its classical expression for probability distributions of random variables. In the second case, by assuming $\sigma = |\psi\rangle\langle\psi|$, we have $\sigma^{\frac{1}{2}} = \sigma$, and hence

$$F(\rho, \sigma) = \sqrt{\langle\psi|\rho|\psi\rangle}, \tag{1.40}$$

the fidelity becomes the square root of the overlap between the two states. The fidelity is widely used in the context of quantum communication to characterize how well a quantum channel preserves information [5].

As commented above, the fidelity is not a metric on density operators because it does not fulfill the triangle inequality. Still it can be turned into a metric by using again the implications of Ullman's theorem (see [5] for details). Indeed the angle

$$\Theta(\rho, \sigma) \equiv \arccos F(\rho, \sigma), \tag{1.41}$$

is non-negative, symmetric in its inputs, equals zero if and only if $\rho = \sigma$, and verifies the triangle inequality. Henceforth, it is a proper metric on density operators.

1.2 Prototypical Systems

In most of the situations of interest in open quantum systems, the reduced system interacting with its surroundings can be described by some simple canonical models [26]. Those models capture the essence of the physical behavior of real systems, while making the calculations simpler, and in some cases analytically tractable. In this section, we introduce and review some important characteristics of two ubiquitous canonical models: the qubit system and the harmonic oscillator. These two models are important not only at a fundamental level, but also because experimental techniques have been developed in the last decade in order to provide in vivo control at the single particle level [27].

The qubit system describes systems with only two discrete accessible states. This implies that its Hilbert space $\mathcal{H}$ can be reduced to dimension $N = 2$, like a spin-$\frac{1}{2}$ particle. This situation arises, among others, in the case of photons with vertical or horizontal polarization, when a particle passes through a two-slit configuration in a Young interferometer, in the case of an atom interacting with a field nearly resonant to one of its atomic transitions, or for a low-energy particle trapped in a double well potential. On the other hand, the harmonic oscillator (or bosonic mode) model describes very accurately the electromagnetic field, and is well suited in general to account for potentials within a quadratic approximation e.g. close to a minimum of the potential. The harmonic oscillator model is usually employed e.g. to describe the vibrational modes of atoms bound in a molecule, ions trapped by electromagnetic fields, or the phonons generated in crystalline media [2].

1.2.1 The Qubit System

The qubit system is probably the simplest object in quantum theory. Its Hilbert space can be expanded by just two state vectors, which we denote $\{|0\rangle, |1\rangle\}$, often called *computational basis states* in quantum information contexts [5]. The two basis states may have different energy. We hence write the Hamiltonian of the system as

$$\hat{H} = E|1\rangle\langle 1| = \frac{E}{2}\left(\mathbb{1} + \hat{\sigma}_Z\right), \tag{1.42}$$

where we set to zero the energy of the (ground) state $|0\rangle$, and keep E as the energy gap between the two basis states. Furthermore, in the second equality we introduced the three Pauli operators $\hat{\sigma}_i$ for $i = X, Y, Z$, which are hermitian, unitary, traceless, and have eigenvalues ± 1. Moreover they fulfill the commutation relations $[\hat{\sigma}_i, \hat{\sigma}_j] = 2i\epsilon_{ijk}\hat{\sigma}_k$, being ϵ_{ijk} the Levi-Civita symbol, which is zero if two of the three indices are equal, takes the value 1 for even permutations of XYZ, and -1 for the odd ones. Importantly, any observable of the qubit system can be written as a combination of the three Pauli operators $\hat{\sigma}_i$ and the identity $\mathbb{1}$.

We also introduce the raising and lowering operators:

$$\begin{aligned} \hat{\sigma} &\equiv \frac{1}{2}\left(\hat{\sigma}_X + i\hat{\sigma}_Y\right) = |0\rangle\langle 1|, \\ \hat{\sigma}^\dagger &\equiv \frac{1}{2}\left(\hat{\sigma}_X - i\hat{\sigma}_Y\right) = |1\rangle\langle 0|, \end{aligned} \tag{1.43}$$

which promote jumps between the qubit energy levels, $\hat{\sigma}^\dagger|0\rangle = |1\rangle$ and $\hat{\sigma}|1\rangle = |0\rangle$. The raising and lowering operators fulfill the anti-commutation relation $\{\hat{\sigma}, \hat{\sigma}^\dagger\} \equiv \hat{\sigma}\hat{\sigma}^\dagger + \hat{\sigma}^\dagger\hat{\sigma} = \mathbb{1}$, and verify $\hat{\sigma}\hat{\sigma} = \hat{\sigma}^\dagger\hat{\sigma}^\dagger = 0$.

The most general pure state of the qubit system can be written as the superposition

$$|\psi\rangle = c_0|0\rangle + c_1|1\rangle, \tag{1.44}$$

where c_0 and c_1 are complex coefficients such that $0 \leqslant |c_i|^2 \leqslant 1$ and $|c_0|^2 + |c_1|^2 = 1$. As the Hilbert space dimension of the qubit system is just $N = 2$, $|\psi\rangle$ can be viewed just as a unit vector in a two-dimensional complex vector space. A convenient geometrical representation of such set of states is the so-called Bloch sphere in which any state $|\psi\rangle$ is represented as a point in the surface (see Fig. 1.4). Here the eigenstates of the operator $\hat{\sigma}_Z$, i.e. the basis states $\{|0\rangle, |1\rangle\}$, are represented as the north and shout

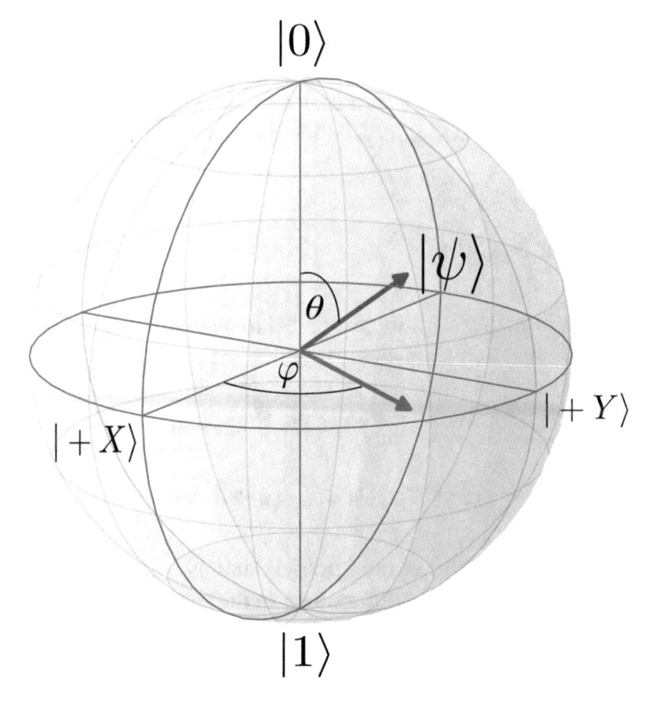

Fig. 1.4 Bloch sphere for a qubit system. The green vector pointing the surface represents the pure state $|\psi\rangle$, as specified by the angles θ and φ, while the orange vectors is its projection in the XY plane

poles of the sphere, while the x and y axes correspond to the eigenstates of $\hat{\sigma}_X$ and $\hat{\sigma}_Y$ respectively

$$|\pm X\rangle = \frac{1}{\sqrt{2}}\left(|0\rangle \pm |1\rangle\right), \quad |\pm Y\rangle = \frac{1}{\sqrt{2}}\left(|0\rangle \pm i|1\rangle\right). \tag{1.45}$$

Indeed, any two arbitrary orthonormal states $\langle\psi|\psi_\perp\rangle = 0$, are represented by antipode points on the Bloch sphere surface. Taking spherical coordinates we can associate the above introduced coefficients c_i to the polar and azimuthal angles, $\theta \in [0, \pi]$ and $\varphi \in [0, 2\pi)$, which allows us to rewrite Eq. (1.44) as

$$|\psi\rangle = \cos(\theta/2)|0\rangle + e^{i\phi}\sin(\theta/2)|1\rangle. \tag{1.46}$$

Pure states are represented as points in the surface of the Bloch sphere, while the inner volume corresponds to mixed states. A general state of the qubit system can be generally written as

$$\rho = \frac{1}{2}\left(\mathbb{1} + \vec{r}\cdot\vec{\sigma}\right), \tag{1.47}$$

where $\vec{r} = r(\sin\theta\cos\varphi, \sin\theta\sin\varphi, \cos\theta)$ is a real three-dimensional vector with $0 \leqslant r \leqslant 1$, and $\vec{\sigma} = (\hat{\sigma}_X, \hat{\sigma}_Y, \hat{\sigma}_Z)$. Notice also that the maximally mixed state $\mathbb{1}/2$ corresponds to the center of the sphere.

An important state of any quantum system is the thermal equilibrium or Gibbs state, which is generically defined for any quantum system with Hamiltonian $\hat{H}$ as

$$\rho_{\text{th}}(\beta) \equiv \frac{e^{-\beta\hat{H}}}{Z(\beta)}, \tag{1.48}$$

where $\beta \equiv 1/k_B T$ is the inverse temperature of the system, and $Z(\beta) = \text{Tr}[e^{-\beta\hat{H}}]$ is the partition function. Indeed the following relations are verified

$$\langle\hat{H}\rangle_{\rho_{\text{th}}} = -\partial_\beta \ln Z, \qquad S(\rho_{\text{th}}) = k_B(\ln Z + \beta\langle\hat{H}\rangle_{\rho_{\text{th}}}). \tag{1.49}$$

It can be furthermore shown that the Gibbs state maximizes the von Neumann entropy for a fixed value of the mean energy, and analogously that it is the state minimizing the mean energy for a fixed value of the entropy. In the case of the qubit system the thermal equilibrium state reads

$$\rho_{\text{th}}(\beta) = \frac{1}{Z}\left(|0\rangle\langle 0| + e^{-\beta E}|1\rangle\langle 1|\right), \tag{1.50}$$

with the partition function $Z = 1 + e^{-\beta E}$. The average energy of the qubit system in the Gibbs state reads

$$\langle\hat{H}\rangle_{\text{th}} = E\,\frac{e^{-\beta E}}{Z} = \frac{E}{1 + e^{-\beta E}}, \tag{1.51}$$

where $0 \leqslant \langle\hat{H}\rangle_{\text{th}} \leqslant E$, and the dispersion is given by

$$\sigma^2_{\text{th}}(\hat{H}) = E^2\,\frac{e^{-\beta E}}{(1 + e^{-\beta E})^2}. \tag{1.52}$$

This corresponds to the thermal energy fluctuations for a classical two-level system, fulfilling the following thermodynamical relation

$$\sigma^2_{\text{th}}(\hat{H}) = k_B T^2\,\partial_T\langle\hat{H}\rangle_{\text{th}}. \tag{1.53}$$

An important property of the qubit system extensively used in quantum thermodynamic contexts is that, given the Hamilton operator in (1.42), *any* mixture $\rho = p_0|0\rangle\langle 0| + p_1|1\rangle\langle 1|$ can be seen as a Gibbs state for an unequivocally inverse temperature defined by the detailed balance relation

$$\frac{p_0}{p_1} \equiv e^{\beta E} \quad \Leftrightarrow \quad \beta \equiv \frac{1}{\hbar\omega}\ln\frac{p_0}{p_1}. \tag{1.54}$$

The definition of such an effective temperature for a qubit system has proven very useful in recent studies on quantum thermal machines and theoretical investigations in quantum work extraction scenarios, where it has been sometimes called *virtual temperature* (see e.g. Ref. [28]).

The time evolution of an isolated qubit system is also rather simple. From the Hamiltonian in Eq. (1.42) it turns out that only the elements of the density operator non-diagonal in the computational basis evolve in time, acquiring a time-dependent phase factor $e^{iEt/\hbar}$. In the Bloch sphere representation, this implies that the azimuthal angle, θ, is kept constant during the evolution, together with r, the distance of the system state to the center of the sphere. The only effect is hence a rotation in the XY-plane at angular velocity $E/\hbar$. However, arbitrary rotations on the Bloch sphere become possible if we introduce an extra *classical field* interacting with our qubit system.

1.2.2 Manipulation of Qubits by Classical Fields

Consider the situation in which an external classical field interacts with a qubit system as described above. This setup can be used to prepare arbitrary states of the qubit system, provided it is completely isolated from the environment and we have a precise control over the field parameters. Physically, this situation arises in nuclear magnetic resonance (NMR) experiments when spin$-\frac{1}{2}$ nuclei are put on a constant magnetic field and perturbed by radio frequency pulses, or in the case of an isolated two-level atom interacting with a classical time-dependent electric field [2]. The interaction of the external field with the atomic dipole introduces an extra term in the Hamiltonian $\hat{H}$ of Eq. (1.42) of the form

$$\hat{H}_{\mathrm{f}}(t) = \frac{\hbar \Omega_R}{2} \left(\hat{\sigma}^{\dagger} e^{-i(\omega_{\mathrm{f}} t + \varphi_0)} + \hat{\sigma}\; e^{i(\omega_{\mathrm{f}} t + \varphi_0)} \right), \tag{1.55}$$

where ω_{f} is the frequency of the field, φ_0 its phase, and Ω_R is sometimes called the *classical* Rabi frequency [2]. This Hamiltonian results after eliminating terms oscillating at fast frequencies, $\pm 2\omega_{\mathrm{f}}$, a procedure which is usually known as the Rotating Wave Approximation (RWA).

The inclusion of extra Hamiltonian terms non-diagonal in the computational basis, changes drastically the time-evolution of the qubit system, making possible the exchange of energy between the qubit and the external field. In order to illustrate the dynamics, let us adopt an interaction frame with respect to $H_0 \equiv \hbar \omega_{\mathrm{f}} |1\rangle\langle 1|$, which represents a rotating frame at frequency ω_{f}. In such interaction picture the time-dependences are eliminated from the total Hamiltonian $\hat{H} = \hat{H}_0 + \hat{H}_{\mathrm{f}}$:

$$\hat{H}_I = \hbar \Delta_{\mathrm{f}} \hat{\sigma}^{\dagger} \hat{\sigma} + \hbar \frac{\Omega_R}{2} \left(\hat{\sigma}^{\dagger} e^{-i\varphi_0} + \hat{\sigma}\; e^{i\varphi_0} \right), \tag{1.56}$$

with $\Delta_{\rm f} = E/\hbar - \omega_f$ the detuning between the atomic transition and field frequencies. It can be indeed rewritten in the more suggestive form

$$\hat{H}_I \equiv \frac{\hbar\Delta_{\rm f}}{2}\mathbb{1} + \frac{\hbar\Omega'_R}{2}\,\vec{n}\cdot\vec{\sigma}, \quad \text{with} \;\; \Omega'_R = \sqrt{\Delta_{\rm f}^2 + \Omega_R^2} \tag{1.57}$$

the Rabi frequency, and the vector

$$\vec{n} = \frac{1}{\Omega'_R}(\Omega_R\cos(\varphi_0), \Omega_R\sin(\varphi_0), \Delta_{\rm f}). \tag{1.58}$$

This Hamiltonian can be interpreted by looking at the qubit state in the Bloch sphere as a pseudo-spin. Hence the second term in Eq. (1.57) describes the Larmor precession of the pseudo-spin at angular frequency Ω'_R, around the axis defined by the vector $\vec{n}$, while the first term is a constant with no effect on the dynamics. Therefore, the dynamics of the two-level atom state in the Bloch sphere can be viewed as the combination of two movements: a circular motion of frequency Ω'_R around the axis $\vec{n}$ in which the distance r to the center is maintained (unitary evolution), combined with a rotation in the XY plane at frequency $\omega_{\rm f} \sim E/\hbar$. It is worth noticing that the Z component of the vector $\vec{n}$ is proportional to the detuning between the two-level atom and the field frequencies, $\Delta_{\rm f}$, in contrast to the X and Y components, which are proportional to Ω_R. This implies that a large detuning, $\Delta_{\rm f} \gg \Omega_R$, will only imply rotations in the XY plane, making all the Z axis points invariant, corresponding to no-energy exchange between the two-level atom and the field. On the other hand, only when nearly resonant frequencies are reached, $\Delta_{\rm f} \ll \Omega_R$, the vector $\vec{n}$ is contained in the equatorial plane, and the field is able to induce flips between the basis states $|0\rangle$ and $|1\rangle$, which corresponds to the situation in which the two-level atom and the field exchange quanta of energy $E \simeq \hbar\omega_{\rm f}$.

1.2.3 *The Harmonic Oscillator*

The second canonical quantum extensively adopted in this PhD thesis is the one dimensional harmonic oscillator. This corresponds for instance to a particle of mass m moving in one dimension, trapped in a quadratic (attractive) potential $\hat{V} = m\omega^2\hat{x}^2/2$. The Hamiltonian of the model when including the kinetic energy of the particle reads

$$\hat{H} = \frac{\hat{p}^2}{2m} + \frac{m\omega^2}{2}\hat{x}^2, \tag{1.59}$$

where $\hat{x}$ and $\hat{p}$ (with $[\hat{x}, \hat{p}] = i\hbar$) are respectively the position and momentum operators of the particle. The harmonic oscillator model, apart of being ubiquitous in quantum physics as argued above, is also one of the few models for which an exact analytical solution is known. The eigenvalues and eigenstates of the Hamiltonian

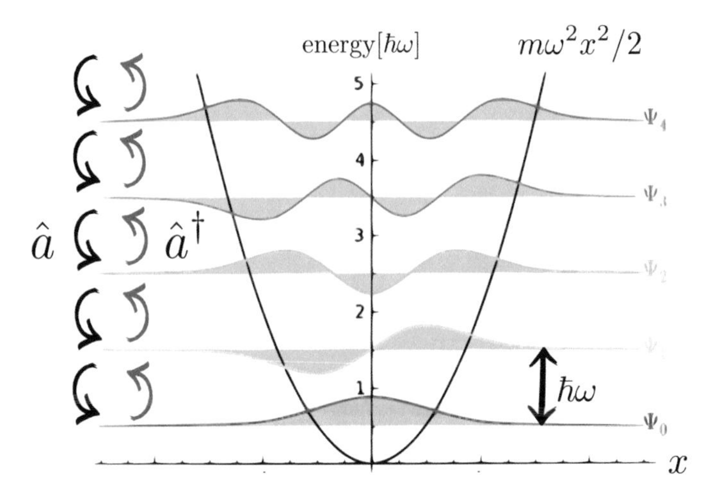

Fig. 1.5 Picture of the first four energy levels and wavefunctions of the quantum harmonic oscillator model. The infinite set of energy levels are equally spaced with energy gap $\hbar\omega$. The annihilation and creation operators, $\hat{a}$ and $\hat{a}^\dagger$, produce respectively jumps down and up in the energy levels ladder

(1.59), $\hat{H}|\psi_n\rangle = E_n|\psi_n\rangle$, are

$$E_n = \hbar\omega\left(n + \frac{1}{2}\right), \tag{1.60}$$

$$\Psi_n(x) = \frac{1}{\sqrt{2^n n!}}\left(\frac{m\omega}{\pi\hbar}\right)^{\frac{1}{4}} e^{-\frac{m\omega^2 x^2}{2\hbar}} H_n\left(\sqrt{\frac{m\omega}{\hbar}}x\right), \tag{1.61}$$

for $n = 0, 1, 2, \ldots$, and where we introduced the Hermite polynomials $H_n(z) = (-1)^n e^{z^2}\partial_z^n(e^{-z^2})$. We hence obtain an infinite number of equally spaced (quantized) energy levels spanning the Hilbert space of the harmonic oscillator. The wavefunction corresponding to the nth energy level is a Gaussian function on x modulated by a polynomial with n zeros or nodes (see Fig. 1.5).The eigenstates of $\hat{H}$ are called Fock states, and will be denoted by $\{|0\rangle, |1\rangle, |2\rangle, \ldots, |n\rangle, \ldots\}$. As an orthonormal basis, the Fock states fulfill $\langle n|k\rangle = \delta_{k,n}$, and $\sum_n |n\rangle\langle n| = \mathbb{1}$. It is also worth noticing that the ground state of the model, the state $|0\rangle$, is characterized by a non-zero energy, $E_0 = \hbar\omega/2$, usually called the zero point energy, or vacuum fluctuation energy.

It is very useful at this point to introduce the following annihilation and creation (or simply ladder) operators

$$\hat{a} = \sqrt{\frac{m\omega}{2\hbar}}\left(\hat{x} + \frac{i}{m\omega}\hat{p}\right), \quad \hat{a}^\dagger = \sqrt{\frac{m\omega}{2\hbar}}\left(\hat{x} - \frac{i}{m\omega}\hat{p}\right), \tag{1.62}$$

which promote jumps in the energy ladder of the harmonic oscillator,

$$\hat{a}|n\rangle = \sqrt{n}|n-1\rangle, \qquad \hat{a}^\dagger|n\rangle = \sqrt{n+1}|n+1\rangle, \tag{1.63}$$

and $a|0\rangle = 0$. As can be easily appreciated, those operators remove or add an energy quantum $\hbar\omega$ to the oscillator, and fulfill the commutation relation $[a, a^\dagger] = \mathbb{1}$. We can furthermore introduce the *number operator* as $\hat{N} \equiv \hat{a}^\dagger \hat{a}$, whose action over the Fock states is simply $\hat{N}|n\rangle = n|n\rangle$, and fulfill

$$[\hat{N}, \hat{a}] = -\hat{a}, \qquad [\hat{N}, \hat{a}^\dagger] = \hat{a}^\dagger. \tag{1.64}$$

The Hamiltonian (1.59) can be rewritten in the canonical form

$$\hat{H} = \hbar\omega\left(\hat{N} + \frac{1}{2}\right). \tag{1.65}$$

In this above canonical form, the constant term corresponding to the zero-point energy, $E_0 = \hbar\omega/2$, is often neglected by redefining the energy origin. The energy corresponding to the Fock states $\{|n\rangle\}$, c.f. Eq. (1.60), can be hence attributed to n quanta $\hbar\omega$ [2]. The Hamiltonian (1.65) may describe as well an electromagnetic field mode of angular frequency ω in a (completely isolated) cavity, or more generally, one of the components of the electromagnetic field propagating in free space. In this case the excitations of the oscillator are called photons, $\hat{N}$ the photon number operator, and the states $|n\rangle$, the photon number states. In light of the intimate connexion between the harmonic oscillator model and the nature of the electromagnetic field, we will often employ during this thesis the characteristic nomenclature used for fields to refer as well to the general harmonic oscillator model, and vice-versa.

Any generic pure state of the harmonic oscillator model can be written as a superposition of Fock states

$$|\psi\rangle = \sum_{n=0}^{\infty} c_n |n\rangle, \tag{1.66}$$

with complex coefficients c_n, such that $0 \leqslant |c_n|^2 \leqslant 1$ and $\sum_n |c_n|^2 = 1$. The time evolution of such state under the action of the Hamiltonian (1.65) is

$$|\psi\rangle(t) = e^{-\frac{i}{\hbar}\hat{H}t}|\psi\rangle = \sum_{n=0}^{\infty} c_n e^{-i\omega(n+1/2)t}|n\rangle, \tag{1.67}$$

that is, each of the Fock states appearing in the superposition $|\psi\rangle$, acquires a time dependent phase. It is worth noticing that while the number of quanta in the Fock states $|n\rangle$ is always well defined, the mean value of the position (and momentum) operator vanishes, $\langle n|\hat{x}|n\rangle = 0$, indicating that Fock states do not behave in a classical way.

The thermal equilibrium state can be again defined for the harmonic oscillator model from Eq. (1.48) as a mixture of Fock states with Boltzmann weights

$$\rho_{\text{th}} = \sum_{n=0}^{\infty} \frac{e^{-\beta\hbar\omega(n+\frac{1}{2})}}{Z(\beta)} |n\rangle\langle n|, \tag{1.68}$$

where $\beta = 1/k_B T$ represents again the inverse temperature, and now the partition function reads $Z(\beta) = e^{\beta\hbar\omega/2}/(e^{\beta\hbar\omega} - 1)$. The mean energy in ρ_{th} using Eq. (1.65) reproduces Planck's law $\langle\hat{H}\rangle_{\text{th}} = \hbar\omega(\langle\hat{N}\rangle_{\text{th}} + 1/2)$, where $\langle\hat{N}\rangle_{\text{th}} = (e^{\beta\hbar\omega} - 1)^{-1}$ is the mean number of quanta. Furthermore we see that in the limit $T \to \infty$ (or equivalently $\beta \to 0$) we obtain

$$\langle\hat{H}\rangle_{\text{th}} \simeq k_B T, \qquad \langle\hat{N}\rangle_{\text{th}} \simeq \frac{k_B T}{\hbar\omega}, \tag{1.69}$$

approaching the value predicted by the classical Maxwell–Boltzmann distribution [10]. The variance in the number of quanta can be calculated yielding

$$\sigma_{\text{th}}^2(\hat{N}) = \frac{e^{\beta\hbar\omega}}{(e^{\beta\hbar\omega} - 1)^2}. \tag{1.70}$$

It is now easy to check that the state ρ_{th} shows equilibrium energy fluctuations verifying, like in the qubit system case,

$$\sigma_{\text{th}}^2(\hat{H}) = k_B T^2 \,\partial_T \langle\hat{H}\rangle_{\text{th}} \tag{1.71}$$

as corresponds to the canonical ensemble [29].

1.2.4 Coherent States

In contrast to Fock states, the *coherent* or Glauber states [30] (also Gaussian states in more general contexts [6]) reproduce much better the classical behavior required by the correspondence principle [10]. They are usually denoted as $|\alpha\rangle$, and can be defined as the (right) eigenstates of the annihilation operator, that is $a|\alpha\rangle = \alpha|\alpha\rangle$ (or equivalently $\langle\alpha|a^\dagger = \alpha^*\langle\alpha|$), with complex eigenvalue α. In the Fock states basis they can be written as the superpositions

$$|\alpha\rangle = e^{-\frac{1}{2}|\alpha|^2} \sum_{n=0}^{\infty} \frac{\alpha^n}{\sqrt{n!}} |n\rangle, \tag{1.72}$$

with a Poissonian probability distribution $p_\alpha(n) = \exp(-|\alpha|^2)|\alpha|^{2n}/n!$ for the photon number. The distribution is peaked around the mean number of quanta $\langle\hat{N}\rangle_\alpha = \langle\alpha|\hat{N}|\alpha\rangle = |\alpha|^2$, with variance given by $\sigma_\alpha^2 = \langle\hat{N}^2\rangle_\alpha - \langle\hat{N}\rangle_\alpha^2 = |\alpha|^2 = \langle N\rangle_\alpha$. This implies that, as the mean number of quanta increases, the distribution becomes more peaked around its mean value, presenting smaller relative energy fluctuations, corresponding to the classical case. Indeed coherent states can be obtained by considering the radiation emitted by a classical oscillating electric current [31, 32]. Another prominent physical example is the highly coherent monochromatic light emitted by a laser.

An important property of coherent states is that they also correspond to minimum uncertainty states for position and momenta

$$\sigma_\alpha(\hat{x})\sigma_\alpha(\hat{p}) = \frac{\hbar}{2} \tag{1.73}$$

where $\sigma_\alpha(\hat{x}) = \sqrt{\hbar/2m\omega}$ and $\sigma_\alpha(\hat{p}) = \sqrt{\hbar m\omega/2}$. Furthermore the set of coherent states fulfill the completeness relation $\int \alpha|\alpha\rangle\langle\alpha| = \pi\,\mathbb{1}$, while being non-orthogonal $|\langle\alpha|\alpha'\rangle|^2 = \exp(-|\alpha - \alpha'|^2)$ [31].

The time evolution of a coherent state $|\alpha\rangle$ can be easily calculated from Eq. (1.67) to be $|\alpha\rangle(t) = e^{-i\omega t/2}|\alpha e^{-i\omega t}\rangle$, meaning that the harmonic oscillator always remains in a coherent state. Its wave function in the position representation reads

$$\psi_\alpha(x) = \left(\frac{m\omega}{\pi\hbar}\right)^{\frac{1}{4}} e^{-|\alpha|^2/2} e^{\frac{m\omega x^2}{2\hbar}} e^{(\sqrt{m\omega/\hbar}x - \alpha e^{-i\omega t}/\sqrt{2})^2}. \tag{1.74}$$

It corresponds to a minimum uncertainty Gaussian wake packet which maintains the same shape during its evolution, while its centroid follows a classical oscillatory trajectory between the two turning points of the potential [32].

It is also convenient to introduce at this point the field quadratures of the harmonic oscillator, corresponding to the operators

$$\hat{X}_\varphi \equiv \frac{1}{\sqrt{2}}(ae^{-i\varphi} + a^\dagger e^{i\varphi}) \quad \text{and} \quad \hat{P}_\varphi \equiv \hat{X}_{\varphi+\pi/2}. \tag{1.75}$$

They are dimensionless versions of the position and momenta operators, rotated by an angle φ in phase space. They fulfill the commutation relation $[\hat{X}_\varphi, \hat{P}_\varphi] = i$, as well as the uncertainty principle $\sigma(\hat{X}_\varphi)\sigma(\hat{P}_\varphi) \geqslant \frac{1}{2}$, in accordance with Eq. (1.1). With the help of the field quadratures, we may consider an 'optical' phase space in terms of the quadratures $\hat{X}_0$ and $\hat{P}_0$, instead of using the original position and momentum operators, $\hat{x}$ and $\hat{p}$, used to define the quantum-mechanical phase space. In the optical phase space, coherent states can be represented by filled circles centered around the point $(Re(\alpha), Im(\alpha))$ with diameter $1/\sqrt{2}$, as the uncertainty in any field quadrature turns out to be equal, $\sigma_\alpha(\hat{X}_\varphi) = \sigma_\alpha(\hat{P}_\varphi) = 1/\sqrt{2}$. Their time evolution, as calculated above, corresponds to a uniform clockwise circular motion of the filled circle with respect to the origin of coordinates at angular velocity ω (see Fig. 1.6). A mere general representation of states in the quantum optics phase space would be obtained with quasi-probability distributions as discussed in Ref. [33].

Notice that the vacuum state $|0\rangle$ is also a coherent state with $\alpha = 0$. It can be indeed represented in optical phase space as a filled circle with the same diameter as any other coherent state, centered at the origin of coordinates. This picture suggests

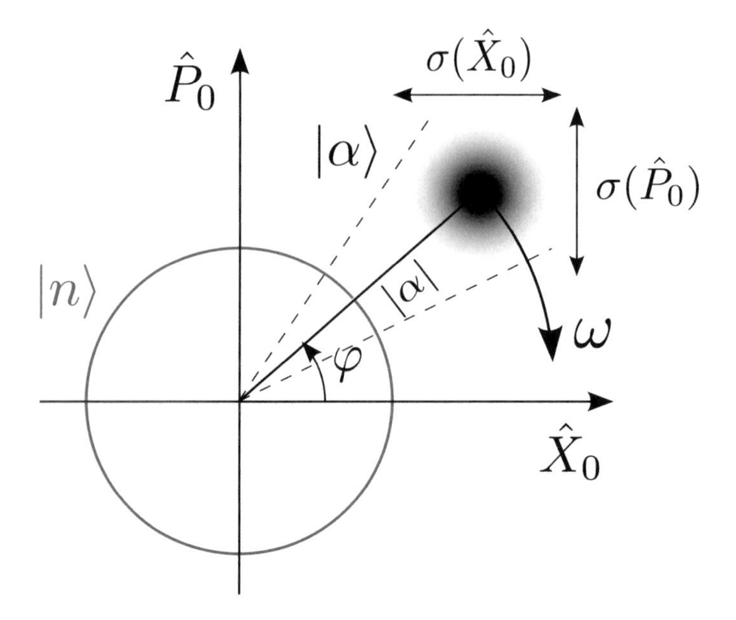

Fig. 1.6 Representation of a Fock state $|n\rangle$ (blue color) and a coherent state $|\alpha\rangle$ (black color) in optical phase space. Fock states are represented by hollow circles centered at the origin of coordinates, being its uncertainty in any field quadrature negligible. In the other hand, coherent states with $\alpha = |\alpha|e^{i\varphi}$, are represented by filled circles stressing the presence of equal uncertainties in the quadrature fields. The later are displaced from the origin of coordinates a distance $|\alpha|$ and rotate clockwise around the origin of coordinates at angular velocity ω (see text)

that coherent states are just vacuum states displaced in phase space a distance $|\alpha|$ in the direction determined by the angle θ defined by the complex number $\alpha = |\alpha|e^{i\theta}$. The above idea can be made mathematically precise by defining a *displacement* operator

$$\hat{\mathfrak{D}}(\alpha) \equiv e^{\alpha\hat{a}^\dagger - \alpha^*\hat{a}}, \tag{1.76}$$

which is unitary, $\hat{\mathfrak{D}}(\alpha)\hat{\mathfrak{D}}(\alpha)^\dagger = \hat{\mathfrak{D}}(\alpha)^\dagger\hat{\mathfrak{D}}(\alpha) = \mathbb{1}$, and fulfills $\hat{\mathfrak{D}}(\alpha)^\dagger = \hat{\mathfrak{D}}(-\alpha)$. Any coherent state can be generated applying the displacement operator to the vacuum state $\hat{\mathfrak{D}}(\alpha)|0\rangle = |\alpha\rangle$ [31]. Moreover, the action of the displacement operator over the annihilation and creation operators turns out to be

$$\hat{\mathfrak{D}}(\alpha)\, a\, \hat{\mathfrak{D}}(\alpha)^\dagger = a - \alpha, \qquad \hat{\mathfrak{D}}(\alpha)\, a^\dagger\, \hat{\mathfrak{D}}(\alpha)^\dagger = a^\dagger - \alpha^*. \tag{1.77}$$

Using the above properties one can also prove that

$$\begin{aligned} \hat{\mathfrak{D}}(\alpha)\hat{\mathfrak{D}}(\beta) &= \hat{\mathfrak{D}}(\alpha+\beta)\, e^{(\alpha\beta^* - \alpha^*\beta)/2}, \\ \hat{\mathfrak{D}}(\alpha)|\beta\rangle &= e^{(\alpha\beta^* - \alpha^*\beta)/2}\, |\alpha+\beta\rangle. \end{aligned} \tag{1.78}$$

These relationships play an important role in quantum optics. They have been used e.g. in the development of field measurement techniques such as *homodyne* detection [2].

Once the displacement operator has been introduced in Eq. (1.76), we may apply it to arbitrary states. Of particular relevance is the so-called *coherent thermal state*, defined as

$$\rho_{\rm d} \equiv \hat{\mathfrak{D}}(\alpha)\rho_{\rm th}(\beta)\hat{\mathfrak{D}}(\alpha)^\dagger = \frac{e^{-\beta\hat{\mathfrak{D}}(\alpha)\hat{H}\hat{\mathfrak{D}}(\alpha)^\dagger}}{Z(\beta)}. \tag{1.79}$$

This state corresponds to the thermal equilibrium state of a displaced field with modified Hamiltonian $\hat{H}_{\rm d} \equiv \hat{\mathfrak{D}}(\alpha)\hat{H}\hat{\mathfrak{D}}(\alpha)^\dagger = \hbar\omega(\hat{A}^\dagger\hat{A} + 1/2)$, being $\hat{A} = a - \alpha$. The state $\rho_{\rm d}$ has the same entropy as the Gibbs state $\rho_{\rm th}$ but modified mean energy

$$\langle \hat{H}\rangle_{\rho_{\rm d}} = {\rm Tr}[\hat{\mathfrak{D}}(\alpha)^\dagger \hat{H}\hat{\mathfrak{D}}(\alpha)\rho_{\rm th}(\beta)] = \langle \hat{H}\rangle_{\rho_{\rm th}} + \hbar\omega|\alpha|^2, \tag{1.80}$$

with the addition of a term scaling with $|\alpha|^2$. In the above equation we used the cyclic property of the trace and the relations (1.77). In a similar way we can calculate the energy fluctuations in the state

$$\sigma^2_{\rho_{\rm d}}(\hat{H}) = \sigma^2_{\rho_{\rm th}}(\hat{H}) + |\alpha|^2(\hbar\omega + 2\langle\hat{H}\rangle_{\rho_{\rm th}}), \tag{1.81}$$

leading to a stretched photon number distribution with respect to the thermal equilibrium case. In analogy to coherent states with respect to vacuum, the coherent thermal state has the same shape as the Gibbs state, displaced a distance $|\alpha|$ in direction θ.

1.2.5 Squeezed States

A second class of states of great importance in quantum optics and quantum information, are the so-called *squeezed states*. In order to introduce their definition and properties it is convenient to turn back to the generalized Heisenberg's uncertainty relation in Eq. (1.1). The fluctuations of two conjugate observables $\hat{A}$ and $\hat{B}$ in any state ρ of the quantum system under consideration are linked by

$$\sigma(\hat{A})\,\sigma(\hat{B}) \geqslant \frac{1}{2}|\langle[\hat{A},\hat{B}]\rangle_\rho|, \tag{1.82}$$

where $\sigma^2(\hat{A}) = \langle\hat{A}^2\rangle_\rho - \langle\hat{A}\rangle^2_\rho$ denotes the variance of the observable $\hat{A}$ in the state ρ. A state is called a squeezed state if it satisfies [31]

$$\sigma^2(\hat{A}) < \frac{1}{2}|\langle[\hat{A},\hat{B}]\rangle_\rho|, \quad \text{or} \quad \sigma^2(\hat{B}) < \frac{1}{2}|\langle[\hat{A},\hat{B}]\rangle_\rho|, \tag{1.83}$$

that is, a squeezed state verifies that the variance in the statistics of some observable is below the standard levels of quantum noise, at the expense of increasing the fluctuations of the conjugate observable. If the state of the system furthermore verifies the equality in Eq. (1.82), then it is called an *ideal squeezed state*, a nomenclature originally introduced in Ref. [34]. In the case of the harmonic oscillator, we usually deal with *quadrature squeezing*, which is defined by

$$\sigma^2(\hat{X}_\varphi) < \frac{1}{2}, \quad \text{or} \quad \sigma^2(\hat{P}_\varphi) < \frac{1}{2}. \tag{1.84}$$

It follows from that definition that Fock and coherent states are not squeezed. Notice that the concept of squeezing may be applied as well to any other two conjugate observables of the harmonic oscillator other than the field quadratures [32].

Squeezed states may be mathematically characterized by means of the following *squeezing operator*

$$\hat{\mathbb{S}}(\xi) \equiv e^{\frac{1}{2}(\xi^* \hat{a}^2 - \xi \hat{a}^{\dagger 2})}, \tag{1.85}$$

where $\xi \equiv re^{i\theta}$ is a complex number with $r \geqslant 0$, usually called the *squeezing parameter*, and $\theta \in [0, 2\pi]$. The squeezing operator may act on any state of the harmonic oscillator and is intimately connected to the second order coherences generated by the operators $\hat{a}^2$ and $\hat{a}^{\dagger 2}$, describing two quanta annihilation and creation processes. Squeezing must be henceforth considered as a quantum effect [32] with no classical analogue [10].

The squeezing operator in Eq. (1.85) has many similarities with the displacement operator introduced in Eq. (1.76). It is also unitary $\hat{\mathbb{S}}(\xi)\hat{\mathbb{S}}^\dagger(\xi) = \hat{\mathbb{S}}^\dagger(\xi)\hat{\mathbb{S}}(\xi) = \mathbb{1}$, with $\hat{\mathbb{S}}^\dagger(\xi) = \hat{\mathbb{S}}(-\xi)$. Furthermore it acts on the annihilation and creation operators as

$$\begin{aligned} \hat{R} &\equiv \hat{\mathbb{S}}(\xi)\, \hat{a}\, \hat{\mathbb{S}}^\dagger(\xi) = \hat{a} \cosh(r) + \hat{a}^\dagger \sinh(r) e^{i\theta} \\ \hat{R}^\dagger &\equiv \hat{\mathbb{S}}(\xi)\, \hat{a}^\dagger\, \hat{\mathbb{S}}^\dagger(\xi) = \hat{a}^\dagger \cosh(r) + \hat{a} \sinh(r) e^{-i\theta} \end{aligned} \tag{1.86}$$

which defines a canonical Bogoliubov–Valatin transformation, mapping the annihilation and creation operators to the new operators $\hat{R}$ and $\hat{R}^\dagger$ as given in Eq. (1.86), with $[\hat{R}, \hat{R}^\dagger] = \mathbb{1}$.

The most simple example of a squeezed state follows from applying the operator (1.85) to the vacuum state $|0\rangle$, to get

$$|\xi\rangle \equiv \hat{\mathbb{S}}(\xi)|0\rangle = \frac{1}{\sqrt{\cosh(r)}} \sum_{n=0}^{\infty} (-\tanh(r))^n \frac{\sqrt{(2n)!}}{2^n n!} |2n\rangle, \tag{1.87}$$

called the squeezed vacuum state. Notice that this state is an infinite superposition of even-quanta Fock states, with non-zero mean number of quanta $\langle \hat{N} \rangle_\xi = \sinh(r)^2$ and variance

$$\sigma_\xi^2(\hat{N}) = 2\sinh^2(r)\cosh^2(r) = 2\langle \hat{N} \rangle_\xi (\langle \hat{N} \rangle_\xi + 1),$$

showing photon bunching and super-Poissonian statistics [35, 36]. The squeezed vacuum state also verifies $\hat{R}|\xi\rangle = 0$, and can be regarded as the vacuum (ground) state of the Bogoliubov mode defined by a Hamiltonian of the form $\hat{H}_B = \hbar\omega(\hat{R}^\dagger \hat{R} + 1/2)$. The crucial property of the state (1.87) is the modification of their variances in the field quadratures along the angle $\theta/2$ with respect to the vacuum state

$$\sigma_\xi^2(\hat{X}_{\theta/2}) = \frac{e^{-2r}}{2}, \quad \sigma_\xi^2(\hat{P}_{\theta/2}) = \frac{e^{2r}}{2}. \tag{1.88}$$

As we can see in the above equation, the fluctuations in the field quadrature $\hat{X}_{\theta/2}$ are 'squeezed' by an exponential factor depending on the squeezing parameter r, while the ones in $\hat{P}_{\theta/2}$ are increased by the inverse multiplicative factor. They still correspond to a minimum uncertainty state (or ideal squeezed state), that is $\sigma_\xi(\hat{X}_{\theta/2})\sigma_\xi(\hat{P}_{\theta/2}) = 1/2$. The squeezed vacuum (1.87) can be represented in optical phase space as a filled ellipse centered at the origin, minor axis in the $\theta/2$ direction and major axis in the $(\theta + \pi)/2$ one, with respective widths given by the variance in Eq. (1.88). Furthermore, the time evolution generated by the Hamiltonian (1.65) will produce clockwise rotation of the ellipse at uniform angular frequency ω.

The above introduced squeezed vacuum state may be generalized by applying the displacement operator in Eq. (1.76) as

$$|\alpha, \xi\rangle = \hat{\mathcal{D}}(\alpha)|\xi\rangle = \hat{\mathcal{D}}(\alpha)\hat{\mathcal{S}}(\xi)|0\rangle, \tag{1.89}$$

where again $\xi = re^{i\theta}$ and $\alpha = |\alpha|e^{i\varphi}$. This family of squeezed displaced states, including the vacuum squeezed state $|\xi\rangle$, are all ideal squeezed states with the same variances in the field quadratures as in Eq. (1.88), independently of the field amplitude α. Indeed squeezing is a macroscopic quantum effect which may be present in high intensity fields [35], a prediction which has been recently demonstrated in the laboratory [37] (see also Ref. [38]). The photon number distribution in squeezed coherent states is peaked around $\langle \hat{N} \rangle_{\alpha,\xi} = |\alpha|^2 + \sinh^2(r)$ with variance $\sigma_{\alpha,\xi}(\hat{N}) = |\alpha \cosh(r) - \alpha^* \sinh(r)e^{i\theta}| + 2\cosh^2(r)\sinh^2(r)$ [10]. It can show both sub-Poissonian or super-Poissonian statistics depending on the squeezing and displacement angles θ and φ, and hence display anti-bunching or bunching effects [36]. The representation of squeezed displaced states in optical phase space can be easily obtained by just shifting the ellipse representing the squeezed vacuum state $|\xi\rangle$ a distance $|\alpha|$ along the direction defined by φ [32]. Squeezed displaced states have been studied since the 60s by authors interested in the generalization of the minimum uncertainty states in different systems [34, 39–45] (for reviews see [35, 36]).

Another interesting example of a squeezed state to particular relevance for this thesis is the *squeezed thermal state*, resulting from the application of the squeezing operator (1.85) to the thermal equilibrium state [47, 48]

$$\rho_{\mathrm{sq}} = \hat{\mathcal{S}}(\xi)\, \rho_{\mathrm{th}}\, \hat{\mathcal{S}}^\dagger(\xi) = \hat{\mathcal{S}}(\xi)\, \frac{e^{-\beta \hat{H}}}{Z(\beta)}\, \hat{\mathcal{S}}^\dagger(\xi), \tag{1.90}$$

with again $\xi = re^{i\theta}$, to be compared with Eq. (1.79). In analogy to the displaced thermal state, the squeezed thermal state has the same entropy as the Gibbs state but an increased mean energy

$$\langle \hat{H} \rangle_{\rho_{\text{sq}}} = \text{Tr}[\hat{H}\hat{\mathbb{S}}(\xi)\, \rho_{\text{th}}\, \hat{\mathbb{S}}(\xi)^{\dagger}] = \hbar\omega \left(\langle \hat{N} \rangle_{\text{th}} \cosh(2r) + \sinh(r)^2 + \frac{1}{2} \right),$$

which increases exponentially with r. Its fluctuations are also larger than in the thermal equilibrium case, and given by

$$\sigma_{\text{sq}}(\hat{H}) = (\hbar\omega)^2 \left(\cosh(4r)(\langle \hat{N} \rangle_{\text{th}}^2 + \langle \hat{N} \rangle_{\text{th}}) + \sinh^2(2r)/2 \right). \tag{1.91}$$

Moreover, in this case its variances in the field quadratures $\hat{X}_{\theta/2}$ and $\hat{P}_{\theta/2}$ read

$$\begin{aligned} \sigma_{\rho_{\text{sq}}}(\hat{X}_{\theta/2})^2 &= e^{-2r} \sigma_{\rho_{\text{th}}}(\hat{X}_{\theta/2})^2 = e^{-2r} \left(\langle \hat{N} \rangle_{\text{th}} + \frac{1}{2} \right), \\ \sigma_{\rho_{\text{sq}}}(\hat{P}_{\theta/2})^2 &= e^{2r} \sigma_{\rho_{\text{th}}}(\hat{P}_{\theta/2})^2 = e^{2r} \left(\langle \hat{N} \rangle_{\text{th}} + \frac{1}{2} \right), \end{aligned} \tag{1.92}$$

again producing the squeezing of the fluctuations in the first quadrature at expenses of increasing in the conjugate one. The will turn on the interesting properties of the squeezed thermal states from a quantum nonequilibrium thermodynamics point of view in Part IV of this thesis.

The first experimental generation of squeezed light was performed in 1985 using four-wave mixing techniques in an optical cavity with sodium atoms [49, 50]. Other pioneering experimental realizations of squeezing states include four-wave mixing in optical fibers [51] and degenerate parametric down conversion by using a nonlinear-optical crystal of MgO : LiNbOs inside an optical cavity [52]. The development of quantum state reconstruction methods and related theoretical tools, allowed then to comprehensively analyze and reconstruct squeezed states of light (see e.g. Fig. 1.7 and Ref. [46]). Squeezed thermal states have been also generated in the laboratory in a variety of physical systems of interest, its first experimental realization being throughout a Josephson-parametric amplifier operated at microwave frequencies [53].

Squeezing nowadays constitutes a central tool in quantum optics, with several applications in quantum metrology, quantum communication and quantum information processing [54, 55]. Some examples are the use of squeezed states of light in gravitational-wave detection [34, 56, 57], as well as in quantum teleportation scenarios [58], quantum computation with continuous variable systems [59], secure quantum key distribution protocols [60], or quantum imaging [61]. Furthermore, squeezing has been also successfully generated and detected in non-optical systems such as trapped atoms [62, 63], Bose–Einstein condensates [64, 65], collective spin-wave excitations (magnons) [66], or motional degrees of freedom in optomechanical setups [67, 68].

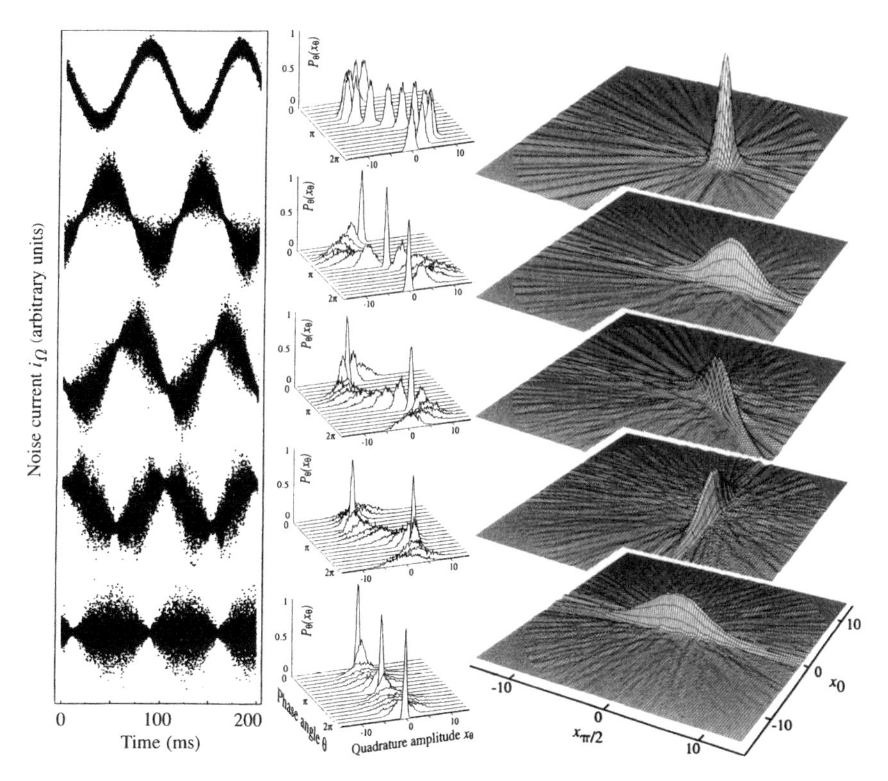

Fig. 1.7 Experimental results from generation and detection of different states of squeezed light by using a lithium-niobate optical parametric oscillator, pumped by a frequency-doubled continuous-wave Nd : YAG laser. In the left column the noise traces in the photocurrent $i_\Omega(t)$ when measuring the field quadratures are shown. In the center column the quadrature distributions are plotted, which can be interpreted as the time evolution of wave packets during one oscillation period. Finally in the third column it is shown one of the reconstructed quasi-probability distributions (the Wigner function [33]) of the generated states. From top to bottom: coherent state, phase-squeezed state, state squeezed in the $\theta = 48°$-quadrature, amplitude squeezed state, and squeezed vacuum state. Image obtained from Ref. [46]

1.3 Quantum Measurement

As commented previously, quantum theory ascribes an intrinsic probabilistic nature to measurement results of any observable. Even when we have maximal knowledge about the state of a system, that is, a pure state, the observables of the system are not completely determined, and if we measure it, different random results are obtained. In this section we introduce the formalism of quantum measurement. First we will introduce ideal quantum measurements to then move to the case of indirect measurements, i.e. measurements that are performed with the help of an ancillary system which is correlated with the system to be measured. We will use along the section the formalism of density operators introduced above and the related quantities used to measure purity, uncertainty or correlations in composite systems.

1.3.1 Ideal Measurements

Ideal quantum measurements constitute the simplest case of measurements in quantum theory, in which no extra sources of noise coming from the measuring apparatus are considered. They were introduced by Von Neumann [69] and Lüders [70] and are given in terms of *projectors*. Imagine we want to measure some observable given by an hermitian operator

$$\hat{O} = \sum_k o_k \hat{\Pi}_k, \tag{1.93}$$

where o_k are the real eigenvalues of $\hat{O}$, which we consider to be discrete. The operators $\hat{\Pi}_k$ are called the *projectors* onto the eigenspace of $\hat{O}$ with eigenvalue o_k. If the eigenvalues are non-degenerate we recover the expression introduced at the beginning of Sect. 1.1, that is, the projectors $\hat{\Pi}_k = |o_k\rangle\langle o_k|$ have rank-1. Otherwise a new quantum number has to be introduced in order to take into account the degeneracy, the projector then reading $\hat{\Pi}_k = \sum_{n=1}^{d_k} |o_k, n\rangle\langle o_k, n|$, where d_k is the degeneracy of the eigenvalue o_k. Notice that the sum of all degeneracies gives us the dimension of the system Hilbert space, $\sum_k d_k = \dim(\mathcal{H})$. In any case the projectors always fulfill orthonormal relations $\hat{\Pi}_k \hat{\Pi}_n = \delta_{k,n} \hat{\Pi}_k$ and the completeness relation $\sum_k \hat{\Pi}_k = \hat{\mathbb{1}}$, i.e. they provide a resolution of the identity operator.

The *projection postulate* states that we can choose any arbitrary observable $\hat{O}$ with a corresponding set of projectors $\{\hat{\Pi}_k\}$ to measure our system. The state of the system prior to measurement can be described by a density operator ρ. As a result of the measurement we will obtain an eigenvalue o_k, with probability $P_k = \mathrm{Tr}[\rho \hat{\Pi}_k]$, leaving the system after measurement in state

$$\rho'_k = \frac{\hat{\Pi}_k \, \rho \, \hat{\Pi}_k}{P_k}, \tag{1.94}$$

i.e. the state is projected onto the eigenspace $\hat{\Pi}_k$ of $\hat{O}$. Then any posterior measurement of the observable $\hat{O}$ would produce the same result o_k, leaving unaltered the system so far, as can be easily checked iterating Eq. (1.94). We call ρ'_k the *conditional* state of the system after the *selective* measurement, which implies knowledge about the result, as labeled by index k. If, on the contrary, we know that a measurement has been performed but we ignore the result (or forgot it), then the final state after measurement is given by averaging over all possible outcomes of the measurement

$$\rho' = \sum_k P_k \rho'_k = \sum_k \hat{\Pi}_k \, \rho \, \hat{\Pi}_k, \tag{1.95}$$

to which we refer as the *unconditional* state after the *unselective* measurement. Different observers can ascribe different density operators to the same system depending on their knowledge in a compatible way [6].

A *selective* measurement, in general, tends to reduce the entropy of a state ρ, as we obtain information about it. Indeed if the projectors $\hat{\Pi}_k$ are rank-1, the *conditional* state after measurement is pure, $\rho'_k = \hat{\Pi}_k = |o_k\rangle\langle o_k|$ and then $S(\rho) \geqslant S(\rho'_k) = 0$. On the other hand, for an *unselective* measurement the entropy of the state cannot decrease, $S(\rho') \geqslant S(\rho)$, holding the equality if and only if the state of the system is unaltered by the measurement [5]. The proof of the above statement follows from Klein's inequality, by noticing that

$$\begin{aligned} S(\rho') - S(\rho) &= -\operatorname{Tr}[\rho' \ln \rho'] - S(\rho) = -\sum_k \operatorname{Tr}[\Pi_k \rho \Pi_k \ln \rho'] - S(\rho) \\ &= -\operatorname{Tr}[\rho \ln \rho'] - S(\rho) = D(\rho || \rho') \geqslant 0, \end{aligned} \tag{1.96}$$

where we have used $[\hat{\Pi}_k, \rho'] = 0$, the cyclic property of the trace and $\sum_k \hat{\Pi}_k^2 = \mathbb{1}$. This result can be understood from the fact that an *unselective* quantum measurement disturbs the system in a random way, as we have no information about the result of the measurement, and hence we lose information about its state. From the above expression it is also clear that $S(\rho') = S(\rho)$ if and only if $\rho' = \rho$, i.e. the measurement does not change the state of the system. For this to happen we need $[\hat{O}, \rho] = 0$. Hence in contrast to unitary evolutions, measurements can change the entropy of a system, leading, in general, to irreversible processes in which information is lost.

To give a simple example of an ideal measurement, let us consider a spin-$\frac{1}{2}$ particle. We denote as $\{|0\rangle, |1\rangle\}$ the two spin eigenstates on the z-axis. The system is prepared in a state of the form

$$\rho = p_g |0\rangle\langle 0| + c_{ge} |0\rangle\langle 1| + c_{ge}^* |1\rangle\langle 0| + p_e |1\rangle\langle 1|, \tag{1.97}$$

where $p_e + p_g = 1$ and $|c_{ge}|^2 \leqslant p_e p_g$, corresponding to a pure state when the equality is reached. Now imagine we perform a measurement of the spin on the z-axis, as given by the operator $\hat{\mathbb{S}}_z = -\frac{\hbar}{2}\hat{\Pi}_0 + \frac{\hbar}{2}\hat{\Pi}_1$, with projectors $\hat{\Pi}_0 = |0\rangle\langle 0|$ and $\hat{\Pi}_1 = |1\rangle\langle 1|$. This corresponds to the prototypical measurement of spin first realized in 1922 in the Stern–Gerlach experiment (see Fig. 1.8). As a result of the measurement we obtain spin $-\hbar/2$ with probability p_g, collapsing the state of the particle to $|0\rangle$, or spin $\hbar/2$ with probability p_e, then collapsing the state to $|1\rangle$. If the ensemble of all the outputs is considered we get the following *unconditional* state after measurement:

$$\rho' = p_g |0\rangle\langle 0| + p_e |1\rangle\langle 1|, \tag{1.98}$$

which is simply a mixture of states $|0\rangle$ and $|1\rangle$. Here the coherences in the $\{|0\rangle, |1\rangle\}$ basis present in the initial state ρ, have 'disappeared' by effect of the measurement. This means that the system has been disturbed in a random way (collapsing either to

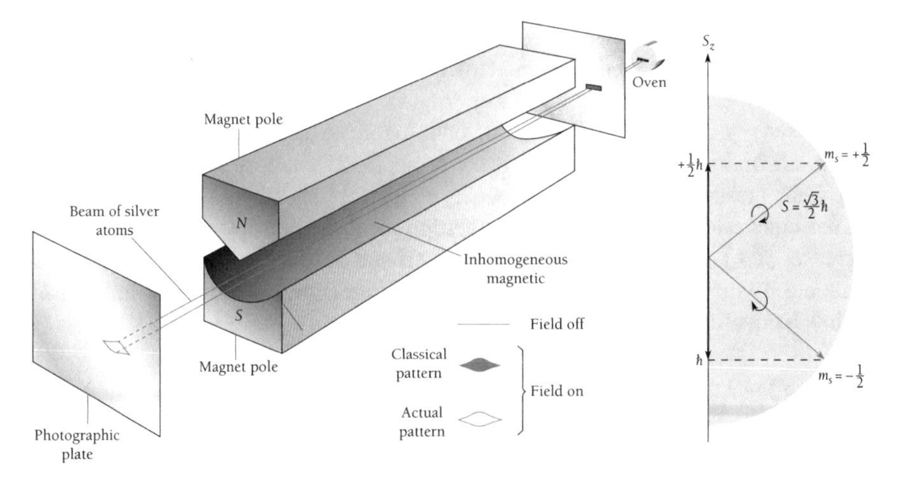

Fig. 1.8 Configuration of the Stern–Gerlach experiment showing the existence of the spin and the influence of quantum measurement on the state of the system being measured. Pictures from Ref. [71]

$|0\rangle$ or $|1\rangle$) and we lose information if we does not have access to the result of each measurement.

1.3.2 Generalized Measurements

In the previous section we introduced projective measurements as the simplest case of quantum measurements. However, in real experiments this ideal description is often not adequate, for example when the apparatus performing the measurement introduces some noise, or when we verify that the final state of our system is not an eigenstate of the observable that we are measuring [6]. Furthermore, one never measures directly the system of interest. It usually interacts with another systems, such as the environment or the meter, from which we finally collect the information. Then if we want to generalize the description of measurements to account for more general situations, we must consider composite systems. One of the major problems in the interpretation of quantum measurements is that, even if we include in our description a chain of other systems interacting one with the next and broadcasting progressively the information we want to retrieve, at some stage we always have to cut the chain introducing an ideal measurement. This is the so-called Heisenberg's cut [72]. In practice, we will introduce the Heisenberg's cut by simply adding a second stage to our measurement process, that is, our system of interest interacts with another ancillary system on which we assume to perform a projective measurement. Considering a chain with only one extra element is appropriate when the ancillary system undergoes a rapid decoherence process, which will yield results negligibly different from those obtained by adding further stages in the measurement process [6].

Let's consider the quantum system which we want to measure, $\mathcal{S}$, with associated Hilbert space $\mathcal{H}_\mathcal{S}$, and an ancillary system $\mathcal{A}$ (with $\mathcal{H}_\mathcal{A}$) representing the measuring apparatus. As in the introduction to composite systems in Sect. 1.1.5, we assume system and ancilla to be completely independent of each other at the initial instant of time, $t = 0$, i.e. the global state of the composite system is a product state $\rho_{\mathcal{S}\mathcal{A}}(0) = \rho_\mathcal{S} \otimes \rho_\mathcal{A}$, where $\rho_\mathcal{A}$ is a generic known state of the meter. Then for some arbitrary period of time, τ, system and meter interact via the Hamiltonian

$$\hat{H}(t) = \hat{H}_\mathcal{S} + \hat{H}_\mathcal{A} + \hat{H}_{\text{int}}(t), \tag{1.99}$$

where $\hat{H}_\mathcal{S}$ is the Hamiltonian of system alone, $\mathcal{H}_\mathcal{A}$ is the Hamiltonian of the ancilla, and $\hat{H}_{\text{int}}(t)$ represents some time-dependent interaction between system and ancilla, which becomes zero outside the time interval $[0, \tau]$. Hence the composite system follows unitary evolution as given by Eq. (1.6), that is

$$\hat{U} \equiv \hat{U}(\tau, 0) = T_+ \exp\left(-\frac{i}{\hbar}\int_0^\tau dt\, \hat{H}(t)\right), \tag{1.100}$$

building up correlations between system and ancilla. The state of the composite system after evolution is

$$\rho_{\mathcal{S}\mathcal{A}}(\tau) = \hat{U}\,(\rho_\mathcal{S} \otimes \rho_\mathcal{A})\,\hat{U}^\dagger, \tag{1.101}$$

which in general is an entangled state. Then an ideal measurement of some observable $\hat{R}_\mathcal{A} = \sum_k r_k \hat{\Pi}_k^\mathcal{A}$, with $\hat{\Pi}_k^\mathcal{A}$ projectors on the eigen-spaces corresponding to eigenvalues r_k, is performed on the ancilla (here $\hat{R}_\mathcal{A}$ acts only on the ancilla's Hilbert space $\mathcal{H}_\mathcal{A}$). This measurement may take some time as well, during which we assume the evolution of system and ancilla is negligible, and hence consider the measurement as instantaneous. In Fig. 1.9 we provide a sketch of this generalized measurement scheme.

The probability of obtaining the result r_k on the ancillary system, according to the projection postulate, is then given by

$$P_k = \text{Tr}[(\hat{\mathbb{1}}_\mathcal{S} \otimes \hat{\Pi}_k^\mathcal{A})\rho_{\mathcal{S}\mathcal{A}}(\tau)] = \text{Tr}[\hat{U}^\dagger(\hat{\mathbb{1}}_\mathcal{S} \otimes \hat{\Pi}_k^\mathcal{A})\hat{U}(\rho_\mathcal{S} \otimes \rho_\mathcal{A})], \tag{1.102}$$

where the trace is performed over the whole Hilbert space $\mathcal{H}_\mathcal{S} \otimes \mathcal{H}_\mathcal{A}$. Analogously, the global state of system and ancilla after obtaining outcome k in the measurement reads

$$\rho_{\mathcal{S}\mathcal{A}}^{\prime(k)}(\tau) = (\hat{\mathbb{1}}_\mathcal{S} \otimes \hat{\Pi}_k^\mathcal{A})\; \rho(\tau)\; (\hat{\mathbb{1}}_\mathcal{S} \otimes \hat{\Pi}_k^\mathcal{A}) \;/\; P_k, \tag{1.103}$$

according to Eq. (1.95), while its *unconditional* counterpart should read $\rho'_{\mathcal{S}\mathcal{A}}(\tau) = \sum_k P_k \rho_{\mathcal{S}\mathcal{A}}^{\prime(k)}(\tau)$. Given the spectral decomposition $\rho_\mathcal{A} = \sum_n q_n |\phi_n\rangle\langle\phi_n|_\mathcal{A}$, and

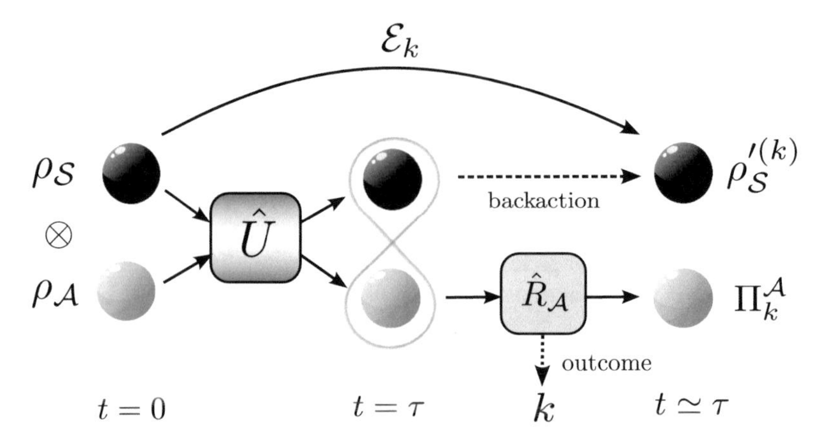

Fig. 1.9 Schematic representation of the generalized measurement process introduced in the text. System and ancilla are prepared in an uncorrelated state, and interact by means of the unitary evolution operator $\hat{U}$ from $t = 0$ to $t = \tau$, building up correlations between them. Then the observer performs quasi-instantaneously a local ideal measurement of the observable $\hat{R}_\mathcal{A}$ on the ancilla, obtaining outcome k, and leading to back-action on the system

assuming for simplicity the ancillary projectors to be rank-1, i.e. $\hat{\Pi}_k^\mathcal{A} = |r_k\rangle\langle r_k|_\mathcal{A}$, the above Eq. (1.102) may be rewritten as follows

$$P_k = \text{Tr}_\mathcal{S}[\sum_n \hat{M}_{k,n}^\dagger \hat{M}_{k,n}\ \rho_\mathcal{S}], \tag{1.104}$$

where we have introduced the *measurement operators*:

$$\hat{M}_{k,n} = \langle r_k|_\mathcal{A} \hat{U}\ |\phi_n\rangle_\mathcal{A} \sqrt{q_n}, \tag{1.105}$$

acting on the system Hilbert space $\mathcal{H}_\mathcal{S}$. Notice that $\hat{M}_{k,n}$ actually depend on two indices corresponding to the measurement result r_k, and the initial pure state $|\phi_n\rangle$ of the ancilla, as given by the spectral decomposition of $\rho_\mathcal{A}$. However, since we can rewrite the ancilla density operator as an arbitrary mixture of (non-orthonormal) pure states, the term *measurement operator* is often restricted to the case in which the ancilla starts in a pure state, i.e. $\hat{M}_k = \langle r_k|_\mathcal{A} \hat{U}\ |\phi\rangle_\mathcal{A}$. In this case the operators are unequivocally associated to the results of the measurement. In any case we can always define the positive operators

$$\hat{F}_k = \sum_n \hat{M}_{kn}^\dagger \hat{M}_{kn} \leqslant \hat{\mathbb{1}}_\mathcal{S}, \tag{1.106}$$

These are the so-called *effects* of the measurement corresponding to outcome k, as they completely determine the statistics of the measurement results. It is worth noticing that $\sum_k \hat{F}_k = \sum_{k,n} \hat{M}_{k,n}^\dagger \hat{M}_{k,n} = \hat{\mathbb{1}}_\mathcal{S}$, i.e. the effects constitute a resolution of the identity in the system Hilbert space $\mathcal{H}_S$, which implies $\sum_k P_k = 1$. The set $\{\hat{F}_k; k\}$ is usually called a *Positive-Operator-Valued Measure* (POVM) [6].

The measurement operators are also useful to account for the local disturbance of the measurement on the system. The reduced state of the system after the *selective* measurement conditioned to outcome k, taking the trace in Eq. (1.103) over the ancilla degrees of freedom, reads

$$\rho_S'^{(k)} = \sum_n \frac{\hat{M}_{k,n}\, \rho_S\, \hat{M}_{k,n}^\dagger}{P_k} \equiv \frac{\mathcal{E}_k(\rho_S)}{P_k}, \tag{1.107}$$

to be compared to Eq. (1.95). We call the mapping

$$\mathcal{E}_k(\rho_S) = \sum_n \hat{M}_{k,n} \rho_S \hat{M}_{k,n}, \tag{1.108}$$

transforming positive operators into positive operators, a *quantum operation*. Notice that $\mathcal{E}_k$ in general does not preserve the trace, unless there is only one possible outcome in the measurement ($P_k = 1$). From Eq. (1.107) we immediately find that the *unconditional* system state after measurement is

$$\rho_S' = \sum_k P_k\, \rho_S'^{(k)} = \sum_{k,n} \hat{M}_{k,n}\, \rho_S\, \hat{M}_{k,n}^\dagger \equiv \mathcal{E}(\rho_S), \tag{1.109}$$

and we will refer to $\mathcal{E}(\rho_S) = \sum_k \mathcal{E}_k(\rho_S)$ as a *quantum map*, which preserves the trace and hence transforms physical states into physical states of the system. In Chap. 2 we will turn to this important class of maps, also called *Completely Positive and Trace Preserving* (CPTP) maps. On the other hand, after obtaining result k in the measurement, the ancillary system collapses to the pure state $\Pi_k^{\mathcal{A}}$, and hence the corresponding *unconditional* ancillary state is

$$\rho_{\mathcal{A}}' = \sum_k P_k |r_k\rangle\langle r_k|_{\mathcal{A}}. \tag{1.110}$$

1.3.3 Classes of Measurements

The measurement scheme presented in the previous section did not assume any form of the interaction Hamiltonian leading to the global evolution $\hat{U}$. This allows to include measurements that give only partial information about the system, that disturb the system minimally or even that do not correspond to observables of the system. Such generalized measurements are essential to account for the effects of both classical and quantum noise in the measurement apparatus by means of the initial state of the ancilla, a general mixed state diagonal in a different basis than the eigenstates $\{|r_k\rangle\}$ of $\hat{R}_{\mathcal{A}}$.

In the following we will describe some important classes of measurements by imposing extra conditions on the form of the *effects* or the *measurement operators* (for a more complete classification of measurements see [6]). However, it is important to first introduce the so-called *polar decomposition theorem*. This theorem states that any operator can be expressed as a product of a positive operator and a unitary one. For instance, for the measurement operators:

$$\hat{M}_k = \hat{U}_k \hat{N}_k, \tag{1.111}$$

where the unitary part is $\hat{U}_k \hat{U}_k^\dagger = \hat{U}_k^\dagger \hat{U}_k = \hat{\mathbb{1}}$ and we have introduced the positive and Hermitian operator $\hat{N}_k = \hat{N}_k^\dagger$. The theorem is really relevant as it gives us a picture of a quantum measurement as a combination of two processes. The first process determined by the action of the positive operator $\hat{N}_k$ is irreversible and responsible for the back-action on the system associated with the information gathering process [6]. Indeed it completely determines the statistics of the measurement

$$P_k = \mathrm{Tr}_{\mathbb{S}}[\hat{M}_k^\dagger \hat{M}_k \rho_{\mathbb{S}}] = \mathrm{Tr}_{\mathcal{S}}[(\hat{N}_k)^2 \rho_{\mathbb{S}}], \tag{1.112}$$

and extracts information about the system corresponding to its eigenbasis, hence disturbing it when $[\hat{N}_k, \rho_{\mathbb{S}}] \neq 0$. The second process of the measurement consist on the unitary operation $\hat{U}_k$:

$$\rho_{\mathbb{S}}^{\prime(k)} = \hat{U}_k \frac{\hat{N}_k \rho_{\mathbb{S}} \hat{N}_k}{P_k} \hat{U}_k^\dagger, \tag{1.113}$$

which can be viewed as a reversible *feedback process* depending on the specific result of the measurement. The positive operators $\{\hat{N}_k\}$ can change the eigenvalues of $\rho_{\mathbb{S}}$, and therefore its entropy. In contrast, the unitary operation does not provide any further information about the system [13].

1.3.3.1 Efficient Measurements

A first class of measurements are the so-called *efficient measurements*. They correspond to the case in which the operations $\mathcal{E}_k$ are defined in terms of a single measurement operator

$$\mathcal{E}_k(\rho_{\mathbb{S}}) = \hat{M}_k \rho_{\mathbb{S}} \hat{M}_k^\dagger, \tag{1.114}$$

and results when the ancillary system starts in a pure state, as we showed in Sect. 1.3.2. Hence efficient measurements transform pure states into pure states and any noise arising in the process can be interpreted as quantum noise [6].

The term efficient makes reference to the fact that this kind of measurements are the only ones that produce, on average, an information gain [73, 74]

$$S(\rho_{\mathbb{S}}) \geqslant \sum_k P_k S(\rho_{\mathbb{S}}^{\prime(k)}), \tag{1.115}$$

which implies a refining of the observer's state of knowledge. However the entropy of the state conditioned to outcome k is not necessarily lower than the entropy of the initial state for all the outcomes. Furthermore, the entropy of the *unconditional* system state after measurement, $\rho_{\mathbb{S}}' = \sum_k P_k \rho_{\mathbb{S}}^{\prime(k)}$, can be either greater or lower than the entropy of the initial state $\rho_{\mathbb{S}}$. In the later case, the reduction of entropy in the system is always compensated by an increase in the entropy of the ancilla, the total entropy of the composite system being always a non-decreasing quantity.

1.3.3.2 Bare Measurements

Bare measurements are a subclass of efficient measurements. They are defined from the polar decomposition theorem introduced in Eq. (1.111) as the ones in which the unitary part of the decomposition is absent, $\hat{U}_k = \hat{\mathbb{1}}_{\mathbb{S}}$ up to an arbitrary phase. In this case we have that $\hat{M}_k = \sqrt{\hat{F}_k} = \hat{N}_k$ and the measurement disturbs the system state only 'minimally', in the sense that the average fidelity between a pure initial state and the *conditional* state after measurement is maximized [6].[4]

As *bare* measurements are given by positive Hermitian operators, they do not disturb the system state if $[\hat{M}_k, \rho_{\mathbb{S}}] = 0$. Furthermore, the measurement operators represent different observables of the system, being Hermitian operators acting on the Hilbert space $\mathcal{H}_{\mathbb{S}}$, from which partial information is obtained in the measurement process. Finally *bare* measurements also satisfy

$$S(\rho_{\mathbb{S}}') \geqslant S(\rho_{\mathbb{S}}) \geqslant \sum_k P_k S(\rho_{\mathbb{S}}^{\prime(k)}). \tag{1.116}$$

As for ideal measurements, ignoring the result of the measurement always produces a loss of information about the system state. Indeed, Eq. (1.116) can be extended to all measurements for which the unitary operators $\hat{U}_k$ are equal (i.e. not depending on the measurement outcome k) and to any other measure of mixedness of the state [6].

1.3.3.3 Complete Measurements

Complete measurements extract all the information contained in the initial state of the system, and hence further measurements do not provide new insight about it. This condition can be made formal by requiring that the *conditional* states associated to any outcome k do not depend on $\rho_{\mathbb{S}}$. This implies that the *operations* associated to

[4]However as pointed in Ref. [13] the disturbance of a measurement can be defined in other different ways.

the measurements are of the form [6]

$$\mathcal{E}_k(\rho_{\mathcal{S}}) = \sum_{n,m} |\psi_n^{(k)}\rangle\langle\phi_m^{(k)}|\,\rho_{\mathcal{S}}|\phi_m^{(k)}\rangle\langle\psi_n^{(k)}|, \tag{1.117}$$

$\{|\psi_n^{(k)}\rangle\}$ and $\{|\phi_m^{(k)}\rangle\}$ being arbitrary states of $\mathcal{H}_{\mathcal{S}}$ (possibly non normalized). The probability of obtaining outcome k is then

$$P_k = \sum_n |\langle\psi_n^{(k)}|\psi_n^{(k)}\rangle|^2 \sum_m \langle\phi_m^{(k)}|\,\rho_{\mathcal{S}}|\phi_m^{(k)}\rangle, \tag{1.118}$$

and the effects of the measurement are

$$\hat{F}_k = \sum_{n,m} \langle\psi_n^{(k)}|\psi_n^{(k)}\rangle|\phi_m^{(k)}\rangle\langle\phi_m^{(k)}|, \tag{1.119}$$

with $\sum_k \hat{F}_k = \hat{\mathbb{1}}_{\mathcal{S}}$. It is easy to see that in this case

$$\rho_{\mathcal{S}}^{\prime(k)} = \frac{\mathcal{E}_k(\rho_{\mathcal{S}})}{P_k} = \frac{\sum_n |\psi_n^{(k)}\rangle\langle\psi_n^{(k)}|}{\sum_n |\langle\psi_n^{(k)}|\psi_n^{(k)}\rangle|^2}, \tag{1.120}$$

which is independent of $\rho_{\mathcal{S}}$. Notice that here any initial state of the system is transformed into $\rho_{\mathcal{S}}^{\prime(k)}$ only depending on the outcome k. On the other side the unconditional state of the system reads

$$\rho_{\mathcal{S}}' = \mathcal{E}(\rho_{\mathcal{S}}) = \sum_k \mathcal{E}_k(\rho_{\mathcal{S}}) \equiv \sum_k a_k \hat{A}_k, \tag{1.121}$$

with $\hat{A}_k = \sum_n |\psi_n^{(k)}\rangle\langle\psi_n^{(k)}|$ and $a_k = \sum_m \langle\phi_m^{(k)}|\,\rho_{\mathcal{S}}|\phi_m^{(k)}\rangle$, which may still depend on the initial state $\rho_{\mathcal{S}}$ through the quantities a_k.

1.3.3.4 Non-demolition Measurements

A *quantum non-demolition measurement* (QND measurement), as introduced by Braginsky and Khalili [75], is defined as a measurement for which the probability distribution of some observable of the system, $\hat{O}_{\mathcal{S}}$, does not change during the measurement process. For the case of a *unselective* measurement this implies [1]

$$\mathrm{Tr}[(\hat{O}_{\mathcal{S}})^l \rho_{\mathcal{S}}] = \mathrm{Tr}[(\hat{O}_{\mathcal{S}})^l \rho_{\mathcal{S}}'] = \sum_{k,n} \mathrm{Tr}[(\hat{O}_{\mathcal{S}})^l \hat{M}_{k,n}\rho_{\mathcal{S}}\hat{M}_{k,n}^\dagger], \tag{1.122}$$

where l is any integer. This means that all the moments of $\hat{O}_{\mathcal{S}}$ are the same before and after the measurement, and hence the complete distribution of its eigenvalues.

We can rewrite the above equation as

$$\sum_{k,n} \mathrm{Tr}[\hat{M}_{k,n}^{\dagger}(\hat{O}_{\mathcal{S}})^l \hat{M}_{k,n}\rho_{\mathcal{S}}] - \mathrm{Tr}[(\hat{O}_{\mathcal{S}})^l \rho_{\mathcal{S}}] =$$
$$= \sum_{k,n} \mathrm{Tr}[\hat{M}_{k,n}^{\dagger}[(\hat{O}_{\mathcal{S}})^l, \hat{M}_{k,n}]\, \rho_{\mathcal{S}}] = 0, \tag{1.123}$$

where we have used $\sum_{k,n} \hat{M}_{k,n}^{\dagger}\hat{M}_{k,n} = \sum_k \hat{F}_k = \hat{\mathbb{1}}_{\mathcal{S}}$. The above condition Eq. (1.122) [or equivalently Eq. (1.123)] must be satisfied for all initial states $\rho_{\mathcal{S}}$. Hence a sufficient condition for a QND measurement is

$$[\hat{O}_{\mathcal{S}}, \hat{M}_{k,n}] = 0 \ \forall \ k, n, \tag{1.124}$$

which are sometimes also called *back-action-evading measurements*. By following the general measurement scheme of Sect. 1.3.2 this condition can be translated into $[\hat{O}_{\mathcal{S}} \otimes \hat{\mathbb{1}}_{\mathcal{A}}, \hat{U}] = 0$, $\hat{U}$ being the unitary evolution operator in Eq. (1.100), coupling system and ancilla. This can be achieved by requiring both

$$[\hat{H}_{\mathcal{S}}, \hat{O}_{\mathcal{S}}] = 0, \quad \text{and} \quad [\hat{H}_{\mathrm{int}}, \hat{O}_{\mathcal{S}} \otimes \hat{\mathbb{1}}_{\mathcal{A}}] = 0, \tag{1.125}$$

which ensures that the statistics of the system observable $\hat{O}_{\mathcal{S}}$ are not disturbed by the interaction with the ancilla. Here the observable $\hat{O}_{\mathcal{S}}$ defines the so-called *pointer observable*, which determines the basis states (or *pointer basis*) which are robust against the measurement process [76].

1.3.3.5 Projective Measurements

Finally, we stress that ideal measurements introduced in Sect. 1.3.1 are a particular case of the above generalized measurements, also called *projective measurements*. They correspond to the case in which the measurement operators are projectors onto the eigenspaces of some observable $\hat{O}_{\mathcal{S}}$ on $\mathcal{H}_{\mathcal{S}}$:

$$\hat{M}_k = \hat{F}_k = \hat{\Pi}_k^{\mathcal{S}}, \tag{1.126}$$

with the orthogonality $\hat{\Pi}_k^{\mathcal{S}}\hat{\Pi}_l^{\mathcal{S}} = \delta_{k,l}\hat{\Pi}_k^{\mathcal{S}}$ and completeness relations $\sum_k \hat{\Pi}_k^{\mathcal{S}} = \hat{\mathbb{1}}_{\mathcal{S}}$. Again the observable $\hat{O}_{\mathcal{S}}$ is generally determined from the specific form of the interaction Hamiltonian $\hat{H}_{\mathrm{int}}(t)$ in Eq. (1.99) which couples system and ancilla during the first part of the measurement process. In the case of a strong interaction we can approximate $\hat{H}(t) \simeq \hat{H}_{\mathrm{int}}(t)$, and the *pointer observable* is the one satisfying $[\hat{H}_{\mathrm{int}}(t), \hat{O}_{\mathcal{S}}] = 0$.

We notice that *projective measurements* are simultaneously *efficient*, *bare*, and *non-demolition* measurements. However they are only *complete* when the projectors

$\hat{\Pi}_k^{\mathrm{S}}$ onto eigenspaces of the system observable $\hat{O}_{\mathrm{S}}$ are rank-1. The latter subclass of *projective measurements* is called by some authors, *von Neumann measurements* [6].

1.4 Classical and Quantum Correlations

Correlations are the main characterizing signature of multipartite systems. Furthermore, it has been a long standing question in quantum theory the distinction between classical and quantum correlations from a theoretical point of view, but also in the search of new applications and the development of quantum technologies. In this section we review different indicators of classical and quantum correlations present in bipartite or multipartite systems and introduce specific quantifiers which will be especially useful in Part II of this thesis.

1.4.1 Entanglement

The phenomenon of entanglement is a consequence of the superposition principle when applied to composite systems. It yields a rich and striking phenomenology which led Scrödinger to consider it as "The characteristic trait of quantum physics" [77]. It constitutes a key quantum resource for the development of quantum communication and quantum computation, responsible of applications such as superdense coding, quantum teleportation, quantum error-correction algorithms or key-distribution protocols for quantum cryptography [5, 78].

We already introduced entanglement in the case of pure states through the Schmidt decomposition theorem in Sect. 1.1.5. In the more general context of the density operator formalism, we say that a state ρ_{AB} of a composite quantum system AB (with Hilbert space $\mathcal{H}_A \otimes \mathcal{H}_B$) is *entangled* if it cannot be written in the form

$$\rho_{AB} = \sum_k p_k \, \rho_A^{(k)} \otimes \rho_B^{(k)} \tag{1.127}$$

where $\rho_A^{(k)}$ and $\rho_B^{(k)}$ are local states of systems A and B respectively, and $0 \leqslant p_k \leqslant 1$ with $\sum_k p_k = 1$. The states of the form in Eq. (1.127) are called *separable* states and can always be prepared by distant observers following instructions from a common source [10]. On the other hand, for two systems to be entangled it is required some kind of interaction, which makes them lose their local identity. In this case there exist properties of the composite system that cannot be reconstructed by means of local operations on the subsystems, even if the local observers are allowed to communicate classically. This makes entanglement a notion of the quantumness of correlations, as under the framework of Local Operations and Classical Communication (LOCC), only separable states in the form of Eq. (1.127) can be generated [78]. Any other state contains some degree of entanglement.

The Schmidt decomposition not only allows to characterize entanglement in bipartite pure quantum states, but also leads to a measure of entanglement, the so-called *entropy of entanglement*. Indeed, if the global state is pure, $\rho_{AB} = |\Psi\rangle\langle\Psi|$, it follows that the reduced states, $\rho_A = \mathrm{Tr}_B[\rho_{AB}]$ and $\rho_B = \mathrm{Tr}_A[\rho_{AB}]$, have exactly the same eigenvalues, which implies, as we already pointed in Sect. 1.1.6, that they have the same von Neumann entropy. The entropy of entanglement is precisely defined as this quantity:

$$E(|\Psi\rangle\langle\Psi|) = S(\mathrm{Tr}_A[|\Psi\rangle\langle\Psi|]) = S(\mathrm{Tr}_B[|\Psi\rangle\langle\Psi|]) \geqslant 0, \tag{1.128}$$

which is only zero in the case of a separable state in the form $|\Psi\rangle = |\psi\rangle_A \otimes |\phi\rangle_B$.

However, in the case of mixed states the situation becomes much more complicate and a remarkable theoretical effort has been devoted to quantifying entanglement. In this context, many indicators with its own advantages and disadvantages have been proposed. In the following we will review some of them, but let us first state a general set of desirable conditions which a measure of entanglement in mixed states, $E(\rho_{AB})$, should fulfill [25, 78]:

1. If the state ρ_{AB} is separable then $E(\rho_{AB}) = 0$. That is, separable states have zero entanglement.
2. Local unitary operators cannot modify the amount of entanglement

$$E(\rho_{AB}) = E(\hat{U}_A \otimes \hat{U}_B \rho_{AB} \hat{U}_A^\dagger \otimes \hat{U}_B^\dagger) \tag{1.129}$$

   where $\hat{U}_A$ and $\hat{U}_B$ are arbitrary unitary operators acting on $\mathcal{H}_A$ and $\mathcal{H}_B$ respectively.
3. LOCC operations cannot increase the entanglement

$$E(\rho_{AB}) \geqslant E\left(\sum_i p_i \hat{M}_i^{\mathcal{A}} \otimes \hat{M}_i^{\mathcal{B}} \, \rho_{AB} \, \hat{M}_i^{\mathcal{A}\dagger} \otimes \hat{M}_i^{\mathcal{B}\dagger}\right) \tag{1.130}$$

   for any set of measurement operators $\{\hat{M}_i^{\mathcal{A}}\}$ and $\{\hat{M}_i^{\mathcal{B}}\}$ on subsystems A and B respectively. Notice that they can share a common index, meaning that the operations in subsystems A and B may be correlated.
4. When the state ρ_{AB} is pure, the measurement of entanglement reduces to the entropy of entanglement in Eq. (1.128).

The following measures of entanglement fulfill conditions (1–3) and some of them also condition (4). Some of them provide specific operational interpretations of entanglement while others are more easily computable. Whether we use one or the other depends mainly on the specific situation for which entangled states are needed.

- *Logarithmic Negativity*: This measure is based on the Positive Partial Transpose (PPT) criterion, which constitutes a strong condition for the separability of a quan-

tum state [79]. It asserts that if ρ_{AB} is separable then, the partial transposed matrix $\rho_{AB}^{T_A}$, where the indices of one of the two subsystems (here A) have been exchanged, is a 'legal' density operator (i.e. it is non-negative and has unit trace). The PPT criterion is in general a necessary condition for separability, but it is also a sufficient condition for the case of composite systems consisting of two two-level systems and the composition of a two-level system and a three-level system [78].
Logarithmic negativity captures the violation of the PPT criterion. It is defined as

$$E_{\mathcal{N}}(\rho_{AB}) \equiv \ln \mathrm{Tr}[\sqrt{\tilde{\rho}_{AB}^{\dagger}\tilde{\rho}_{AB}}] \tag{1.131}$$

being $\tilde{\rho}_{AB} = \rho_{AB}^{T_A}$ (or equivalently $\rho_{AB}^{T_B}$), the partial transpose of the density operator ρ_{AB}. Notice that for separable states, $\rho_{AB}^{T_A}$ is a legal density operator and hence $E_{\mathcal{N}}(\rho_{AB}) = \ln \mathrm{Tr}[\rho_{AB}^{T_A}] = 0$, thus satisfying condition (1). Furthermore it also fulfills conditions (2) and (3), while representing an additive and easily computable quantity [80]. In particular, we will employ the logarithmic negativity as a measure of entanglement between a pair of dissipative harmonic oscillators in Chaps. 5 and 6, for which a closed expression exists when the global state ρ_{AB} is Gaussian (see below).

- *Entanglement of formation*: This is an extension of the entropy of entanglement to mixed states. It is defined as [81]

$$E_F(\rho_{AB}) \equiv \min \sum_i p_i S(\rho_A^{(i)}), \tag{1.132}$$

$\rho_A^{(i)} = \mathrm{Tr}_B[|\Psi^{(i)}\rangle\langle\Psi^{(i)}|]$ being the local reduced state of subsystem A in the decomposition of the global state as a mixture of pure states $\rho_{AB} = \sum_i p_i |\Psi^{(i)}\rangle\langle\Psi^{(i)}|$. Notice that, as long as the decomposition of ρ_{AB} is not unique, we need to take the minimum over all possible decompositions. This measure satisfies conditions (1–4) and represents the asymptotic ratio n/m at which two observers (Alice and Bob) can create n copies of ρ_{AB} in the LOCC framework by using m copies of maximally entangled pairs (e.g. Bell states) in the limit of n and m large. Closely related to the concept of *entanglement of formation* is the so-called *entanglement of distillation*, $E_D(\rho_{AB})$, which involves the opposite process: here Alice and Bob obtain maximally entangled states from many copies of the state ρ_{AB} by using the LOCC framework. The logarithmic negativity introduced above provides and upper bound to the entanglement of distillation, i.e. $E_D(\rho_{AB}) \leqslant E_{\mathcal{N}}(\rho_{AB})$.
Even if the computation of the entanglement of formation from its definition is usually hard due to the minimization procedure, it can be easily evaluated in some specific cases. For instance in the case of a pair of two-level systems it can be computed by using another entanglement measure, the so-called *concurrence*, $C(\rho_{AB})$ [82]. An example for continuous variable systems, are two-mode symmetric Gaussian states, for which an expression of the entanglement of formation has been derived [83, 84].

- *Relative entropy of entanglement*: This is another important measure of entanglement, which satisfies conditions (1–4) and is defined by means of the relative entropy

$$E_R(\rho_{AB}) \equiv \min_{\sigma_{AB} \in \mathcal{S}} D(\rho_{AB} || \sigma_{AB}) \tag{1.133}$$

where $\mathcal{S}$ denote the set of separable states. This measure quantifies entanglement as the distinguishability of ρ_{AB} from the set of non-entangled states, i.e. the more the state is entangled the easier is to distinguish it from a separable state [85]. It can be shown that:

$$E_D(\rho_{AB}) \leqslant E_R(\rho_{AB}) \leqslant E_F(\rho_{AB}), \tag{1.134}$$

i.e. the relative entropy of entanglement, $E_R(\rho_{AB})$, is bounded from above by the entanglement of formation, $E_F(\rho_{AB})$, and constitutes an upper bound for the entanglement of distillation, $E_D(\rho_{AB})$ [25]. Finally, an advantage of the relative entropy of entanglement with respect to other measures is that it can be easily extended to multipartite systems.

Of particular interest to this thesis is the expression of the *logarithmic negativity* for bipartite Gaussian states, which will be used in Chaps. 5 and 6 of Part II in order to calculate the entanglement between a pair of dissipative harmonic oscillators. This expression has been obtained in Refs. [80, 86], to which we refer for details on the derivation, and is based on the properties of the so-called *covariance matrix*. Indeed, all the information about the correlations in bipartite continuous-variable systems AB is encoded in this matrix, which we denote V_{AB}, and whose entries correspond to the average values of the ten covariances of position $\hat{x}_k$ and momenta $\hat{p}_k$ operators for $k = A, B$ in the global Gaussian state ρ_{AB} (see Sect. 1.2.4)

$$[V_{AB}]_{ij} = \langle (\hat{X}_i - \langle \hat{X}_i \rangle_{\rho_{AB}})(\hat{X}_j - \langle \hat{X}_j \rangle_{\rho_{AB}}) \rangle_{\rho_{AB}}, \quad i, j = 1, 2, 3, 4 \tag{1.135}$$

and where $\hat{X}_1 = \hat{x}_A$, $\hat{X}_2 = \hat{p}_A$, $\hat{X}_3 = \hat{x}_B$, and $\hat{X}_4 = \hat{p}_B$. It can be rewritten in block form as

$$V_{AB} = \begin{pmatrix} \alpha & \gamma \\ \gamma^t & \beta \end{pmatrix}, \tag{1.136}$$

where α, β, γ are (2×2) blocks: $\alpha(\beta)$ contains the variances (1.135) of subsystem $A(B)$, and γ contains correlations of both subsystems. The minimum symplectic eigenvalue of the partially transposed covariance matrix, $\tilde{V}_{AB} = V_{AB}^{T_A}$ corresponding to time inversion in one subsystem, is given by

$$\nu_- = \sqrt{\frac{1}{2}(a + b - 2g - \sqrt{(a + b - 2g)^2 - 4s}}), \tag{1.137}$$

with $a = 4\det(\alpha)/\hbar^2$, $b = 4\det(\beta)/\hbar^2$, $g = 4\det(\gamma)/\hbar^2$ and $s = 16\det V_{AB}/\hbar^4$. The expression of the logarithmic negativity can be shown to depend only on this quantity [80, 86]

$$E_{\mathcal{N}}(\rho_{AB}) = \max\{0, -\log \nu_-\}. \tag{1.138}$$

1.4.2 Mutual Information

We are now in position to introduce another of the fundamental quantities needed to characterize correlations and thermodynamics in open quantum systems. Let us consider again the composite system AB, with global state ρ_{AB} and reduced states ρ_A and ρ_B, respectively. The quantum mutual information (or simply the mutual information) between subsystems A and B is defined as the distance

$$I(A:B) \equiv D(\rho_{AB} \,||\, \rho_A \otimes \rho_B) \geqslant 0, \tag{1.139}$$

quantifying the distinguishability of the actual state of the composite system ρ_{AB} with respect to the completely uncorrelated state $\rho_A \otimes \rho_B$. Mutual information is hence a measure of the total correlations present in the composite system, being zero if and only if $\rho_{AB} = \rho_A \otimes \rho_B$. It is related to the von Neumann entropies of the subsystems by the general relation

$$S(\rho_{AB}) = S(\rho_A) + S(\rho_B) - I(A:B), \tag{1.140}$$

from which subadditivity of von Neumann entropy is derived (see Sect. 1.1.6). From the above Eq. (1.140) we deduce that the mutual information corresponds to the information about the composite system (contained in ρ_{AB}) which is unaccessible by local measurements in the subsystems (as characterized by ρ_A and ρ_B). It indeed corresponds to the total amount of correlations (both classical and quantum) between the subsystems [87].

The quantum mutual information is the extension to the quantum domain of the classical mutual information between random variables X and Y, taking values x and y according to the probability distribution $p(x, y)$. The latter is defined as

$$\begin{aligned} I(x:y) &\equiv \sum_{x,y} p(x,y)\,(\ln p(x,y) - \ln p(x)p(y)) = \\ &= H(Y) - H(Y|X) = H(X) - H(X|Y) = \\ &= H(X) + H(Y) - H(X,Y), \end{aligned} \tag{1.141}$$

where $p(x) = \sum_y p(x, y)$ and $p(y) = \sum_x p(x, y)$ are marginal probability distributions, and we denoted again the Shannon information of $p(x, y)$ as $H(X, Y) = -\sum_x p(x, y) \ln p(x, y)$ and equivalently for the marginals of variables X and Y. Further, we also introduced the classical conditional entropy

$$H(Y|X) = H(X,Y) - H(X) = \sum_x p(x) H(p(y|x)), \tag{1.142}$$

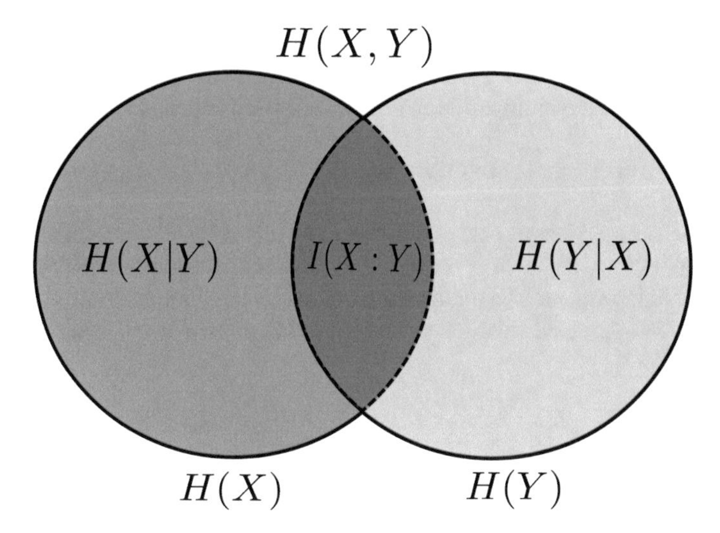

Fig. 1.10 Information diagram illustrating the different Shannon information measures introduced in the main text. The area of each circle represent the Shannon informations $H(X)$ and $H(Y)$ respectively. The Shannon entropy of the joint distribution, $H(X, Y)$, is the union of the circles (delimited by the solid black line), while the mutual information, $I(X : Y)$, corresponds to its intersection (delimited by the dashed black line)

where $p(y|x) = p(x, y)/p(x)$ is the conditional probability distribution of y given x (see the diagram of Fig. 1.10). However, there exist important differences between quantum and classical mutual information. One way to see these differences is to rewrite the quantum mutual information in Eq. (1.140) as

$$I(A : B) = S(\rho_A) + S(\rho_B) - S(\rho_{AB}) = S(\rho_A) - S(\rho_B|\rho_A), \tag{1.143}$$

where we have introduced the conditional quantum entropy $S(\rho_B|\rho_A) \equiv S(\rho_{AB}) - S(\rho_A)$. The latter would be the analogous to the classical conditional entropy, $H(Y|X) \geqslant 0$. However, in the quantum case the conditional entropy $S(\rho_B|\rho_A)$ can be negative [88] in sharp contrast to the classical situation. This is due to the fundamental difference between the quantum conditional entropy as defined by $S(\rho_B|\rho_A)$ and its classical counterpart, $H(Y|X)$. The latter represents the average uncertainty in the value of Y when we know the value of X. In the quantum case this concept would involve measurements on the subsystems that may disturb the quantum states [89]. Hence, one may define a different version of the conditional entropy based on this concept of information acquisition, which would lead to a different notion of mutual information. In this context, different indicators of the quantumness of the correlations alternative to entanglement have been proposed, such as the so-called *quantum discord* [90] in order to distinguish between the classical and quantum contents of the mutual information [87] (see below).

Quantum mutual information can be easily generalized to multipartite systems. Consider a composite system in state ρ which can be split into subsystems $1, 2, \ldots, M$. Then the multipartite mutual information is defined as the distance:

$$I(1:2:\ldots:M) \equiv D(\rho \,||\, \rho_1 \otimes \rho_2 \otimes \cdots \otimes \rho_M), \tag{1.144}$$

where $\rho_i = \mathrm{Tr}_{j \neq i}[\rho]$ denotes the partial trace over all subsystems except subsystem i, and the state $\rho_1 \otimes \rho_2 \otimes \cdots \otimes \rho_M$ corresponds to the case in which the subsystems are completely independent. The multipartite mutual information measures the amount of correlations among all subsystems and the relation (1.140) still holds in the form

$$I(1:2:\ldots:M) = \sum_{i=1}^{M} S(\rho_i) \;-\; S(\rho). \tag{1.145}$$

However, in the multipartite case there exist many other partial mutual informations measuring the correlations in some subset of systems, or conditional mutual informations stressing the amount of correlations between some subsystems given some knowledge about other ones.

1.4.3 Quantum Discord

The separation of correlations into classical and quantum parts by means of entanglement has been questioned in the last decade, motivated in part by the discovery of quantum exponential speedups in some computational tasks with vanishingly small entanglement [91]. This led some authors to consider quantum correlations beyond entanglement, introducing the so-called quantum discord as a quantifier of the quantumness of correlations [87, 90].

As we commented in the previous section, the classical notion of mutual information unfolds into two different concepts when extending it to the quantum case. Together with the quantum mutual information defined in Eqs. (1.139) and (1.140), which represents the intersection of uncertainties between the two subsystems A and B of the composite system AB (see Fig. 1.10), we may define a different mutual information based on information acquisition through (ideal) measurements on one of the subsystems. Consider again the state ρ_{AB} with marginals ρ_A and ρ_B. If we measure subsystem B by means of a set of projectors $\{\hat{\Pi}_k^B\}$ (corresponding to some observable of subsystem B), then the information we obtain about subsystem A reads

$$\begin{aligned} J(A:B)_{\{\hat{\Pi}_k^B\}} &\equiv S(\rho_A) - S(A|\{\hat{\Pi}_k^B\}) \\ &= S(\rho_A) - \sum_k P_k S(\rho_A^{\prime(k)}) \geqslant 0, \end{aligned} \tag{1.146}$$

where $P_k = \mathrm{Tr}[(\hat{\mathbb{1}}_A \otimes \hat{\Pi}_k^B)\rho_{AB}]$ is the probability of obtaining outcome k in the measurement, and

$$\rho_A'^{(k)} = \mathrm{Tr}_B[\rho_{AB}'^{(k)}] = \mathrm{Tr}_B[(\hat{\mathbb{1}}_A \otimes \hat{\Pi}_k^B)\rho_{AB}(\hat{\mathbb{1}}_A \otimes \hat{\Pi}_k^B)]/P_k, \tag{1.147}$$

is the *conditional* state of subsystem A after measurement in B (see Sect. 1.3.2). This version of the mutual information, as the average information gain (entropy decrease) in one subsystem by performing measurements in the other, does not coincide, in general, with the definition in Eq. (1.139), representing the total amount of correlations. As we mentioned above, this is in sharp contrast with the classical case, in which the two definitions of mutual information are equivalent due to Bayes' rule.

When the mutual information in Eq. (1.146) is maximized over all possible measurements in subsystem B, we obtain the classical part of the correlations in the state ρ_{AB} [87]:

$$J(A:B) = \max_{\{\hat{\Pi}_k^B\}} J(A:B)_{\{\hat{\Pi}_k^B\}}. \tag{1.148}$$

This is the maximum amount of information that can be obtained on average about one subsystem by measuring the other one. The *quantum discord* is defined as the difference [90]:

$$\begin{aligned}\delta(A:B) &\equiv I(A:B) - J(A:B) = \\ &= \min_{\{\hat{\Pi}_k^B\}} S(A|\{\hat{\Pi}_k^B\}) + S(\rho_B) - S(\rho_{AB}) \geqslant 0,\end{aligned} \tag{1.149}$$

which hence corresponds to the quantum part of the correlations. Notice that the maximization in Eq. (1.148) translates into a minimization in Eq. (1.149), equivalent to finding the measurement which disturbs least the global state ρ_{AB} while extracting the maximum amount of information about A [90]. Furthermore discord is asymmetric in A and B, and zero if and only if

$$\delta(A:B) = 0 \;\Leftrightarrow\; \rho_{AB} = \sum_k (\hat{\mathbb{1}}_A \otimes \hat{\Pi}_k^B)\rho_{AB}(\hat{\mathbb{1}}_A \otimes \hat{\Pi}_k^B), \tag{1.150}$$

meaning that all the mutual information can be locally recovered without disturbance in the global state $\rho_{AB} = \sum_k P_k \rho_{AB}'^{(k)}$. This corresponds to a different notion of classicality in correlations, not based on separability, but on the disturbance of the global state when acquiring local information. Interestingly quantum discord reduces to entanglement in the case of pure states. On the other hand, quantum correlations in mixed states can be present even in separable states (see Fig. 1.11). We also mention that discord is bounded from above by the von Neumann entropy of the measured subsystem, that is $\delta(A:B) \leqslant S(\rho_B)$ [92].

Quantum discord has attracted increasing attention in the quantum information, quantum computation and quantum foundations communities, raising the question of how to distinguish between classical and quantum correlations, also in view of

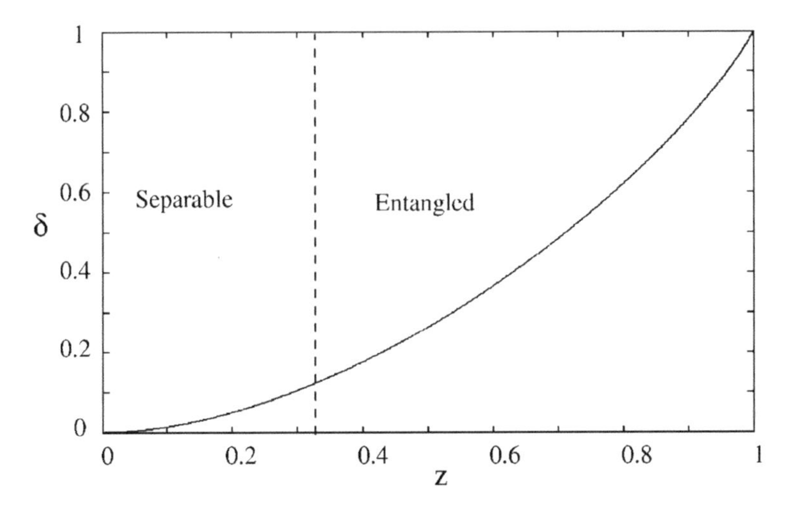

Fig. 1.11 Value of the discord for Werner states $\frac{1-z}{4}\hat{\mathbb{1}} + z|\psi\rangle\langle\psi|$ (for $z \in [0, 1]$), with $|\psi\rangle = (|00\rangle + |11\rangle)/\sqrt{2}$. Discord does not depend on the basis of measurement in this case because both $\hat{\mathbb{1}}$ and $|\psi\rangle$ are invariant under local rotations. Picture taken from Ref. [90]

applications (see [89] for a recent review). Discord has been shown to be responsible of computational speedup in some quantum protocols not using entanglement [93, 94], to spotlight critical points associated with quantum phase transitions [95], or to be a resource for quantum state merging [96, 97], remote state preparation [98], entanglement distribution [99], and device-dependent quantum cryptography [100], among others.

Albeit discord has been originally introduced using (local) rank-1 orthogonal projectors [90] (here denoted as $\{\hat{\Pi}_k^B\}$), it can be easily extended to generalized local measurements (see e.g. [89]). Indeed, the class of measurements minimizing Eq. (1.149) are not orthogonal rank-1 projectors but a class of genuine rank-1 POVMs, whose *effects* are proportional to projectors, not necessarily orthogonal between them [93] and extremal [101], that is, their elements are linearly independent [102]. Nevertheless, in the case of two qubits (or two-level systems) it has been shown [103] that orthogonal-projectors are optimal for rank-2 states and that they provide the correct value for rank-3 and rank-4 states up to minimal corrections.

Even if it is difficult in general to obtain an analytical expression for the quantum discord when optimizing over generalized measurements, a closed expression for the case of Gaussian states has been reported [104, 105]. We introduce here the relevant expression for the *Gaussian discord* in the case of two-mode states, minimized for the set of single-mode generalized Gaussian measurements [104, 105]

$$\mathcal{D}(\rho) = f(\sqrt{b}) - f(\nu_+) - f(\nu_-) + f\left(\frac{\sqrt{a} + 2\sqrt{ab} + 2\sqrt{g}}{1 + 2\sqrt{b}}\right). \tag{1.151}$$

This expression is valid for squeezed thermal states of the form $\hat{\mathbb{S}}\rho_A^{\text{th}} \otimes \rho_B^{\text{th}}\hat{\mathbb{S}}^\dagger$ with $\hat{\mathbb{S}} \equiv e^{r(\hat{a}^\dagger\hat{b}^\dagger - \hat{a}\hat{b})}$ the two-mode squeezing operator ($\hat{a}$ and $\hat{b}$ are the corresponding annihilation operators of modes A and B), and ρ_i^{th} with $i = A, B$, arbitrary Gibbs states for each mode. In Eq. (1.151) the quantities a, b, g and s correspond to the determinants of the blocks of the covariance matrix V_{AB} [Eqs. (1.135) and (1.136)], while $\nu_\pm$ are the maximum and minimum symplectic eigenvalues of V_{AB} (see Sect. 1.4.1)

$$\nu_\pm^2 = \frac{1}{2}\left(a + b + 2g \pm \sqrt{(a + b + 2g)^2 - 4s}\right), \tag{1.152}$$

and the function $f(x)$ is defined as

$$f(x) = \left(x + \frac{1}{2}\right)\ln\left(x + \frac{1}{2}\right) - \left(x - \frac{1}{2}\right)\ln\left(x - \frac{1}{2}\right). \tag{1.153}$$

Those expressions will be used extensively along Part II of the thesis to quantify the quantum correlations shared between interacting harmonic oscillators in a composite Gaussian state.

Appendix

A.1 Proof of the Micro-Reversibility Principle

In Sect. 1.1.4 of this chapter we claimed that the microreversibility principle for non-autonomous systems can be generalized to the case in which the system Hamiltonian is not invariant under the action of the time-reversal operator, that is $[\hat{\Theta}, \hat{H}(\lambda)] \neq 0$. Here we provide a proof which is essentially the proof presented by Campisi et al. in Ref. [18], while the key point is the observation that the condition $[\hat{\Theta}, \hat{H}(\lambda)] = 0$ is not needed if one defines the time-reversed unitary evolution $\hat{U}_{\tilde{\Lambda}}(t, 0)$ as the one governed by the Hamiltonian

$$\hat{H}_R(\tilde{\lambda}) \equiv \hat{\Theta}\hat{H}(\tilde{\lambda})\hat{\Theta}^\dagger. \tag{A.1}$$

Following Ref. [18] one may discretize $\hat{U}_\Lambda(\tau - t, 0)$ as a time-ordered product in a large number of steps of duration $\epsilon = t/N$ as

$$\hat{U}_{\tilde{\Lambda}}(\tau - t, 0) = \lim_{N\to\infty} e^{-\frac{i}{\hbar}\hat{H}_R(\tilde{\lambda}(\tau - N\epsilon))\epsilon}\, e^{-\frac{i}{\hbar}\hat{H}_R(\tilde{\lambda}(\tau-(N-1)\epsilon))\epsilon} \ldots$$
$$\ldots e^{-\frac{i}{\hbar}\hat{H}_R(\tilde{\lambda}(\epsilon))\epsilon}\, e^{-\frac{i}{\hbar}\hat{H}_R(\tilde{\lambda}(0))\epsilon}. \tag{A.2}$$

Then using $\tilde{\lambda}(t) = \lambda_{\tau - t}$, i.e. in the time-reversed dynamics the control parameter takes the inverse sequence of values, we have

$$\hat{U}_{\tilde{\Lambda}}(\tau - t, 0) = \lim_{N\to\infty} e^{-\frac{i}{\hbar}\hat{H}_R(\lambda(N\epsilon))\epsilon} e^{-\frac{i}{\hbar}\hat{H}_R(\lambda((N-1)\epsilon))\epsilon} \ldots$$

$$\ldots e^{-\frac{i}{\hbar}\hat{H}_R(\lambda(\tau-\epsilon))\epsilon} e^{-\frac{i}{\hbar}\hat{H}_R(\lambda(\tau))\epsilon}. \quad \text{(A.3)}$$

Now inserting the product $\hat{\Theta}^\dagger\hat{\Theta} = \mathbb{1}$ in between any two exponentials in the decomposition we may calculate

$$\hat{\Theta}^\dagger \hat{U}_{\tilde{\Lambda}}(\tau - t, 0)\hat{\Theta} = \lim_{N\to\infty} \hat{\Theta}^\dagger e^{-\frac{i}{\hbar}\hat{H}_R(\tilde{\lambda}(N\epsilon))\epsilon}\hat{\Theta}^\dagger\hat{\Theta} e^{-\frac{i}{\hbar}\hat{H}_R(\tilde{\lambda}((N-1)\epsilon))\epsilon}$$

$$\hat{\Theta}^\dagger\hat{\Theta} \ldots \hat{\Theta}^\dagger\hat{\Theta} e^{-\frac{i}{\hbar}\hat{H}_R(\tilde{\lambda}(\tau-\epsilon))\epsilon}\hat{\Theta}^\dagger\hat{\Theta} e^{-\frac{i}{\hbar}\hat{H}_R(\tilde{\lambda}(\tau))\epsilon}\hat{\Theta}.$$

At this point, we calculate the action of the time-reversal operation on any exponential, which leads to

$$\hat{\Theta}^\dagger e^{-\frac{i}{\hbar}\hat{H}_R(\lambda)\epsilon}\hat{\Theta} = e^{\frac{i}{\hbar}\hat{\Theta}^\dagger\hat{H}_R(\lambda)\hat{\Theta}\epsilon} = e^{\frac{i}{\hbar}\hat{H}(\lambda)\epsilon}, \quad \text{(A.4)}$$

where we have used that ϵ is a real number and, crucially, the definition (A.1). Using Eq. (A.4) we hence obtain

$$\hat{\Theta}^\dagger \hat{U}_{\tilde{\Lambda}}(\tau - t, 0)\hat{\Theta} = \lim_{N\to\infty} e^{\frac{i}{\hbar}\hat{H}(\lambda(N\epsilon))\epsilon} e^{\frac{i}{\hbar}\hat{H}(\lambda((N-1)\epsilon))\epsilon} \quad \text{(A.5)}$$

$$\ldots e^{\frac{i}{\hbar}\hat{H}(\lambda(\tau-\epsilon))\epsilon} e^{\frac{i}{\hbar}\hat{H}(\lambda(\tau))\epsilon}$$

$$= \lim_{N\to\infty} \Big[e^{-\frac{i}{\hbar}\hat{H}(\lambda(\tau))\epsilon} e^{-\frac{i}{\hbar}\hat{H}(\lambda(\tau-\epsilon))\epsilon}$$

$$\ldots e^{-\frac{i}{\hbar}\hat{H}(\lambda((N+1)\epsilon))\epsilon} e^{-\frac{i}{\hbar}\hat{H}(\lambda(N\epsilon))\epsilon}\Big]^\dagger,$$

and then $\hat{\Theta}^\dagger \hat{U}_{\tilde{\Lambda}}(\tau - t, 0)\hat{\Theta} = \hat{U}_\Lambda(\tau, t)$, that is, we recover the micro-reversibility principle in Eq. (1.15).

References

1. H.-P. Breuer, F. Petruccione, *The Theory of Open Quantum Systems* (Oxford University Press, New York, 2002)
2. S. Haroche, J.M. Raimond, *Exploring the Quantum: Atoms, Cavities and Photons*, Oxford Graduate Texts (Oxford University Press, Oxford, 2006)
3. A. Rivas, S.F. Huelga, *Open Quantum Systems : An Introduction* (Springer, Berlin, 2012)
4. W.H. Zurek, Decoherence, einselection, and the quantum origins of the classical. Rev. Mod. Phys. **75**, 715–775 (2003)
5. M.A. Nielsen, I.L. Chuang, *Quantum Computation and Quantum Information* (Cambridge University Press, Cambridge, 2000)
6. H.M. Wiseman, G.J. Milburn, *Quantum Measurement and Control* (Cambridge University Press, Cambridge, 2010)
7. M. Born, Zur Quantenmechanik der Stovorgänge [On the quantum mechanics of collisions], Z. Phys. **37**, 863–867 (1926), [English translation in Ref. [106], pp. 52–62]

8. W. Heisemberg, Über den anschaulichen Inhalt der quantentheoretischen Kinematik und Mechanik [The physical content of quantum kinematics and mechanics], Z. Phys. **43**, 172–198 (1927), [English translation in Ref. [106], pp. 62–87]
9. H.P. Robertson, The uncertainty principle. Phys. Rev. **34**, 163–164 (1929)
10. G. Auletta, M. Fortunato, G. Parisi, *Quantum Mechanics into a Modern Perspective* (Cambridge University Press, New York, 2009)
11. J. Von Neumann, Wahrscheinlichkeitstheoretischer Aufbau der Quantenmechanik [Probabilistic structure of quantum mechanics], Göttinger Nachrichten **10**, 245–272 (1927), [English translation in Ref. [107], pp. 208–235]
12. M. Schlüter, L.J. Sham, Density functional theory. Phys. Today **35**, 36–43 (1982)
13. K. Jacobs, *Quantum Measurement Theory and its Applications* (Cambridge University Press, Cambridge, 2014)
14. T. Young, *A Course of Lectures on Natural Philosophy and the Mechanical Arts* (William Savage, London, 1907)
15. T. Juffmann, A. Milic, M. Mullneritsch, P. Asenbaum, A. Tsukernik, J. Tuxen, M. Mayor, O. Cheshnovsky, M. Arndt, Realtime single-molecule imaging of quantum interference. Nat. Nano. **7**, 297–300 (2012)
16. E. Schrödinger, An undulatory theory of the mechanics of atoms and molecules. Phys. Rev. **28**, 1049 (1926)
17. D. Andrieux, P. Gaspard, Quantum work relations and response theory. Phys. Rev. Lett. **100**, 230404 (2008)
18. M. Campisi, P. Hänggi, P. Talkner, Colloquium: quantum fluctuation relations: foundations and applications. Rev. Mod. Phys. **83**, 771–791 (2011)
19. F. Haake, *Quantum Signatures of Chaos*, 3rd edn. Springer Series in Synergetics (Springer, Berlin, 2010)
20. D. Reeb, M.M. Wolf, An improved Landauer principle with finite-size corrections. New J. Phys. **16**, 103011 (2014)
21. A. Einstein, B. Podolsky, N. Rosen, Can quantum-mechanical description of physical reality be considered complete? Phys. Rev. **47**, 777–780 (1935)
22. J. Bell, *Speakable and Unspeakable in Quantum Mechanics* (Cambridge University Press, Cambridge, 1987)
23. S. Kullback, R.A. Leibler, On information and sufficiency. Ann. Math. Statist. **22**, 79–86 (1951)
24. T.M. Cover, J.A. Thomas, *Elements of Information Theory* (Wiley, Hoboken, 2006)
25. V. Vedral, The role of relative entropy in quantum information theory. Rev. Mod. Phys. **74**, 197 (2002)
26. M. Schlosshauer, *Decoherence and the Quantum-to-Classical Transition* (Springer, Berlin, 2008)
27. S. Haroche, Nobel lecture: controlling photons in a box and exploring the quantum to classical boundary. Rev. Mod. Phys. **85**, 1083 (2013)
28. N. Brunner, N. Linden, S. Popescu, P. Skrzypczyk, Virtual qubits, virtual temperatures, and the foundations of thermodynamics. Phys. Rev. E **85**, 051117 (2012)
29. J.W. Gibbs, *Elementary Principles in Statistical Mechanics* (Charles Scribner's Sons, New York, 1902)
30. R.J. Glauber, Coherent and incoherent states of the radiation field. Phys. Rev. **161**, 2766 (1963)
31. M.O. Scully, M.S. Zubairy, *Quantum Optics* (Cambridge University Press, Cambridge, 1997)
32. C.C. Gerry, P.L. Knight, *Introductory Quantum Optics* (Cambridge University Press, Cambridge, 2005)
33. W.P. Schleich, *Quantum Optics in Phase Space* (Wiley-VCH, Berlin, 2001)
34. C.M. Caves, Quantum-mechanical noise in an interferometer. Phys. Rev. D **23**, 1693 (1981)
35. D.F. Walls, Squeezed states of light. Nature **306**, 141–146 (1983)
36. R. Loudon, P.L. Knight, Squeezed light. J. Mod. Opt. **34**, 709–759 (1987)
37. H. Vahlbruch, M. Mehmet, S. Chelkowski, B. Hage, A. Franzen, N. Lastzka, S. Goßler, K. Danzmann, R. Schnabel, Observation of squeezed light with 10-dB quantum-noise reduction. Phys. Rev. Lett. **100**, 033602 (2008)

38. H. Vahlbruch, M. Mehmet, K. Danzmann, R. Schnabel, Detection of 15 dB squeezed states of light and their application for the absolute calibration of photoelectric quantum efficiency. Phys. Rev. Lett. **117**, 110801 (2016)
39. D.W. Robinson, The ground state of the Bose gas. Comm. Math. Phys. **1**, 159–174 (1965)
40. D. Stoler, Equivalence classes of minimum uncertainty packets. Phys. Rev. D **1**, 3217 (1970)
41. D. Stoler, Equivalence classes of minimum-uncertainty packets. II. Phys. Rev. D **4**, 1925 (1971)
42. E.Y.C. Lu, New coherent states of the electromagnetic field. Lett. Nuovo Cim. **2**, 1241 (1971)
43. E.Y.C. Lu, Quantum correlations in two-photon amplification. Lett. Nuovo Cim. **3**, 585 (1972)
44. H.P. Yuen, Two-photon coherent states of the radiation field. Phys. Rev. A **13**, 2226 (1976)
45. J.N. Hollenhorst, Quantum limits on resonant-mass gravitationalradiation detectors. Phys. Rev. D **19**, 1669 (1979)
46. G. Breitenbach, S. Schiller, J. Mlynek, Measurement of the quantum states of squeezed light. Nature **387**, 471–475 (1997)
47. H. Fearn, M.J. Collett, Representations of squeezed states with thermal noise. J. Mod. Opt. **35**, 553–564 (1988)
48. M.S. Kim, F.A.M. de Oliveira, P.L. Knight, Properties of squeezed number states and squeezed thermal states. Phys. Rev. A **40**, 2494 (1989)
49. R.E. Slusher, L.W. Hollberg, B. Yurke, J.C. Mertz, J.F. Valley, Observation of squeezed states generated by four-wave mixing in an optical cavity. Phys. Rev. Lett. **55**, 2409 (1985)
50. R.E. Slusher, L.W. Hollberg, B. Yurke, J.C. Mertz, J.F. Valley, Erratum: observation of squeezed states generated by four- wave mixing in an optical cavity. Phys. Rev. Lett. **56**, 788 (1986)
51. R.M. Shelby, M.D. Levenson, S.H. Perlmutter, R.G. DeVoe, D.F. Walls, Broad-band parametric deamplification of quantum noise in an optical fiber. Phys. Rev. Lett. **57**, 691 (1986)
52. L.-A. Wu, H.J. Kimble, J.L. Hall, H. Wu, Generation of squeezed states by parametric down conversion. Phys. Rev. Lett. **57**, 2520 (1986)
53. B. Yurke, P.G. Kaminsky, R.E. Miller, E.A. Whittaker, A.D. Smith, A.H. Silver, R.W. Simon, Observation of 4.2- K equilibrium-noise squeezing via a Josephson-parametric amplifier. Phys. Rev. Lett. **60**, 764 (1988)
54. V. Giovannetti, S. Lloyd, L. Maccone, Quantum-enhanced measurements: beating the standard quantum limit. Science **306**, 1330 (2004)
55. E.S. Polzik, The squeeze goes on. Nature **453**, 45–46 (2008)
56. K. Goda, O. Miyakawa, E.E. Mikhailov, S. Saraf, R. Adhikari, K. McKenzie, R. Ward, S. Vass, A.J. Weinstein, N. Mavalvala, A quantum-enhanced prototype gravitational-wave detector. Nat. Phys. **4**, 472–476 (2008)
57. T.L.S. Collaboration, A gravitational wave observatory operating beyond the quantum shot-noise limit. Nat. Phys. **7**, 962–965 (2011)
58. A. Furusawa, J.L. Sorensen, S.L. Braunstein, C.A. Fuchs, H.J. Kimble, E.S. Polzik, Unconditional quantum teleportation. Science **282**, 706–709 (1998)
59. S. Lloyd, L. Braunstein, Quantum computation over continuous variables. Phys. Rev. Lett. **82**, 1784 (1999)
60. M. Hillery, Quantum cryptography with squeezed states. Phys. Rev. A **61**, 022309 (2000)
61. G. Brida, M. Genovese, I.R. Berchera, Experimental realization of sub-shot-noise quantum imaging. Nat. Photon. **4**, 227–230 (2010)
62. D.M. Meekhof, C. Monroe, B.E. King, W.M. Itano, D.J. Wineland, Generation of nonclassical motional states of a trapped atom. Phys. Rev. Lett. **76**, 1796 (1996)
63. D.M. Meekhof, C. Monroe, B.E. King, W.M. Itano, D.J. Wineland, Erratum: generation of nonclassical motional states of a trapped atom. Phys. Rev. Lett. **77**, 2346 (1996)
64. C. Orzel, A.K. Tuchman, M.L. Fenselau, M. Yasuda, M.A. Kasevich, Squeezed states in a Bose-Einstein condensate. Science **291**, 2386–2389 (2001)
65. J. Estève, C. Gross, A. Weller, S. Giovanazzi, M.K. Oberthaler, Squeezing and entanglement in a Bose-Einstein condensate. Nature **455**, 1216 (2008)

66. J. Zhao, A.V. Bragas, D.J. Lockwood, R. Merlin, Magnon squeezing in an antiferromagnet: reducing the spin noise below the standard quantum limit. Phys. Rev. Lett. **93**, 107203 (2004)
67. E.E. Wollman, C.U. Lei, A.J. Weinstein, J. Suh, A. Kronwald, F. Marquardt, A.A. Clerk, K.C. Schwab, Quantum squeezing of motion in a mechanical resonator. Science **349**, 952–955 (2015)
68. J.-M. Pirkkalainen, E. Damskägg, M. Brandt, F. Massel, M.A. Sillanpää, Squeezing of quantum noise of motion in a micromechanical resonator. Phys. Rev. Lett. **115**, 243601 (2015)
69. J. Von Neumann, *Mathematical Foundations of Quantum Mechanics*, translated by R. T. Geyer (Princeton University Press, Princeton, 1955)
70. G. Lüders, Über die Zustandsänderung durch den Meßprozeß [Concerning the state-change due to the measurement process], Ann. Phys. (Leipzig) **8**, 322 (1951), [English translation by K. A. Kirkpatrick, Ann. Phys. (Leipzig) **15**, 633, (2006)]
71. A. Beiser, *Concepts of Modern Physics*, 6th edn. (McGraw-Hill, International Ed, Boston, 2003)
72. W. Heisenberg, *The Physical Principles of Quantum Theory*, translated by C. Eckart, F. C. Hoyt. (University of Chicago Press, Chicago, 1930)
73. M.A. Nielsen, Characterizing mixing and measurement in quantum mechanics. Phys. Rev. A **63**, 022114 (2001)
74. C.A. Fuchs, K. Jacobs, Information-tradeoff relations for finitestrength quantum measurements. Phys. Rev. A **63**, 062305 (2001)
75. V.B. Braginsky, F.Y. Khalili, *Quantum Measurement* (Cambridge University Press, Cambridge, 1992)
76. W.H. Zurek, Pointer basis of quantum apparatus: Into what mixture does the wave packet collapse? Phys. Rev. D **24**, 1516–1525 (1981)
77. E. Schrödinger, Die gegenwärtige Situation in der Quantenmechanik [The present situation in quantum mechanics], Naturwissenschaften **23**, 844 (1935), [English translation in Ref. [106], pp. 152–168]
78. R. Horodecki, P. Horodecki, M. Horodecki, K. Horodecki, Quantum entanglement. Rev. Mod. Phys. **81**, 865–942 (2009)
79. A. Peres, Separability criterion for density matrices. Phys. Rev. Lett. **77**, 1413 (1996)
80. G. Vidal, R.F. Werner, Computable measure of entanglement. Phys. Rev. A **65**, 032314 (2002)
81. C.H. Bennett, D.P. DiVincenzo, J.A. Smolin, W.K. Wootters, Mixed-state entanglement and quantum error correction. Phys. Rev. A **54**, 3824 (1996)
82. W.K. Wootters, Entanglement of formation of an arbitrary state of two qubits. Phys. Rev. Lett. **80**, 2245 (1998)
83. G. Giedke, M.M. Wolf, O. Krüger, R.F. Werner, I. Cirac, Entanglement of formation for symmetric Gaussian states. Phys. Rev. Lett. **91**, 107901 (2003)
84. M.M. Wolf, G. Giedke, O. Krüger, R.F. Werner, J.I. Cirac, Gaussian entanglement of formation. Phys. Rev. A **69**, 052320 (2004)
85. V. Vedral, M.B. Plenio, M.A. Rippin, P.L. Knight, Quantifiying entanglement. Phys. Rev. Lett. **78**, 2275 (1997)
86. G. Adesso, A. Serafini, F. Illuminati, Quantification and scaling of multipartite entanglement in continuous variable systems. Phys. Rev. Lett. **93**, 220504 (2004)
87. L. Henderson, V. Vedral, Classical, quantum and total correlations. J. Phys. A: Math. Gen. **34**, 6899 (2001)
88. M. Horodecki, J. Oppenheim, A. Winter, Partial quantum information. Nature **436**, 673–679 (2005)
89. K. Modi, A. Brodutch, H. Cable, T. Paterek, V. Vedral, The classical-quantum boundary for correlations: discord and related measures. Rev. Mod. Phys. **84**, 1655–1707 (2012)
90. H. Ollivier, W.H. Zurek, Quantum discord: a measure of the quantumness of correlations. Phys. Rev. Lett. **88**, 017901 (2001)
91. E. Knill, R. Laflamme, Power of one bit of quantum information. Phys. Rev. Lett. **81**, 5672 (1998)

92. N. Li, S. Luo, Classical and quantum correlative capacities of quantum systems. Phys. Rev. A **84**, 042124 (2011)
93. A. Datta, A. Shaji, C.M. Caves, Quantum discord and the power of one qubit. Phys. Rev. Lett. **100**, 050502 (2008)
94. B.P. Lanyon, M. Barbieri, M.P. Almeida, A.G. White, Experimental quantum computing without entanglement. Phys. Rev. Lett. **101**, 200501 (2008)
95. T. Werlang, C. Trippe, G.A.P. Ribeiro, G. Rigolin, Quantum correlations in spin chains at finite temperatures and quantum phase transitions. Phys. Rev. Lett. **105**, 095702 (2010)
96. D. Cavalcanti, L. Aolita, S. Boixo, K. Modi, M. Piani, A. Winter, Operational interpretations of quantum discord. Phys. Rev. A **83**, 032324 (2011)
97. V. Madhok, A. Datta, Interpreting quantum discord through quantum state merging. Phys. Rev. A **83**, 032323 (2011)
98. B. Dakić et al., Quantum discord as resource for remote state preparation. Nat. Phys. **8**, 666–670 (2012)
99. T.K. Chuan, J. Maillard, K. Modi, T. Paterek, M. Paternostro, M. Piani, Quantum discord bounds the amount of distributed entanglement. Phys. Rev. Lett. **109**, 070501 (2012)
100. S. Pirandola, Quantum discord as a resource for quantum cryptography. Sci. Rep. **4**, 06956 (2014)
101. S. Hamieh, R. Kobes, H. Zaraket, Positive-operator-valued measure optimization of classical correlations. Phys. Rev. A **70**, 052325 (2004)
102. G.M. D'Ariano, P.L. Presti, P. Perinotti, Classical randomness in quantum measurements. J. Phys. A: Math. Gen. **38**, 5979–5991 (2005)
103. F. Galve, G.L. Giorgi, R. Zambrini, Orthogonal measurements are almost sufficient for quantum discord of two qubits. Europhys. Lett. **96**, 40005 (2011)
104. G. Adesso, A. Datta, Quantum versus classical correlations in Gaussian states. Phys. Rev. Lett. **105**, 030501 (2010)
105. P. Giorda, M.G.A. Paris, Gaussian quantum discord. Phys. Rev. Lett. **105**, 020503 (2010)
106. J.A. Wheeler, W.H. Zurek (eds.), *Quantum Theory and Measurement* (Princeton University Press, Princeton, 1983)
107. J. Von Neumann, in *John Von Neumann Collected Works (6 volume set)*, ed. by A.H. Traub (Pergamon Press, Oxford, 1961)

Chapter 2
Open Quantum Systems Dynamics

In the previous chapter we reviewed a number of fundamental concepts and elements needed to build a satisfactory description of open quantum systems. We are now in position to properly focus on the dynamical evolution of open quantum systems and its main properties. In the framework of open quantum systems the interaction of a system with its surroundings induces a noise affecting the evolution of the system of interest. This noise appears as a result of neglecting or averaging over the complete isolated evolution of system plus environment, which allows us to obtain an approximate effective description of the open system dynamics which is mathematically tractable.

Along this chapter, we will see that there exist different approaches to describe the dynamics of open quantum systems, which involve different levels of generality and approximations, and may result useful in different contexts. As a field developed by many different communities one may also find, as pointed in Ref. [1], that the kind of tools and approximations involved in the description of e.g. optical systems [2–4], may greatly differ from those employed in condensed matter [5], quantum information theory [6], or statistical physics [7–9]. The purpose of this chapter is not to provide a review of these many different approaches, but to introduce the main tools we will employ along the thesis. Modern expositions providing a unifying view while covering the most important methods in open quantum system theory, can be found in excellent specialized textbooks in the matter (see e.g. [1, 10, 11]).

Here we primarily differentiate between two main approaches to describe the evolution of open quantum systems. The first one corresponds to the general formalism of *quantum maps and operations*, a powerful tool which has received increased attention in the context of quantum computation and quantum information [6]. This general framework provides a mathematical description which can be derived from few physically motivated axioms. It is based on discrete state changes where the explicit reference to time plays a very secondary role. Its range of applicability is huge, including systems interacting weakly as well as strongly with their

G. Manzano Paule, *Thermodynamics and Synchronization in Open Quantum Systems*, Springer Theses, https://doi.org/10.1007/978-3-319-93964-3_2

surroundings, or being suddenly measured. The second approach to describe open quantum systems dynamics is provided by both *master equations* and *stochastic differential equations*. They give rise to a continuous-time description of the dynamics useful to understand the processes involved in the system evolution and their properties. However, the development of such approaches often requires more specific approximations. In the following we will present and discuss both dynamical descriptions, highlighting the specific concepts and techniques which will be useful in the following chapters of this thesis.

This chapter starts in Sect. 2.1 by introducing the general framework of quantum maps and operations, together with fundamental related concepts such as positivity or complete positivity. We will introduce an important tool in the quantum maps and operations formalism, the Kraus sum decomposition, and discuss its implications, establishing connections to quantum measurement theory and the environmental modelings of the maps. We then move to the time-continuous approach for open quantum systems, introducing the concept of quantum master equations in Sect. 2.2. We will focus on a particular and important case of open quantum system dynamics, the one following a Markov process, and derive the master equation governing its evolution from both theoretical considerations and microscopic derivations. In Sect. 2.3 we present different examples of Markovian master equations concerning the open system dynamics of prototypical systems (qubit and harmonic oscillator) in different dissipative situations. Next, we discuss in Sect. 2.4 the extension of the master equation formalism to the case of many-body systems where different mechanisms of dissipation may be present, developing a master equation approach for the case of coupled dissipative oscillators. Finally, we review in Sect. 2.5 the formalism of quantum trajectories, where measurement and environmental action meet in a unique framework giving rise to the stochastic Schrödinger equation, complementing the master equation approach.

2.1 Quantum Maps and Operations

Noise in classical systems is usually described by the theory of stochastic processes, in which the state of a system is allowed to change into other states by following some probability rules. If, for instance, the states of the system are a discretized set, following the exposition in Ref. [6], one can associate an input probability vector $\vec{p} = (p_1, p_2, \ldots, p_N)$ to the initial probabilities of the system to be in its different N states. Then after some time in which the system interacts with the environment, we will have some output probabilities $\vec{p}'$, that in the simplest case are linearly related to the input ones by

$$\vec{p}' = M\, \vec{p}, \tag{2.1}$$

where M is a $N \times N$ matrix whose elements are conditional probabilities, usually called the *evolution matrix*. This evolution matrix must fulfill two important properties in order to guarantee that the components of the vector $\vec{p}'$ are well defined probabilities. The first one is known as *positivity*, meaning that all the entries of M are non-negative. The second is known as *completeness*, requiring that the columns of M sum up to one. Of course, the values of the conditional probabilities depend on the specific nature of the interaction with the environment, implying that one will need in general some modeling of the underlying processes to determine M.

In order to characterize noisy processes in quantum systems the previous formalism must be generalized. In first place we replace the probability vectors such as $\vec{p}$ by density operators, ρ. Hence, analogously to Eq. (2.1), we introduce the map

$$\rho' = \mathcal{E}(\rho) \tag{2.2}$$

providing an output state ρ' when the input state is ρ. $\mathcal{E}$ is a *quantum map* and captures any dynamical change in the state of a quantum system. The simplest example of a quantum map is given by a unitary transformation (see Sect. 1.1 of the previous chapter), for which $\rho' = \mathcal{E}(\rho) = \hat{U}\rho\hat{U}^\dagger$, and corresponds to the case for which the system does not decohere into the environment (but at most interacts with an ideal external driver). Other examples of quantum maps are unconditional measurements, both ideal or generalized. In that case $\rho' = \mathcal{E}(\rho) = \sum_{k,n} \hat{M}_{k,n}\rho\hat{M}^\dagger_{k,n}$, where $\hat{M}_{k,n}$ are the measurement operators introduced in Sect. 1.3. However, the theory of quantum maps and operations can be used to describe more general situations [6].

Following Sect. 1.3.2 in Chap. 1 a natural way of interpreting Eq. (2.2) is considering it as the result of partial tracing over the environment degrees of freedom after some unitary interaction between our system of interest and its surroundings. If we assume that system and environment start in some product state $\rho \otimes \rho_E$, where the subscript E denotes the environment degrees of freedom, after an arbitrary transformation $\hat{U}$ our system is described by

$$\rho' = \mathcal{E}(\rho) \equiv \mathrm{Tr}[\hat{U}(\rho \otimes \rho_E)\hat{U}^\dagger]. \tag{2.3}$$

This is one way to provide a definition of quantum maps [6]. Here we assumed system and environment as initially uncorrelated, which in principle may limit the applicability of the formalism. However, it must be stressed that in many situations of physical interest the initial experimental preparation of the state ρ ideally implies the destruction of all the previously generated correlations between system and environment, in accordance with Eq. (2.3). We will turn to this question below.

A second way of defining quantum maps, more interesting from an operational point of view, is by direct imposition on Eq. (2.2) of a set physically motivated constraints, in analogy to the case of classical stochastic processes. In order to ensure that $\mathcal{E}$ describes a physical process transforming well defined density operators on well defined density operators, we must require the following conditions:

1. The quantum map must preserve the trace, that is $\mathrm{Tr}[\mathcal{E}(\rho)] = \mathrm{Tr}[\rho] = 1$, to return a physical output density operator.
2. The quantum map is required to be convex linear. This means that, for an ensemble of density operators $\{\rho_i\}$ randomly chosen with probabilities p_i with $\sum_i p_i = 1$,

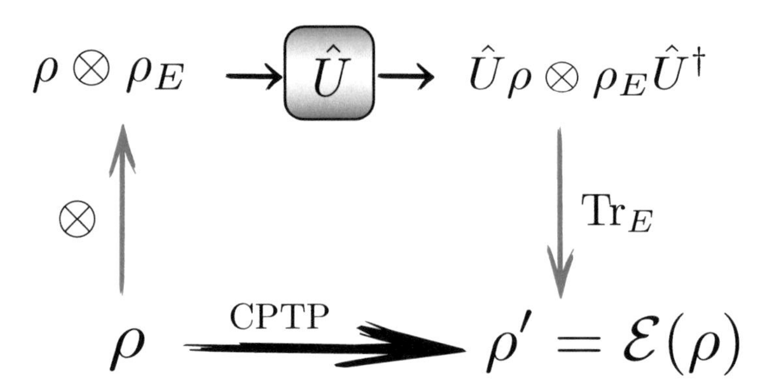

Fig. 2.1 Schematic representation of the equivalence between the definition of a quantum map as a result of partial tracing over the environmental degrees of freedom after an interaction between the system of interest and its surroundings, c.f. Eq. (2.3), and in terms of requiring it to be a linear CPTP map (red arrow)

we have

$$\mathcal{E}\left(\sum_i p_i \rho_i\right) = \sum_i p_i \mathcal{E}(\rho_i). \tag{2.4}$$

This requirement can be physically motivated by application of Bayes' theorem [10].

3. The quantum map $\mathcal{E}$ must be *completely positive*. This condition is stronger than positivity, the latter meaning that $\mathcal{E}$ maps positive operators onto positive operators. Complete positivity implies that if we enlarge the Hilbert space of our system of interest by including an extra ancillary system of arbitrary dimension d_A, but which do not interact with the system of interest at any stage, the map $(\mathcal{E} \otimes \mathbb{1}_A)$ acting on the global system, must be also positive (see Ref. [1] for a transparent discussion of this point). We stress that, while complete positivity implies positivity, the contrary is not always true, a statement which can be proven by considering the partial transposition operation.

Quantum maps obeying those three general requirements are commonly called *completely positive and trace preserving* (CPTP) maps or also *quantum channels*. In addition to CPTP maps, we may define a more general class of *quantum operations* by relaxing the first assumption regarding the preservation of the trace, while maintaining intact the second and the third requirements. In such case we replace the condition $\mathrm{Tr}[\rho'] = \mathrm{Tr}\mathcal{E}(\rho)$, by $0 \leqslant \mathrm{Tr}[\mathcal{E}(\rho)] \leqslant 1$, which implies that the operator resulting from the application of the mapping $\mathcal{E}(\rho)$ must be normalized in order to represent a legal density operator. This generalization allows to include general selective quantum measurements into the framework, in which case $\mathcal{E}$ is related to some measurement result, and the quantity $\mathrm{Tr}[\mathcal{E}(\rho)]$ must be regarded as the probability to obtain such result (Fig. 2.1).

2.1.1 Properties of CPTP Maps

Some interesting properties of CPTP maps are the following:

- CPTP maps are contractive, that is, any CPTP map $\mathcal{E}$ causes a contraction on the space of density operators. This property can be properly expressed in mathematical terms by using the trace distance (see Sect. 1.1.7 in Chap. 1) as

$$T(\rho, \sigma) \geqslant T(\mathcal{E}(\rho), \mathcal{E}(\sigma)). \tag{2.5}$$

  It is worth noticing that the last inequality is fulfilled as well when replacing the trace distance by the quantum relative entropy [12], $D(\rho||\sigma)$ as introduced in Sect. 1.1.6 of Chap. 1. This result, called the Uhlmann inequality, is of central importance both in the context of quantum information processing, and when considering the thermodynamics of open quantum systems [13]. With respect to the fidelity, $F(\rho, \sigma)$, the above equation is also fulfilled inverting the direction of the inequality.
- Any CPTP map $\mathcal{E}$ has at least an invariant state (or fixed point) π, such that

$$\mathcal{E}(\pi) = \pi, \tag{2.6}$$

  a result which follows from Schauder's fixed point theorem [6]. Furthermore, if $\mathcal{E}$ is strictly contractive, i.e.

$$T(\rho, \sigma) > T(\mathcal{E}(\rho), \mathcal{E}(\sigma)), \tag{2.7}$$

  for any ρ and σ, then the map has a unique fixed point, which is called the *stationary state*.
- CPTP maps fulfilling $\mathcal{E}(\mathbb{1}) = \mathbb{1}$, that is, for which the identity operator is a fixed point of the dynamics, are called *unital maps* or *bistochastic maps*. They constitute an important class of quantum CPTP maps, including unitary evolution as well as ideal projective measurements as special cases. Unital maps exhibit special thermodynamic properties, as we will see in Part III of the thesis. An important property of unital maps is that they can never decrease the von Neumann entropy of any state ρ:

$$S(\rho) \leqslant S(\mathcal{E}(\rho)). \tag{2.8}$$

- Any quantum CPTP map of the form

$$\mathcal{E}(\rho) = p\rho_0 + (1-p)\mathcal{E}'(\rho), \tag{2.9}$$

  where ρ_0 is a density operator, $\mathcal{E}'(\rho)$ is another CPTP map, and $0 \leqslant p \leqslant 1$, is strictly contractive, and therefore it has a unique fixed point $\pi = \rho_0$. The physical meaning of the above map is that with probability p it replaces the state ρ by the predefined ρ_0, and with probability $1-p$ it applies the map $\mathcal{E}'(\rho)$.

- The set of CPTP maps possesses a semi-group structure under concatenation. If $\mathcal{E}_1$ and $\mathcal{E}_2$ are two CPTP maps, the concatenation

$$\Omega(\rho) \equiv \mathcal{E}_2 \circ \mathcal{E}_1(\rho) = \mathcal{E}_2(\mathcal{E}_1(\rho)), \tag{2.10}$$

is also a CPTP map. However notice that the operation $\circ$ is not commutative, i.e. in general $\mathcal{E}_2 \circ \mathcal{E}_1(\rho) \neq \mathcal{E}_1 \circ \mathcal{E}_2(\rho)$.
- A CPTP map $\mathcal{E}$ is invertible, that is, one can guarantee the existence of another CPTP map $\mathcal{E}^{-1}$ such that

$$\mathcal{E}^{-1} \circ \mathcal{E} = \mathbb{1}, \tag{2.11}$$

if and only if it is a reversible transformation given by a unitary mapping, $\mathcal{E}(\rho) = \hat{U}_{\mathcal{S}} \rho \hat{U}_{\mathcal{S}}^{\dagger}$ with $\hat{U}_{\mathcal{S}}$ unitary.

2.1.2 *Kraus Operator-Sum Representation*

An important theorem due to Karl Kraus [14] states that the map $\mathcal{E}$ satisfies the three requirements introduced in the above section (both for the case of CPTP maps or for general quantum operations) if and only if it can be written as

$$\mathcal{E}(\rho) = \sum_k \hat{M}_k \rho \hat{M}_k^{\dagger}. \tag{2.12}$$

Here the countable set of operators $\{\hat{M}_k\}$ are called the Kraus operators, mapping the input Hilbert space into the output Hilbert space, and fulfill $\sum_k \hat{M}_k^{\dagger} \hat{M}_k \leqslant \mathbb{1}$ (reaching the equality in the case of CPTP maps). The form in Eq. (2.12) is usually called the *Kraus operator-sum representation* or decomposition of the map $\mathcal{E}$, and the above theorem is usually referred to as the *representation theorem*.

The operator-sum representation provides a useful way to write a CPTP map (or a quantum operation) without having to consider the specific properties of the environment, which are just encoded in the form of the Kraus operators $\hat{M}_k$. Indeed, many different environments can result in the same dynamical representation. This is an important feature, as it can greatly simplify the calculations and provide theoretical insights [6]. The Kraus representation also provides a physical interpretation of the process (2.2) analogous to classical stochastic maps. The map $\mathcal{E}$ is understood as the application of a number of physical operations on the system

$$\mathcal{E}(\rho) = \sum_k \mathcal{E}_k(\rho), \quad \text{with} \quad \mathcal{E}_k(\rho) = \hat{M}_k \rho \hat{M}_k^{\dagger}, \tag{2.13}$$

occurring with probability $P_k = \mathrm{Tr}[\mathcal{E}_k(\rho)]$. Each operations transforms the initial state ρ into

$$\rho_k' = \frac{\mathcal{E}_k(\rho)}{P_k}, \tag{2.14}$$

thus the map $\mathcal{E}$ randomly replaces ρ by ρ_k' with probability P_k. This is a very similar picture as in the case of noisy communication channels [6]. Furthermore, it is worth noticing that the Kraus operator-sum representation in not unique. If we have operator elements $\{\hat{M}_k\}_{k=1}^K$ and $\{\hat{F}_j\}_{j=1}^J$ corresponding to two quantum operations $\mathcal{E}$ and $\mathcal{F}$ respectively, and we append zero operators to the shorter set until $K = J$, it follows that $\mathcal{E} = \mathcal{F}$ if and only if

$$\hat{M}_k = \sum_j u_{kj} \hat{F}_j \tag{2.15}$$

for u_{kj} the entries of a $K \times K$ unitary matrix. This freedom implies that all quantum operations $\mathcal{E}$ acting on a Hilbert space $\mathcal{H}$ of dimension $\dim(\mathcal{H}) = N$ can be generated by an operator-sum representation of at most N^2 Kraus operators. Among all possible Kraus representations there exists a canonical form with an orthogonality relation between the Kraus operators $\mathrm{Tr}[\hat{M}_k^\dagger \hat{M}_l] \sim \delta_{k,l}$.

It is finally worth recalling here that complete positivity requires that the map $\mathcal{E}$ is physically meaningful for any arbitrary initial state ρ. Following the representation theorem, this implies that the Kraus operators appearing in Eq. (2.12) do not depend on the input state ρ, which is also equivalent to the first definition of quantum maps provided in Eq. (2.3) for an initial product state between system and environment [1]. On the contrary, maps of the form $\mathcal{E}(\rho) \equiv \mathrm{Tr}_E[\hat{U} \rho_{\mathrm{tot}} \hat{U}^\dagger]$ with $\mathrm{Tr}_E[\rho_{\mathrm{tot}}] = \rho$ and ρ_{tot} an arbitrary correlated state may be written as in (2.12), but complete positivity is not guaranteed for any initial state. This can be understood by simply noticing that the initial correlations between system and environment will be encoded in the Kraus operators $\hat{M}_k$ of the mapping, which are related to the specific state ρ sharing that correlations, and not to any arbitrary state. Henceforth it can happen that for some initial input states the dynamics $\sum_k \hat{M}_k \rho \hat{M}_k^\dagger$ is not completely positive. However, it has been shown in Refs. [15, 16] that the Kraus operator-sum representation in Eq. (2.12) is still valid for any quantum evolution by allowing the Kraus operators to explicitly depend on the initial state ρ, i.e. $\hat{M}_k = \hat{M}_k(\rho)$. This kind of more general evolution is not included in the framework of CPTP maps presented here.

2.1.3 Environmental Models

Notice the similarities of the above introduced framework of quantum CPTP maps and the general quantum measurements introduced in Sect. 1.3.2. Indeed, using the Kraus representation, any CPTP map can be viewed as an *efficient measurement* for which the environment plays the role of the ancillary system, starting in some pure state $|\phi_0\rangle_E$ and being found to be in some state of the basis $\{|\phi_k\rangle_E\}$ after the global interaction $\hat{U}$. In this case the Kraus operators would read

$$\hat{M}_k \equiv \langle \phi_k|_E \, \hat{U} \, |\phi_0\rangle_E \, . \tag{2.16}$$

Here the operator $\hat{U}$ must be defined by its action as $\hat{U}\,|\psi\rangle\,|\phi_0\rangle_E \equiv \sum_k \hat{M}_k\,|\psi\rangle\,|\phi_k\rangle_E$, for any arbitrary state of the system $|\psi\rangle$. This imposes that the first column elements of $\hat{U}$ when represented in the basis $\{|\phi_k\rangle_E\}$ correspond to the set of Kraus operators $\{\hat{M}_k\}$, while the rest of the elements can be arbitrarily chosen such that $\hat{U}$ is unitary [6]. This kind of environmental model is sometimes called the *Stinespring representation* of the map, and is always well defined even in the case of an environment starting in some mixed state ρ_E via *purification* [17, 18].

Finding an environmental model for the case of quantum operations not preserving the trace follows in a similar manner, with the difference that now one can associate to the initial state of the environment a probability, therefore including a source of classical noise in the description

$$\hat{M}_k \equiv \sqrt{p_0}\,\langle \phi_k|_E \, \hat{U} \, |\phi_0\rangle_E \, , \tag{2.17}$$

$0 \leqslant p_0 \leqslant 1$ being the probability that the environment is initially in $|\phi\rangle_0$. Now it can be easily checked that $\sum_k \hat{M}_k^\dagger \hat{M}_k = p_0 \hat{\mathbb{1}} \leqslant \hat{\mathbb{1}}$. Alternatively one can also associate the operation $\mathcal{E}$ to the result of a measurement performed in the system [6].

Finally, we stress that the unitary freedom in the selection of the Kraus operator-sum representation can be easily identified in the environmental representation introduced above by introducing a final local unitary transformation acting only on the environmental degrees of freedom, $\hat{U}_E$. Clearly, the introduction of this unitary process does not influence the dynamics of the system of interest, which is still described by the same CPTP map (or quantum operation) $\mathcal{E}$. However, alternatively to (2.16) [or (2.17) for quantum operations] we can define a new representation given by elements

$$\hat{F}_k \equiv \langle \phi_k|_E \, (\hat{\mathbb{1}} \otimes \hat{U}_E)\hat{U} \, |\phi_0\rangle_E = \langle \phi_k'|_E \, \hat{U} \, |\phi_0\rangle_E \, , \tag{2.18}$$

where we introduced a new set of orthonormal basis elements for the environment $\{|\psi_k'\rangle_E \equiv \hat{U}_E^\dagger \, |\phi_k\rangle_E\}$. We hence see that the action of the local unitary $\hat{U}_E$ can be included in the selection of a different environmental basis. Now it can be easily shown that the relation (2.15) between the old and new Kraus operators is

$$\hat{F}_k = \sum_j \langle \phi_k|_E \, \hat{U}_E \, |\phi_j\rangle_E \;\; \langle \phi_j|_E \, \hat{U} \, |\psi_0\rangle_E = \sum_j u_{kj}\hat{M}_j, \tag{2.19}$$

with $u_{kj} \equiv \langle \phi_k|_E \, \hat{U}_E \, |\psi_k\rangle_E$ entries of a unitary matrix. From the above reasoning we hence see that the unitary freedom of the Kraus operator-sum representation is equivalent to the selection of a specific environmental basis which, following Sect. 1.3.2, corresponds to the basis where measurements are performed.

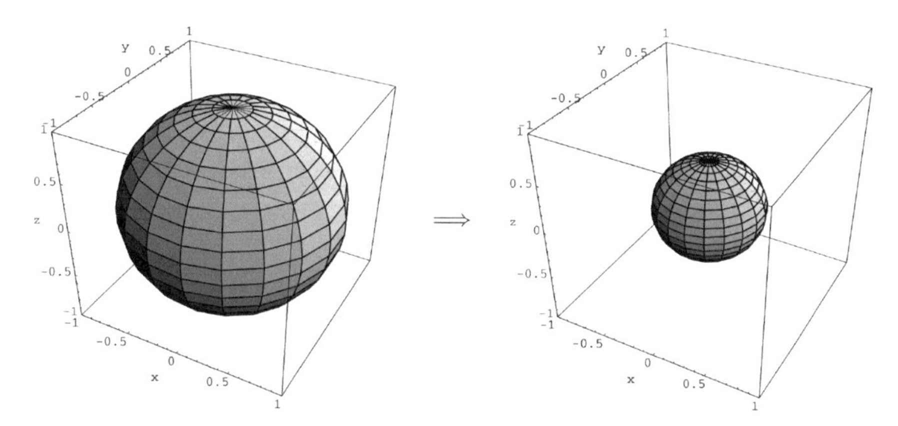

Fig. 2.2 Effect of the depolarizing channel on the Bloch sphere of a qubit system. All pure states in the surface of the Bloch sphere are mapped to mixed states in the inner sphere. The entire sphere contracts uniformly with p, in this case $p = 0.5$. The picture has been obtained from Ref. [6]

2.1.4 *Some Examples of CPTP Maps*

Here we give two simple examples of CPTP maps. The first one is the so-called *depolarizing channel*, operating on finite dimensional quantum systems

$$\mathcal{E}(\rho) = (1-p)\rho + p\frac{\hat{\mathbb{1}}}{N}\mathrm{Tr}[\rho], \tag{2.20}$$

where $0 \leqslant p \leqslant 1$ is called the probability of error, and N is the dimension of the Hilbert space, $\mathcal{H}$. This CPTP quantum map transform the initial state ρ into the maximally mixed state $\mathbb{1}/N$ with probability p and has no effect with probability $1-p$. A Kraus operator-sum representation of the map is given by [18]

$$\mathcal{E}(\rho) = \hat{M}_0\rho\hat{M}_0^\dagger + \sum_{ij=1}^{N}\hat{M}_{ij}\rho\hat{M}_{ij}^\dagger, \tag{2.21}$$

with $\hat{M}_0 = \sqrt{1-p}\ \mathbb{1}$, and $\hat{M}_{ij} = \sqrt{p/N}\ \ |\psi_i\rangle\langle\psi_j|$ being $\{|\psi_i\rangle\}_{i=1}^N$ an arbitrary orthonormal basis of $\mathcal{H}$. This gives us a decomposition of the map into operations leaving the state untouched with probability $1-p$ (operator $\hat{M}_0$), or transforming it into the pure state $|\psi_i\rangle$ with probability $p\,\langle\psi_j|\,\rho\,|\psi_j\rangle/N$ (operator $\hat{M}_{ij}$). It can be easily checked that the depolarizing channel is strictly contractive, as it is of the form (2.9). Therefore, it has a unique fixed point (steady state) $\pi = \frac{\hat{\mathbb{1}}}{N}\mathrm{Tr}[\rho]$, to which any initial state converges after a large sequence of successive applications of the map. In the case of a qubit system, a single application of the map can be visualized on the Bloch sphere (see Sect. 1.2.1 in Chap. 1) by representing all the output states when the input are pure states (that is, the surface of the sphere), as shown in Fig. 2.2.

Our second example is the *generalized amplitude damping channel* for qubit systems. It describes e.g. the relaxation process of a spin due to its coupling to other spins in a surrounding lattice (or more general spin environments) when they are in thermal equilibrium, a relevant situation in NMR quantum computation [6]. A Kraus operator-sum representation of such CPTP map is defined by four Kraus operators of the form

$$\hat{M}_0 = \sqrt{p}\begin{pmatrix} 1 & 0 \\ 0 & \sqrt{1-\lambda} \end{pmatrix}, \qquad \hat{M}_1 = \sqrt{p}\begin{pmatrix} 0 & \sqrt{\lambda} \\ 0 & 0 \end{pmatrix},$$
$$\hat{M}_2 = \sqrt{1-p}\begin{pmatrix} \sqrt{1-\lambda} & 0 \\ 0 & 1 \end{pmatrix}, \quad \hat{M}_3 = \sqrt{1-p}\begin{pmatrix} 0 & 0 \\ \sqrt{\lambda} & 0 \end{pmatrix}, \tag{2.22}$$

in the basis defined by $|0\rangle = \begin{pmatrix} 1 \\ 0 \end{pmatrix}$ and $|1\rangle = \begin{pmatrix} 0 \\ 1 \end{pmatrix}$.

The two parameters appearing in Eq. (2.22), $0 \leqslant p \leqslant 1$ and $0 \leqslant \lambda \geqslant 1$ can be associated to physical processes producing this map. For instance in the case of a two-level atom with energy spacing E weakly interacting in the dipole approximation with a thermal radiation reservoir at inverse temperature $\beta = 1/k_B T$, we can make the identifications

$$p \equiv \frac{n_{\text{th}} + 1}{2n_{\text{th}} + 1}, \qquad \lambda \equiv 1 - e^{-t/\tau_R}, \tag{2.23}$$

where $n_{\text{th}} = (e^{\beta E} - 1)^{-1}$ is the mean number of thermal photons with energy E (see Sect. 1.2.3), t parametrizes time, and $\tau_R \propto (2n_{\text{th}} + 1)^{-1}$ is the relaxation time. In such case, the operator $\hat{M}_1$ ($\hat{M}_3$) describes a jump process where the atom emits (absorbs) an energy quantum E to (from) the environment, and the operator $\hat{M}_0$ ($\hat{M}_2$) a monitoring process in which the amplitude of state $|1\rangle$ ($|0\rangle$) decreases, while the coherences in the energy basis are damped.

The generalized amplitude damping channel is also strictly contractive, being its unique fixed point (stationary state) $\pi = p\,|0\rangle\,\langle 0| + (1-p)\,|1\rangle\,\langle 1|$. Using the identification (2.23) for p, the stationary state π is easily shown to correspond to the Gibbs thermal equilibrium state at inverse temperature β:

$$p = \frac{1}{Z}, \quad 1 - p = \frac{e^{-\beta E}}{Z}, \quad \text{with } Z = 1 + e^{-\beta E}, \tag{2.24}$$

as introduced in Sect. 1.2.1. Therefore, many successive applications of the map (or also when $t \to \infty$) produce the complete thermalization of the two-level system to the reservoir's temperature. We notice that the generalized amplitude damping channel can be further generalized to describe the effect of a squeezed thermal reservoir (e.g. a continuum of light modes in a squeezed thermal state) on the qubit system [19].

2.2 Markovian Master Equations

So far we provided a characterization of open quantum system dynamics mainly focused on discrete transformations, where time does not explicitly enter in the description, and the specific time evolution occurring in the larger Hilbert space of the global system (open system plus its environment) played only a secondary role. Here we will turn our perspective to a continuous-time description of open quantum systems based on the development of differential equations for the density operator, a method usually called the *master equation approach*. This approach requires a more careful look at the global picture, as we will shortly see. In this section we show how one can deduce master equations describing the evolution of an open quantum system both by deriving it from the above quantum maps and operations formalism, or considering specific models for the environment and its interaction with the system of interest.

Let us start by discussing continuity in time of quantum CPTP maps and operations inspired by the more detailed discussion presented in Ref. [1]. We have previously seen that the concatenation of CPTP maps provides a well defined CPTP map, however we will now see that the converse statement is not always true. Consider a CPTP map $\mathcal{E}$ describing the dynamics of an open quantum system from some initial instant of time t_0 to some posterior instant t_2. We pursue the splitting of the map as a concatenation of two CPTP maps, $\mathcal{E} \equiv \mathcal{E}_2 \circ \mathcal{E}_1$, where $\mathcal{E}_1$ describes the evolution of the open system from the initial time t_0 to t_1, and $\mathcal{E}_2$ describes it from time t_1 to t_2. As we have previously seen, the CPTP map $\mathcal{E}$ can be viewed as the result of tracing the environmental degrees of freedom after a global unitary evolution of system and environment

$$\mathcal{E}(\rho(t_0)) = \mathrm{Tr}_E[\hat{U}(\rho(t_0) \otimes \rho_E)\hat{U}^\dagger] = \rho(t_2), \tag{2.25}$$

where initially, system and environment must be completely uncorrelated. However, we may in principle do the same for the maps $\mathcal{E}_1$ and $\mathcal{E}_2$ as we want they to be also CPTP. By splitting the global unitary evolution as $\hat{U} = \hat{U}_2\hat{U}_1$ we can indeed provide such environmental representation for $\mathcal{E}_1$, but for the case of $\mathcal{E}_2$ we would obtain

$$\mathcal{E}_2(\rho(t_1)) = \mathrm{Tr}_E[\hat{U}_2\rho_{\mathrm{tot}}(t_1)\hat{U}_2^\dagger] = \rho(t_2). \tag{2.26}$$

Here the state $\rho_{\mathrm{tot}}(t_1) = \hat{U}_1(\rho(t_0) \otimes \rho_E)\hat{U}_2^\dagger$ in general contains correlations, implying that it cannot be written in the required product state form $\rho_{\mathrm{tot}}(t_1) \neq \rho(t_1) \otimes \rho_E(t_1)$, with $\rho_E(t_1)$ the reduced state of the environment at time t_1. From the above reasoning it follows that one cannot in general deduce a continuous-time evolution, as we cannot in general split a generic CPTP map $\mathcal{E}$ in a concatenation of many CPTP maps describing infinitesimal time-steps of the dynamics. The evolutions which allow time divisibility $\mathcal{E} \equiv \mathcal{E}_2 \circ \mathcal{E}_1$ as discussed above, are sometimes called *Markovian evolutions*, a convention that we will adopt in this thesis, and are analogous to the classical evolutions fulfilling the Chapman–Kolmogorov equation, introduced in the context of Markov processes [1]. The underlying physical idea behind Markovian evolution in both classical and quantum cases, is that the dynamical effects arising in the open system as a result of the interaction with its surroundings can be considered to be uncorrelated from one infinitesimal instant of time to the next. However, other notions of Markovianity have been introduced in the context of open quantum systems, e.g. based on measures of the back-flow of information between system and environment [20] (for reviews see Refs. [21, 22]).

From the above discussion, it becomes clear that we may need to introduce extra conditions in order to ensure a description of CPTP dynamics in terms of differential equations. Consider the splitting of time into a sequence of small time steps of duration τ. If we denote $\rho(t)$ the density operator of the open system at an arbitrary time t, after one of those small time steps we have $\rho(t+\tau) = \mathcal{E}_t(\rho(t))$ being $\mathcal{E}_t$ a CPTP map, which may depend both on t and τ. From this coarse-grained description, a first-order differential equation (or master equation) can be mathematically defined as

$$\frac{d\rho(t)}{dt} = \lim_{\tau\to 0} \frac{\rho(t+\tau) - \rho(t)}{\tau} = \mathcal{L}_t(\rho(t)), \tag{2.27}$$

with $\mathcal{L}_t \equiv \lim_{\tau\to 0}\left(\mathcal{E}_t - \hat{\mathbb{1}}\right)/\tau$, provided the limit $\tau \to 0$ is well defined (smooth evolution) [1]. However, as we have seen previously, to define the CPTP maps $\mathcal{E}_t$, we must require that the global state of system and environment at any time t

$$\rho_{\text{tot}}(t) = \rho(t) \otimes \rho_E(t) + \delta\rho_{\text{corr}}(t) \approx \rho(t) \otimes \rho_E(t), \tag{2.28}$$

where $\delta\rho_{\text{corr}}(t)$ is a traceless term containing the correlations between system and environment due to its prior interaction and $\rho_E(t)$ is the reduced state of the environment at time t.

Equation (2.28) may be fulfilled under a variety of physical circumstances. The most common situation is to consider the environment a large system in some steady state with small fluctuations $\rho_E(t) = \rho_E + \delta\rho_E(t)$ and with levels spanning a wide energy range $\hbar\Delta\omega$ [23] (see also Sect. 2.2.2 below). In this case the time scale associated to the two-time correlations of the terms $\delta\rho_E(t)$ and $\delta\rho_{\text{corr}}(t)$ is very short, of the order $\tau_c = \hbar/\Delta\omega$, while the time scale associated to the variation of the system density operator is $\tau_\rho = \hbar^2/g^2\tau_c$, g being the magnitude of the coupling between system and environment. Therefore, if $\tau_c \ll \tau_\rho$, which is the case when the coupling is weak, $g \ll \hbar/\tau_c = \Delta\omega$, we can always choose a coarse-grained time τ such that $\tau_c \ll \tau \ll \tau_\rho$. Then the influence of $\delta\rho_E(t)$ and $\delta\rho_{\text{corr}}(t)$ in $\rho(t+\tau)$, developed only during the interval $[t, t+\tau_c]$, is negligible (for details see Ref. [23]). This corresponds to Markov conditions, as the environment becomes effectively memoryless (see Sect. 2.2.2 below). It is important to notice that, although Eq. (2.28) is fulfilled, system and environment will continuously create classical and quantum correlations between them, but those correlations are not significantly affecting the evolution of the open system. A second situation, described by the so-called collisional models, consists of a system that interacts sequentially with independent parts of the environment at random times. In such case, if the environment is sufficiently big, the probability that the system of interest interacts more than one time with the same part of the environment becomes negligible, and we can assume that Eq. (2.28) is true. This kind of situation is usually engineered in cavity quantum electrodynamics, where an electromagnetic field mode is trapped in a micro-cavity and interacts with a sequence of atoms crossing the cavity and acting as its environment [23]. As important examples we mention the derivation of the one-atom maser master equation [24], and the bosonic collisional model developed in Sect. 2.3.

2.2.1 The Lindblad Form

The more general form for a Markovian master equation [see Eq. (2.27)] reads [1]

$$\frac{d\rho}{dt} = -\frac{i}{\hbar}[\hat{H}, \rho] + \sum_{k=1}^{K} \gamma_k(t) \left(\hat{L}_k \rho \hat{L}_k^\dagger - \sum_k \{\hat{L}_k^\dagger \hat{L}_k, \rho\} \right) = \mathcal{L}_t(\rho), \tag{2.29}$$

where $\hat{H} \equiv \hat{H}(t)$ is a time-dependent Hermitian (Hamiltonian-like) operator, $\hat{L}_k \equiv \hat{L}_k(t)$ are called the *Lindblad operators*, and $\gamma_k(t) \geqslant 0 \ \forall \ k, t$, are positive time-dependent rates. In the above equation, the first term is reminiscent of the Liouville–von Neumann equation (see Sect. 1.1 in Chap. 1) and describes a reversible unitary evolution in the system of interest. In contrast, the second term containing the Lindblad operators introduces an irreversible component, which is decomposed in a set of K different processes, each of them occurring with respective rate γ_k. Indeed the Lindblad master equation (2.29) provides a physical picture of the evolution consisting of a smooth dynamics punctuated by different irreversible transformations, occurring at rates γ_k. This interpretation will be developed in more detail when we introduce the *quantum trajectory* formalism in Sect. 2.5.

The problem of finding the most general form of a CPTP master equation was investigated by Gorini, Kossakowski and Sudarshan [9], and by Lindblad [8]. They first derived Eq. (2.29) for the case of time-homogeneous equations, that is, when the generator $\mathcal{L}_t$ is time-independent. In this case the CPTP maps $\mathcal{E}_t(\rho(t_0)) = \exp(\mathcal{L}(t - t_0))\rho(t_0)$ form a one-parameter semigroup, i.e. they satisfy the divisibility condition $\mathcal{E}_t \mathcal{E}_s = \mathcal{E}_{t+s}$, and $\mathrm{Tr}[\hat{\mathcal{O}} \mathcal{E}_t(\rho(t_0))]$ is a continuous function of t for any density operator $\rho(t_0)$ and Hermitian operator $\hat{\mathcal{O}}$ [11]. From now on, we will focus on this simpler case.

As in the Kraus operator-sum decomposition, the Lindblad operators $\{\hat{L}_k\}_{k=0}^{K}$ in Eq. (2.29) are not unique. They obey the same unitary freedom relation Eq. (2.15), replacing the Kraus operators $\hat{M}_k$ by $\sqrt{\gamma_k}\hat{L}_k$. Analogously, by requiring the Lindblad operators to be linearly independent, a Lindblad master equation can be derived with at most N^2 elements, N being the dimension of the open system Hilbert space. In addition, Eq. (2.29) is invariant under the transformation

$$\hat{L}_k \to \hat{L}_k + l_k, \quad \hat{H} \to \hat{H} - \frac{i}{2}\sum_{k=1}^{K} \gamma_k \left(l_k^* \hat{L}_k - l_k \hat{L}_k^\dagger \right) + r, \tag{2.30}$$

where l_k are arbitrary complex coefficients, and r is a real number. The latter property implies that it is always possible to choose traceless Lindblad operators [10].

Two further important properties of the evolution generated by a completely positive semigroup $\mathcal{E}_t = e^{\mathcal{L}(t-t_0)}$ (with time-independent generator $\mathcal{L}$) are:

- There exists always one invariant state π, such that $\mathcal{L}(\pi) = 0$. This is a consequence of the properties of CPTP maps, which, as stated before, have always at least one fixed point $\mathcal{E}_t(\pi) = \pi$. If the invariant state π is unique we call it *steady state*, and the semigroup $\mathcal{E}_t$ is strictly contractive (see Sect. 2.1.1), implying

$$\lim_{t\to\infty} \mathcal{E}_t(\rho(t_0)) = \pi. \tag{2.31}$$

- If the set of Lindblad operators $L \equiv \{\hat{L}_{k=1}^K\}$ is self-adjoint, that is, the adjoint of any operator $\hat{L}_k$ is also in the set L, and all the elements $[\hat{L}_k, \hat{A}] \neq 0$ for any arbitrary operator $\hat{A}$ except $\hat{A} = \hat{\mathbb{1}}$, hence the semigroup is relaxing [25] (see also [1]). In such case we say that the Lindblad operators $\{\hat{L}_k\}_{k=1}^K$ come in pairs, a condition which will be proven very useful in discussing the thermodynamics generated from Lindblad master equations in Part III of this thesis.

2.2.2 The Born–Markov Master Equation

In many situations of interest a Markovian master equation in Lindblad form, Eq. (2.29), can be obtained from microscopic models taking into account the global dynamics of system and environment, and then tracing over the environmental degrees of freedom. This kind of approach requires however to perform approximations in order to guarantee that the dynamics is well described by a Markovian stochastic process. Here we will sketch a general microscopic derivation of the generator of quantum dynamical semigroups for an open system continuously interacting with its surroundings in the weak coupling limit.

We start with the Hamiltonian

$$\hat{H}_{\text{tot}}(t) = \hat{H} + \hat{H}_E + \hat{H}_{\text{int}}(t), \tag{2.32}$$

where $\hat{H}$ is the Hamiltonian of the open system, $\hat{H}_E$ is the environment Hamiltonian, and $\hat{H}_{\text{int}}(t)$ represents the interaction between them. We will assume for simplicity that $\hat{H}_{\text{tot}}$ is time-independent, and that the global system is closed, following a unitary evolution given by the Liouville–von Neumann equation starting at $t = 0$. In the interaction picture with respect to $\hat{H}_0 \equiv \hat{H} + \hat{H}_E$, the global evolution reads

$$\frac{d\rho^I_{\text{tot}}(t)}{dt} = -\frac{i}{\hbar}[\hat{H}^I_{\text{int}}(t), \rho^I_{\text{tot}}(t)], \tag{2.33}$$

where $\hat{H}^I_{\text{int}}(t) = e^{\frac{i}{\hbar}\hat{H}_0 t}\hat{H}_{\text{int}}e^{-\frac{i}{\hbar}\hat{H}_0 t}$, and $\rho^I_{\text{tot}}(t) = e^{\frac{i}{\hbar}\hat{H}_0 t}\rho_{\text{tot}}(t)e^{-\frac{i}{\hbar}\hat{H}_0 t}$ are the total Hamiltonian and the global density operator at time t respectively in the interaction picture (see Sect. 1.1). For ease of notation we will neglect from now on the superscript I denoting interaction picture operators.

Equation (2.33) may be rewritten in the integral form

$$\rho_{\text{tot}}(t) = \rho_{\text{tot}}(0) - \frac{i}{\hbar}\int_{s=0}^{t} \text{ds } [\hat{H}_{\text{int}}(s), \rho_{\text{tot}}(s)]. \tag{2.34}$$

If we introduce again Eq. (2.34) into Eq. (2.33) and take the trace over the environment, we get

$$\frac{d\rho(t)}{dt} = -\frac{i}{\hbar}\text{Tr}_E\left([\hat{H}_{\text{int}}, \rho_{\text{tot}}(0)]\right) - \frac{1}{\hbar^2}\int_0^t ds\ \text{Tr}_E\left([\hat{H}_{\text{int}}(t), [\hat{H}_{\text{int}}(s), \rho_{\text{tot}}(s)]]\right),$$

where $\rho(t) = \text{Tr}_E[\rho_{\text{tot}}(t)]$ is the density operator of the open system (in interaction picture). This is still an exact equation. However, in order to proceed we need to introduce some approximations. First, we assume that system and environment are initially uncorrelated, that is $\rho_{\text{tot}}(0) \approx \rho(0) \otimes \rho_E$, where $\rho(0)$ is the initial density operator of the open system, and ρ_E the environment density operator. In addition, we assume that the interaction Hamiltonian (in interaction picture) verifies $\text{Tr}_E\left([\hat{H}_{\text{int}}, \rho_{\text{tot}}(0)]\right) = 0$. This second condition is not restrictive as one can always redefine $\hat{H}_0$ such that it is verified, by including an extra Hamiltonian term only acting on the system Hilbert space [11]. This implies that the first term in the above equation can be neglected.

Furthermore, an important approximation should be taken at this point, called the *Born* approximation. It assumes that the open system only affects very weakly the state of the reservoir during the evolution, so that we can replace

$$\rho_{\text{tot}}(s) \approx \rho(s) \otimes \rho_E, \tag{2.35}$$

inside the integral term in the above equation. It is important to notice that this approximation does not imply that we neglect the correlations built up between system and environment, but only that they do not affect appreciably the reduced system dynamics. Furthermore we will perform a first Markov-like approximation by replacing $\rho(s)$ by $\rho(t)$, which requires the integrand to be only non-zero in a small region around $s \sim t$. Implementing the three above approximations we obtain an integro-differential equation which is local in time

$$\frac{d\rho(t)}{dt} = -\frac{1}{\hbar^2}\int_0^t ds\ \text{Tr}_E\left([\hat{H}_{\text{int}}(t), [\hat{H}_{\text{int}}(s), \rho(t) \otimes \rho_E]]\right), \tag{2.36}$$

called the *Redfield equation*, which is still not strictly Markovian [10, 11]. We need to perform a further Markov approximation by substituting s by $t-s$ inside the integrand of Eq. (2.36), and letting the upper limit of the integral go to infinity [10]. Doing this, we obtain a Markovian master equation with time-independent coefficients [10, 11]

$$\frac{d\rho(t)}{dt} = -\frac{1}{\hbar^2}\int_0^\infty ds\, \mathrm{Tr}_E\left([\hat{H}_{\mathrm{int}}(t), [\hat{H}_{\mathrm{int}}(t-s), \rho(t)\otimes\rho_E]]\right), \tag{2.37}$$

which gives us the evolution of the open system density operator ρ with a limited resolution on a coarse-grained time axis. The various approximations leading to Eq. (2.37) are usually termed the *Born–Markov approximation*. They can be physically justified in the case of a large environment with a continuous energy spectrum over a wide range $\hbar\Delta\omega$ as we explained at the beginning of the section. The crucial point is the separation of the time scales, $\tau_\rho \gg \tau_c$, where τ_ρ is the characteristic time scale of the open system dynamics in the interaction picture, and τ_c the characteristic decay time for the environment correlation functions [10], which also characterizes the generation of correlations between system and environment [23]. However, it is important to stress that Eq. (2.37) is not necessarily the generator of a dynamical semigroup and therefore is not guaranteed that it can be written in Lindblad form (2.29) [7].

In order to ensure that Eq. (2.37) describes a CPTP dynamics we usually need to perform a final *secular approximation*, consisting of a kind of rotating wave approximation (RWA) in which one averages over rapidly oscillating terms. At this point we assume without loss of generality the interaction Hamiltonian (in Schrödinger picture) to be of the form

$$\hat{H}_{\mathrm{int}} = \hbar\sum_{i=1}^{I}\hat{A}_i\otimes\hat{E}_i, \tag{2.38}$$

where $\hat{A}_i$ and $\hat{E}_i$ are $\forall i = 1, 2, \ldots, I$ Hermitian operators acting on the system and environment degrees of freedom respectively. We then proceed by decomposing $\hat{H}_{\mathrm{int}}$ into eigenoperators of the system Hamiltonian, whose spectral decomposition reads $\hat{H} = \sum_l \epsilon_l \hat{\Pi}_{\epsilon_l}$. We define

$$\hat{A}_i(\omega) \equiv \sum_{l,l'}\delta(\epsilon_{l'} - \epsilon_l - \hbar\omega)\,\hat{\Pi}_{\epsilon_l}\hat{A}_i\hat{\Pi}_{\epsilon_{l'}}\,, \tag{2.39}$$

where $\delta(\epsilon' - \epsilon - \hbar\omega)$ is the Dirac delta function selecting all possible transitions in the system spectrum with a fixed energy difference $\hbar\omega$. Those operators fulfill

$$[\hat{H}, \hat{A}_i(\omega)] = -\hbar\omega\hat{A}_i(\omega), \qquad [\hat{H}, \hat{A}_i^\dagger(\omega)] = \hbar\omega\hat{A}_i^\dagger(\omega),$$

with $\hat{A}_i^\dagger(\omega) = \hat{A}_i(-\omega)$, and $[\hat{H}, \hat{A}_i^\dagger(\omega)\hat{A}_i(\omega)] = 0$. Furthermore they obey the completeness relation $\sum_\omega \hat{A}_i(\omega) = \hat{A}_i$. In terms of those operators the interaction Hamiltonian (2.38) in interaction picture reads

$$\hat{H}_{\rm int}(t) = \hbar \sum_{i,\omega} e^{-i\omega t}\, \hat{A}_i(\omega) \otimes \hat{E}_i(t), \tag{2.40}$$

being $\hat{E}_i(t) = e^{\frac{i}{\hbar}\hat{H}_0 t} \hat{E}_i e^{-\frac{i}{\hbar}\hat{H}_0 t}$ the environment operators in interaction picture. In many cases of interest, by making explicit also the eigenoperator decomposition of the reservoir observables $\hat{E}_i$ (with respect to $\hat{H}_E$), one can directly perform a RWA in the Hamiltonian (2.40) by neglecting rapidly oscillating terms, leading to a Markovian master equation (2.37) which can be directly written in Lindblad form. This is the case when considering e.g. radiation-matter interaction throughout the Jaynes–Cummings–Paul model [26, 27].

Otherwise the RWA can be performed inside the integrand of Eq. (2.37) as follows. Introducing Eq. (2.40) into the master equation (2.37) we arrive at

$$\frac{d\rho}{dt} = \sum_{\omega,\omega'} \sum_{i,j} e^{i(\omega'-\omega)t} \Gamma_{ij}(\omega) \left(\hat{A}_j(\omega)\rho \hat{A}_i^\dagger(\omega') - \hat{A}_i^\dagger(\omega')\hat{A}_j(\omega)\rho \right) + \text{ h.c.}, \tag{2.41}$$

where h.c. denotes the hermitian conjugate of the first term, and we have defined

$$\Gamma_{ij}(\omega) \equiv \int_0^\infty ds e^{i\omega s} \mathrm{Tr}_E[\hat{E}_i(t)\hat{E}_j(t-s)\rho_E], \tag{2.42}$$

the one-sided Fourier transform of the *environment correlation functions*, $\mathrm{Tr}_E[\hat{E}_i(t)\hat{E}_j(t-s)\rho_E]$ with an associated characteristic decay time $\tau_c \sim \hbar/\Delta\omega$. In some cases the correlation functions are homogeneous in time, meaning that $\mathrm{Tr}_E[\hat{E}_i(t)\hat{E}_j(t-s)\rho_E] = \mathrm{Tr}_E[\hat{E}_i(s)\hat{E}_j(0)\rho_E]$ and ensuring that the functions $\Gamma_{ij}(\omega)$ are time independent. This is the case of environments such that $[\hat{H}_E\ , \rho_E] = 0$, the thermal equilibrium reservoir being the most important and paradigmatic case. However, more exotic states of the environment may present time-varying coefficients, as in the case of a squeezed thermal reservoir. For those cases one may split Eq. (2.42) into time-dependent and time-independent parts [10]. Assuming homogeneous correlation functions, we can now easily perform the *secular approximation* by neglecting in Eq. (2.41) the terms with $\omega' \neq \omega$. This approximation is justified when the time scale of the intrinsic (isolated) system is small compared to the time scale associated to the interaction with the reservoir, $\tau_s \sim 1/|\omega' - \omega| \ll \tau_\rho$. In this case the non-secular terms ($\omega' \neq \omega$) oscillate very rapidly during the time over which $\rho(t)$ varies (in interaction picture), and can be neglected [10]. This implies that Eq. (2.41) transforms into

$$\frac{d\rho}{dt} = \sum_{\omega} \sum_{i,j} \Gamma_{ij}(\omega) \left(\hat{A}_j(\omega)\rho \hat{A}_i^\dagger(\omega) - \hat{A}_i^\dagger(\omega)\hat{A}_j(\omega)\rho \right) + \text{h.c.}$$

Finally, it is convenient to decompose the Fourier transforms of the reservoir correlation functions into two terms $\Gamma_{ij}(\omega) = \frac{1}{2}\gamma_{ij}(\omega) + i\chi_{ij}(\omega)$, where the first one is always positive, and the second one are the entries of an Hermitian matrix

$$\gamma_{ij}(\omega) \equiv \Gamma_{ij}(\omega) + \Gamma^*_{ji}(\omega) = \int_{-\infty}^{\infty} ds e^{i\omega s}\langle \hat{E}_i(s)\hat{E}_j(0)\rangle \tag{2.43}$$
$$\chi_{ij}(\omega) \equiv \frac{1}{2i}\left(\Gamma_{ij}(\omega) - \Gamma^*_{ij}(\omega)\right).$$

Rearranging terms we obtain the master equation in interaction picture

$$\frac{d\rho(t)}{dt} = -\frac{i}{\hbar}[\hat{H}_{LS}, \rho(t)] + \mathcal{D}(\rho(t)), \tag{2.44}$$

where we defined the Hermitian operator

$$\hat{H}_{LS} \equiv \sum_{\omega}\sum_{i,j} \hbar\, \chi_{ij}(\omega)\hat{A}_i^\dagger(\omega)\hat{A}_j(\omega), \tag{2.45}$$

often called the *Lamb shift Hamiltonian*, which introduces a (small) renormalization on the unperturbed energy levels of the open system induced by the coupling to the environment; $[\hat{H}, \hat{H}_{LS}] = 0$. The second term in Eq. (2.44) is a super-operator usually called the *dissipator*, which takes the form

$$\mathcal{D}(\rho) \equiv \sum_{\omega}\sum_{i,j} \gamma_{ij}(\omega)\left(\hat{A}_j(\omega)\rho\hat{A}_i^\dagger(\omega) - \frac{1}{2}\{\hat{A}_i^\dagger(\omega)\hat{A}_j(\omega)\, ,\, \rho\}\right). \tag{2.46}$$

The Markovian master equation in Eq. (2.44) describes a CPTP dynamics, as can be directly seen by rewriting it in Lindblad form. This can be done by simply diagonalizing the matrices $\gamma_{ij}(\omega)$, which are positive by virtue of Bochner's theorem [10]. Turning back to the Schrödinger picture Eq. (2.44) reads

$$\frac{d\rho(t)}{dt} = -\frac{i}{\hbar}[\hat{H} + \hat{H}_{LS}, \rho(t)] + \mathcal{D}(\rho(t)), \tag{2.47}$$

which is the final form for the CPTP Markovian master equation. In practical applications the Lamb shift Hamiltonian $\hat{H}_{LS}$ is usually neglected, as the weak coupling limit ensures that the energy shift is small compared to the eigenvalues $\{\epsilon_l\}$ of $\hat{H}$. One can also include a *counter-term* in the original system Hamiltonian in order to cancel the Lamb shift contribution. However a proper calculation of the Lamb shift Hamiltonian requires the use of renormalization theory and relativistic quantum mechanics [11].

2.3 Dissipative Qubits and Harmonic Oscillators

Here we present some relevant examples of Markovian master equations with a wide range of applicability in the study of open quantum systems. They are based on the prototypical systems introduced in Sect. 1.2, namely, the qubit system and the quantum harmonic oscillator. In the first example we discuss the relaxation of a qubit system in the presence of a bosonic reservoir (harmonic oscillators) in the optical regime, where a CPTP Markovian master equation is obtained even in the case of a squeezed reservoir. We start with an interaction Hamiltonian in the RWA, and show that the result is the same when the whole interaction term is considered after the *secular approximation* (see the above Sect. 2.2.2). In the second example we consider an harmonic oscillator interacting with the bosonic reservoir now from a collisional approach, which is well suited for quantum information and quantum thermodynamic studies (see e.g. Refs. [28, 29]). When the RWA or the secular approximation is carried out, we obtain a very similar master equation than for the qubit case. However, in some important regimes this approximation fails leading to a non-CPTP Markovian master equation, as in the case of the so-called *quantum Brownian motion*, which we discuss as a third example. Other relevant (simple) examples of Markovian master equations can be found in textbooks covering open quantum systems, such as the one-atom maser [24], the spin-boson model [30], quantum dots interacting with fermionic reservoirs [11], or dynamical models of quantum measurements [10, 11].

2.3.1 *Qubit Relaxation in a Bosonic Environment*

Let us start by considering a qubit system (two-level atom) with Hamilton operator $\hat{H} = \hbar\omega\hat{\sigma}^\dagger\hat{\sigma}$, $\hat{\sigma} \equiv |0\rangle\langle 1|$ being its lowering operator and satisfying $\{\hat{\sigma}, \hat{\sigma}^\dagger\} = \mathbb{1}$. The system interacts with a bosonic reservoir (e. g. electromagnetic radiation). The reservoir is described as an infinite collection of uncorrelated bosonic modes (harmonic oscillators) spanning a continuous frequency spectrum with Hamiltonian $\hat{H}_E = \sum_k \hbar\Omega_k \hat{b}_k^\dagger \hat{b}_k$, where $[\hat{b}_k, \hat{b}_{k'}^\dagger] = \hat{\mathbb{1}}_E \delta_{k,k'}$, and $\hat{b}_k$ and $\hat{b}_k^\dagger$ being the annihilation and creation operators for mode k. The qubit and the reservoir modes interact via the Jaynes–Cummings Hamiltonian, which in interaction picture reads

$$\hat{H}_{\text{int}} = \sum_k \hbar g_k (\hat{\sigma}\, \hat{b}_k^\dagger e^{-i(\omega-\Omega_k)t} + \hat{\sigma}^\dagger\, \hat{b}_k e^{i(\omega-\Omega_k)t}) \tag{2.48}$$

where g_k is the coupling strength between the qubit and the mode k in the reservoir. The interaction (2.48) results from performing the RWA on the Hamiltonian describing the interaction between the two-level atom and the radiation field in the dipole approximation (see e.g. [2, 24]). Introducing Eq. (2.48) into the Redfield equation (2.36)

$$\frac{d\rho(t)}{dt} = \Gamma_1(\hat{\sigma}\rho(t)\hat{\sigma}^\dagger - \hat{\sigma}^\dagger\hat{\sigma}\rho(t)) + \Gamma_2(\hat{\sigma}^\dagger\rho(t)\hat{\sigma} - \hat{\sigma}\hat{\sigma}^\dagger\rho(t))$$
$$+ \Gamma_3\, 2\hat{\sigma}\rho(t)\hat{\sigma} + \text{h.c.}, \tag{2.49}$$

where we have used $\hat{\sigma}^2 = 0$, and defined the following four coefficients, similar to the ones in Eq. (2.42):

$$\Gamma_1 \equiv \int_0^t ds \sum_k g_k^2 e^{i(\omega-\Omega_k)(t-s)}(N_{\Omega_k}+1),$$
$$\Gamma_2 \equiv \int_0^t ds \sum_k g_k^2 e^{-i(\omega-\Omega_k)(t-s)} N_{\Omega_k},$$
$$\Gamma_3 \equiv \int_0^t ds \sum_k g_k^2 e^{-i(\omega-\Omega_k)(t+s)} M^*_{\Omega_k}, \tag{2.50}$$

with $N_{\Omega_k} \equiv \text{Tr}[\hat{b}_k^\dagger \hat{b}_k \rho_E]$ and $M_{\Omega_k} \equiv \text{Tr}[\hat{b}_k^2 \rho_E]$. Next, we take the continuous limit by defining the *spectral density* of the reservoir

$$J(\Omega) \equiv \sum_k g_k^2 \delta(\Omega - \Omega_k), \tag{2.51}$$

that characterize the number of modes in the reservoir interacting with the system with a given strength. The spectral density allows us to replace the sum in Eq. (2.50) by an integral over the reservoir frequencies:

$$\Gamma_1 = \int_0^t ds \int_0^\infty d\Omega\, J(\Omega) e^{i(\omega-\Omega)(t-s)}(N_{\Omega_k}+1),$$

and analogously for coefficients Γ_2 and Γ_3. In the above expression $\hat{N}_\Omega$ is the number operator of the reservoir mode with frequency Ω. Taking the upper limit $t \to \infty$ in the above integrals (Markov approximation) and using $\int_0^\infty dse^{-ixs} = \pi\delta(x) - iP(1/x)$, P being the Cauchy principal value, we can split the coefficients into two terms $\Gamma_i \equiv \frac{1}{2}\gamma_i + i\chi_i$, where

$$\gamma_1 = \int_0^\infty d\Omega J(\Omega)\, 2\pi\delta(\omega - \Omega)(N_\Omega + 1) = 2\pi\, J(\omega)(N_\omega + 1),$$
$$\gamma_2 = \int_0^\infty d\Omega J(\Omega)\, 2\pi\delta(\omega - \Omega) N_\Omega\rangle_{\rho_E} = 2\pi\, J(\omega) N_\omega,$$
$$\gamma_3 = \int_0^\infty d\Omega J(\Omega)\, 2\pi\delta(\omega - \Omega) M^*_\Omega = 2\pi\, J(\omega) M^*_\omega, \tag{2.52}$$

and the coefficients χ_i only enter in the Lamb-shift Hamiltonian (see below). In the above equations we see how the Markov approximation implies the selection of the resonant frequency in the reservoir as the dominant contribution for the damping coefficients γ_i.

Introducing all the coefficients back in Eq. (2.49) and rearranging terms we finally obtain

$$\begin{aligned}\dot{\rho} =& -\frac{i}{\hbar}[\hat{H}_{LS}, \rho] + \gamma_0 M_\omega^* \hat{\sigma}\rho\hat{\sigma} + \gamma_0 M_\omega \hat{\sigma}^\dagger \rho \hat{\sigma}^\dagger \\ &+ \gamma_0 (N_\omega + 1)\left(\hat{\sigma}\rho\hat{\sigma}^\dagger - \frac{1}{2}\{\hat{\sigma}^\dagger\hat{\sigma}, \rho\}\right) \\ &+ \gamma_0 N_\omega \left(\hat{\sigma}^\dagger\rho\hat{\sigma} - \frac{1}{2}\{\hat{\sigma}\hat{\sigma}^\dagger, \rho\}\right). \end{aligned} \tag{2.53}$$

The above Markovian master equation equation describes the relaxation of the qubit system in contact with a (generalized) bosonic reservoir characterized by the damping coefficient $\gamma_0 \equiv 2\pi J(\omega)$, which depends on the density of states at the qubit frequency ω, the mean number of quanta in the reservoir's resonant mode, $\langle \hat{N}_\omega \rangle_{\rho_E}$, and its second-order coherences $\langle \hat{M}_\omega \rangle_{\rho_E}$. In the derivation we implicitly assumed $\langle \hat{b}_k \rangle_{\rho_E} = 0$ for all modes k in the reservoir. The Lamb-shift Hamiltonian introduces a renormalization of the qubit frequency ω given by

$$\hat{H}_{LS} = \hbar\Omega_{LS}\hat{\sigma}^\dagger\hat{\sigma}, \quad \text{with} \quad \Omega_{LS} \equiv P\int_0^\infty d\Omega\, J(\Omega)\frac{2\langle \hat{N}_\Omega \rangle_{\rho_E} + 1}{\Omega - \omega}, \tag{2.54}$$

which depends on the mean number of quanta in the reservoir [31].

It is important to notice that we could also derived the master equation (2.53) by considering the complete dipole approximation without the RWA in the interaction between the qubit and the environment, $\hat{H}_{\text{int}} = \sum_k \hbar g_k (\hat{\sigma} e^{-i\omega t} + \hat{\sigma}^\dagger e^{i\omega t})(\hat{b}_k e^{-i\Omega_k t} + \hat{b}_k^\dagger e^{i\Omega_k t})$. This corresponds to a Hamiltonian of the form (2.38) with a single term, where it is easy to identify two system eigenoperators $\{\hat{A}(\omega)\}$ corresponding to frequencies $\pm\omega$ (see Sect. 2.2.2)

$$\hat{A}(\omega) = \hat{\sigma}, \quad \hat{A}(-\omega) = \hat{\sigma}^\dagger, \tag{2.55}$$

and the reservoir operator $\hat{B} = \sum_k g_k(\hat{b}_k e^{-i\Omega_k t} + \hat{b}_k^\dagger e^{i\Omega_k t})$. Calculating the reservoir correlation functions $\Gamma_{ij}(\omega)$ (in this case $i = j = 1$ as the interaction Hamiltonian only contains a single term), and splitting it in homogeneous and non-homogeneous in time parts [10], the above Markovian master equation (2.53) is recovered.

Let us now consider some particular cases of Eq. (2.53). Probably the most natural case to begin with is assuming that the reservoir is in thermal equilibrium, $\rho_E = e^{-\beta \hat{H}_E}/Z_E$, at inverse temperature $\beta = 1/k_B T$. The mean number of quanta in the thermal reservoir with frequency ω reads $N_\omega = (e^{\beta\hbar\omega - 1})^{-1} \equiv n_{\text{th}}$, and $M_\omega = 0$ (see Sect. 1.2.3). Turning back to the Schrödinger picture, the master equation (2.53) then reads

$$\dot{\rho} = -\frac{i}{\hbar}[\hat{H}, \rho] + \gamma_0(n_{\rm th}+1)\left(\hat{\sigma}\rho\hat{\sigma}^\dagger - \frac{1}{2}\{\hat{\sigma}^\dagger\hat{\sigma}, \rho\}\right)$$
$$+ \gamma_0 n_{\rm th}\left(\hat{\sigma}^\dagger\rho\hat{\sigma} - \frac{1}{2}\{\hat{\sigma}\hat{\sigma}^\dagger, \rho\}\right) \equiv -\frac{i}{\hbar}[\hat{H}, \rho] + \mathcal{L}_{\rm th}(\rho), \tag{2.56}$$

where we have neglected the frequency shift introduced by the Lamb-shift Hamiltonian. This equation is written in Lindblad form by identifying the two Lindblad operators: $\hat{L}_\downarrow = \sqrt{\gamma_0(n_{\rm th}+1)}\,\hat{\sigma}$ and $\hat{L}_\uparrow = \sqrt{\gamma_0 n_{\rm th}}\,\hat{\sigma}^\dagger$. Here the Hamiltonian term describes the phase evolution of the qubit system while the superoperator $\mathcal{L}_{\rm th}$ accounts for the dissipative effects induced by the thermal reservoir, consisting of two processes. The first one in the upper line of Eq. (2.56) associated to $\hat{L}_\downarrow$, is the spontaneous and stimulated emission of quanta $\hbar\omega$ from the qubit to the reservoir by performing a jump from its excited state $|1\rangle$ to the ground state $|0\rangle \propto \hat{L}_\downarrow |1\rangle$, and occurring at a rate $\gamma_\downarrow = \gamma_0(n_{\rm th}+1)$. In analogy, the second line in (2.56) associated to $\hat{L}_\uparrow$ corresponds to the stimulated absorption of quanta $\hbar\omega$ from the reservoir while producing a jump in the qubit from the ground $|0\rangle$ to the excited level $|1\rangle \propto \hat{L}_\uparrow |0\rangle$, a process which occurs at rate $\gamma_\uparrow = \gamma_0 n_{\rm th}$. The interplay of these two processes, whose rates are related by a detailed balance relation $\gamma_\downarrow = e^{-\beta\hbar\omega}\gamma_\uparrow$, implies the thermalization of the qubit in the long time run, $\pi = e^{-\beta\hat{H}}/Z$, with $[\hat{H}, \pi] = \mathcal{L}_{\rm th}(\pi) = 0$. It is easy to see that, in the thermal equilibrium state, the populations of the two levels are such that the two processes become equally probable. In the limiting case in which the temperature of the reservoir vanishes $T \to 0$, we have $n_{\rm th} \to 0$, and spontaneous emission becomes dominant. As a consequence, the steady state of the qubit tends to the ground state $\pi \to |0\rangle\langle 0|$.

An important generalization of the above situation results from including the coherent driving of the qubit by a nearly resonant classical field, a model which is known as *resonance fluorescence* (the case of an isolated qubit system driven by a classical field has been considered in Sect. 1.2.2). If the intensity of the classical field is weak compared to the qubit frequency ω, it can be neglected in the derivation of the interaction picture master equation, and simply added at the end, when turning back to the Schödinger picture [11]. In such case we obtain the following master equation in Schrödinger picture

$$\dot{\rho} = -\frac{i}{\hbar}[\hat{H} + \hat{H}_{\rm f}(t), \rho] + \gamma_0(n_{\rm th}+1)\left(\hat{\sigma}\rho\hat{\sigma}^\dagger - \frac{1}{2}\{\hat{\sigma}^\dagger\hat{\sigma}, \rho\}\right)$$
$$+ \gamma_0 n_{\rm th}\left(\hat{\sigma}^\dagger\rho\hat{\sigma} - \frac{1}{2}\{\hat{\sigma}\hat{\sigma}^\dagger, \rho\}\right), \tag{2.57}$$

where $\hat{H}_{\rm f}(t) = \hbar\Omega_R\left(\hat{\sigma}^\dagger e^{-i\omega_{\rm f}t} + \hat{\sigma}\, e^{i\omega_{\rm f}t}\right)/2$, being $\omega_{\rm f}$ the frequency of the field and Ω_R the *Rabi frequency*. The master equation Eq. (2.57) is known as the resonance fluorescence master equation, which describes the damping of the Rabi oscillations due to the reservoir. The steady state solution is however no longer diagonal in the $\hat{H}$ eigenbasis. For the strictly resonant case, $\omega_{\rm f} = \omega$, one can obtain the following steady state values for the excited level population and coherence [10]:

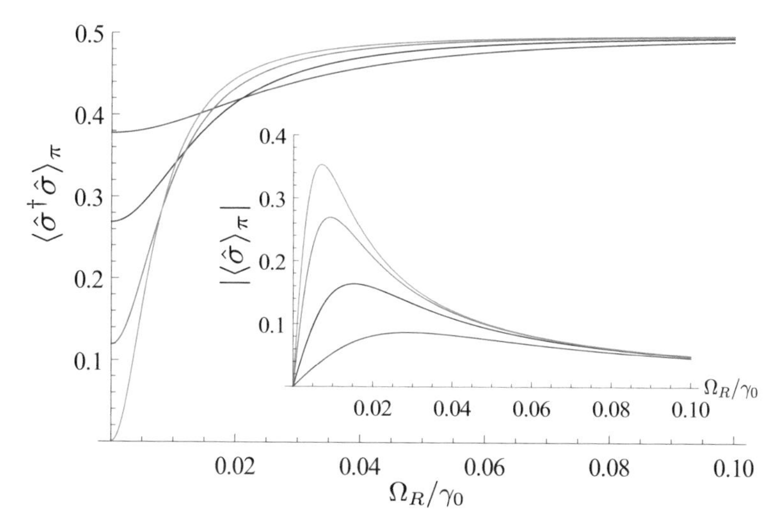

Fig. 2.3 Population of the excited level and coherence (inset plot) in the steady state of the resonance fluorescence master equation, Eq. (2.58), as a function of the ratio between the Rabi frequency Ω_R and the damping rate γ_0 for different values of the (scaled) reservoir's inverse temperature. The different lines correspond to $\beta = \{0.5, 1.0, 2.0, 7.0\}/\hbar\omega$ from bottom to top in the right side of the plot. On the other hand, maximum coherence is attained when $\Omega_R = \gamma_0(2n_{\text{th}} + 1)/\sqrt{2}$

$$\begin{aligned}\langle\hat{\sigma}^\dagger\hat{\sigma}\rangle_\pi &= \frac{\gamma_0^2 n_{\text{th}}(2n_{\text{th}}+1) + \Omega_R^2}{\gamma_0^2(2n_{\text{th}}+1)^2 + 2\Omega_R^2}, \\ \langle\hat{\sigma}\rangle_\pi &= \frac{i\gamma_0\Omega_R e^{-i\omega t}}{\gamma_0^2(2n_{\text{th}}+1)^2 + 2\Omega_R^2},\end{aligned} \tag{2.58}$$

which are plotted in Fig. 2.3. Pumping from the coherent classical field increases the excited level population $\langle\hat{\sigma}^\dagger\hat{\sigma}\rangle_\pi \geqslant \langle\hat{\sigma}^\dagger\hat{\sigma}\rangle_{\text{th}}$ and the presence of coherence in the steady state produces the rotation of the qubit state in the XY-plane of the Bloch sphere at constant frequency ω. Notice that when $\Omega_R \to 0$ we recover the thermal equilibrium state.

The thermal reservoir case can be further generalized by considering a squeezed thermal reservoir, for which

$$\rho_E = \prod_k \hat{\mathbb{S}}_k \left(e^{-\beta\hat{H}_E}/Z\right)\hat{\mathbb{S}}_k^\dagger, \tag{2.59}$$

$\hat{\mathbb{S}}_k(r_k, \theta_k)$ being the squeezing operator for mode k with e parameters r_k and θ_k (see Sect. 1.2.5). In this situation, the expectation values change to $N_\omega = \cosh(2r)n_{\text{th}} + \sinh^2(r)$, and $M_\omega = -(2n_{\text{th}}+1)\sinh(r)\cosh(r)e^{i\theta}$, r and θ being the squeezing parameters of the resonant mode [10]. In this case, the extra terms in Eq. (2.53) come into play, generating a transient enhancement of the qubit coherence [terms $\langle\hat{\sigma}\rangle_\rho(t)$ and $\langle\hat{\sigma}^\dagger\rangle_\rho(t)$] before its final suppression, and modifying the populations of the steady state as if it were at a higher temperature [19, 32]. The master equation Eq. (2.53) can be rewritten as

$$\dot{\rho} = -\frac{i}{\hbar}[\hat{H}, \rho] + \gamma_0(n_{\text{th}} + 1)\left(\hat{R}\rho\hat{R}^\dagger - \frac{1}{2}\{\hat{R}^\dagger\hat{R}, \rho\}\right) + \gamma_0 n_{\text{th}}\left(\hat{R}^\dagger\rho\hat{R} - \frac{1}{2}\{\hat{R}\hat{R}^\dagger, \rho\}\right), \tag{2.60}$$

with $\hat{R} = \cosh(r)\hat{\sigma} - \sinh(r)e^{i\theta}\hat{\sigma}^\dagger$. We can easily identify here the Lindblad operators $\hat{L}_- \equiv \sqrt{\gamma_0(n_{\text{th}} + 1)}\hat{R}$, and $\hat{L}_+ \equiv \sqrt{\gamma_0 n_{\text{th}}}\hat{R}^\dagger$, associated to correlated jumps between the ground and the excited states of the qubit.

2.3.2 *Bosonic Collisional Model*

Now we move to the case in which our system of interest is an harmonic oscillator, represented for instance by an electromagnetic field mode in a cavity or more generally by a bosonic mode. The Hamilton operator of the system reads $\hat{H} = \hbar\omega\hat{a}^\dagger\hat{a}$, where $[\hat{a}, \hat{a}^\dagger] = \mathbb{1}$ are respectively the annihilation and creation operators of the mode. The environment is considered again to be composed by a bosonic reservoir with Hamiltonian $\hat{H}_E = \sum_k \hbar\Omega_k\hat{b}_k^\dagger\hat{b}_k$, with $[\hat{b}_k, \hat{b}_{k'}^\dagger] = \mathbb{1}_E\delta_{k,k'}$ (see the previous example). Let us assume the interaction Hamiltonian between the system and the reservoir (in interaction picture) to be in the RWA

$$\hat{H}_{\text{int}} = \sum_k \hbar g_k i\left(\hat{a}\,\hat{b}_k^\dagger e^{-i(\omega-\Omega_k)t} - \hat{a}^\dagger\,\hat{b}_k e^{i(\omega-\Omega_k)t}\right), \tag{2.61}$$

to be compared with Eq. (2.48). As in the previous example, this interaction is a good approximation in the weak coupling regime for optical frequencies, which implies that the characteristic frequency of the system is much larger than the decay rate [11]. Following the same steps as in the qubit system example, we obtain a master equation similar to Eq. (2.53) for the relaxation of the harmonic oscillator in a general environment. However, in this case we develop a collisional model which will provide a more intuitive picture of the dynamical evolution. This represents a generalization of the model we recently reported in Ref. [33] for the study of the thermodynamical features of the squeezed thermal reservoir (see Chap. 10).

In the collisional model, the system bosonic mode interacts at random times, given by some rate $\mathcal{R}$, with a generic mode k of the bosonic environment once at a time during some small interval τ. It is convenient to introduce the Hamiltonian of a single reservoir's mode k, $\hat{H}_E(\Omega_k) = \hbar\Omega_k\hat{b}_k^\dagger\hat{b}_k$. In each collision we assume that the bosonic mode interacts with a different reservoir mode, which may have a different frequency, depending on the reservoir *density of states*, $\vartheta(\Omega_k)$, which characterizes the number of modes with a given frequency Ω_k. Let us specify the interaction Hamiltonian (2.61) to account for the interaction with a single mode in the reservoir, $\hat{H}_{\text{int}} = i\hbar g_k(\hat{a}\hat{b}_k^\dagger e^{-i\Delta_k t} - \hat{a}^\dagger \hat{b}_k e^{i\Delta_k t})$, with $\Delta_k = \omega - \Omega_k$. Assuming again weak coupling, $g_k \tau \ll 1 \; \forall k$, the unitary evolution governing a single collision occurring at time t, reads, in the interaction picture:

$$\hat{U}_I(t+\tau, t) = \hat{T}_+ \exp\left(-\frac{i}{\hbar}\int_t^{t+\tau} dt_1 \hat{H}_{\text{int}}(t_1)\right), \tag{2.62}$$

where $\hat{T}_+$ is the time-ordering operator. The evolution of the two-mode (total) density matrix can be expanded up to second order in the coupling using the Dyson series (see Sect. 1.1):

$$\rho_{\text{tot}}(t+\tau, t) \simeq \rho_{\text{tot}}(t) - \frac{i}{\hbar}\int_t^{t+\tau} dt_1 [\hat{H}_{\text{int}}(t_1), \rho_{\text{tot}}(t)] \\ -\frac{1}{\hbar^2}\int_t^{t+\tau} dt_2 \int_t^{t_2} dt_1 [\hat{H}_{\text{int}}(t_2), [\hat{H}_{\text{int}}(t_1), \rho_{\text{tot}}(t)]]. \tag{2.63}$$

The first order commutator reads

$$[\hat{H}_{\text{int}}(t_1), \rho_{\text{tot}}(t)] = i\hbar g_k \left([\hat{a}\hat{b}_k^\dagger, \rho_{\text{tot}}(t)]e^{-i\Delta_k t_1} - \text{h.c.}\right), \tag{2.64}$$

and the second-order one

$$[\hat{H}_{\text{int}}(t_2), [\hat{H}_{\text{int}}(t_1), \rho_{\text{tot}}(t)]] = -\hbar^2 g_k^2 ([\hat{a}^\dagger \hat{b}_k, [\hat{a}^\dagger \hat{b}_k, \rho_{\text{tot}}(t)]] \\ e^{i\Delta_k(t_1+t_2)} - [\hat{a}\,\hat{b}_k^\dagger, [\hat{a}^\dagger \hat{b}_k, \rho_{\text{tot}}(t)]]e^{i\Delta_k(t_1-t_2)} + \text{h.c.}). \tag{2.65}$$

The reduced evolution in the system and in the reservoir mode, can be obtained by partial tracing Eq. (2.63) over the corresponding degrees of freedom. We also assume $\rho_{\text{tot}}(t) = \rho(t) \otimes \rho_E^{(k)}$, i.e. the system mode always interacts with a 'fresh' reservoir mode k in the same state. The master equation can be constructed from the following coarse-grained derivative for the system mode. During some small interval of time $\delta t \ll \mathcal{R}^{-1}$ (but $\delta t \gg \tau$), for which at most one interaction occurs:

$$\rho(t+\delta t) = \mathcal{R}\delta t\; \rho(t+\tau) \; + \; (1 - \mathcal{R}\delta t)\rho(t), \tag{2.66}$$

where $\rho(t) = \text{Tr}_E[\rho_{\text{tot}}(t)]$. That is, with probability $\mathcal{R}\delta t$ the system interacts with a bosonic mode in the reservoir and with the complementary probability, $1 - \mathcal{R}\delta t$, it is unaltered (we stress that we are working in the interaction picture). Therefore we have

$$\dot{\rho}(t) \simeq \frac{1}{\delta t}[\rho(t+\delta t) - \rho(t)] \;=\; \mathcal{R}[\rho(t+\tau) - \rho(t)].$$

This is valid when the reservoir modes always have the same frequency Ω_k, but if we want to take into account that the reservoir contains many frequencies, the above equation should be averaged over the reservoir density of states:

$$\dot{\rho}(t) \simeq \mathcal{R} \sum_k \vartheta(\Omega_k)[\rho(t+\tau) - \rho(t)]. \tag{2.67}$$

Performing the time integrals, the partial trace, and defining the following reservoir expectation values:

$$\begin{aligned} \langle \hat{b}_k \rangle_{\rho_E^{(k)}} &= D_{\Omega_k} & \langle \hat{b}_k^\dagger \rangle_{\rho_E^{(k)}} &= D_{\Omega_k}^* \\ \langle \hat{b}_k^2 \rangle_{\rho_E^{(k)}} &= M_{\Omega_k} & \langle \hat{b}_k^{\dagger 2} \rangle_{\rho_E^{(k)}} &= M_{\Omega_k}^* \\ \langle \hat{b}_k^\dagger \hat{b}_k \rangle_{\rho_E^{(k)}} &= N_{\Omega_k} & \langle \hat{b}_k \hat{b}_k^\dagger \rangle_{\rho_E^{(k)}} &= N_{\Omega_k} + 1 \end{aligned} \tag{2.68}$$

we obtain the following preliminary form for the master equation

$$\begin{aligned} \dot{\rho} = & -\frac{i}{\hbar}[\Delta\hat{H}, \rho] + [\epsilon^* \hat{a} - \epsilon \hat{a}^\dagger, \rho] \\ & + \Gamma_e \left(\hat{a}\rho\hat{a}^\dagger - \frac{1}{2}\{\hat{a}^\dagger\hat{a}, \rho\} \right) + \Gamma_a \left(\hat{a}^\dagger\rho\hat{a} - \frac{1}{2}\{\hat{a}\hat{a}^\dagger, \rho\} \right) \\ & - \Gamma_s \left(\hat{a}^\dagger\rho\hat{a}^\dagger - \frac{1}{2}\{\hat{a}^{\dagger 2}, \rho\} \right) - \Gamma_s^* \left(\hat{a}\rho\hat{a} - \frac{1}{2}\{\hat{a}^2, \rho\} \right), \end{aligned} \tag{2.69}$$

where we identified the decay factors charactering the time scales of emission/ absorption processes and squeezing:

$$\begin{aligned} \Gamma_e &\equiv \mathcal{R}\tau^2 \int_0^\infty d\Omega J(\Omega) \mathrm{sinc}^2(\tau\Delta/2)(N_\Omega + 1) \\ \Gamma_a &\equiv \mathcal{R}\tau^2 \int_0^\infty d\Omega J(\Omega) \mathrm{sinc}^2(\tau\Delta/2) N_\Omega \\ \Gamma_s &\equiv \mathcal{R}\tau^2 \int_0^\infty d\Omega J(\Omega) \mathrm{sinc}^2(\tau\Delta/2) M_\Omega e^{i\Delta(2t+\tau)}, \end{aligned} \tag{2.70}$$

together with the driving amplitude-like coefficient

$$\epsilon \equiv \mathcal{R}\tau \sum_k g_k \vartheta(\Omega_k) \mathrm{sinc}(\tau\Delta_k/2) D_{\Omega_k} e^{i\Delta_k(t+\tau/2)}, \tag{2.71}$$

and the reservoir-induced frequency shift

$$\begin{aligned} \Delta\hat{H} = \mathcal{R} \int_0^\infty d\Omega J(\Omega) \frac{\tau}{\Delta} \{ & \hat{a}^\dagger\hat{a}(\mathrm{sinc}(\tau\Delta/2)\cos(\tau\Delta/2) - 1) \\ & + 1 - \mathrm{sinc}(\tau\Delta/2) \left(2N(\Omega)(\cos(\tau\Delta/2) - 1) + e^{i\tau\Delta/2}\right)\}, \end{aligned} \tag{2.72}$$

whose first term induces a renormalization of the mode frequency, and the second one shifts the zero-point energy. In the above equations we have introduced the reservoir *spectral density*, defined here as $J(\Omega) = \sum_k g_k^2 \vartheta(\Omega_k)\delta(\Omega - \Omega_k)$, and took the continuum limit. As a consequence, we drop the subscripts k in the reservoir frequencies Ω_k and related quantities, e.g. $\Delta \equiv \omega - \Omega$. Notice that the three integrals in Eq. (2.70) are weighted by the function $\mathrm{sinc}^2(\tau\Delta/2)$. As this factor is highly peaked around $\Delta = 0$ (that is $\Omega = \omega$), it acts as a Dirac delta function ($\delta(\tau\Delta/2)$) when integrating over the reservoir frequencies, meaning that the effect of detuned modes in the reservoir is very weak in comparison with the resonant ones [24]. This implies:

$$\begin{aligned} \Gamma_e &\simeq \mathcal{R}\tau^2 J(\omega)(N_\omega + 1) \equiv \gamma_0(N_\omega + 1) \\ \Gamma_a &\simeq \mathcal{R}\tau^2 J(\omega)N_\omega \equiv \gamma_0 N_\omega \\ \Gamma_s &\simeq \mathcal{R}\tau^2 J(\omega)M_\omega \equiv \gamma_0 M_\omega \end{aligned} \tag{2.73}$$

and we obtain an effective decay rate $\gamma_0 = \mathcal{R}\tau^2 J(\omega)$ characterizing the global system-reservoir interaction dynamics, proportional to the density of resonant modes in the reservoir. We note that this approximation is justified when the time-scale of the intrinsic (isolated) system dynamics $\tau_s \sim 1/\omega$ is small compared with the interaction time between system and environment, $\tau_s \ll \tau$, in analogy to the secular approximation introduced in Sect. 2.2.

We end up with the following Markovian master equation in the interaction picture

$$\begin{aligned} \dot{\rho} = &-\frac{i}{\hbar}[\Delta\hat{H}, \rho] + [\epsilon^* \hat{a} - \epsilon\hat{a}^\dagger, \rho] \\ &+ \gamma_0(N_\omega + 1)\left(\hat{a}\rho\hat{a}^\dagger - \frac{1}{2}\{\hat{a}^\dagger\hat{a}, \rho\}\right) + \gamma_0 N_\omega\left(\hat{a}^\dagger\rho\hat{a} - \frac{1}{2}\{\hat{a}\hat{a}^\dagger, \rho\}\right) \\ &- \gamma_0 M_\omega^*\left(\hat{a}\rho\hat{a} - \frac{1}{2}\{\hat{a}^2, \rho\}\right) - \gamma_0 M_\omega\left(\hat{a}^\dagger\rho\hat{a}^\dagger - \frac{1}{2}\{\hat{a}^{\dagger 2}, \rho\}\right). \end{aligned} \tag{2.74}$$

As we can see the structure of the equation is the same as in the qubit case, Eq. (2.53), replacing the lower and raising operators of the qubit system by the creation and annihilation operators of the cavity mode. However, we see that here the terms in the last two lines retain the anticommutator part previously vanishing due to $\hat{\sigma}^2 = 0$, and its sign has changed due to the different relative phase introduced in the interaction Hamiltonian (2.61). Furthermore in this case we considered the reservoir to have non-zero initial averages $\langle b_k\rangle_{\rho_E}$, which implies the inclusion of a driving-like term with amplitude ϵ.

By particularizing the reservoir state we can obtain different versions of the master equation. We just consider here the case of a squeezed thermal reservoir, as it will be be of particular interest in later chapters. The reservoir density operator is in this case $\rho_E = \prod_k \hat{\mathbb{S}}_k \rho_{\mathrm{th}} \hat{\mathbb{S}}_k^\dagger$, $\hat{\mathbb{S}}_k(r_k, \theta_k)$ being the squeezing operator for mode k and $\rho_{\mathrm{th}} = e^{-\beta\hat{H}_E}/Z_E$ the equilibrium thermal (Gibbs) state (see Sect. 1.2.5). We hence obtain the master equation coefficients

$$N_\omega = \cosh(2r)n_{\text{th}} + \sinh^2(r), \quad M_\omega = -\sinh(r)\cosh(r)(2n_{\text{th}} + 1)e^{i\theta},$$

and $\epsilon = 0$. In the above expressions r and θ are the squeezing parameters of the resonant mode in the reservoir, however one may alternatively assume that all the modes in the reservoir have the same squeezing parameters [33]. The master equation (2.74) then can be rewritten in the Schrödinger picture as

$$\begin{aligned} \dot{\rho} = & -\frac{i}{\hbar}[\hat{H}, \rho] + \gamma_0(n_{\text{th}} + 1)\left(\hat{R}\rho\hat{R}^\dagger - \frac{1}{2}\{\hat{R}^\dagger\hat{R}, \rho\}\right) \\ & + \gamma_0 n_{\text{th}}\left(\hat{R}^\dagger\rho\hat{R} - \frac{1}{2}\{\hat{R}\hat{R}^\dagger, \rho\}\right), \end{aligned} \tag{2.75}$$

where we have defined $\hat{R} \equiv \cosh(r)\hat{a} + \sinh(r)e^{i\theta}\hat{a}^\dagger = \hat{\mathbb{S}}\hat{a}\hat{\mathbb{S}}^\dagger$, and $\hat{\mathbb{S}}(r, \theta) = \exp(r(\hat{a}^2 e^{-i\theta} + \hat{a}^{\dagger 2}e^{i\theta})/2)$ the squeezing operator acting on the system. As in the case of the qubit, it is easy to check that the above master equation is in Lindblad form (see Sect. 2.3.1). However, in contrast to the previous case, the steady state of the dynamics is no longer diagonal in the system energy eigenbasis. This stationary state is in fact the squeezed thermal state

$$\pi = \hat{\mathbb{S}}\frac{e^{-\beta\hat{H}}}{Z}\hat{\mathbb{S}}^\dagger. \tag{2.76}$$

The different properties of this anomalous relaxation process are analyzed from a thermodynamical point of view in Chap. 10. It is also interesting to notice that Eq. (2.75) can be mapped to the case of a Bogotified mode with Hamilton operator $\hat{H}_B \equiv \hbar\omega\hat{R}^\dagger\hat{R}$ in weak contact with a traditional thermal reservoir. This property can be easily checked by introducing the squeezed frame $\rho \to \hat{\mathbb{S}}^\dagger\rho\hat{\mathbb{S}}$ and $\hat{H} \to \hat{\mathbb{S}}\hat{H}\hat{\mathbb{S}}^\dagger = \hat{H}_B$.

2.3.3 Quantum Brownian Motion

In the preceding examples the fast evolution of the coherent inner system dynamics, as compared with the relaxation characteristic time scales, allowed us to perform either the RWA or the secular approximation to obtain CPTP Markovian master equations. This condition is usually satisfied by optical systems, but not in other scenarios such as in solid state physics. A paradigmatic case in which the RWA cannot be performed is the *quantum Brownian motion*, as described by the Caldeira–Legget model [34]. In the weak-coupling limit and for high temperatures a Markovian master equation can be derived for this model, but otherwise the non-Markovian character of the dynamical evolution needs to be addressed with more powerful techniques, such as generalized quantum Langevin equations, or the Feynman–Vernon influence functional [10, 35].

Consider a single Brownian particle of mass m and position and momentum operators $\hat{x}$ and $\hat{p}$, trapped in a potential $V(\hat{x})$. The corresponding Hamiltonian reads

$$\hat{H} = \frac{\hat{p}^2}{2m} + V(\hat{x}). \tag{2.77}$$

The environment of the particle is modeled as a large reservoir of harmonic oscillators in thermal equilibrium (bath) with Hamiltonian

$$\hat{H}_E = \sum_n \hbar\Omega_k \left(\hat{b}_k^\dagger \hat{b}_k + \frac{1}{2}\right) = \sum_k \frac{\hat{\Pi}_k}{2M_k} + \frac{M_k}{2}\Omega_k^2 \hat{Q}_k, \tag{2.78}$$

where $[\hat{b}, \hat{b}_{k'}^\dagger] = \hat{\mathbb{1}}_E \delta_{kk'}$ are annihilation and creation operators, and $[\hat{\Pi}_k, \hat{Q}_{k'}] = i\hbar\hat{\mathbb{1}}_E\delta_{kk'}$ canonical momentum and position operators. Notice that, in contrast to previous examples, we explicitly consider here the masses of the reservoir harmonic oscillators and include the zero point energy term (see Sect. 1.2.3). We stress that both models of the environment are equivalent and that we adopt this approach here just for historical reasons.

In this model, the particle position $\hat{x}$ and that of the bath's oscillators $\hat{X}_k$ are coupled through the interaction Hamiltonian

$$\hat{H}_{\text{int}} = -\sum_k g_k \hat{x}\hat{Q}_k = -\hat{x}\hat{B}_k, \tag{2.79}$$

where g_k represents the coupling strength of the particle to the harmonic oscillator k in the reservoir, and we identify $\hat{B} = \sum_k g_k \hat{Q}_k$ as the global bath operator coupled to the open system. Furthermore we may include a *counter-term* which compensates for the renormalization of frequencies appearing later in the form of a Lamb-shift Hamiltonian term (see Sect. 2.2.2)

$$\hat{H}_{\text{c-t}} = \hat{x}^2 \sum_k \frac{g_k^2}{2M_k\Omega_k^2}, \tag{2.80}$$

which acts only on the Hilbert space of the particle.

Assuming a factorized initial condition, weak coupling between the Brownian particle and the bath, and that the bath is at a high temperature, we can obtain a Markovian master equation which cannot be written in Lindblad form, and is not completely positive (CP). Let us start from the Born–Markov master equation derived in Sect. 2.2.2 in Schrödinger picture

$$\frac{d\rho(t)}{dt} = -\frac{i}{\hbar}[\hat{H} + \hat{H}_{\text{c-t}}, \rho(t)] - \frac{1}{\hbar^2}\int_0^{\infty} d\tau \, \text{Tr}_E([\hat{H}_{\text{int}}, [\hat{H}_{\text{int}}(-\tau), \rho(t) \otimes \rho_E]]), \tag{2.81}$$

with $\rho_E = e^{-\beta \hat{H}_E}/Z_E$, $\beta = 1/k_B T$ being the inverse temperature of the bath. In the above expression $\hat{H}_{\text{int}}(t)$ is the interaction Hamiltonian in interaction picture. In the following, operators with an explicit time dependence are expressed in the interaction picture (see Sect. 1.1). After rearranging terms in Eq. (2.81) we obtain the following most convenient form of the master equation [10]

$$\frac{d\rho(t)}{dt} = -\frac{i}{\hbar}[\hat{H} + \hat{H}_{\text{c-t}}, \rho(t)] + \frac{1}{\hbar^2}\int_0^{\infty} d\tau \left(i\frac{\xi(\tau)}{2}[\hat{x}, \{\hat{x}(-\tau), \rho(t)\}] - \frac{\nu(\tau)}{2}[\hat{x}, [\hat{x}, \rho(t)]] \right), \tag{2.82}$$

where we defined the *dissipation* and *noise kernels* respectively as

$$\xi(\tau) \equiv i\text{Tr}_E([\hat{B}, \hat{B}(-\tau)]) = 2\hbar \sum_k \frac{g_k^2}{2M_k\Omega_k} \sin(\Omega_k \tau), \tag{2.83}$$

$$\nu(\tau) \equiv \text{Tr}_E(\{\hat{B}, \hat{B}(-\tau)\}) = 2\hbar \sum_k \frac{g_k^2}{2M_k\Omega_k} (2\hat{b}_k^\dagger \hat{b}_k + 1) \cos(\Omega_k \tau).$$

As in the previous examples, we may consider now the continuous limit by introducing the spectral density function of the reservoir, which in this case is defined as

$$J(\Omega) \equiv \sum_k \frac{g_k^2}{2M_k\Omega_k} \delta(\Omega - \Omega_k). \tag{2.84}$$

The spectral density is assumed to be a continuous and smooth function function of Ω. The dissipation and noise kernels can be hence rewritten as

$$\xi(\tau) = 2\hbar \int_0^{\infty} d\Omega \, J(\Omega) \, \sin(\Omega\tau), \tag{2.85}$$

$$\nu(\tau) = 2\hbar \int_0^{\infty} d\Omega \, J(\Omega) \, \coth(\beta\hbar\omega/2) \, \cos(\Omega\tau). \tag{2.86}$$

From the above equations it becomes clear that the specific form of the spectral density $J(\Omega)$ may strongly affect the properties of the master equation through the noise and dissipation kernels. The usual approach assumes the so-called *Ohmic* spectral density, depending linearly on Ω for low frequencies, i.e. in the vicinity of the system frequency ω. Furthermore, one assumes a high-frequency cutoff $\Lambda \gg \omega$ in order to account for the system frequency renormalization induced by the interaction of the

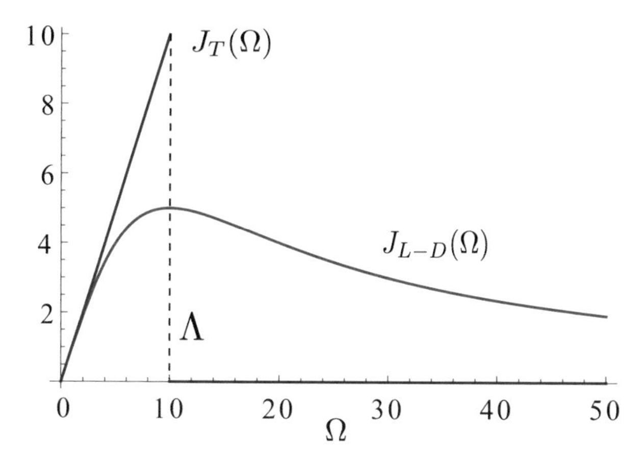

Fig. 2.4 Plots of the Lorentz–Drude $J_{L-D}(\Omega)$ and triangular $J_T(\Omega)$ spectral density functions in Eqs. (2.87) and (2.88) respectively as a function of the environment frequencies Ω. The cutoff frequency $\Lambda = 10$ corresponds to the dotted line, and the spectral densities are scaled in units of $2m\gamma_0/\pi$

system with far detuned oscillators of the environment [10]. An example of a spectral density function fulfilling the above required characteristics is

$$J_{L-D}(\Omega) = \frac{2m\gamma_0}{\pi}\Omega \frac{\Lambda^2}{\Lambda^2 + \Omega^2}, \tag{2.87}$$

which contains a Lorentz–Drude cutoff function. Another simpler example is given by a triangular function with a sharp cutoff at frequency Λ

$$J_T(\Omega) = \frac{2m\gamma_0}{\pi}\Omega\, \Theta(\Lambda - \Omega), \tag{2.88}$$

where $\Theta(x)$ is the Heaviside step function taking the value 1 for $x > 1$ and 0 for $x \leqslant 0$ (see Fig. 2.4). This Ohmic spectral density gives rise to a frequency independent damping at rate γ_0, which is usually determined phenomenologically.

Taking into account the above considerations and approximating $\hat{x}(-\tau) \approx \hat{x} - \tau\hat{p}/m$, the master equation in Eq. (2.82) becomes

$$\begin{aligned} \frac{d\rho(t)}{dt} = & -\frac{i}{\hbar}[\hat{H} + \hat{H}_{\text{c-t}}, \rho(t)] + \frac{iR}{2\hbar^2}[\hat{x}, [\hat{x}, \rho(t)]] \\ & - \frac{1}{2\hbar^2}\left(i\Gamma\,[\hat{x}, \{\hat{p}, \rho(t)\}] + D\,[\hat{x}, [\hat{x}, \rho(t)]] - F\,[\hat{x}, [\hat{p}, \rho(t)]]\right), \end{aligned} \tag{2.89}$$

where we have introduced the coefficients

$$R \equiv \int_0^\infty d\tau\, \xi(\tau) = 2\hbar \sum_k \frac{g_k^2}{2M_k\Omega_k^2}, \qquad F \equiv \int_0^\infty d\tau\, \frac{\tau}{m}\, \xi(\tau), \tag{2.90}$$
$$\Gamma \equiv \int_0^\infty d\tau\, \tau\, \xi(\tau) = 2m\hbar\gamma_0, \qquad D \equiv \int_0^\infty d\tau\, \nu(\tau) = 4m\gamma_0/\beta,$$

and we have used the properties $\int_0^\infty d\tau \sin(\Omega\tau) = P(1/\Omega)$, P being the Cauchy principal value, and $\int_0^\infty d\tau\, \tau \sin(\Omega\tau) = -\partial_\Omega \int_0^\infty d\tau \cos(\Omega\tau) = -\pi\, \partial_\Omega\delta(\Omega)$ [10]. The (Lamb-shift) term accompanying the coefficient R in Eq. (2.89) can be rewritten as $[\hat{x}, [\hat{x}, \rho]] = [\hat{x}^2, \rho]$ and hence cancels with the counter-term $\hat{H}_{\text{c-t}}$ in Eq. (2.80). Furthermore, we stress that the coefficient Γ is related to the damping of the particle motion, while the terms D and F, which are temperature dependent, describe fluctuations induced by the thermal bath. The first one, D, leads to diffusion in momentum and the second one, F, to the so-called anomalous diffusion [11]. We also stress that Eq. 2.89 can be obtained from the Redfield master equation [Eq. (2.36) in Sect. 2.2.2] leading to the same form (2.89) but with time-dependent coefficients, by taking the asymptotic expressions, $t \to \infty$.

The anomalous diffusion coefficient F depends also on the cutoff frequency Λ, and hence requires a specific shape for the spectral density to be calculated. Taking the Ohmic spectral density with Lorentz–Drude cutoff in Eq. (2.87) one obtains in the high temperature limit $k_B T \gtrsim \hbar\Lambda$ [10]

$$F \approx \frac{4\gamma_0}{\beta\Lambda} \tag{2.91}$$

and hence the anomalous diffusion term differs from the momentum diffusion by a factor ω/Λ. As this factor is very small (recall that we assume $\omega \ll \Lambda$) the anomalous diffusion term can be neglected in the high-temperature limit, leading to the Caldeira–Legget master equation [34]

$$\frac{d\rho(t)}{dt} = -\frac{i}{\hbar}[\hat{H}, \rho(t)] - \frac{i\gamma_0}{\hbar}[\hat{x}, \{\hat{p}, \rho(t)\}] - \frac{2m\gamma_0}{\beta\hbar^2}[\hat{x}, [\hat{x}, \rho(t)]]. \tag{2.92}$$

This corresponds to a non CP Markovian master equation which cannot be written in Lindblad form. The first dissipative term describes the loss of the particle's kinetic energy, and the second one the gain [11]. However, a minimal modification allows us to obtain a CP master equation from Eq. (2.92). It consists in adding a further term, namely $-(\gamma_0\beta/8m)[\hat{p}, [\hat{p}, \rho]]$, which is small in the high-temperature limit, $\hbar\omega \ll k_B T$, and hence can be safely included in this regime. By defining a single Lindblad operator of the form [10]

$$\hat{L} \equiv \sqrt{\frac{4m}{\beta\hbar^2}}\hat{x} + i\sqrt{\frac{\beta}{4m}}\hat{p}, \tag{2.93}$$

we may rewrite Eq. (2.92) in Lindblad form as

$$\frac{d\rho(t)}{dt} = -\frac{i}{\hbar}[\hat{H}, \rho(t)] + \hat{L}\rho(t)\hat{L}^{\dagger} - \frac{1}{2}\{\hat{L}^{\dagger}\hat{L}, \rho(t)\}, \tag{2.94}$$

describing CPTP Markovian dynamics.

2.4 Open Many-Body Systems

In the previous section we have seen various examples of dissipative quantum systems whose dynamics can be expressed in the form of a Markovian master equation. However, the three considered cases were single quantum systems (qubit system or harmonic oscillator) coupled to a reservoir which plays the role of the environment. In this section we consider the case of a many-body system, consisting of different prototypical interacting subsystems, dissipating and decohering into the environment. This is a natural and necessary extension of the single open system case, which allows the study of more general configurations. We are interested in the emergence of collective quantum phenomena in realistic (non-isolated) extended systems, such as sub-radiance and super-radiance [23, 36, 37], subdecoherence [38], synchronization [39–42], or quantum phase transitions [43–48]. Open many-body systems are present in a wide range of situations of physical, chemical, or biological interest, and can be often controlled or simulated in the laboratory with high precision (see the reviews [49, 50] on quantum simulation and references therein). Some relevant examples also include the possibility of controlling and engineering dissipation, as recently reported in trapped ion configurations [51, 52], cold quantum gases [51, 53, 54], or Josephson junction arrays [55].

When considering many-body systems the modeling of dissipation and decoherence processes results more involved than in the single body case. One may for instance consider that each unit of the composite system is coupled to a totally independent environment, such as in the initial models of decoherence in quantum computers [56], in the first studies of entanglement dynamics in composite open systems [57, 58], or as in cavity optical modes [4, 5]. However, another possibility is to consider the different bodies in the open system to be coupled to the same environment, a situation which may lead to different time-scales for decoherence and dissipation depending on the relation between the properties of the environment and the spatial extension of the many-body system [38, 59–63]. This latter possibility, usually called *common* or *collective* dissipation and decoherence, has attracted much attention due to their potential applications to quantum information and quantum computing [64–67].

2.4.1 Common Versus Independent Environmental Action

In the next, we provide a simple model to illustrate the differences that may arise in the environmental action on a composite quantum system depending on the interplay between the spatial scale of the open system and the properties of the environment. Consider for instance two quantum systems in a spatially extended configuration, characterized by Hamiltonians $\hat{H}_i$ with associated Hilbert spaces $\mathcal{H}_i$ for $i = 1, 2$. We assume that the two subsystems do not interact directly, and then the total Hamiltonian of the system under consideration reads

$$\hat{H} = \sum_{i=1,2} \hat{H}_i. \tag{2.95}$$

We further assume that the composite system interacts with an environment with Hamilton operator $\hat{H}_E$. We model the coupling between system and environment with a general term of the form

$$\hat{H}_{\text{int}} = \sum_{i=1,2} \sum_{k} \lambda_{ik} \, \hat{A}_i \otimes \hat{E}_k, \tag{2.96}$$

where i runs over the two subsystems. We have introduced system Hermitian operators $\hat{A}_i$ acting on $\mathcal{H}_i$, and orthogonal Hermitian operators $\hat{E}_k$, such that $[\hat{E}_k, \hat{E}_l] = 0$ for $k \neq l$ (they could be for instance operators of different bosonic modes in a thermal bath), affecting each subsystem throughout the different coupling strengths λ_{ik}. These coupling terms would in general depend on the different (classical) positions in space of each subsystem, denoted as $\vec{r}_i$. Notice also that, for simplicity, we have assumed a single operator of each subsystem coupled to the environment.

As a first approximation, we may introduce a spatial scale ξ_E above which the couplings λ_{ik} start to depart from each other depending on i, as assumed in Ref. [59], for non-interacting environmental operators $\hat{E}_k$. This means that one can compare the distance between the system components with the distance ξ_E. If the two subsystems are close enough in space, $|\vec{r}_1 - \vec{r}_2| \ll \xi_E$, then the couplings to the baths are similar, $\lambda_k \equiv \lambda_{1k} \approx \lambda_{2k}$ and the two subsystems feel the same environmental action

$$\hat{H}_{\text{int}} \approx \sum_i \hat{A}_i \otimes \sum_k \lambda_k \hat{E}_k = \sum_i \hat{A}_i \otimes \hat{E}, \tag{2.97}$$

where $\hat{E} = \sum_k \lambda_k \hat{E}_k$. On the other hand, if $|r_1 - r_2| \gtrsim \xi_E$, each subsystem is coupled to different operators $\hat{E}_k$ of the environment and they feel a different noise (see the schematic representation of Fig. 2.5). However, this simplified description turns out to be incorrect or insufficient in many cases of interest, as it becomes clear when considering structured environments [68].

The differences between common or independent environmental action can be better understood by considering the Born–Markov master equation reported in Sect. 2.2.2 within the secular approximation

$$\frac{d\rho}{dt} = \sum_{\omega} \sum_{i,j} \Gamma_{ij}(\omega) \left(\hat{A}_j(\omega) \rho \hat{A}_i^{\dagger}(\omega) - \hat{A}_i^{\dagger}(\omega) \hat{A}_j(\omega) \rho \right) + \text{h.c.}, \tag{2.98}$$

Fig. 2.5 Schematic representation of a two-body quantum system dissipating and decohering into the environment. In the simplest case, depending on the spacial distance between the two systems $|\vec{r}_1 - \vec{r}_2|$ and the spatial scale ξ_E, the systems would feel an independent or a correlated noise

where here the indices $i, j = 1, 2$ refer to the two subsystems of the composite open system. Here we introduced the environment correlation functions

$$\Gamma_{ij}(\omega) \equiv \int_0^\infty ds e^{i\omega s} \sum_k \lambda_{ik}^* \lambda_{jk} \mathrm{Tr}_E[\hat{E}_k(t)\hat{E}_k(t-s)\rho_E], \tag{2.99}$$

which explicitly depend on the coupling constants $\{\lambda_{ik}\}$ between subsystem i and environmental operator k. The correlation functions for $i = j$ correspond to self-dissipation of each subsystem by direct contact with the environment, while the terms $i \neq j$ are cross-dissipative terms indirectly coupling the dynamics of the two subsystems. Recall that the environment operators $\hat{E}(t)$ correspond to the interaction picture with respect to $\hat{H} + \hat{H}_E$.

A more rigorous analysis of the transition between common or independent environmental action focus on the behavior of the correlation functions (2.99). It has been shown that, for isotropic dispersion of the bath modes (e.g. electromagnetic radation in free space), a spatial scale ξ_E arises, such that [61, 63, 68]

$$\begin{aligned} \Gamma_{ij}(\omega) &\approx \Gamma(\omega)\ \forall i, j \quad \text{when } |\vec{r}_i - \vec{r}_j| \ll \xi_E, \\ \Gamma_{ij}(\omega) &\approx \delta_{ij}\Gamma(\omega), \quad \text{when } |\vec{r}_i - \hat{r}_j| \gtrsim \xi_E. \end{aligned} \tag{2.100}$$

That is, for intrasystem distances greater than ξ_E, the only terms that survive in the master equation (2.98) are those describing the dissipation or decoherence of each subsystem as if it were coupled to an independent separate environment, while for distances smaller than ξ_E, the environment induces an indirect coupling between the subsystems. On the other hand, for non-isotropic dispersion relations (as those appearing in some structured environments), the transition between the two limit cases of common or independent noise can be more intricate. In such cases the use of the spatial scale ξ_E is not justified, as collective dissipation can ideally arise above any distance [68].

2.4.2 Coupled Dissipative Harmonic Oscillators

We have seen that different forms of dissipation can arise in spatially extended systems, a case which is not usually treated in textbooks. Therefore we devote this section to the derivation of the master equation for a set of coupled harmonic oscillators dissipating into a thermal environment (thermal bath). Our aim is to obtain Markovian master equations for both cases of common and independent dissipation, which we will call *common bath* (CB) and *separate baths* (SB) respectively. The dynamical modeling which we develop here will be used in Part II of the thesis.

Consider an arbitrary set of N harmonic oscillators with different natural frequencies, unit masses, and arbitrary coupling between them. The Hamiltonian of such system reads:

$$\hat{H} = \frac{1}{2}\sum_{i=1}^{N}\left(\hat{p}_i^2 + \omega_i^2\hat{x}_i^2\right) + \sum_{i<j}^{N}\lambda_{ij}\hat{x}_i\hat{x}_j \tag{2.101}$$

where $\hat{x}_i$ and $\hat{p}_j$ are the canonical position and momenta operators of the harmonic oscillators, satisfying $[\hat{x}_i, \hat{p}_j] = i\hbar\delta_{ij}$. It is convenient to express Eq. (2.101) in matrix form:

$$\hat{H} = \frac{1}{2}\left(\mathbf{p}^T\mathbb{1}\,\mathbf{p} + \mathbf{x}^T\mathcal{H}\,\mathbf{x}\right) \tag{2.102}$$

where $\mathbf{x}^T = (\hat{x}_1, \ldots, \hat{x}_N)$ and $\mathcal{H}$ contains the topological properties of the set, i.e. the (squared) natural frequencies of oscillators in the diagonal elements, $\mathcal{H}_{ii} = \omega_i^2$, and the coupling strengths in the off-diagonal ones $\mathcal{H}_{ij} = \lambda_{ij}$.

Following the previous discussion, the environment is considered as either consisting of N independent bosonic thermal baths (SB case), or just by a single bath (CB case)

$$\hat{H}_E^{(SB)} = \sum_{k=1}^{N}\sum_{\alpha=1}^{\infty}\left(\frac{\hat{\Pi}_\alpha^{2\,(k)}}{2M_\alpha^{(k)}} + \frac{M_\alpha^{(k)}}{2}\tilde{\Omega}_\alpha^{2\,(k)}\hat{Q}_\alpha^{2\,(k)}\right) \tag{2.103}$$

$$\hat{H}_E^{(CB)} = \frac{1}{2}\sum_{\alpha=1}^{\infty}\left(\frac{\hat{\Pi}_\alpha^2}{M_\alpha} + \frac{M_\alpha}{2}\tilde{\Omega}_\alpha^2\hat{Q}_\alpha^2\right) \tag{2.104}$$

where $[\hat{Q}_\alpha^{(k)}, \hat{\Pi}_{\alpha'}^{(l)}] = i\hbar\delta_{\alpha,\alpha'}\delta_{k,l}\hat{\mathbb{1}}$, $\hat{Q}_\alpha^{(k)}$ and $\hat{\Pi}_\alpha^{(k)}$ being the corresponding position and momenta operators for the α bosonic mode in the i-th thermal bath. Here and in the following, we will use greek subscripts when labeling the reservoir(s) bosonic modes. The coupling between the open system and the environment is bilinear in the position, and, while for separate baths each harmonic oscillator is connected to a different bath, in the case of the common bath, coupling is present through the system center of mass:

$$\hat{H}_{\text{int}}^{(SB)} = \sum_i \sum_{\alpha=1}^{\infty} g_\alpha^{(i)} \left(\hat{x}_i \otimes \hat{Q}_\alpha^{(i)} \right), \tag{2.105}$$

$$\hat{H}_{\text{int}}^{(CB)} = \sum_i \hat{x}_i \otimes \sum_{\alpha=1}^{\infty} g_\alpha \hat{Q}_\alpha. \tag{2.106}$$

Notice that the SB interaction term may be written as $\hat{H}_{\text{int}}^{(SB)} = \sum_i \hat{A}_i \otimes \hat{B}_i$ where system and bath operators $\hat{A}_i \equiv \hat{x}_i$ and $\hat{B}_i \equiv \sum_\alpha g_\alpha^{(i)} \hat{Q}_\alpha^{(i)}$ can be identified for each term in the sum over the oscillators. On the other hand, in the CB case the interaction Hamiltonian can be written as a single product $\hat{H}_{\text{int}}^{(CB)} = \hat{A} \otimes \hat{B}$ by identifying $\hat{A} \equiv \sum_i \hat{x}_i$ and $\hat{B} \equiv \sum_\alpha g_\alpha \hat{Q}_\alpha$. In any case the constants $g_\alpha^{(i)}$ model the coupling of the reservoir modes with the open system, and can be related to the spectral density function of the baths $J_i(\Omega) \equiv \sum_\alpha (g_\alpha^{(i)2}/\tilde{\Omega}_\alpha)\delta(\Omega - \tilde{\Omega}_\alpha)$.

A simpler picture of this many-body open system can be obtained by considering the normal mode basis of the set of oscillators, which is obtained by diagonalization of its Hamiltonian $\hat{H}$. This problem reduces to the diagonalization of the matrix $\mathcal{H}$ in Eq. (2.102), which can be formally done by introducing the canonical transformation

$$\hat{x}_i = \sum_{j=1}^{N} f_{ij}\hat{X}_j, \qquad \hat{p}_i = \sum_{j=1}^{N} f_{ij}\hat{P}_j. \tag{2.107}$$

This transformation can be alternatively expressed in matrix form as $\mathbf{x} = f\ \mathbf{X}$, and $\mathbf{p} = f\ \mathbf{P}$. We stress that the change of basis matrix, f, must be orthogonal ($f^T = f^{-1}$) since it is a canonical transformation. If we now substitute the new set of coordinates in the original Hamiltonian (2.102) we have

$$\mathbf{x}^T\ \mathcal{H}\ \mathbf{x} = \mathbf{X}^T\ f^T\ \mathcal{H} f\ \mathbf{X} = \mathbf{X}^T\ \Omega\ \mathbf{X}, \tag{2.108}$$

$$\mathbf{p}^T\ \mathbb{1}\ \mathbf{p} = \mathbf{P}^T\ f^T\ \mathbb{1} f\ \mathbf{P} = \mathbf{P}^T\ \mathbb{1}\ \mathbf{P}, \tag{2.109}$$

and then the diagonal matrix $\Omega = f^T \mathcal{H} f$ contains the squared normal modes frequencies Ω_i^2. In the normal mode basis, the complete form of the open system Hamiltonian is transformed into a set of uncoupled harmonic oscillators with frequencies Ω_i, that is

$$\hat{H} = \frac{1}{2}\sum_{i}^{N}\left(\hat{P}_i^2 + \Omega_i^2\,\hat{Q}_i^2\right). \tag{2.110}$$

and the interaction Hamiltonians for SB and CB hence take the new forms

$$\hat{H}_{\text{int}}^{(SB)} = \sum_{i,j=1}^{N}\sum_{\alpha=1}^{\infty} f_{ij}\, g_\alpha^{(i)}\left(\hat{X}_j \otimes \hat{Q}_\alpha^{(i)}\right), \tag{2.111}$$

$$\hat{H}_{\text{int}}^{(CB)} = \sum_{j=1}^{N}\sum_{\alpha=1}^{\infty} \kappa_j\, g_\alpha\left(\hat{X}_j \otimes \hat{Q}_\alpha\right). \tag{2.112}$$

Examining the above equations we observe that in the SB case we can redefine the bath operators such that again $\hat{H}_{\text{int}}^{(SB)} = \sum_j \hat{A}_j \otimes \hat{B}_j'$ with $\hat{B}_j' \equiv \sum_i \sum_\alpha f_{ij} g_\alpha^{(i)} \hat{Q}_\alpha^{(i)}$, meaning that each normal mode couples to a separate bosonic reservoir with redefined canonical positions for each reservoir mode. On the contrary, for the CB case we obtain that each normal mode of the system couples with different strengths to the thermal reservoir through the coefficients $\kappa_j = \sum_i f_{ij}$, that takes into account all the topological (geometry, coupling strengths, frequencies...) characteristics of the open system. In Fig. 2.6 we provide a schematic representation of the SB and CB cases in terms of the normal modes of the open system.

2.4.2.1 Separate Baths Master Equation

We are now in position to derive Markovian master equations for CB and SB in the normal mode basis. Consider first the case of SB, for which we use the interaction Hamiltonian $\hat{H}_{\text{int}}^{(SB)}$ in Eq. (2.111). Assuming an initial product state for system plus reservoir(s), and using Born–Markov approximations in Sect. 2.2.2 we have that the evolution of the system reduced density matrix is, in the Schrödinger picture:

$$\begin{aligned}\frac{d\rho}{dt} = -\frac{i}{\hbar}\left[\hat{H}, \rho\right] - \frac{1}{\hbar^2}\int_0^t d\tau \sum_{l,k}[C_{lk}(\tau)(\hat{A}_l\hat{A}_k(-\tau)\rho - \hat{A}_k(-\tau)\rho\hat{A}_l) \\ + C_{kl}(-\tau)(\rho\hat{A}_k(-\tau)\hat{A}_l - \hat{A}_l\rho\hat{A}_k(-\tau))],\end{aligned} \tag{2.113}$$

$\hat{A}_k(\tau)$ denoting interaction picture operators. We have introduced here the correlation functions $C_{lm}(\tau) = \text{Tr}[\hat{B}_l(\tau)\hat{B}_m\rho_E]$ with $\rho_E = e^{-\beta\hat{H}_E}/Z_E$ the thermal equilibrium (Gibbs) state at inverse temperature $\beta = 1/k_B T$. Let us define

$$C_{lk}^C(\tau) \equiv C_{lk}(\tau) - C_{kl}(-\tau), \qquad C_{lk}^A(\tau) \equiv C_{lk}(\tau) + C_{kl}(-\tau),$$

which allows us to rewrite the master equation (2.113) as

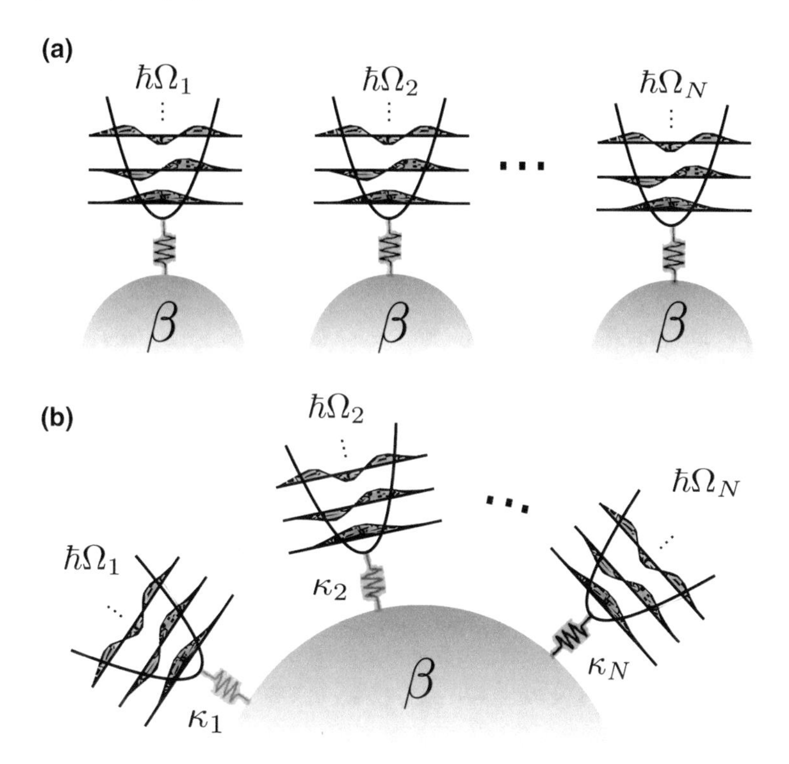

Fig. 2.6 Schematic representation of the dissipative couplings of the N normal modes of the open system with frequencies $\Omega_1, \ldots, \Omega_N$ in **a** the SB situation, where each normal mode is equally coupled to an equivalent thermal bath at the same inverse temperature β, and **b** the CB case, in which all the normal modes couple to the common bath with different strengths, κ_i for $i = 1, \ldots, N$, depending on the natural frequencies and couplings of the original oscillators

$$\frac{d\rho}{dt} = -\frac{i}{\hbar}\left[\hat{H}, \rho\right] - \frac{1}{\hbar^2}\int_0^t d\tau \sum_{l,k}\left(\frac{C^C_{lk}(\tau)}{2}[\hat{A}_l, \{\hat{A}_k(-\tau), \rho\}] + \frac{C^A_{lk}(\tau)}{2}[\hat{A}_l, [\hat{A}_k(-\tau), \rho]]\right). \quad (2.114)$$

As long as we pursue a master equation in the normal mode basis, we must compute the system and bath operators by using the interaction Hamiltonian in Eq. (2.111). The correlation functions read

$$C_{lk}(\tau) = \sum_i f_{li}f_{ki}C_i(\tau), \quad \text{with}$$
$$C_i(\tau) = \hbar\sum_\alpha \frac{g_\alpha^{2(i)}}{2\tilde{\Omega}_\alpha^{(i)}}\left(e^{i\tilde{\Omega}_\alpha^{(i)}\tau} + 2\hat{n}_\alpha^{(i)}\cos(\tilde{\Omega}_\alpha^{(i)}\tau)\right), \quad (2.115)$$

where $\hat{n}_\alpha^{(i)}$ is the number operator of mode α of the i-th thermal bath, and we notice that $C_{lk}(\tau) = C_{kl}(\tau)$. With the help of these expressions we can compute the correlation functions

$$C^C_{lk}(\tau) = \sum_i f_{li} f_{ki} \left(C_i(\tau) - C_i(-\tau)\right) = \sum_i f_{li} f_{ki} i\xi_i(\tau), \tag{2.116}$$

$$C^A_{lk}(\tau) = \sum_i f_{li} f_{ki} \left(C_i(\tau) + C_i(-\tau)\right) = \sum_i f_{li} f_{ki} \nu_i(\tau), \tag{2.117}$$

where we introduced the dissipation and noise kernels (see the Brownian motion model in Sect. 2.3), that could be expressed in terms of the spectral density function as

$$\xi_i(\tau) \equiv \hbar \int d\Omega\, J_i(\Omega) \sin(\Omega\tau), \tag{2.118}$$

$$\nu_i(\tau) \equiv \hbar \int d\Omega\, J_i(\Omega) \cos(\Omega\tau) \coth\left(\frac{\beta\hbar\Omega}{2}\right). \tag{2.119}$$

Introducing the correlation functions in the expression for the master equation (2.114), and using that the free evolution of the system operators is given by $\hat{A}_k(-\tau) = \hat{X}_k \cos(\Omega_k \tau) - \frac{\hat{P}_k}{\Omega_k} \sin(\Omega_k \tau)$, we can rearrange terms to obtain

$$\frac{d\rho}{dt} = -\frac{i}{\hbar}[\hat{H}, \rho] + \sum_{lk} \mathcal{D}_{lk}[\rho]. \tag{2.120}$$

Here we have introduced the super-operator $\mathcal{D}_{lk}[\rho](t)$ defined as

$$\begin{aligned}\mathcal{D}_{lk}[\rho](t) \equiv\ & -\frac{1}{2\hbar^2}\left(-iR_{lk}(t)\,[\hat{X}_l, \{\hat{X}_k, \rho\}] + i\Gamma_{lk}(t)\,[\hat{X}_l, \{[\hat{P}_k, \rho\}]\right.\\ & \left. + D_{lk}(t)\,[\hat{X}_l, [\hat{X}_k, \rho]] - F_{lk}(t)\,[\hat{X}_l, [\hat{P}_k, \rho]]\right),\end{aligned}$$

with the time-dependent coefficients

$$R_{lk}(t) \equiv \int_0^t d\tau \sum_i f_{li} f_{ki} \cos(\Omega_k \tau)\, \xi_i(\tau), \tag{2.121}$$

$$\Gamma_{lk}(t) \equiv \int_0^t d\tau \sum_i f_{li} f_{ki} \frac{\sin(\Omega_k \tau)}{\Omega_k}\, \xi_i(\tau), \tag{2.122}$$

$$D_{lk}(t) \equiv \int_0^t d\tau \sum_i f_{li} f_{ki} \cos(\Omega_k \tau)\, \nu_i(\tau), \tag{2.123}$$

$$F_{lk}(t) \equiv \int_0^t d\tau \sum_i f_{li} f_{ki} \frac{\sin(\Omega_k \tau)}{\Omega_k}\, \nu_i(\tau). \tag{2.124}$$

By taking the Markovian limit, i.e. $t \to \infty$ in the time integrals appearing in Eqs. (2.121)–(2.124), and integrating over the bath frequencies, we obtain time-independent coefficients. The frequency renormalization of the normal modes is given by the coefficient

$$R_{lk} = -\frac{\hbar}{2}\sum_i f_{li}f_{ki}\left[P\left(\frac{J_i(\Omega)}{\Omega-\Omega_k}\right)+P\left(\frac{J_i(\Omega)}{\Omega+\Omega_k}\right)\right], \tag{2.125}$$

P denoting the principal value of Cauchy, which will be neglected since it can be incorporated into the original Hamiltonian $\hat{H}$.

In addition we obtain

$$\Gamma_{lk} = \frac{\hbar\pi}{2}\sum_i f_{li}f_{ki}\,\frac{J_i(\Omega_k)}{\Omega_k} = \delta_{lk}\frac{\hbar\pi}{2}\frac{J(\Omega_k)}{\Omega_k}, \tag{2.126}$$

$$\begin{aligned} D_{lk} &= \frac{\hbar\pi}{2}\sum_i f_{li}f_{ki}\,J_i(\Omega_k)\coth\left(\frac{\beta\hbar\Omega_k}{2}\right) \\ &= \delta_{lk}\frac{\pi}{2}J(\Omega_k)\coth\left(\frac{\beta\hbar\Omega_k}{2}\right), \end{aligned} \tag{2.127}$$

$$\begin{aligned} F_{lk} &= \sum_i f_{li}f_{ki}\int_0^\infty d\tau\,\frac{\sin(\Omega_k\tau)}{\Omega_k}\nu_i(\tau) \\ &= \delta_{lk}\int_0^\infty d\tau\,\frac{\sin(\Omega_k\tau)}{\Omega_k}\,\nu(\tau), \end{aligned} \tag{2.128}$$

where in the second equality we have assumed identical thermal baths with the same spectral density $J_i(\Omega) = J(\Omega)$. Using the time-independent coefficients in Eqs. (2.126)–(2.128), we finally obtain the following Markovian master equation for separate baths in the normal mode basis

$$\begin{aligned} \frac{d\rho}{dt} = -\frac{i}{\hbar}[\hat{H},\rho] - \frac{1}{2\hbar^2}\sum_k (i\Gamma_k\,[\hat{X}_l,\{[\hat{P}_k,\rho\}] + D_k\,[\hat{X}_l,[\hat{X}_k,\rho]], \\ - F_k\,[\hat{X}_l,[\hat{P}_k,\rho]]), \end{aligned} \tag{2.129}$$

with damping coefficients $\Gamma_k \equiv \Gamma_{kk}$, momentum diffusion coefficients $D_k \equiv D_{kk}$, and anomalous diffusion $F_k \equiv F_{kk}$, as given by Eqs. (2.126)–(2.128). We notice that the dissipative terms in the above master equation describe the independent damping, diffusion and anomalous diffusion processes obtained previously in the quantum Brownian motion model in Sect. 2.3.

It is also important to stress that the Markovian master equation Eq. (2.129), as the Brownian motion master equation (2.89) in Sect. 2.3, is not in Lindblad form, and hence complete positivity is not guaranteed, meaning that the evolution may be unphysical for some specific initial conditions and values of the parameters. In order to obtain a CPTP dynamics we would need to perform a further approximation in Eq. (2.129), which we call strong RWA, and is analogous to the secular approximation introduced in the previous section.

2.4.2.2 Common Bath Master Equation

Consider now the case of CB, and hence the interaction Hamiltonian $\hat{H}_{\text{int}}^{(CB)}$ in Eq. (2.111). Following the same lines than in the SB case, the master equation (2.113) now reads

$$\frac{d\rho}{dt} = -\frac{i}{\hbar}\left[\hat{H}, \rho\right] - \frac{1}{\hbar^2}\int_0^t d\tau [C(\tau)(\hat{A}\hat{A}(-\tau)\rho - \hat{A}(-\tau)\rho\hat{A}) + C(-\tau)(\rho\hat{A}(-\tau)\hat{A} - \hat{A}\rho\hat{A}(-\tau))], \tag{2.130}$$

where we have a single operator for the open system and the bath, $\hat{A} = \sum_j \kappa_j \hat{X}_j$ and $\hat{B} = \sum_\alpha g_\alpha \hat{Q}_\alpha$, leading to the bath correlation function

$$C(\tau) \equiv \text{Tr}[\hat{B}(\tau)\hat{B}\rho_E] = \hbar \sum_\alpha \frac{g_\alpha^2}{2}\tilde{\Omega}_\alpha \left(e^{i\tilde{\Omega}_\alpha \tau} + 2\hat{n}_\alpha \cos(\tilde{\Omega}_\alpha \tau)\right). \tag{2.131}$$

Defining as in the previous case $C^C(\tau) \equiv C(\tau) - C(-\tau) = i\xi(\tau)$ and $C^A(\tau) \equiv C(\tau) + C(-\tau) = \nu(\tau)$, we obtain similar expressions for the dissipation and noise kernels of the bath

$$\xi(\tau) = \hbar \int d\Omega\, J(\Omega) \sin(\Omega\tau), \tag{2.132}$$

$$\nu(\tau) = \hbar \int d\Omega\, J(\Omega) \cos(\Omega\tau) \coth\left(\frac{\beta\hbar\Omega}{2}\right), \tag{2.133}$$

with the spectral density function $J(\Omega) \equiv \sum_\alpha (g_\alpha^2/\tilde{\Omega}_\alpha)\delta(\Omega - \tilde{\Omega}_\alpha)$.

Using the above expressions and rearranging terms we arrive to the master equation for the common bath in the normal mode basis:

$$\frac{d\rho}{dt} = -\frac{i}{\hbar}[\hat{H}, \rho] - \frac{1}{2\hbar^2}\sum_{ij}(-iR_{ij}(t)\, [\hat{X}_i, \{\hat{X}_j, \rho\}] + i\Gamma_{lk}(t)\, [\hat{X}_l, \{[\hat{P}_k, \rho\}] + D_{ij}(t)\, [\hat{X}_i, [\hat{X}_j, \rho]] \;-\; F_{ij}(t)\, [\hat{X}_i, [\hat{P}_j, \rho]]), \tag{2.134}$$

where the coefficients are now defined by

$$R_{ij}(t) \equiv \int_0^t d\tau\ \kappa_i\kappa_j \cos(\Omega_j\tau)\ \xi(\tau), \tag{2.135}$$

$$\Gamma_{ij}(t) \equiv \int_0^t d\tau\ \kappa_i\kappa_j \frac{\sin(\Omega_j\tau)}{\Omega_j}\ \xi(\tau), \tag{2.136}$$

$$D_{ij}(t) \equiv \int_0^t d\tau\ \kappa_i\kappa_j \cos(\Omega_j\tau)\ \nu(\tau), \tag{2.137}$$

$$F_{ij}(t) \equiv \int_0^t d\tau\ \kappa_i\kappa_j \frac{\sin(\Omega_j\tau)}{\Omega_j}\ \nu(\tau), \tag{2.138}$$

to be compared with those of Eqs. (2.121)–(2.124). In the Markovian limit $t \to \infty$ these coefficients read

$$\Gamma_{ij} = \kappa_i\kappa_j\ \frac{\hbar\pi}{2}\frac{J(\Omega_j)}{\Omega_j}, \quad D_{ij} = \kappa_i\kappa_j\ \frac{\hbar\pi}{2} J(\Omega_j) \coth\left(\frac{\beta\hbar\Omega_j}{2}\right),$$
$$F_{ij} = \kappa_i\kappa_j \int_0^\infty d\tau \frac{\sin(\Omega_j\tau)}{\Omega_j}\ \nu(\tau), \tag{2.139}$$

and we may again neglect the frequency renormalization terms R_{ij} leading to a Lamb-shift-like Hamiltonian. Examining Eq. (2.139) we note that, in contrast to the SB case, for CB we have no longer independent channels of dissipation for each normal mode of the open system, but they become coupled by means of the master equation coefficients with $i \neq j$. This fact is expected from the arguments presented in Sect. 2.4.1. We also notice that the self-dissipation channels ($i = j$) in the Markovian limit are the same than in the SB case, except for the appearance of the effective couplings κ_i^2, modifying the strength of the environmental dissipative action over each normal mode. Therefore, when the symmetries present in the (original) open system lead to vanishing effective couplings $\kappa_i = 0$, the corresponding normal mode (i) is protected from the environmental action, a situation which is not possible in the case of identical SB. Those features are going to be exploited in Part II of this thesis, where we analyze in detail the emergence of dynamical effects such as synchronization phenomena, or the characteristics of the evolution of quantum correlations between pairs of oscillators of the open system under dissipation.

2.5 Quantum Trajectories

In this section we introduce the formalism of quantum trajectories, also called the quantum jump approach, which allows us the introduction of a detailed stochastic description of the evolution of open quantum systems, more complete than the master equation approach previously studied. The concept of a quantum trajectory is closely related to the Kraus operator decomposition of CPTP maps introduced in Sect. 2.1.2, and consequently to the generalized measurement framework presented in Sect. 1.3.2.

A quantum trajectory can be considered as the path which the (pure) state of an open quantum system follows over time conditioned to a "record" of events occurring during the evolution. This path is generated by a continuous sequence of random quantum operations modifying the state of the system, and generally differing from one trajectory to another, much as the evolution generated by a large succession of selective quantum measurements occurring at infinitesimal intervals of times.

The quantum trajectory formalism was first introduced in the context of quantum optics by different groups [3, 69–73], both as a physical process resulting from continuous monitoring of an open quantum system which undergoes *quantum jumps* induced by environmental action, and also as a powerful simulation tool allowing numerical computation of complicated problems where the master equation approach is intractable (for reviews see [74, 75]). The formalism can be applied to any open quantum system whose evolution is described by a Markovian master equation in Lindblad form, which, as we will see below, can be *unraveled* by splitting the average dynamics in a set of random quantum processes generating the stochastic trajectories.

Quantum trajectory methods are ubiquitous in atomic physics and quantum optics, and they have been applied to tackle laser cooling [72, 76, 77], cascaded quantum systems [78], or the quantum delta-kicked rotor [79] among others [74, 75]. The quantum jumps in which the formalism is based can be traced back to the origins of quantum theory in the ideas of Bohr, but its real existence has been also the subject of strong criticism from long time ago, as manifested e.g. by Schrödinger in 1952 in a paper entitled "Are there quantum jumps?" [80, 81]. The first observations of quantum jumps had to wait for the development of single particle experiments, which were finally performed in ion trapped setups in the middle 1980s [82–84]. The more recent development of novel quantum non-demolition measurements allowed the precise generation, record, and manipulation of quantum trajectories in real time with applications in quantum state generation and control [11, 85–90].

2.5.1 Continuous Measurements and Quantum Jumps

Consider as in Sect. 2.2 the splitting of time into a sequence of small time steps where generalized measurements are performed. The measurement at any time step can be described by a CPTP map $\mathcal{E}$ with a prescribed Kraus representation of M measurement operators, namely $\{\hat{M}_m\}_{m=0}^{M-1}$. Let us assume that $\rho(t)$ is the density operator of the system at an arbitrary time t. Hence, after one time step of small duration dt the (unconditional) state of the system, averaging over possible results, changes to

$$\rho(t+dt) = \sum_k \mathcal{E}_k(\rho(t)) = \sum_m \hat{M}_m(dt)\rho(t)\hat{M}_m^\dagger(dt). \tag{2.140}$$

We want this concatenation of maps to describe a continuous evolution, which implies

$$\lim_{dt\to 0} \frac{\rho(t+dt)-\rho(t)}{dt} \equiv \dot{\rho}(t) = \text{finite}, \tag{2.141}$$

when considering that the time dt during which the measurement is performed is infinitesimal. Therefore we need the state $\rho(t+dt)$ to be only infinitesimally different from $\rho(t)$, which can be done by setting the following form for the measurement (Kraus) operators [11]

$$\hat{M}_0(dt) = \mathbb{1} - dt\left(\hat{R}/2 + \frac{i}{\hbar}\hat{H}\right), \tag{2.142}$$

$$\hat{M}_k(dt) = \sqrt{dt}\,\hat{L}_k, \quad \text{for } k = 1, 2, \ldots, M-1. \tag{2.143}$$

Here $\hat{H}$ and $\hat{R}$ are Hermitian but otherwise arbitrary operators, and we have also introduced the generic set of operators $\{\hat{L}_k\}$, which are only required to obey

$$\sum_k \hat{L}_k^\dagger \hat{L}_k = \hat{R}. \tag{2.144}$$

Under these conditions, we obtain that the trace preserving condition is fulfilled

$$\sum_{m=0}^{M-1} \hat{M}_m(dt)^\dagger \hat{M}_m(dt) = \mathbb{1} - dt\hat{R} + dt\sum_k \hat{L}_k^\dagger \hat{L}_k + \mathcal{O}(dt^2) = \mathbb{1},$$

up to first order in dt. Moreover, the above Eq. (2.140) then reads

$$\begin{aligned} \rho(t+dt) &= \hat{M}_0(dt)\rho(t)\hat{M}_0(dt)^\dagger + dt\sum_k \hat{L}_k\rho(t)\hat{L}_k^\dagger \\ &= \rho(t) - \frac{i}{\hbar}[\hat{H},\rho(t)]dt + \sum_k \hat{L}_k\rho(t)\hat{L}_k^\dagger - \frac{1}{2}\{\hat{L}_k^\dagger\hat{L}_k,\rho(t)\}, \end{aligned} \tag{2.145}$$

implying that Eq. (2.141) reproduces the Lindblad form [c.f. Eq. (2.29) in Sect. 2.2.1].

Let us now focus on the action of the measurement (or Kraus) operators in Eqs. (2.142) and (2.143) over the system state. We assume the state of the system at time t to be pure and given by $\sigma_t = |\psi\rangle_t\,\langle\psi|_t$. The result of the measurement at time t occurring during the infinitesimal time interval dt takes the different values $m = 0$ or $m = k$ for $k = 1, \ldots, M-1$ with probabilities

$$\mathcal{P}_0(t) = \text{Tr}[\hat{M}_0^\dagger \hat{M}_0 \sigma_t] = 1 - dt\langle\hat{R}\rangle_t, \tag{2.146}$$

$$\mathcal{P}_k(t) = \text{Tr}[\hat{M}_k^\dagger \hat{M}_k \sigma_t] = dt\langle\hat{L}_k^\dagger \hat{L}_k\rangle_t, \tag{2.147}$$

where it can be easily verified that $\mathcal{P}_0(t) + \sum_k \mathcal{P}_k(t) = 1\ \forall t$. Notice from the above expressions that, as long as dt is very small, in almost all of the infinitesimal time intervals we will get as outcome of the measurement $m = 0$, while the other outcomes

$m = k$ will occur very rarely. In the first case the state of the system changes as

$$|\psi\rangle^{(0)}_{t+dt} = \frac{\hat{M}_0(dt)}{\mathcal{P}_0(t)}|\psi\rangle_t = |\psi\rangle_t - dt\left(\frac{i}{\hbar}\hat{H} + \frac{\hat{R} - \langle\hat{R}\rangle_t}{2}\right)|\psi\rangle_t\,, \tag{2.148}$$

which corresponds to a smooth evolution. It is important to notice that this evolution is not unitary due to the presence of the operator $\hat{R}$. Indeed the probability of obtaining two consecutive $m = 0$ results would be $\mathcal{P}_0(t + dt) = \mathcal{P}_0(t) + dt^2(\sigma^2(\hat{R})_t - i\langle[\hat{H}, \hat{R}]\rangle_t)/\hbar$, which increases with the fluctuations in the operator $\hat{R}$. For instance, in the relevant case in which $\hat{H}$ and $\hat{R}$ commute, $\mathcal{P}_0(t + dt) = \mathcal{P}_0(t)$ unless $|\psi\rangle_t$ is a superposition of energy eigenstates, meaning that no evolution occurs at all, c.f. Eq. (2.148). Otherwise the above operation will produce the progressive damping of the (superposition) states.

On the other hand, when the outcome of the measurement results to be $m = k$, the back-action induces a change in the system given by

$$|\psi\rangle^{(k)}_{t+dt} = \frac{\hat{L}_k}{\sqrt{\langle\hat{L}_k^\dagger\hat{L}_k\rangle_t}}|\psi\rangle_t\,, \tag{2.149}$$

which corresponds to an abrupt change usually referred to as a *quantum jump*. In the relevant case in which $[\hat{H}, \hat{L}_k] = \pm\hbar\omega\hat{L}_k$, this jump corresponds to the sudden absorption or emission of a quantum of energy $\hbar\omega$ by the system. Notice that those equations are easily extended to the case of mixed states σ_t by applying the generalized measurement formalism introduced in Sect. 1.3.2, where the arbitrary form for the state of the system conditioned to a measurement result was given in Eq. (1.107). Here we will still maintain the assumption of a pure quantum state $|\psi\rangle_t$ for reasons that will become obvious later.

In conclusion, we obtained that the conditioned system evolution when the continuous measurement is performed, consists of large periods of smooth evolution given by Eq. (2.148) and associated to the outcome $m = 0$, while at some random times a jump associated to outcome k occurs (with rate $\mathcal{P}_k(dt)/dt$), and the state of the system suddenly changes according to Eq. (2.149). In the continuous measurement language the outcome $m = 0$ is regarded as the *null result* of the measurement, while outcomes $m = k$ are called *detections* [11]. A simple physical example of continuous measurement is given by photodetection of the light emitted by a damped single-mode cavity in a thermal reservoir at zero temperature (see Sect. 2.3.2), which may be described by only two outcomes $m = \{0, 1\}$ [3]. In such case the null measurement $m = 0$ corresponds to no photons being detected, and the outcome $m = 1$ describes the clicks on the detector when a photon is emitted by the cavity at some (rare) random times (see Fig. 2.7). The corresponding measurement operators are $\hat{M}_0 = \mathbb{1} - dt(i\omega + \gamma_0/2)\hat{N}$ inducing the damping of superpositions in the Fock basis, and $\hat{L}_1 = \sqrt{\gamma_0}\hat{a}$, which produces the annihilation of one photon in the cavity when the detector clicks. Here γ_0 is the exponential decay rate of photons in the cavity.

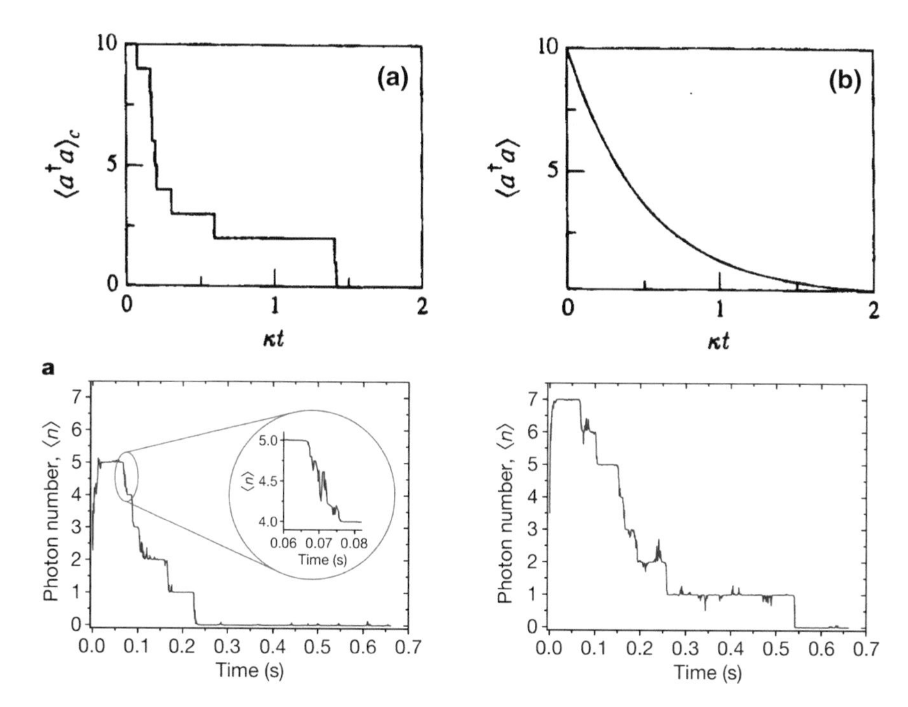

Fig. 2.7 In the top panels we can appreciate **a** a sample trajectory of the mean number of photons in a single-mode cavity initially prepared in the Fock state $|10\rangle$, and **b** the average over 10,000 trajectories reproducing the exponential decay predicted by the master equation formalism (here $2\kappa \equiv \gamma_0$). The bottom panel shows experimental sample trajectories for a damped microwave field in an ultrahigh-Q Fabry–Perot cavity cooled at $0.8K$ and sensed by circular Rydberg atoms of rubidium. The microwave field is prepared to contain $n = 5$ (left) and $n = 7$ (right) photons. From Refs. [3] (top) and [91] (bottom)

2.5.2 *Stochastic Schrödinger Equation*

The above description of the selective continuous measurement can be used to construct a stochastic equation of motion generating trajectories for the system evolution, conditioned to the measurement outcomes obtained at any time during the dynamics. In order to construct such equation we need first to introduce the number of detections corresponding to each outcome k different from the null result up to time t during the dynamical evolution, which we denote by $N_k(t)$. We then define the infinitesimal stochastic increment $dN_k(t)$ which obeys

$$dN_k(t)dN_l(t) = \delta_{kl}dN_k(t), \tag{2.150}$$

$$E[dN_k(t)] = \langle \hat{M}_k^\dagger(dt)\hat{M}_k(dt)\rangle_t = dt\langle \hat{L}_k^\dagger \hat{L}_k\rangle_t, \tag{2.151}$$

where $E[dN_k(t)]$ denotes the classical expectation value of the stochastic quantity $dN_k(t)$ over many realizations of the process. The quantities $dN_k(t)$ are stochastic numbers taking either the value zero or one when a jump of type k is detected (notice that no more than one increment $dN_k(t)$ can be one at the same time). Their classical expectation value coincides with the probability of detecting a quantum jump of type k, that is $\mathcal{P}_k$ in Eq. (2.146). Notice that this probability is of the order of the time-step, dt, and hence the stochastic sequence of jumps corresponds to a Poisson process.

The stochastic increments $dN_k(t)$ allow us to write the conditioned evolution of the system during the interval $[t, t+dt]$ as an Itô stochastic differential equation. Considering the increment $d\,|\psi\rangle \equiv |\psi\rangle_{t+dt} - |\psi\rangle_t$, and assuming that the system starts in a pure state $|\psi\rangle_0$ at the initial time t_0, we can construct the evolution as the sum of two components depending on the value of the increments $dN_k(t)$

$$\begin{aligned} d\,|\psi\rangle_t = &\; dt[1 - \sum_k dN_k(t)] \left(-\frac{i}{\hbar}\hat{H} + \frac{\langle \hat{R}\rangle_t - \hat{R}}{2} \right) |\psi\rangle_t \\ &+ \sum_k dN_k(t) \left(\frac{\hat{L}_k}{\sqrt{\langle \hat{L}_k^\dagger \hat{L}_k\rangle_t}} - \mathbb{1} \right) |\psi\rangle_t . \end{aligned} \tag{2.152}$$

This means that, when none of the increments are one, that is, if no jump occurs during dt, only the first term survives and the system evolves according to the smooth change in Eq. (2.148). On the other hand, if some of the stochastic increments takes the value 1, meaning that a jump of type k has been detected during dt, the evolution is just given by the corresponding sharp change in Eq. (2.149).

The above stochastic differential equation can be further simplified by using the rule $dN_k(t)dt = \mathcal{O}(dt^2)$, meaning that any of the stochastic increments are at least of order dt. We then obtain

$$\begin{aligned} d\,|\psi\rangle_t = &\; dt \left(-\frac{i}{\hbar}\hat{H} + \frac{\langle \hat{R}\rangle_t - \hat{R}}{2} \right) |\psi\rangle_t \\ &+ \sum_k dN_k(t) \left(\frac{\hat{L}_k}{\sqrt{\langle \hat{L}_k^\dagger \hat{L}_k\rangle_t}} - \mathbb{1} \right) |\psi\rangle_t . \end{aligned} \tag{2.153}$$

which is usually known as a Stochastic Schrödinger equation. The solutions of this equation are the so-called *quantum trajectories*, which can be fully characterized by the initial state $|\psi\rangle_0$, and the measurement record $\mathcal{R} = \{(k_1, t_1), (k_2, t_2), \ldots, (k_j, t_j), \ldots, (k_J, t_J)\}$ containing the type of the jumps k_j occurred during the evolution, together with the times t_j at which they have been detected.

The solutions of the stochastic Schrödinger equation (SSE) (2.153) can be constructed with the help of the measurement operators introduced in the previous section. Using the record $\mathcal{R}$ of the J jumps detected during the evolution between t_0 and t, we can calculate the probability of a trajectory to occur as

$$P(\mathcal{R}, t) = \langle \psi_0 | \hat{T}_{\mathcal{R}}^\dagger(t, t_0) \hat{T}_{\mathcal{R}}(t, t_0) | \psi_0 \rangle \tag{2.154}$$

where we defined the following *trajectory operator* generating the correct measurement record

$$\hat{T}_{\mathcal{R}}(t, t_0) \equiv \hat{U}_{\text{eff}}(t, t_J)\hat{L}_{k_J} \ \dots \ \hat{U}_{\text{eff}}(t_2, t_1)\hat{L}_{k_1}\hat{U}_{\text{eff}}(t_1, t_0). \tag{2.155}$$

In the expression of the trajectory operator above we introduced

$$\hat{U}_{\text{eff}}(t, s) \equiv \exp[-\frac{i}{\hbar}\left(\hat{H} - i\hbar\hat{R}\right)(t - s)] \tag{2.156}$$

which corresponds to a non-unitary *drift* operator governing the effective smooth evolution of the system when no jumps are detected. This operator can be obtained by concatenating the operators of the null evolution $\hat{M}_0$ in Eq. (2.142). Following this notation, the solution of the SSE (2.153) can be finally written as

$$|\psi\rangle_t(\mathcal{R}) = \frac{\hat{T}_{\mathcal{R}}(t, t_0)}{\sqrt{P(\mathcal{R}, t)}}|\psi\rangle_0 . \tag{2.157}$$

If we now take the classical average over trajectories, we recover the density operator describing the unconditional state evolution

$$E[|\psi\rangle\langle\psi|_t(\mathcal{R})] = \rho(t) \tag{2.158}$$

which obeys the same Markovian master equation in Lindblad form

$$\frac{d\rho(t)}{dt} = -\frac{i}{\hbar}[\hat{H}, \rho(t)]dt + \sum_k \hat{L}_k\rho(t)\hat{L}_k^\dagger - \frac{1}{2}\{\hat{L}_k^\dagger\hat{L}_k, \rho(t)\}. \tag{2.159}$$

It must be stressed that, despite we assumed the initial state of the system to be pure, the same results can be obtained for the case of an initial mixed state ρ_0. In such case the SSE derived in Eq. (2.153) transforms into a *stochastic master equation* (SME) of the form [73]

$$\begin{aligned} d\sigma_t = & -\frac{i}{\hbar}[\hat{H}, \sigma_t]dt - \left(\frac{1}{2}\{\hat{R}, \sigma_t\} - \langle\hat{R}\rangle_t\sigma_t\right)dt \\ & + \sum_k dN_k(t)\left(\frac{\hat{L}_k\sigma_t\hat{L}_k^\dagger}{\langle\hat{L}_k^\dagger\hat{L}_k\rangle_t} - \sigma_t\right), \end{aligned} \tag{2.160}$$

where now $\sigma_t = \hat{T}_{\mathcal{R}}(t, t_0)\sigma_0\hat{T}_{\mathcal{R}}(t, t_0)^\dagger/P(\mathcal{R}, t)$ being the probability of the trajectory $P(\mathcal{R}, t) = \text{Tr}[\hat{T}_{\mathcal{R}}(t, t_0)^\dagger\hat{T}_{\mathcal{R}}(t, t_0)\sigma_0]$, and $\rho(t) = E[\sigma_t]$. This is an alternative way to introduce a stochastic differential equation for the system evolution more suited to describe situations in which extra sources of noise are considered, arising e.g. by an inefficient detection of the apparatus [92].

2.5.3 Master Equation Unraveling

In the previous section we have seen that a continuous measurement scheme, expressed in terms of a CPTP map equipped with a Kraus representation leads to quantum trajectories generated by the stochastic differential equation (2.153), whose average converges to a Markovian master equation in Lindblad form. Here we will take the other way around, i.e. we start with a generic Lindblad master equation and develop the corresponding trajectory description.

Consider a quantum system weakly interacting with its surroundings and whose dynamical evolution is given by

$$\frac{d\rho(t)}{dt} = -\frac{i}{\hbar}[\hat{H}, \rho(t)]dt + \sum_{k=1}^{K} \hat{L}_k \rho(t) \hat{L}_k^\dagger - \frac{1}{2}\{\hat{L}_k^\dagger \hat{L}_k, \rho(t)\}, \tag{2.161}$$

$\hat{H}$ being the Hamilton operator of the open system. As long as the dynamics is CP, we can always split the evolution into an infinite sequence of CPTP maps, $\mathcal{E}$, governing the evolution along infinitesimal time intervals dt as defined in Eq. (2.140). In addition, the selection of a specific set of Kraus operators $\{\hat{M}_m(dt)\}_{m=1}^{M}$ for the map provides, following Sect. 2.1.2, a physical interpretation of the dynamics consisting in different random operations induced by the environment, and occurring with probabilities $p_m(dt) = \langle \hat{M}_m^\dagger(dt)\hat{M}_m(dt)\rangle_t$. We call this selection of the Kraus operators for the maps an *unraveling* of the master equation (2.161), which is in turn related to a specific monitoring scheme of the environment generating the quantum trajectories (see Sect. 2.1.3).

One option is to choose a Kraus representation for the map consisting of $M = K + 1$ operators, $\hat{M}_0(dt)$ and $\{\hat{M}_k(dt)\}_{k=1}^{K}$ as the ones introduced in Eqs. (2.142) and (2.143). This will of course give us the SSE in Eq. (2.153), or equivalently the SME in Eq. (2.160) if the initial state of the system is not considered to be pure. However, exploiting the symmetries of the master equation (2.161) may lead to different unravelings corresponding to other measurement schemes. In Sect. 2.2.1 we have seen that any master equation in Lindblad form is invariant under the transformation (2.30), which leads to rewrite Eq. (2.161) as

$$\frac{d\rho(t)}{dt} = -\frac{i}{\hbar}[\hat{H}', \rho(t)]dt + \sum_{k=1}^{K} \hat{L}'_k \rho(t) \hat{L}'^\dagger_k - \frac{1}{2}\{\hat{L}'^\dagger_k \hat{L}'_k, \rho(t)\}, \tag{2.162}$$

with the new operators

$$\hat{H}' = \hat{H} - \frac{i\hbar}{2}\sum_{k=1}^{K}\left(l_k^* \hat{L}_k - l_k \hat{L}_k^\dagger\right) + \hbar r, \quad \hat{L}'_k = \hat{L}_k + l_k, \tag{2.163}$$

for l_k arbitrary complex coefficients, and r a real number. Applying now the identification in Eqs. (2.142) and (2.143) for the new defined Hamiltonian and Lindblad operators we obtain:

$$\hat{M}_0(dt) = \mathbb{1} - dt\left(\frac{i}{\hbar}(\hat{H} + \hbar r) + \frac{1}{2}\sum_k \hat{L}_k^\dagger \hat{L}_k + 2\hat{L}_k l_k^* + |l_k|^2\right),$$
$$\hat{M}_k(dt) = \sqrt{dt}\left(\hat{L}_k + l_k\right). \tag{2.164}$$

The corresponding probabilities for the operations to occur are then

$$\mathcal{P}_0(dt) = 1 - dt\sum_k \langle \hat{L}_k^\dagger \hat{L}_k + |l_k|\hat{X}_k + |l_k|^2\rangle_t, \tag{2.165}$$

$$\mathcal{P}_k(dt) = dt\, \langle \hat{L}_k^\dagger \hat{L}_k + |l_k|\hat{X}_k + |l_k|^2\rangle_t, \tag{2.166}$$

where we defined $\hat{X}_k \equiv \hat{L}_k e^{-i\varphi_k} + \hat{L}_k^\dagger e^{i\varphi_k}$ by expressing $l_k = |l_k|e^{i\varphi_k}$. Following the same lines as in the previous sections we can obtain a new SSE in the form

$$d\,|\psi\rangle_t = dt\left(-\frac{i}{\hbar}(\hat{H} + \hbar r) + \frac{\langle \hat{R}\rangle_t - \hat{R}}{2} + \sum_k |l_k|(\frac{\langle \hat{X}_k\rangle}{2} - \hat{L}_k)\right)|\psi\rangle_t$$
$$+ \sum_k dN_k(t)\left(\frac{\hat{L}_k + l_k}{\sqrt{\langle \hat{L}_k^\dagger \hat{L}_k + |l_k|\hat{X}_k + |l_k|^2\rangle_t}} - \mathbb{1}\right)|\psi\rangle_t\,. \tag{2.167}$$

where again $\hat{R} \equiv \sum_k \hat{L}_k^\dagger \hat{L}_k$, and the stochastic increments obey the same properties as in Eq. (2.152) with the replacement $\hat{L}_k \to \hat{L}_k + l_k$.

Equation (2.167) describes different trajectories from the previously derived SSE (2.153) for the same open system dynamics as given by the master equation (2.161). This shows how the quantum trajectory formalism depends on the details of the environmental modeling, which may be interpreted in terms of different protocols to monitor the system evolution. In the above general scheme, we may identify the observable being measured in the open system by looking at the classical expectation value for the stochastic increments [11]

$$\frac{E[dN_k(t)]}{dt} = \langle \hat{L}_k^\dagger \hat{L}_k + |l_k|\hat{X}_k + |l_k|^2\rangle_t \tag{2.168}$$

describing the "clicks" of the detector in the continuous measurement interpretation. This becomes clearer for the damped cavity field mode. In such case the transformation $\hat{L}_1' = \hat{L}_1 + l = \sqrt{\gamma_0}(\hat{a} + \alpha)$ where $\alpha = |\alpha|e^{i\varphi}$, can be seen as a displacement of the cavity field mode. This can be performed by means of *homodyne detection* of the cavity output light, which consists in mixing of the output light with a strong coherent field $|\alpha\rangle$ in a low-reflectivity beam-splitter before it is detected [11]. In such case the output signal would be

$$E[dN_1(t)]/dt = \gamma_0(\langle\hat{a}^\dagger\hat{a}\rangle_t + |\alpha|^2) + \sqrt{2\gamma_0}|\alpha|\langle\hat{X}_\varphi\rangle_t, \tag{2.169}$$

where $|\alpha|^2 \gg \langle\hat{a}^\dagger\hat{a}\rangle_t$, that is, the detector measures a constant signal $\gamma_0|\alpha|^2$ plus a second term proportional to the field quadrature in the φ direction $\langle\hat{X}_\varphi\rangle_t$ (see Sect. 1.2.4) and a very small contribution from direct detection of the photons in the cavity, $\gamma_0\langle\hat{a}^\dagger\hat{a}\rangle_t$.

Turning to the general case, it is also important to stress that, when $|l_k| \to \infty$, the expected number of jumps $E[dN_k(t)] = \mathcal{P}_k(t)$ in the time interval dt becomes very large, allowing us to obtain a diffusive limit by approximating the Poisson process by a Gaussian (or Wiener) process. Following Ref. [4] we can formally perform this approximation by defining new stochastic processes $W_k(t)$ for each type of jump

$$|l_k|\, dW_k(t) \equiv dN_k(t) - E[dN_k(t)] \tag{2.170}$$

which correspond to *Wiener* stochastic increments obeying

$$dW_k(t)^2 = dt, \qquad E[dW_k(t)] = 0. \tag{2.171}$$

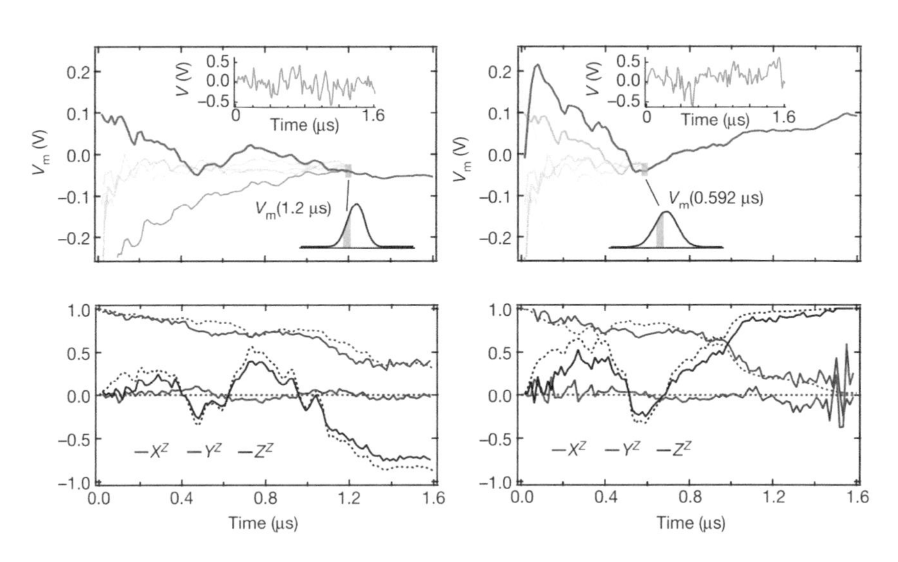

Fig. 2.8 Results from monitorization of a superconducting charge (transmon) qubit coupled to a cooper waveguide cavity. In the top panels individual integrated measurement signals of the amplified cavity field quadrature V_m (green lines) and other measurement traces used to reconstruct the state tomographically (lighter colors) are shown. The inset shows the instantaneous measurement voltage, and the gray region shows the standard deviation. The lower panels display quantum trajectories of the qubit obtained from the measurement analysis (dotted lines) and tomographically reconstructed (solid lines). Left and central columns correspond to Z-measurements in the qubit while the right column has been obtained using ϕ-measurements. Picture from Ref. [89]

Replacing it in the SSE (2.167) and assuming l_k real for simplicity, we obtain in the limit $l_k \to \infty$ the diffusive stochastic Schrödinger equation

$$d\,|\psi\rangle_t = dt\left[-\frac{i}{\hbar}(\hat{H}+\hbar r) - \frac{\hat{R}}{2} + \sum_k \frac{\langle \hat{X}_k\rangle_t}{2}\left(\hat{L}_k - \frac{\langle \hat{X}_k\rangle_t}{2}\right)\right]|\psi\rangle_t$$
$$+ \sum_k dW_k(t)\left(\hat{L}_k - \frac{\langle \hat{X}_k\rangle_t}{2}\right)|\psi\rangle_t\,, \tag{2.172}$$

from which the master equation (2.161) can be again derived by taking the classical expectation value $\rho(t) = E[|\psi\rangle\,\langle\psi|_t]$ and using the properties (2.171). In the continuous limit the point processes defined by counting the different jumps during the evolution are transformed in the monitoring of continuous signals with white noise [11]

$$J_k(t) \equiv \lim_{|l_k|\to\infty} \frac{dN_k(t) - |l_k|^2 dt}{|l_k| dt} = \langle \hat{X}_k\rangle_t + \xi_k(t), \tag{2.173}$$

where $\xi_k(t) = dW_k(t)/dt$ is the white noise term. This corresponds to a continuous measurement of the quantity $\hat{X}_k$. In Fig. 2.8 we show an example of the generated quantum trajectories for the case of continuous monitoring of a qubit system inside a cavity through the detection of one of the cavity field quadratures in a recent experiment [89].

References

1. A. Rivas, S.F. Huelga, *Open Quantum Systems: An Introduction* (Springer, Berlin, 2012)
2. M.O. Scully, M.S. Zubairy, *Quantum Optics* (Cambridge University Press, Cambridge, 1997)
3. H. Carmichael, *An Open Systems Approach to Quantum Optics* (Springer, Berlin, 1993)
4. C. Gardiner, P. Zoller, *Quantum Noise*, 3rd edn. (Springer, Berlin, 2004)
5. U. Weiss, *Quantum Dissipative Systems* (World Scientific, Singapore, 2008)
6. M.A. Nielsen, I.L. Chuang, *Quantum Computation and Quantum Information* (Cambridge University Press, Cambridge, 2000)
7. E.B. Davies, *Quantum Theory of Open Systems* (Academic Press, London, 1976)
8. G. Lindblad, On the generators of quantum dynamical semigroups. Commun. Math. Phys. **48**, 119–130 (1976)
9. V. Gorini, A. Kossakowski, E.C.G. Sudarshan, Completely positive dynamical semigroups of N-level systems. J. Math. Phys. **17**, 821–825 (1976)
10. H.-P. Breuer, F. Petruccione, *The Theory of Open Quantum Systems* (Oxford University Press, New York, 2002)
11. H.M. Wiseman, G.J. Milburn, *Quantum Measurement and Control* (Cambridge University Press, Cambridge, 2010)
12. G. Lindblad, Completely positive maps and entropy inequalities. Commun. Math. Phys. **40**, 147–151 (1975)
13. T. Sagawa, Second law-like inequalities with quantum relative entropy: an introduction, in *Lectures on Quantum Computing, Thermodynamics and Statistical Physics*, ed. by M. Nakahara. Kinki University Series on Quantum Computing, vol. 8 (World Scientific, New Jersey, 2013)

14. K. Kraus, A. Böhm, J.D. Dollard, W.H. Wootters, *States, Effects, and Operations: Fundamental Notions of Quantum Theory*, Lecture Notes in Physics (Springer, Berlin, 1983)
15. D. Salgado, J.L. Sánchez-Gómez, M. Ferrero, Evolution of any finite open quantum system always admits a Kraus-type representation, although it is not always completely positive. Phys. Rev. A **70**, 054102 (2004)
16. D.M. Ton, L.C. Kwek, C.H. Oh, J.-L. Chen, L. Ma, Operatorsum representation of time-dependent density operators and its applications. Phys. Rev. A **69**, 054102 (2004)
17. W.F. Stinespring, Positive functions on C*-algebras. Proc. Am. Math. Soc. **6**, 211–316 (1955)
18. A.S. Holevo, V. Giovannetti, Quantum channels and their entropic characteristics. Rep. Prog. Phys. **75**, 046001 (2012)
19. R. Srikanth, S. Banerjee, Squeezed generalized amplitude damping channel. Phys. Rev. A **77**, 012318 (2008)
20. H.-P. Breuer, E.-M. Laine, J. Piilo, Measure for the degree of non-Markovian behavior of quantum processes in open systems. Phys. Rev. Lett. **103**, 210401 (2009)
21. A. Rivas, S.F. Huelga, M.B. Plenio, Quantum non-Markovianity: characterization, quantification and detection. Rep. Prog. Phys. **77**, 094001 (2014)
22. H.-P. Breuer, E.-M. Laine, J. Piilo, B. Vacchini, Colloquium: non-Markovian dynamics in open quantum systems. Rev. Mod. Phys. **88**, 021002 (2016)
23. S. Haroche, J.M. Raimond, *Exploring the quantum: atoms, cavities and photons, Oxford graduate texts* (Oxford University Press, Oxford, 2006)
24. W.P. Schleich, *Quantum Optics in Phase Space* (Wiley-VCH, Berlin, 2001)
25. H. Spohn, An algebraic condition for the approach to equilibrium of an open N-level system. Lett. Math. Phys. **2**, 33–38 (1977)
26. E.T. Jaynes, F.W. Cummings, Comparison of Quantum and semiclassical radiation theories with applications to the beam maser. Proc. IEEE **51**, 89–109 (1963)
27. H. Paul, Induzierte Emission bei starker Einstrahlung. Ann. Phys. (Leipzig) **11**, 411–412 (1963)
28. V. Scarani, M. Ziman, P. Stelmachovic, N. Gisin, V. Buzek, Thermalizing quantum machines: dissipation and entanglement. Phys. Rev. Lett. **88**, 097905 (2002)
29. B.M. Terhal, D.P. Di Vincenzo, Problem of equilibration and the computation of correlation functions on a quantum computer. Phys. Rev. A **61**, 022301 (2000)
30. M. Schlosshauer, *Decoherence and the Quantum-to-Classical Transition* (Springer, Berlin, 2008)
31. T. Brandes, Quantum dissipation, in *Lectures on Background to Quantum Information*, UMIST-Bradford Lectures on Background to Quantum Information Theory (2003)
32. X.L. Huan, T. Wang, X.X. Yi, Effects of reservoir squeezing on quantum systems and work extraction. Phys. Rev. E **86**, 051105 (2012)
33. G. Manzano, F. Galve, R. Zambrini, J.M.R. Parrondo, Entropy production and thermodynamic power of the squeezed thermal reservoir. Phys. Rev. E **93**, 052120 (2016)
34. A.O. Caldeira, A.J. Leggett, Path integral approach to quantum Brownian motion. Phys. A **121**, 587–616 (1983)
35. P. Hänggi, G.-L. Ingold, Fundamental aspects of quantum Brownian motion. Chaos **15**, 026105 (2005)
36. R.H. Dicke, Coherence in spontaneous radiation processes. Phys. Rev. **93**, 99–110 (1954)
37. M. Gross, S. Haroche, Superradiance: an essay on the theory of collective spontaneous emission. Phys. Rep. **93**, 301–396 (1982)
38. G.M. Palma, K.-A. Suominen, A.K. Ekert, Quantum computers and dissipation. Proc. R. Soc. Lond. A **452**, 567 (1996)
39. S.-B. Shim, M. Imboden, P. Mohanty, Synchronized oscillation in coupled nanomechanical oscillators. Science **316**, 95 (2007)
40. G. Heinrich, M. Ludwig, J. Qian, B. Kubala, F. Marquardt, Collective dynamics in optomechanical arrays. Phys. Rev. Lett. **107**, 043603 (2011)
41. M. Zhang, G.S. Wiederhecker, S. Manipatruni, A. Barnard, P. McEuen, M. Lipson, Synchronization of micromechanical oscillators using light. Phys. Rev. Lett. **109**, 233906 (2012)

42. C.A. Holmes, C.P. Meaney, G.J. Milburn, Synchronization of many nanomechanical resonators coupled via a common cavity field. Phys. Rev. E **85**, 066203 (2012)
43. M.A. Cazalilla, F. Sols, F. Guinea, Dissipation-driven quantum phase transitions in a Tomonaga-Luttinger liquid electrostatically coupled to a metallic gate. Phys. Rev. Lett. **97**, 076401 (2006)
44. M.J. Hartmann, F.G.S.L. Brandão, M.B. Plenio, Strongly interacting polaritons in coupled arrays of cavities. Nat. Phys. **2**, 849–855 (2006)
45. T. Prosen, I. Pižorn, Quantum phase transition in a far-from- equilibrium steady state of an XY spin chain. Phys. Rev. Lett. **101**, 105701 (2008)
46. S. Diehl, A. Tomadin, A. Micheli, R. Fazio, P. Zoller, Dynamical phase transitions and instabilities in open atomic many- body systems. Phys. Rev. Lett. **105**, 015702 (2010)
47. E.G. dalla Torre, E. Demler, T. Giamarchi, E. Altman, Quantum critical states and phase transitions in the presence of nonequilibrium noise. Nat. Phys. **6**, 806–810 (2010)
48. D. Poletti, J.-S. Bernier, A. Georges, C. Kollath, Interaction- induced impeding of decoherence and anomalous diffusion. Phys. Rev. Lett. **109**, 045302 (2012)
49. I. Buluta, F. Nori, Quantum simulators. Science **326**, 108–111 (2009)
50. I.M. Georgescu, S. Ashhab, F. Nori, Quantum simulation. Rev. Mod. Phys. **86**, 153–185 (2014)
51. J.T. Barreiro, M. Müller, P. Schindler, D. Nigg, T. Monz, M. Chwalla, M. Hennrich, C.F. Roos, P. Zoller, R. Blatt, An open-system quantum simulator with trapped ions. Nature **470**, 486–491 (2011)
52. Y. Lin, J.P. Gaebler, F. Reiter, T.R. Tan, R. Bowler, A.S. Sorensen, D. Leibfried, D.J. Wineland, Dissipative production of a maximally entangled steady state of two quantum bits. Nature (London) **504**, 415–418 (2013)
53. P. Schindler, M. Muller, D. Nigg, J.T. Barreiro, E.A. Martinez, M. Hennrich, T. Monz, S. Diehl, P. Zoller, R. Blatt, Quantum simulation of dynamical maps with trapped ions. Nat. Phys. **9**, 361–367 (2013)
54. G. Barontini, R. Labouvie, F. Stubenrauch, A. Vogler, V. Guarrera, H. Ott, Controlling the dynamics of an open many-body quantum system with localized dissipation. Phys. Rev. Lett. **110**, 035302 (2013)
55. R. Labouvie, B. Santra, S. Heun, H. Ott, Bistability in a driven-dissipative superfluid. Phys. Rev. Lett. **116**, 235302 (2016)
56. W.G. Unruh, Maintaining coherence in quantum computers. Phys. Rev. A **51**, 992–997 (1995)
57. A.K. Rajagopal, R.W. Rendell, Decoherence, correlation, and entanglement in a pair of coupled quantum dissipative oscillators. Phys. Rev. A **63**, 022116 (2001)
58. K. Zyczkowski, P. Horodecki, M. Horodecki, R. Horodecki, Dynamics of quantum entanglement. Phys. Rev. A **65**, 012101 (2001)
59. P. Zanardi, Dissipation and decoherence in a quantum register. Phys. Rev. A **57**, 3276–3284 (1998)
60. J.H. Reina, L. Quiroga, N.F. Johnson, Decoherence of quantum registers. Phys. Rev. A **65**, 032326 (2002)
61. R. Doll, M. Wubs, P. Hänggi, S. Kohler, Limitation of entanglement due to spatial qubit separation. Europhys. Lett. **74**, 547–553 (2006)
62. T. Zell, F. Queisser, R. Klesse, Distance dependence of entanglement generation via a bosonic heat bath. Phys. Rev. Lett. **102**, 160501 (2009)
63. D.P.S. McCutcheon, A. Nazir, S. Bose, A.J. Fisher, Longlived spin entanglement induced by a spatially correlated thermal bath. Phys. Rev. A **80**, 022337 (2009)
64. P. Zanardi, M. Rasetti, Noiseless quantum codes. Phys. Rev. Lett. **79**, 3306 (1997)
65. L.-M. Duan, G.-C. Guo, Preserving coherence in quantum computation by pairing quantum bits. Phys. Rev. Lett. **79**, 1953 (1997)
66. D.A. Lidar, I.L. Chuang, K.B. Whaley, Decoherence-free subspaces for quantum computation. Phys. Rev. Lett. **81**, 2594 (1998)
67. A. Beige, D. Braun, B. Tregenna, P.L. Knight, Quantum computing using dissipation to remain in a decoherence-free subspace. Phys. Rev. Lett. **85**, 1762–1765 (2000)
68. F. Galve, A. Mandarino, M.G.A. Paris, C. Benedetti, R. Zambrini, Microscopic description for the emergence of collective dissipation in extended quantum systems. Sci. Rep. **7**, 42050 (2017), arXiv:1606.03390

69. J. Dalibard, Y. Castin, K. Mølmer, Wave-function approach to dissipative processes in quantum optics. Phys. Rev. Lett. **68**, 580–583 (1992)
70. R. Dum, P. Zoller, H. Ritsch, Monte Carlo simulation of the atomic master equation for spontaneous emission. Phys. Rev. A **45**, 4879–4887 (1992)
71. C.W. Gardiner, A.S. Parkins, P. Zoller, Wave-function quantum stochastic differential equations and quantum-jump simulation methods. Phys. Rev. A **46**, 4363–4381 (1992)
72. K. Mølmer, Y. Castin, J. Dalibard, Monte Carlo wave-function method in quantum optics. J. Opt. Soc. Am. B **10**, 524–538 (1993)
73. H.M. Wiseman, G.J. Milburn, Interpretation of quantum jump and diffusion processes illustrated on the Bloch sphere. Phys. Rev. A **47**, 1652–1666 (1993)
74. M.B. Plenio, P.L. Knight, The quantum-jump approach to dissipative dynamics in quantum optics. Rev. Mod. Phys. **70**, 101–144 (1998)
75. A.J. Daley, Quantum trajectories and open many-body quantum systems. Adv. Phys. **63**, 77–149 (2014)
76. P. Marte, R. Dum, R. Taïeb, P.D. Lett, P. Zoller, Quantum wave function simulation of the resonance fluorescence spectrum from one-dimensional optical molasses. Phys. Rev. Lett. **71**, 1335–1338 (1993)
77. Y. Castin, K. Mølmer, Monte Carlo wave-function analysis of 3D optical molasses. Phys. Rev. Lett. **74**, 3772–3775 (1995)
78. H.J. Carmichael, Quantum trajectory theory for cascaded open systems. Phys. Rev. Lett. **70**, 2273–2276 (1993)
79. H. Ammann, R. Gray, I. Shvarchuck, N. Christensen, Quantum delta-kicked rotor: experimental observation of decoherence. Phys. Rev. Lett. **80**, 4111–4115 (1998)
80. E. Schrödinger, Are there quantum jumps? Part I. Br. J. Philos. Sci. **3**, 109–123 (1952)
81. E. Schrödinger, Are there quantum jumps? Part II. Br. J. Philos. Sci. **3**, 233–242 (1952)
82. W. Nagourney, J. Sandberg, H. Dehmelt, Shelved optical electron amplifier: observation of quantum jumps. Phys. Rev. Lett. **56**, 2797–2799 (1986)
83. J.C. Bergquist, R.G. Hulet, W.M. Itano, D.J. Wineland, Observation of quantum jumps in a single atom. Phys. Rev. Lett. **57**, 1699–1702 (1986)
84. T. Sauter, W. Neuhauser, R. Blatt, P.E. Toschek, Observation of quantum jumps. Phys. Rev. Lett. **57**, 1696–1698 (1986)
85. D.J. Wineland, Nobel lecture: superposition, entanglement, and raising Schrödinger's cat. Rev. Mod. Phys. **85**, 1103–1114 (2013)
86. S. Haroche, Nobel Lecture: controlling photons in a box and exploring the quantum to classical boundary. Rev. Mod. Phys. **85**, 1083 (2013)
87. A.N. Vamivakas, C.-Y. Lu, C. Matthiesen, Y. Zhao, S. Falt, A. Badolato, M. Atature, Observation of spin-dependent quantum jumps via quantum dot resonance fluorescence. Nature **467**, 297–300 (2010)
88. J.J. Pla, K.Y. Tan, J.P. Dehollain, W.H. Lim, J.J.L. Morton, F.A. Zwanenburg, D.N. Jamieson, A.S. Dzurak, A. Morello, High-fidelity readout and control of a nuclear spin qubit in silicon. Nature **496**, 334–338 (2013)
89. K.W. Murch, S.J. Weber, C. Macklin, I. Siddiqi, Observing single quantum trajectories of a superconducting quantum bit. Nature **502**, 211–214 (2013)
90. S.J. Weber, A. Chantasri, J. Dressel, A.N. Jordan, K.W. Murch, I. Siddiqi, Mapping the optimal route between two quantum states. Nature **511**, 570–573 (2014)
91. C. Guerlin, J. Bernu, S. Deléglise, C. Sayrin, S. Gleyzes, S. Kuhr, M. Brune, J.-M. Raimond, S. Haroche, Progressive field-state collapse and quantum non-demolition photon counting. Nature **448**, 889–893 (2007)
92. K. Jacobs, D.A. Steck, A straightforward introduction to continuous quantum measurement. Contemp. Phys. **47**, 279–303 (2006)

Chapter 3
Quantum Thermodynamics

In Chaps. 1 and 2 we provided an introduction to fundamental concepts of open quantum systems theory and discussed their dynamical description. As we have previously commented, open quantum systems theory constitutes nowadays a general framework which successfully contributes to the research in many branches dealing with quantum phenomena, such as quantum optics, condensed matter physics, chemical physics, or quantum information science. Its unified and modern view over the dynamics and properties of quantum systems, makes it, in addition, a prominent tool in the development of emerging fields such as quantum thermodynamics.

Quantum thermodynamics can be considered as a rapidly evolving research field, focusing on various thermodynamical aspects of quantum mechanical systems and processes in nonequilibrium situations [1]. Its main subject of study is the behavior of quantities such as heat, work, or entropy in microscopic quantum systems (including individual particles), where thermal and quantum fluctuations may compete, and quantum effects come into play. As an interdisciplinary field, it feeds from different communities and backgrounds, such as stochastic thermodynamics [2, 3], many-body physics [4] or quantum information theory [5], contributing to different issues from the characterization of thermalization and equilibration processes to the analysis of the performance of small quantum thermal machines, passing through the investigation of the link between information and thermodynamics.

The intimate connection between the laws of thermodynamics and their quantum origin comes backs to the very beginning of the quantum theory. Indeed, the consistency with thermodynamical laws was the key point to the introduction of Planck's law for blackbody ration in 1900 [6], and the subsequent quantization of the electromagnetic field in 1905 [7]. Since then, an intense theoretical activity has been devoted to study this connection, some examples being the pioneering introduction of heat engines based on the three-level maser [8, 9], or the different efforts in deriving thermodynamic behavior from quantum mechanics (see e.g. Refs. [10–12]). The interest on a thermodynamic analysis of quantum processes and features has redoubled in the last decade, boosted by the success of stochastic thermodynamics -dealing

G. Manzano Paule, *Thermodynamics and Synchronization in Open Quantum Systems*, Springer Theses, https://doi.org/10.1007/978-3-319-93964-3_3

with small (classical) systems out of equilibrium at the level of single trajectories-, quantum information theory, and the rapid development of experimental techniques for high-precision manipulation and control of quantum systems.

In this chapter we aim to provide an introduction to the emerging framework of quantum thermodynamics while presenting an over-view of some popular topics under current investigation. In particular, two of them will be of major importance to the research developed in this thesis: quantum fluctuation relations, and quantum thermal machines (corresponding to Parts III and IV of the thesis respectively). In both cases, the machinery of open quantum systems theory introduced in the previous chapters will provide us the necessary tools to extend the classical (stochastic) description of small systems to the quantum regime. This however requires to identify classical notions such as work or heat in a purely quantum mechanical scenario, a task which often has been proven difficult or controversial [13–15], as well as the incorporation of information into the nonequilibrium thermodynamic framework [16]. Furthermore, we will provide a brief overview about other current topics in quantum thermodynamics not directly addressed in this thesis, such as thermalization and equilibration processes, and the development of resource theories for quantum thermodynamics.

This chapter begins with a review of some important thermodynamical concepts in Sect. 3.1, such as the laws of thermodynamics, the stochastic description of thermodynamic processes, and the link between information and thermodynamics. Section 3.2 is dedicated to introduce the so-called *fluctuation theorems*, a set of universal relations governing the statistics of different thermodynamical quantities, originally introduced in the classical regime. We discuss the necessary elements commonly assumed in its extension to the quantum regime, and discuss some of the most important known results. In Sect. 3.3 we discuss different small quantum thermal machines composed by elementary quantum systems performing some useful thermodynamic task. We will split them using two main criteria: machines operating in cycles versus continuous operation devices, and machines requiring external driving versus autonomous thermal machines. Finally, in Sect. 3.4, we briefly introduce the reader to some other important topics in quantum thermodynamics: thermalization and equilibration of quantum isolated many-body systems, and the resource theory of thermal operations.

3.1 Principles of Thermodynamics

Thermodynamics was fist developed in the 19th century as a phenomenological theory concerning the relations between macroscopic observables, or *state variables*, such as volume V, pressure p, temperature T, or magnetization M, in systems composed by a large number of degrees of freedom (of the order of $N \sim 10^{24}$). These macroscopic systems are described within the theory by *state functions*, like the internal energy U or the thermodynamic entropy S_{th}, rather than by the Hamilton function (or operator in the quantum mechanical case), providing relations between macroscopic observables in equilibrium. The simplicity of the mathematical description used in thermodynamics is therefore in contrast with the complexity of the processes occurring at the atomic scale and the large number of degrees of freedom needed to describe a macroscopic object. The key point of this simple description comes from the large time required to perform macroscopic measurements, in comparison with the characteristic time scales at which molecular dynamics takes place, detecting only the average behavior of microscopic degrees of freedom through a reduced set of quantities [17].

Thermodynamic systems are usually classified in terms of the allowed exchange processes with their surroundings, the *boundaries* of the system. In this sense a system can be: *completely closed* if neither matter nor energy are exchanged, *closed* if it exchanges energy but not matter, and *open* if it exchanges both of them. Notice the difference with the classification introduced in open quantum systems theory, that considers a system to be *closed* if it is completely isolated, and *open* if it interacts with an environment. In general we will use this second classification during the thesis, while explicitly stressing the meaning of the thermodynamic classification when needed.

A particularly fundamental state of a system is its thermal *equilibrium state*, in which the number of independent state variables becomes a minimum. In such state the internal energy $U(T, V, N_k)$ and the thermodynamic entropy $S_{\text{th}}(T, V, N_k)$ are completely specified by the temperature T, the volume V, and the amounts of chemical constituents N_k of the system. The existence of equilibrium states is taken in thermodynamics as a fundamental fact of experience [12]: after some relaxation time any *completely closed* system (not exchanging energy nor matter with its surroundings) reaches an equilibrium state. This spontaneous evolution is due to *irreversible* processes, such as heat conduction or chemical reactions. Once it is reached, the state variables of a system do not change spontaneously any more, and no macroscopic flows of heat or matter are observed. If two or more systems interact exchanging energy and/or matter they will eventually reach an equilibrium state as defined above, sharing a uniform and common temperature. Furthermore, if two systems A and B are in thermal equilibrium, and a third system C is in thermal equilibrium with system A, then it follows that B and C are also in thermal equilibrium. This transitivity property of equilibrium states is the so-called zeroth law of thermodynamics [18].

Thermodynamic processes are defined via the change in time of one or more state variables of a system. Those changes are produced by both the interaction of the system with its surroundings, and the action of some external agent through the variation in time of some *control parameter* $\lambda(t)$. When the variation of the control parameter is infinitely slow, we say that the process is *quasi-static*, while if it takes a negligible amount of time to occur we say it is *instantaneous*. Furthermore, if no external action occurs, we say that the process is *spontaneous*. A fundamental classification of thermodynamic processes is the following. We call a *reversible process*, a process for which the state of the system is infinitesimally close to an equilibrium state with its surroundings during the entire evolution. In a reversible process, the system and its surroundings will be returned to their original states if the sequence of values adopted by the external parameter, i.e. the so-called *protocol*

$\Lambda = \{\lambda(t) \ \ ||t_{\mathrm{ini}} \leqslant t \leqslant t_{\mathrm{fin}}\}$, is reversed in time. In contrast, *irreversible processes* occurs when the state of the system departs from equilibrium during the time evolution, generating a permanent change in the environment even if the system turns back to its original state by reversing the protocol Λ. Reversible processes are implemented by starting with a thermodynamic system in equilibrium with its surroundings and implementing a quasi-static variation of the control parameter, such that the changes produced by the external agent make the system only depart infinitesimally from the equilibrium state. Intuitively, the system reaches equilibrium more quickly than the control parameter changes. However, notice that reversible processes just correspond to idealized hypothetical processes (implicitly assumed to always exist), but as long as any change of the control parameter occurs at a finite speed, all processes are truly irreversible. It is also worth stressing that, although all reversible processes are quasi-static, the converse it not always true [17].

Furthermore, depending on the type of contact of the system with its environment and the nature of the later, we may classify thermodynamic processes as:

- **Isothermal processes**: here system and the environment can exchange energy and matter but the temperature of the system is maintained constant during the process. Common examples of isothermal processes are melting, evaporation, or other phase changes when occurring at constant pressure.
- **Isochoric processes**: the system can exchange energy but no matter with its surroundings, and the process is performed at constant volume. This is the case e.g. of the heating or cooling of a liquid inside a closed container with non-zero thermal conduction.
- **Adiabatic processes**: in such processes the system cannot exchange energy nor matter with the environment, except for the work performed by the external agent (see below). Adiabatic processes sometimes occur when a physical process takes place so rapidly that there is no time enough to exchange energy with the environment. It is worth mentioning that the word *adiabatic* here has a different meaning than in standard quantum mechanics, where it reefers to a slow perturbation of an isolated quantum system, and is hence more related to the concept of a quasi-static process.

3.1.1 The First Law of Thermodynamics

While conservation of mechanical energy (kinetic plus potential) was well known at the beginning of the 19th century, it was only after the contributions of J.R. von Mayer, J. P. Joule, and H. von Helmholtz, that the equivalence of a plethora of phenomena (mechanical work, heat, electricity, magnetism and chemical transformations) was established through the unique concept of *energy* in the second half of the 19th century [18].

In a thermodynamic process as those introduced above, the changes in the internal energy ΔU of the system between initial and final equilibrium states, can be split into

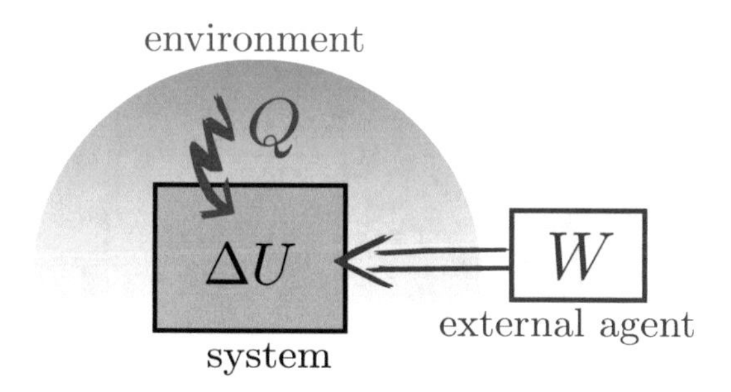

Fig. 3.1 Schematic diagram showing the meaning of the first law of thermodynamics. The changes in internal energy ΔU of the thermodynamic system can be decomposed into two energetic contributions, the work W coming from the control of an external agent (blue arrow), and heat Q exhausted by the environment in an uncontrollable way (red arrow)

two contributions, called *heat* and *work*, which represent two different and process-dependent sources of energy. The *heat* Q represents the energy introduced into the system in an uncontrolled way, associated to the energy exchanged with thermal baths or equilibrium reservoirs. On the other hand, the *work* W represents a controllable (and hence useful) energy source, which is associated to the action of an external agent on the system via a control parameter $\lambda(t)$. The first law of thermodynamics asserts the conservation of energy:

$$\Delta U = Q + W, \tag{3.1}$$

where Q and W, unlike ΔU, are process dependent. A schematic representation of the first law is depicted in Fig. 3.1.

In quantum thermodynamics, the first law can be stated by identifying the internal energy with the *average* energy of a quantum system, as defined by the expectation value of its (time-dependent) Hamiltonian $\hat{H}(\lambda(t))$. If the system is in a generic state $\rho(t)$, at an arbitrary time t, its internal energy reads [1]

$$U(t) \equiv \mathrm{Tr}[\rho(t)\hat{H}(\lambda(t))]. \tag{3.2}$$

Therefore, in any process starting at time t_A in state ρ_A with the control parameter at point λ_A, and ending at time t_B in state ρ_B with the control parameter in λ_B, the change in internal energy becomes

$$\Delta U = U_B - U_A = \mathrm{Tr}[\rho_B \hat{H}(\lambda_B)] - \mathrm{Tr}[\rho_A \hat{H}(\lambda_A)]. \tag{3.3}$$

This change in average energy can be then ascribed to heat and work contributions

$$Q \equiv \int_A^B \delta Q = \int_{t_A}^{t_B} \mathrm{Tr}\left[\frac{d\rho(t)}{dt}\, \hat{H}(\lambda(t))\right] dt, \tag{3.4}$$

$$W \equiv \int_A^B \delta W = \int_{t_A}^{t_B} \mathrm{Tr}\left[\rho(t)\, \frac{d\hat{H}(\lambda(t))}{dt}\right] dt, \tag{3.5}$$

where one recovers Eq. (3.1). In an open quantum system, the work contribution corresponds to the total change in the average energy introduced in the global system (open system + environment) by the external modification of the Hamilton operator. Indeed, by denoting $\rho_{\mathrm{tot}}(t)$ the total density operator at time t in the Schrödinger picture, and the total Hamiltonian as $\hat{H}_{\mathrm{tot}}(t) = \hat{H}(\lambda(t)) + \hat{H}_E + \hat{H}_{\mathrm{int}}$, with time-independent environmental and interaction terms, we have that the change in the total energy is

$$\frac{d}{dt}\mathrm{Tr}[\rho_{\mathrm{tot}}(t)\, \hat{H}_{\mathrm{tot}}(t)] = \mathrm{Tr}\left[\rho_{\mathrm{tot}}(t)\, \frac{d\hat{H}_{\mathrm{tot}}(t)}{dt}\right] = \dot{W}, \tag{3.6}$$

where the first equality follows from $\mathrm{Tr}[\hat{H}_{\mathrm{tot}}\dot{\rho}] = 0$ by exploiting the Liouville–von Neumann equation (1.5) (see Sect. 1.1) and the second one by performing the partial trace over the environmental degrees of freedom. Analogously, an interpretation of the heat Q in Eq. (3.4) can be obtained from $\mathrm{Tr}[\hat{H}_{\mathrm{tot}}\dot{\rho}] = 0$, as it implies

$$\dot{Q} = -\mathrm{Tr}_E[\hat{H}_E\dot{\rho}_E] - \mathrm{Tr}[\hat{H}_{\mathrm{int}}\dot{\rho}_{\mathrm{tot}}], \tag{3.7}$$

and the heat can be interpreted as the energy transferred from the environment (first term) and interaction degrees of freedom (second term). Notice that in the common situation of weak coupling between system and environment the second term can be neglected.

As can be seen from definitions (3.3)–(3.5), the change in internal energy only depends on the initial and final states of the system, while work and heat are process dependent, that is, they depend on the specific path followed by the control parameter from λ_A to λ_B. Henceforth in an infinitesimal process the exact differential $dU = \delta Q + \delta W$, splits into two terms, δQ and δW, which are not, in general, exact differentials [1]. This implies that in a cyclic process in which $\rho_B = \rho_A$, and $\hat{H}(\lambda_B) = \hat{H}(\lambda_A)$, we always have $\Delta U = 0$, but work and heat are in general non-zero, fulfilling $W_{\mathrm{cyc}} = -Q_{\mathrm{cyc}}$.

3.1.2 *The Second Law of Thermodynamics*

The origins of the second law of thermodynamics go back to the pioneering analysis of Sadi Carnot on the power and efficiency of heat engines in 1824 [19], motivated by the spread of the steam engine during the industrial revolution. Carnot identified the flow of heat as a fundamental process required for the generation of work [18]. He introduced the condition of maximum work extraction of an idealized heat engine operating cyclically between a difference of temperatures. This condition corresponds to the case in which all the operations performed by the machine are *reversible*. Under such conditions, assuming no further leaks of heat, a cyclic operation of the machine may reach a maximum efficiency given by

$$\frac{W}{Q_{\text{hot}}} \leqslant \eta_{\text{C}} = 1 - \frac{T_{\text{cold}}}{T_{\text{hot}}}, \tag{3.8}$$

where W refers to the work extracted in a cycle, Q_{hot} is the heat absorbed from the higher temperature reservoir at T_{hot} in the cycle, and $T_{\text{cold}} \leqslant T_{\text{hot}}$ is the temperature of the cold reservoir acting as a heat sink. *Carnot efficiency* establishes a fundamental limitation on the performance of ideal heat engines solely based on the temperature ratio of the reservoirs and independent of the specific model of the engine. However, it should be pointed out that Carnot's analysis was based on the old theory of heat, which considered heat as an indestructible quantity not fulfilling the first law. Its work, while passing mostly unnoticed during decades, was essential to the later introduction of the concept of (thermodynamic) entropy [20] and the formalization of the second law of thermodynamics by Rudolf Clausius and Lord Kelvin (see e.g. Ref. [18]).

As we have seen before, there are many ways to operate a thermodynamic process from some equilibrium state A to some other equilibrium state B. Each of these different paths will provide different values for the work and the heat entering the system during the process. Among all those paths there always exists at least one idealized path, corresponding to a reversible process, for which, as stated before, the system is always infinitesimally close to an equilibrium state. Indeed there exist in general infinitely many reversible paths connecting the two states A and B. The existence of reversible paths leads us to define the thermodynamic entropy, a concept as fundamental and universal as energy, entering the theory as a state function. It has been first defined through its change in a reversible process by Clausius in 1865 [20]

$$\Delta S_{\text{th}} \equiv \int_A^B \frac{\delta Q_{\text{rev}}}{T}, \quad \text{or} \quad \oint \frac{\delta Q_{\text{rev}}}{T} = 0, \tag{3.9}$$

where $\Delta S_{\text{th}} = S_{\text{th}}^B - S_{\text{th}}^A$ is the change in entropy from point A to B, and δQ_{rev} is the heat absorbed by the system in a differential step of the process when it is reversible. Recall that the thermodynamic entropy is a quantity that only depends on the initial and final states of the process and not on the specific reversible path followed. Then the quantity $dS = \delta Q_{\text{rev}}/T$ becomes an exact differential. Notice that if the process is, in addition, isothermal, we just obtain $\Delta S_{\text{th}} = Q/T$.

However, a generic thermodynamic process does not necessarily follow a reversible path [see Fig. 3.2a]. In such situation we have $dS \geqslant \delta Q/T$, or equivalently:

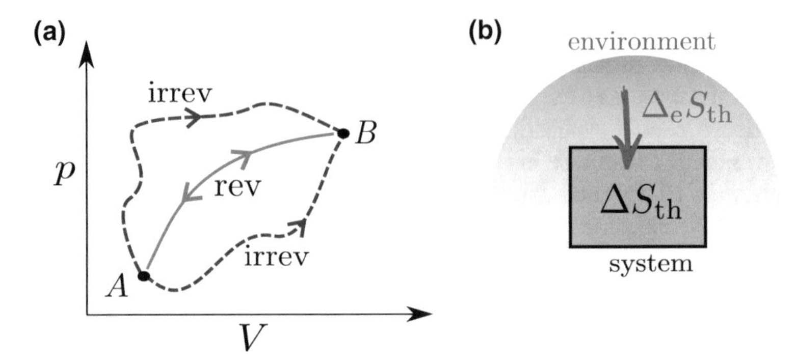

Fig. 3.2 **a** Schematic pressure-volume diagram for processes connecting the two equilibrium states A and B. A reversible process is depicted as the solid green line, while two irreversible processes are represented by the red dashed lines. **b** Schematic picture of the thermodynamic entropy equality (3.11). The changes in thermodynamic entropy of the system, ΔS_{th}, are viewed in part as coming from the environment through energy and matter exchange. The entropy production hence refers to the increase in thermodynamic entropy of the system not accounted by the flow $\Delta_e S_{\text{th}}$

$$\Delta S_{\text{th}} \geqslant \int_A^B \frac{\delta Q}{T}, \qquad \text{or} \qquad \oint \frac{\delta Q}{T} \leqslant 0. \tag{3.10}$$

The above inequalities are usually referred to as the *Clausius inequality* for arbitrary and cyclic processes respectively. Again, for isothermal processes the expressions simplify to $\Delta S_{\text{th}} \geqslant Q/T$. We stress that the equality case is only fulfilled for a reversible process.

A modern formulation of the second law is expressed by introducing a split of ΔS_{th} in Eq. (3.9) into two terms [18]:

$$\Delta S_{\text{th}} = \Delta_i S_{\text{th}} + \Delta_e S_{\text{th}}, \tag{3.11}$$

which are respectively called *entropy production* and *entropy flow*. The entropy flow term $\Delta_e S_{\text{th}}$ corresponds to the changes in entropy of the system due to the exchange of energy and matter with the surroundings [see Fig. 3.2b]. For a system which only exchanges energy in a process connecting the equilibrium states A and B it reads:

$$\Delta_e S_{\text{th}} \equiv \int_A^B \frac{\delta Q}{T}, \tag{3.12}$$

which can be either positive or negative (it has positive sign when heat flows into the system). On the other hand, the entropy production term is the change in entropy of the system due to irreversible processes within the system. This second contribution is non-negative by virtue of the second law (3.10):

$$\Delta_i S_{\text{th}} = \Delta S_{\text{th}} - \int_A^B \frac{\delta Q}{T} \geqslant 0. \tag{3.13}$$

Notice that this equality substitutes the second law inequality. It was first introduced by Clausius who referred to $\Delta_i S_{th}$ as the so-called *uncompensated heat* [20]. In a cyclic process for which the net change of entropy is zero, we have $\Delta_i S_{th} = -\oint \frac{\delta Q}{T} \geqslant 0$, meaning that for the system to return to its initial state, the entropy produced during the irreversible process has to be discarded through the expulsion of heat to the surroundings, hence increasing the entropy of the environment, or more generally *the universe*. Indeed, when the environment is considered to be a thermal reservoir, which maintains its state even if it exchanges energy or matter with the system, the entropy flow term is identified with the entropy decrease in the environment, i.e. $\Delta_e S_{th} = -\Delta S_{th}^E$. Consequently, following Eq. (3.13), the entropy production can be identified with the sum of the thermodynamic entropy changes in system and surroundings, $\Delta_i S_{th} = \Delta S_{th} + \Delta S_{th}^E \geqslant 0$. This leads to the formulation of the second law as the non-decrease of the sum of the changes in entropies of a system and its exterior, or, as summarized by Clausius: "The entropy of the universe approaches a maximum" [18]. Notice the importance of this statement as it leads to the distinction between future and past, and hence the existence of an *arrow of time*.

3.1.3 Statistical Mechanics and Entropy

While 19th century thermodynamics focused on equilibrium properties and reversible transformations, being principally a theory of states, in the 20th century nonequilibrium processes were progressively analyzed and incorporated into the theory [18]. Using the concepts of entropy production and entropy flows, the predictable power of thermodynamics was extended to the description of irreversible process, from thermoelectric phenomena to dissipative structures such as convection patterns in fluids. The development of non-equilibrium thermodynamics has been made possible thanks to the introduction of statistical mechanics, which established a connection between thermodynamic properties and molecular dynamics, providing a physical interpretation of the concept of entropy [17]. Furthermore, the random motion of the molecules induces fluctuations in all thermodynamic quantities, and the interaction with the surroundings makes the system to be continuously subjected to perturbations, leading naturally to work with probabilities. These fluctuations are extremely small in macroscopic equilibrium systems ($\sim 1/\sqrt{N}$) and they can be neglected in normal circumstances. However, this is no longer the case e.g. when approaching the critical point in systems presenting phase transitions, when the size of the systems under description scales down, or when the motion of single molecules can be observed by an intelligent being able to use this information (Maxwell's demon).

The introduction of statistical mechanics comes back to the kinetic theory of gases, created to explain the equilibrium properties of dilute gases from their underlying molecular dynamics. In this formalism the basic concept is the probability distribution, $\rho(x, v, t)$, to find a molecule at position x, with velocity v at time t. When assuming only binary collisions between the gas molecules, this distribution adopts the form of the well-known *Maxwell–Boltzmann distribution* as obtained by

Boltzmann in 1871. In doing so, he also considered the probability $\rho(E)$, to find a molecule with energy E, obtaining

$$\rho(E) \propto \vartheta(E)e^{-E/k_BT}, \tag{3.14}$$

where $\vartheta(E)$, called the *density of states*, is the number of different states in which the molecule has energy E [18]. Nevertheless, the scope of statistical mechanics is broader, being a central topic the explanation of the concept of entropy [17]. In a macroscopic system there exist many microscopic states compatible with the few macroscopic parameters used as state variables in its thermodynamic description. This feature is valid both in classical and quantum frameworks. In an isolated system, it might seem at a first sight that, if the system is in one of such particular states, this state should be maintained forever, evolving according to the Schrödinger equation. However, as we have pointed in Chaps. 1 and 2, no physical system is truly isolated. Even more, if the system is macroscopic, the energy differences between quantum states become extremely small, and the inner interactions will induce transitions between the quantum states [17].

Within classical statistical mechanics this complicated plethora of phenomena is simplified by assuming rapid random microscopic transitions, and takes as a *fundamental postulate* the assignment of equal probabilities to all the permissible *microstates*, as specified by the positions and velocities of all particles in the system $(x, p) \equiv \{\vec{x}_i, \vec{p}_i\}_{i=1}^N$, compatible with a given *macrostate*, that is, the state defined by the macroscopic variables like internal energy U, the volume V or number of particles N. This applies for systems that do not exchange energy or matter with the exterior. The second basis of statistical mechanics is the identification of thermodynamic entropy with the (logarithm of the) number of available microstates $\Omega(U, V, N)$, also due to Boltzmann

$$S_{\text{th}} = k_B \ln \Omega(U, V, N), \tag{3.15}$$

where the Boltzmann constant, k_B, ensures the agreement with the Kelvin scale of temperature [17]. This implies that in the equilibrium state, $\Omega(U, V, N)$ is maximum. This is usually considered as the second fundamental postulate of statistical mechanics.

Statistical mechanics aims to explain macroscopic variables by averaging over ensembles of microscopic states. The probability of the system to be within $[x, x + dx]$ and $[p, p + dp]$ at some particular time reads

$$P(x + dx, p + dp, t) = dxdp\ \rho(x, p, t) \tag{3.16}$$

where $\rho(x, p, t)$ is known as the *phase space density* of the system at time t, and $dxdp$ is a *phase-space* cell. The internal energy of the system can be defined as the ensemble average of the Hamiltonian of the microscopic system $H(x, p)$

$$U = \int dxdp\ \rho(x, p, t) H(x, p), \tag{3.17}$$

to be compared with Eq. (3.2). The number of microstates $\Omega(U, V, N)$ can be hence related to a hypersurface in phase space with energy $H(x, p) = U$ in a constrained region of volume V [21]. Analogously to the internal energy, one can define ensemble averages for any suitable microscopic observable $O(x, q)$ in correspondence to a macroscopic quantity O.

Importantly, Gibbs expressed the entropy of a system as the average of the quantity $s(x, p, t) \equiv -k_B \ln \rho(x, p, t)$, that is

$$\begin{aligned} S(x, p, t) &= \int dxdp\ \rho(x, p, t) s(x, p, t) \\ &= -k_B \int dxdp\ \rho(x, p, t) \ln \rho(x, p, t). \end{aligned} \tag{3.18}$$

Notice that the von Neumann's definition of entropy constitutes an extension of the Gibbs entropy to the quantum case. It is worth stressing here that macroscopic phenomenological thermodynamics can only correctly describe the macrostate of a system when it is in equilibrium, as nonequilibrium states cannot in general be described by a small set of macroscopic measurable variables, henceforth Eq. (3.15) can only describe the thermodynamic entropy in equilibrium states. In contrast, Gibbs entropy can be defined as well for nonequilibrium states, reducing to Boltzmann's expression [Eq. (3.15)] when we assign equal probabilities to all microstates.

3.1.4 Helmholtz and Nonequilibrium Free Energy

A third way to state the second law of thermodynamics in Eq. (3.10) comes from the introduction of the (equilibrium) *Helmholtz free energy* of a system in equilibrium at temperature T, defined as $F \equiv U - TS_{\text{th}}$. In this case the second law establishes a bound on the work that can be extracted in any isothermal process connecting two equilibrium states

$$T\Delta_{\text{i}} S_{\text{th}} = W - \Delta F \geqslant 0, \qquad \text{or} \quad W_{\text{ext}} \equiv -W \leqslant -\Delta F, \tag{3.19}$$

where we have used Eq. (3.9) for constant temperature T, and the first law $Q = \Delta U - W$. This expression can be further extended to nonequilibrium initial and final states, which makes it particularly important for the development of information and quantum thermodynamics [16]. In particular, in the quantum regime we introduce the *nonequilibrium free energy* of a quantum system with Hamiltonian $\hat{H}$ in state ρ, with respect to a thermal reservoir at temperature T as

$$\mathcal{F}(\rho, \hat{H}, T) \equiv U - k_B T\ S(\rho), \tag{3.20}$$

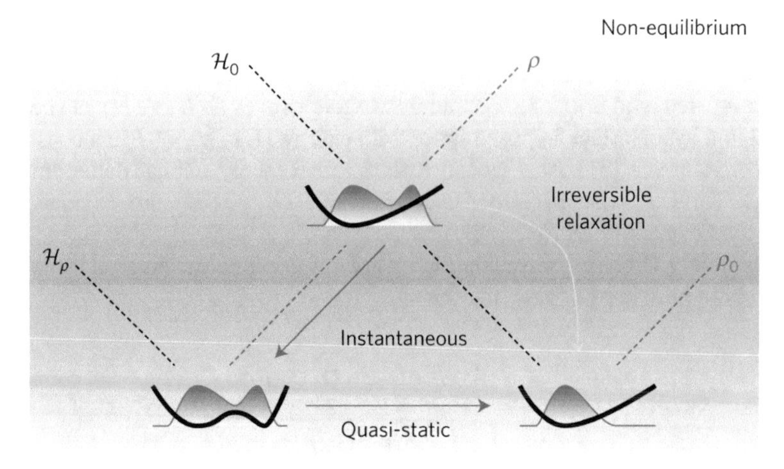

Fig. 3.3 The nonequilibrium state ρ can be transformed into a thermal equilibrium state ρ_0 while extracting a maximal amount of work as given by Eq. (3.21). A protocol which accomplish this task consists of a first instantaneous quench of the Hamilton operator $\hat{H}_0 \to \hat{H}_\rho \equiv k_B T \ln \rho$, followed by a quasi-static isothermal transformation which returns the Hamilton operator back to $\hat{H}_0$ (in the quantum case this will be described by an adiabatic Markovian master equation [24]). This is in contrast to the irreversible relaxation $\rho \to \rho_0$ occurring by directly putting the system in contact with the thermal reservoir. Picture taken from Ref. [16]

where again $U = \mathrm{Tr}[\rho \hat{H}]$ is the (average) internal energy of the quantum system, and $S = -\mathrm{Tr}[\rho \ln \rho]$ is the von Neumann entropy introduced in Sect. 1.1.6. We notice that the nonequilibrium free energy in Eq. (3.20) can be defined as well for classical systems by replacing the von Neumann entropy by the Shannon entropy [16]. It has been shown that the maximum work that can be extracted from a quantum system starting in an arbitrary nonequilibrium state ρ, with Hamiltonian $\hat{H}_0$, by using a thermal reservoir at temperature T, is bounded by the nonequilibrium free energy change [16, 22, 23]

$$W_{\mathrm{ext}} \leqslant \mathcal{F}(\rho, \hat{H}_0, T) - \mathcal{F}(\rho_0, \hat{H}_0, T), \tag{3.21}$$

leaving the system at the end of the process in state ρ_0 at thermal equilibrium with the reservoir. In Fig. 3.3 a generic way of performing this operation is illustrated.

The second law of thermodynamics for isothermal processes (in the sense of the presence of a single thermal reservoir at fixed temperature) connecting two generic nonequilibrium states ρ_A and ρ_B with same Hamiltonian $\hat{H}$ reads [16, 23]

$$T \Delta_{\mathrm{i}} S_{\mathrm{th}} = W - \Delta \mathcal{F} \geqslant 0, \tag{3.22}$$

where W is the work performed during the process, and we have denoted by $\Delta\mathcal{F} = \mathcal{F}(\rho_B, \hat{H}, T) - \mathcal{F}(\rho_A, \hat{H}, T)$ the change in nonequilibrium free energy as defined in Eq. (3.20).

Notice that we have written the thermodynamic entropy as S_{th} using a different notation than the one employed for the Gibbs entropy $S(x, p, t)$, and the von Neumann entropy $S(\rho)$ in Sect. 1.1.6. The von Neumann entropy coincides with the thermodynamic entropy for equilibrium thermal states (or Gibbs states), $\rho_{\text{th}} = \exp(-\beta\hat{H})/Z$ with $Z = \exp(-\beta F)$ the partition function, when multiplied by the Boltzmann constant k_B, that is, $k_B S(\rho_{\text{th}}) = S_{\text{th}}$. However, more care must be taken in nonequilibrium situations. The von Neumann entropy is usually considered to represent the nonequilibrium extension of the thermodynamic entropy in quantum thermodynamics [1, 5] (as well as the Gibbs entropy is widely used in classical statistical mechanics) and, although this identification has been demonstrated to be correct in different nonequilibrium scenarios (see e.g. Refs. [12, 16, 25–27]), its equivalence in arbitrary situations is still a controversial issue in both classical and quantum cases [28–33]. We stress that, for the identification to be correct, one needs to demonstrate the existence of reversible processes connecting the nonequilibrium initial and final states, for which Eq. (3.9) [or equivalently Eq. (3.22)] is fulfilled when using the von Neumann entropy. In Chap. 8 we analyze various expressions for the entropy production (3.13) in open quantum systems based on the von Neumann entropy.

3.1.5 The Third Law of Thermodynamics

Walther Nernst completed the fundamental laws between 1906 and 1912 by noticing that the changes in thermodynamic entropy of all isothermal processes tends to zero when the temperature approaches zero. Planck reformulated this principle in a stronger way by stating that the entropy of any system in equilibrium tends to zero as its temperature approaches zero (when the system has a non-degenerated minimum energy state) [17]

$$S_{\text{th}} \to 0 \quad \text{when} \quad T \to 0K. \tag{3.23}$$

This is known as the third law of thermodynamics, the Nernst heat theorem [18] or Nernst postulate [17]. Notice that the theorem provides an absolute scale for the thermodynamic entropy, making it a well defined quantity for any thermodynamic state.

Furthermore, the third law of thermodynamics has implications in the attainability of the absolute zero temperature. In particular, it sets that no adiabatic process (in the thermodynamic sense) initiated at $T \neq 0$ can reach $T = 0$. This is because the adiabat $S = 0$ coincide with the isotherm $T = 0$, and we have that two adiabats can never cross each other [17]. The third law is hence sometimes formulated as the unattainability of absolute zero temperature for any process in a finite number of operations and at a finite time, as was first pointed by Nernst [34] (see also Ref. [35]).

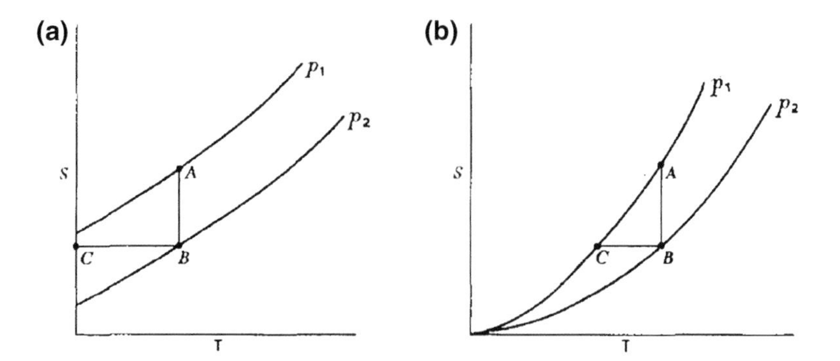

Fig. 3.4 Schematic entropy-temperature diagrams for cooling processes ABC in which **a** absolute zero is attainable and **b** absolute zero is unattainable. The curves p_1 and p_2 represent different values of the external control parameter p (e. g. the pressure in the case of a gas), $S \equiv S_{\text{th}}$ is the thermodynamic entropy and T the temperature. Nernst heat theorem implies situation (**b**) for which an infinite number of cooling processes is needed in order to reach absolute zero temperature. Picture obtained from Ref. [35]

This is indeed probably the most popular formulation of the third law nowadays. In Fig. 3.4 this statement is illustrated in terms of a cooling process represented in the entropy-temperature diagram. It should be further pointed that the unattainability formulation does not implies itself that the thermodynamic entropy tends to zero at $T \rightarrow 0$, but only that the changes in entropy vanish in this regime [35].

In the context of quantum thermodynamics, it has been pointed that the full quantum treatment of matter can shed light into the relation between the above two formulations of the third law [25, 36]. In particular, the Nernst heat theorem has been largely studied in Ising models and lattice systems with generalized ferromagnetic many-body interactions [37, 38] and the role of degeneracy has been discussed. More recently it has been checked for a harmonic oscillator coupled to a general heat bath [39, 40]. On the other hand, the third law has been also explored in its dynamical form in the context of quantum thermal machines (see Sect. 3.3 for an introduction to quantum thermal machines). Four-stroke [41] and swap-based [42] Otto refrigerators, as well as continuously driven [43] and autonomous fridges [44, 45] have been analyzed, and different power laws for the decrease of the refrigeration heat current when the temperature approaches zero have been reported. However, non-Markovian models have been also introduced leading to refrigeration at a constant rate in the limit $T \rightarrow 0$, hence challenging the unattainability principle [46]. Furthermore, other formulations of the third law taking into account e.g. the size of the reservoir, or the energy needed to perform cooling, have been discussed in the context of purification of quantum states and information erasure for quantum information tasks [36, 47–51]. In Chap. 11 we will provide a new formulation based on the Hilbert space dimension of multilevel autonomous quantum fridges.

3.1.6 Thermodynamics and Information

The link between information and thermodynamics is as old as the thermodynamic theory itself, going back to the gedanken (thought) experiment proposed by James Clerk Maxwell in 1871, and today known as *Maxwell's demon* (for a review see Ref. [52]). Maxwell imagined a little intelligent being (the demon) able to acquire information about the positions and velocities of the molecules of two gases at different temperatures, in containers of equal volume separated by a rigid wall equipped with a tiny door which can be opened or closed at will. The demon can control the door to let the fast (hot) particles of the gas at the cold temperature be transferred to the gas at the hot temperature when they approach the wall. Analogously it can also let the slow (cold) particles of the hot gas pass to the cold gas [see Fig. 3.5a]. In this way, he can make the hot gas hotter and the cold gas colder without any work invested, hence challenging the second law of thermodynamics. Maxwell's demon pointed at two important characteristics involving thermodynamic laws, which have been largely investigated. First, the second law of thermodynamics seems to be only a *statistical principle* of large systems which holds almost all the time [52], but events defeating the law can happen at the microscopic scale. Second, the inclusion of information seems to modify the energetic restrictions imposed by the second law [16].

The recognition of the thermodynamic significance of information is due to Léo Szilárd, who proposed another thought experiment in 1929 consisting in a cyclic engine that uses information to perform work, and today known as the *Szilard engine* [54]. In this case the working substance of the engine is a single-molecule gas which starts in a container of given volume V_0, in thermal contact with a reservoir at temperature T [see Fig. 3.5b]. The demon starts by rapidly introducing a piston splitting the container in half, each one of volume $V_0/2$. Subsequently, the demon 'measures' in which half the particle is contained, and moves the piston inducing a reversible expansion to the other side until the volume turns back to the initial value V_0. The piston can then be removed and the cycle starts again. As long as introducing and removing the piston costs no work (as it can be done reversibly [55]), the net effect of the cycle is the extraction of work in the isothermal reversible expansion, which

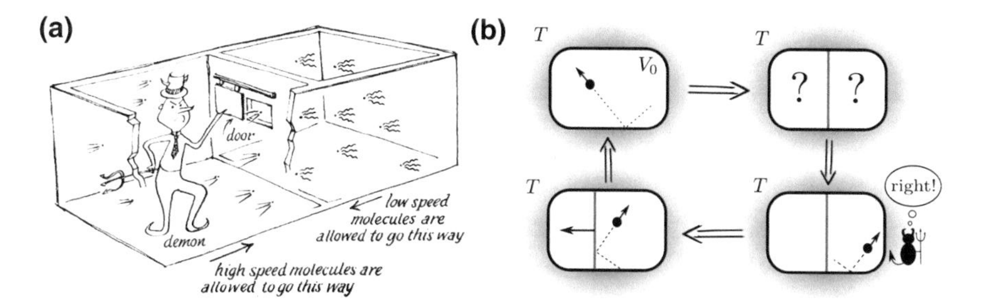

Fig. 3.5 **a** Cartoon representing the Maxwell's demon setting obtained from Ref. [53] and **b** Szilard's engine cycle in four steps starting from the top-left panel (see main text)

equals the heat absorbed from the reservoir, $W_{\rm ext} = Q = k_B T \ln 2$. This produces a decrease of entropy in the reservoir $\Delta S^E_{\rm th} = -Q/T = k_B \ln 2$. This machine hence apparently contradicts Eq. (3.10), or the equivalent formulation: "It is impossible to construct an engine which will work in a complete cycle, and convert all the heat it absorbs from a reservoir into mechanical work" (Ref. [18], p. 103).

The proposal of Leó Szilárd has been analyzed in detail during decades, highlighting different possible drawbacks of the scheme and clarifying the origin of the entropy decrease [53]. Although Szilárd already pointed out that an increase of entropy in the measurement process must compensate the entropy reduction in the reservoir in order to save the second law, he did not mention the role of the demon's memory [52]. A popular resolution of the paradox is due to Charles Bennett [56] who argued that the demon's memory retaining the information about the chamber in which the one-molecule gas is in, must be reset (or erased) to truly close the cycle, while acquisition of information can be done reversibly. In his derivation, Bennett invoked Landauer's erasure principle [57], which sets that the logical erasure of one bit of information, in a system in contact with a thermal reservoir at temperature T, requires a minimum dissipation of heat:

$$Q_{\rm eras} \geqslant k_B T \;\ln 2. \tag{3.24}$$

This minimum amount of heat is known as *Landauer's bound*, and must be compensated by an equal amount of invested work $W_{\rm eras} = Q_{\rm eras}$ if one wants to maintain the internal energy of the working substance constant. Therefore, turning to the Szilard engine, we have that the work extracted in the reversible expansion of the cycle must be spent to reset the memory at the end, implying that the overall gain of work in the whole cycle is (at most) zero. Landauer's principle establishes the connection between logical irreversibility and energy dissipation in computing processes, arguing that information is always stored in physical devices, and consequently it needs to be considered as physical. The increasing ability to control systems at the single particle level has made possible the experimental verification of Landauer's bound [58], while Szilard-like engines are also being implemented nowadays in the laboratory using as working substances a colloidal Brownian particle [59, 60] or a single electron [61, 62]. Furthermore, other devices operating as a Maxwell demon have been recently experimentally realized in a photonic setup [63], or an autonomous version in capacitively coupled single-electron devices [64].

The quantum version of the Szilard engine has been studied in Refs. [65, 66], from where the classical results can be recovered. Furthermore, Landauer's principle can be straightforwardly extended to the quantum domain by considering the erasure of a general quantum state, ρ, to be reset into a *ready-to-measure* fixed pure state, $|0\rangle$, with same internal energy. In this case we obtain $Q_{\rm eras} \geqslant k_B T \;S(\rho)$, where $S(\rho)$ is the von Neumann entropy. Again the dissipated heat can be seen as work externally invested in the operation. Indeed a simple calculus using Eq. (3.21) shows that this is indeed the change in nonequilibrium free energy during the process [16] (see also Refs. [67, 68])

$$W_{\rm eras} \geqslant \Delta \mathfrak{F} = k_B T \;S(\rho). \tag{3.25}$$

We stress that here the meaning of the word "erasure" in the transformation from a high-entropy state ρ to the pure state $|0\rangle$, comes from historical reasons, while it is related to a decrease in the uncertainty about the state (i.e. an increase in the information about the state). It is also worth stressing that Eq. (3.25) must be modified in the case in which the system, say $\mathcal{S}$, to be erased with the transformation $\rho_{\mathcal{S}} \rightarrow |0\rangle_{\mathcal{S}}$, shares correlations with a further ancilla, whose local state $\rho_{\mathcal{A}} \equiv \mathrm{Tr}[\rho_{\mathcal{SA}}]$ is not changed during the process [1]. Notice that this is completely equivalent to the above erasure process from the local point of view, while now Eq. (3.25) generalizes to [69]

$$W \geqslant k_B T \; S(\mathcal{S}|\mathcal{A}), \tag{3.26}$$

that is, work can be extracted from the consumption of quantum correlations if the relative entropy $S(\mathcal{S}|\mathcal{A}) = S(\rho_{\mathcal{SA}}) - S(\rho_{\mathcal{A}})$ is negative, as it is the case for a certain subset of bipartite entangled states (see Ref. [70] and Sect. 1.4). A general protocol obtaining the maximum amount of work is presented in Ref. [69]. The thermodynamics of quantum feedback processes has been also investigated [71, 72] including the role of quantum correlations in different setups [73–76]. Finally we point that the verification of Landauer's principle in a fully quantum system has been recently reported in Ref. [77], and extensions to the case of finite-size reservoirs [51], nonequilibrium reservoirs [78], or probabilistic erasure [79] are currently important topics of research.

3.2 Fluctuation Theorems

In the past decades there has been an increasing interest in applying thermodynamics to small systems of microscopic or even nanoscopic size, and extended more recently to the quantum regime. Individual molecules became accessible to high-precision manipulation and measurement in the laboratory, while simulation techniques for molecular systems were established. As a consequence, small-scale thermodynamics attracts a multidisciplinary interest from biology, chemistry and physics [80]. When the size of the systems under consideration scales down, fluctuations become important, and nonequilibrium situations appear everywhere. In order to describe such situations, the inclusion of fluctuations in the nonequilibrium thermodynamic description is mandatory, a task which has been first accomplished within the framework of *stochastic thermodynamics* [2, 81].

The laws of thermodynamics, as they apply to macroscopic objects, are blurred when considering the random motion of small particles, continuously colliding with the particles of their environmental surroundings, and one may expect they to hold only on average [80]. However, an interesting first law like energy balance can also be stated for individual stochastic trajectories of microscopic objects, and entropy can also be defined at this level. Furthermore, the study of the fluctuations in the microscopic versions of work, heat and entropy, have led to the discovery of universal relations called *fluctuation theorems* (FT) which introduce precise constraints on the statistics followed for those quantities. Among the most popular fluctuations relations are the Jarzynski relation [82, 83], the Crooks work theorem [84], and the integral fluctuation theorem for the total entropy production introduced by Seifert [85], who followed and extended the pioneering work of Evans, Cohen and Morris [86]. They refined our understanding of irreversibility and the second law of thermodynamics, substituting the usual inequalities for average quantities by equalities for the microscopic based ones [80]. Stochastic thermodynamics and fluctuation theorems are introduced in the following sections.

The development of quantum thermodynamics passes through the extension of the different fluctuation theorems to the quantum regime. On the other hand, important issues concerning the nature of quantum work and heat arise when dealing with quantum systems [13, 87, 88], and the trajectory concept is indeed misleading unless quantum measurements are considered (see Sect. 2.5). This motivates the introduction of a framework which we call the *two measurement protocol* (TMP) from which some of the most important fluctuation theorems can be derived both for isolated and open quantum systems [3, 89]. Deriving the fluctuation theorems without the TMP framework as well as developing extensions to nonequilibrium situations in which quantum features or finite-size effects can be naturally incorporated, constitute nowadays important challenges [90–96]. Part III of this thesis is dedicated to present our contributions in this active and promising field of research.

3.2.1 Stochastic Thermodynamics

Stochastic thermodynamic describes the thermodynamic behavior of small systems driven out of equilibrium in a framework which incorporates the fluctuations induced by the environment. The framework combines the stochastic energetics introduced by Sekimoto [97], allowing the formulation of the first law of thermodynamics for stochastic trajectories, and the definition of entropy for fluctuating trajectories [85] leading to different results refining the second law of thermodynamics [2]. It extends statistical mechanics to nonequilibrium situations and it has been satisfactorily applied to isothermal processes followed by microscopic systems from biopolymers to single electron transistors.

Following classical statistical mechanics, we can describe a single particle (or few of them) in phase space by specifying its position and momentum $(x, p) = \{\vec{x}, \vec{p}\}$. The Hamiltonian function of the particle is $H(x, p, \lambda)$, where we introduced the external control parameter $\lambda(t)$, responsible of driving the system out of equilibrium. When a particular protocol for the external parameter is applied, $\Lambda = \{\lambda(t)|t_0 \leqslant t \leqslant \tau\}$, the particle will evolve subjected to the externally applied force while interacting randomly with its surrounding environment, assumed to be a thermal reservoir (or heat bath) in equilibrium at a well defined temperature T. The system evolution is assumed to be described by Langevin dynamics

$$\dot{x} = p/m, \qquad \dot{p} = F(x, \lambda) - \gamma p/m + \xi(t), \tag{3.27}$$

where $F(x, \lambda) = -\partial_x V(x, \lambda) + f(x, \lambda)$ is a systematic force with conservative contribution from the potential $V(x, \lambda)$ and a directly applied non-conservative force $f(x, \lambda)$, m is the mass of the particle, and γ_0 the friction coefficient. The stochastic term $\xi(t)$ describes the thermal noise with $\langle \xi(t) \rangle = 0$ and correlation $\langle \xi(t_1)\xi(t_2) \rangle = 2D\delta(t_1 - t_2)$, D being the diffusion constant. A paradigmatic model is the overdamped regime of the above dynamics, reducing to

$$\dot{x} = \mu F(x, \lambda) + \xi(t), \tag{3.28}$$

where $\mu = 1/\gamma_0$ is the *mobility*. This regime occurs when the particle mass is big $m \gg \gamma_0 \Delta t$, Δt being the resolution of the coarse-grained evolution, or equivalently when the friction is high $\gamma_0 \gg \Delta t/m$. Prototypical systems obeying the overdamped Langevin equation are colloidal microscopic particles, molecular motors, and some magnetic and electric circuits at low intensities. Here it is assumed that the Einstein relation holds, $D = \mu k_B T$, meaning that the noise is not affected by the time-dependent force [2]. For simplicity we considered a single degree of freedom x. In the case of multiple degrees of freedom, x and F become vectors, and D and μ become tensors possibly depending on x [98].

The dynamical evolution generates stochastic *trajectories* in phase-space, denoted as $\gamma_t \equiv (x_t, p_t)$. As long as the interactions with the environment are of random nature, for a given protocol Λ we will obtain different trajectories in different realizations. Therefore, the physical properties of the system at the trajectory level become stochastic variables, which can be described through an appropriate probability distribution. The ensemble of trajectories is characterized by the phase space distribution $\rho(x, p, t)$ denoting the probability to find the particle with position x and momentum p at time t. It evolves according to a Fokker–Planck equation, which in the case of the overdamped colloidal particle reads

$$\partial_t \rho(x, t) = -\partial_x j(x, t), \tag{3.29}$$

with $j(x, t) = \mu F(x, \lambda)\rho(x, t) - D\, \partial_x \rho(x, t)$, the *current*. We recall that in the overdamped regime the momentum of the particle, p, becomes superfluous and has been neglected from the expressions. In some situations, if the system can only occupy discrete states, Eq. (3.29) is replaced by a Markovian master equation of the type

$$\dot{p}_m = \sum_n W_{m,n}\, p_n \tag{3.30}$$

where p_m is the probability of the system for to be in state m, and $W_{m,n}$ are the elements of the so-called *stochastic matrix*, with the property $\sum_m W_{mn} = 0$, which ensures the conservation of probability [81].

The first law of thermodynamics is extended to the trajectory level by decomposing the internal energy change

$$\Delta u[\gamma] \equiv H(x_\tau, p_\tau, \lambda_\tau) - H(x_{t_0}, p_{t_0}, \lambda_{t_0}), \tag{3.31}$$

where $\gamma \equiv \{\gamma_t\}_{t_0}^{\tau}$ denotes the whole trajectory from t_0 to τ, into heat and work contributions as

$$q[\gamma] \equiv \int_{t_0}^{\tau} \big(\partial_x H(x,p,\lambda)\dot{x}_t + \partial_p H(x,p,\lambda)\dot{p}_t\big)dt,$$
$$w[\gamma] \equiv \int_{t_0}^{\tau} \partial_\lambda H(x,p,\lambda)\dot{\lambda}dt, \tag{3.32}$$

fulfilling $\Delta u[\gamma] = w[\gamma] + q[\gamma]$. In isolated systems the dynamics of the particle is governed by the Hamilton equations of motion, which imply $q[\gamma] = 0$ for any trajectory γ, and hence $\Delta u[\gamma] = w[\gamma]$. For open systems governed by the overdamped Langevin equation, Sekimoto [97] identified explicit expressions for the (non exact) differentials of work and heat along trajectories [98]

$$đw = \partial_\lambda V(x,\lambda)\dot{\lambda}dt + f(x,\lambda)dx, \qquad đq = -F(x,\lambda)dx, \tag{3.33}$$

recovering the first law of thermodynamics at the differential level $du = đw + đq = \dot{V}(x,\lambda)dt$. When integrating those expressions, Stratonovich rule has to be used, for which the usual rules of differential calculus apply. For the case of discrete systems, the driving modifies the energy of the states $\epsilon_m(\lambda(t))$ and the trajectory consists of abrupt jumps occurring at random times $\{t_1, t_2, ..., t_J\}$ where the state of the system changes $m(\gamma) = \{m_0 \to m_1, m_1 \to m_2, ..., m_{J-1} \to m_J\}$ [81]. Heat and work are defined as

$$q[\gamma] \equiv \sum_{j=0}^{J-1} \epsilon_{m_{j+1}}(t_{j+1}) - \epsilon_{m_j}(t_{j+1}), \tag{3.34}$$

$$w[\gamma] \equiv \sum_{j=0}^{J} \epsilon_{m_j}(t_{j+1}) - \epsilon_{m_j}(t_j), \tag{3.35}$$

where $t_{J+1} \equiv \tau$ and we have $\Delta u[\gamma] \equiv \epsilon_{m_J}(\tau) - \epsilon_{m_0}(t_0) = q[\gamma] + w[\gamma]$.

When sampling trajectories from some initial phase space density $\rho(x, p, t_0)$ and applying the protocol Λ, the averages of quantities defined at the trajectory level can be recovered by assigning to each trajectory a weight, which for Eq. (3.28) reads

$$p[\gamma|\gamma_{t_0}] = \mathcal{N}\exp\left(-\int_{t_0}^{\tau} dt[(\dot{x} - \mu F)^2/4D + \mu\partial_x F/2]\right), \tag{3.36}$$

where $\mathcal{N}$ is a normalization factor [2]. This allows us to calculate the average of any trajectory dependent quantity given by a functional $O[\gamma_t]$ as

$$\langle O[\gamma_t] \rangle = \int dxdp \int d\gamma_t \; O[\gamma_t] \; p[\gamma|\gamma_{t_0}] \; \rho(x, p, t_0). \qquad (3.37)$$

Such averages over trajectories coincide with the same quantities calculated from the Fokker–Planck equation (3.29) (see details in Ref. [2]).

The last important ingredient in the thermodynamic description is the identification of entropy at the trajectory level and the formulation of the second law. The stochastic or trajectory entropy was introduced by Seifert [85] and reads

$$s_t \equiv -k_B \ln \rho(\gamma_t, t), \qquad (3.38)$$

where $\rho(x, p, t)$ is evaluated along the trajectory γ_t. The definition is similar for the discrete case, where $s_t = -k_B \ln p_{m(\gamma_t)}$ [81]. Notice the similarity in both cases with the Gibbs entropy introduced in Sect. 3.1. In addition, the entropy increase in the medium (the thermal reservoir) during the trajectory γ is given by

$$\Delta s^{\mathrm{m}}[\gamma] \equiv -q[\gamma]/T, \qquad (3.39)$$

which corresponds to (minus) the entropy flow to the system, as introduced in Sect. 3.1. Employing the Fokker–Planck equation (3.29), and the later definitions, the entropy production rate during a stochastic trajectory of the overdamped Langevin equation reads

$$\dot{s}_{\mathrm{i}}[\gamma_t] \equiv \dot{s}_t + \dot{s}^{\mathrm{m}}[\gamma_t] = -\left.\frac{\partial_t \rho(x,t)}{\rho(x,t)}\right|_{\gamma_t} + \left.\frac{j(x,t)}{D\,\rho(x,t)}\right|_{\gamma_t} \dot{x}, \qquad (3.40)$$

which at the ensemble level becomes [2]

$$\dot{S}_{\mathrm{i}} = \int dx \frac{j(x,t)^2}{D\,\rho(x,t)} \geqslant 0. \qquad (3.41)$$

Finally, in the case of discrete systems, the ensemble expression for the entropy production rate is [99]

$$\dot{S}_{\mathrm{i}} = k_B \sum_{m,n} W_{m,n} p_n \ln \frac{W_{m,n} p_n}{W_{n,m} p_m} \geqslant 0, \qquad (3.42)$$

which is zero if and only if the detailed balance condition is satisfied

$$W_{m,n} p_n = W_{n,m} p_m. \qquad (3.43)$$

We stress that the above expression for the entropy production rate is consistent with the identification of the entropy flow entering the system as the heat divided by temperature [81, 99].

3.2.2 Classical Fluctuation Theorems

A major achievement of stochastic thermodynamics has been the formulation of precise equations governing the statistics of thermodynamical fluctuating quantities in processes arbitrarily far from equilibrium. Those relations are known as *fluctuation theorems* (FT) and generalize the second law of thermodynamics in the form of equalities [80]. In the context of the simulation of sheared fluids, Evans, Cohen and Morris [86] proposed in 1993 that the second law may be generalized considering the probabilities of obtaining fluctuations increasing or decreasing the entropy production (see also Ref. [100]). In a nonequilibrium steady state of a thermostatted system, the statement can be formulated as

$$P(-\Delta_{\mathrm{i}}s) = P(\Delta_{\mathrm{i}}s)\, e^{-\Delta_{\mathrm{i}}s}, \tag{3.44}$$

where $P(\Delta_{\mathrm{i}}s)$ is the probability distribution for obtaining a particular value for the entropy production, such that $\langle \Delta_{\mathrm{i}}s \rangle = -Q/T$, the heat transferred to the thermal reservoir divided by its temperature. This would imply that the occurrence of negative entropy production events, although being not forbidden, is exponentially less probable than the occurrence of their positive counterparts. The result is an overall entropy production at the ensemble level which is always positive, $\langle \Delta_{\mathrm{i}}s \rangle \geqslant 0$. Further statistical properties of $P(\Delta_{\mathrm{i}}s)$ can be derived as well from the above fluctuation theorem [101]. The fluctuation theorem in Eq. (3.44) is usually called the steady-state fluctuation theorem, as it applies for systems in steady states observed for long times (the time should be greater than the decorrelation time). It was first proven to hold for deterministic thermostatted systems in Ref. [102], and then extended to Langevin dynamics [103], and general stochastic processes described by Markovian master equations [104]. However, the fluctuation theorem in Eq. (3.44) requires the identification of the entropy production $\Delta_{\mathrm{i}}s$ in the setup of interest, which is not always a simple task. Furthermore, its scope could not surpass steady-state dynamics until the introduction of a general expression for the nonequilibrium entropy of the system at the stochastic level by Seifert [85] (see below).

Nevertheless, other fluctuation theorems were developed in the meanwhile. Probably the most famous one is the Jarzynski equality [82, 83], which had the merit to link the statistics of the work performed in small systems driven out of equilibrium in contact with a thermal reservoir, with its equilibrium properties. The setup considered by Jarzynski is the following. Consider a system that starts in thermal equilibrium with its surroundings at temperature T and can be driven through the variation of a parameter λ controlling the Hamiltonian of the system $H(\lambda)$. The parameter starts in position $\lambda(t_0) \equiv \lambda_A$, and the state of the system is given by the canonical distribution, $\rho_A \equiv e^{-\beta H(\lambda_A)}/Z_A$. Then it is varied, perhaps abruptly, following a specific protocol $\Lambda = \{\lambda(t)\}$ which drives the system arbitrarily far from equilibrium, while still in contact with its surrounding environment (even if the coupling is strong [105]). At some point τ the driving is stopped, acquiring a final fixed value $\lambda(\tau) \equiv \lambda_B$, and the system is let to relax back to equilibrium with the thermal environment, ending at

$\rho_B \equiv e^{-\beta H(\lambda_B)}/Z_B$. During this process, an amount of work $w[\gamma]$ is performed due to the driving, as defined in Eq. (3.32). Repeating the process many times with the same driving protocol Λ, a probability distribution $P(w)$ can be further obtained. Jarzynski equality states that

$$\langle e^{-\beta w} \rangle = e^{-\beta \Delta F_{AB}} = \frac{Z_B}{Z_A}, \tag{3.45}$$

where $\langle e^{-\beta w} \rangle \equiv \int dw P(w) e^{-\beta w}$, and $\Delta F_{AB} = F_B - F_A$ is the difference of Helmholtz free energies between the final and initial equilibrium states.

Equation (3.45) has been obtained using a great variety of deterministic and stochastic evolutions (see Ref. [80] and references therein). In fact, since the equality is also valid for isolated evolution and the work is performed through the degrees of freedom affected by the external parameter lambda, we conclude that Jarzynski equality holds for any reduced dynamics containing those degrees of freedom, either deterministic, stochastic, Markovian or non-Markovian, etc. The only requirement is that the global isolated system is initially in equilibrium at temperature T. It is also worth stressing that Eq. (3.45) remains valid even if the system does not equilibrate with the thermal environment at the end of the protocol, as this final step occurring at constant λ does not contribute to the work [2]. Nonetheless the final thermalization is required to give a specific meaning to Eq. (3.45) in terms of the entropy produced in the process. Jarzynski equality has been experimentally tested reversibly and irreversibly stretching a single molecule of RNA immersed in an aqueous solution at a temperature of 208 to 301K [106].

From Eq. (3.45), and using Jensen's inequality, $\langle e^x \rangle \leqslant e^{\langle x \rangle}$, we obtain

$$\langle w_{\text{diss}} \rangle = \langle w \rangle - \Delta F_{AB} \geqslant k_B T \ln \langle e^{-\beta(w - \Delta F_{AB})} \rangle = 0, \tag{3.46}$$

where again $\langle w \rangle_\gamma \equiv \int dw P(w) w$, and we recover the average form of the second-law inequality, $\langle w_{\text{diss}} \rangle \geqslant 0$, in terms of the work performed to drive a system between two equilibrium states [Eq. (3.19)]. Here if the driving is quasi-static we have reversible conditions, $\langle w \rangle = \Delta F_{AB}$. Jarzynski equality therefore refines Eq. (3.46), establishing a universal property of the statistics of irreversible work [80]. Furthermore, we stress that Jarzynski's result is of remarkable practical utility. Determining the free energy landscape $F(\lambda)$ of a system usually requires the realization of reversible (infinitesimally slow) processes which are difficult to implement. Instead, using Eq. (3.45), one can easily obtain free-energies of equilibrium states by measuring the work performed in arbitrary protocols overcoming experimental difficulties [106–109], that is

$$-k_B T \ln \langle e^{-w/k_B T} \rangle = \Delta F_{AB}. \tag{3.47}$$

The results put forward by Jarzynski were later refined by Crooks in 1999 [84], who derived for stochastic, microscopically reversible dynamics, the fluctuation theorem for the work statistics in Eq. (3.48) below. In the general configuration introduced above for the Jarzynski equality, Crooks considered together with the *forward*

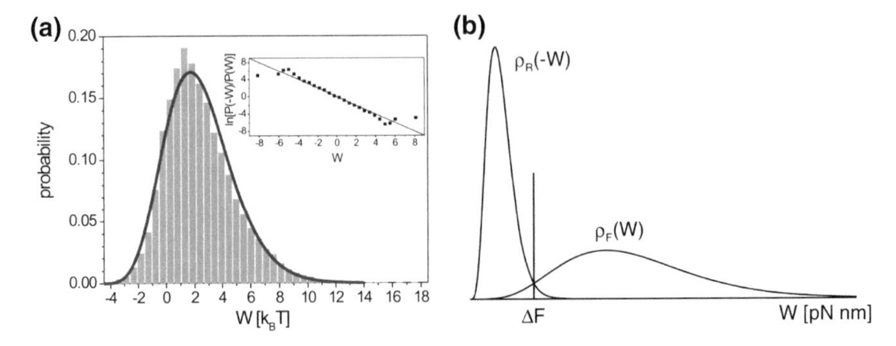

Fig. 3.6 **a** Experimental reconstruction of the (non-Gaussian) work probability distribution of a colloidal particle in a time-dependent nonharmonic potential applied throughout the light pressure of optical tweezers, while using total internal reflection microscopy to determine the trajectories. In the inset the logarithm of the ratio of the probability to find trajectories with work $-w$ to those with work w is plotted. The figure has been obtained from Ref. [110]. **b** Illustration of work probability distributions in the forward process $\rho_F(W) \equiv P(w)$, and in the backward process $\rho_R(-W) \equiv \tilde{P}(-w)$. The two probability distributions intersect at ΔF according to Eq. (3.48). Picture obtained from Ref. [111]

process, taking ρ_A into ρ_B through the driving protocol Λ, also the *backward* (or reverse) process, in which the time-reversed protocol $\tilde{\Lambda} \equiv \{\lambda(\tau + t_0 - t)\}$ is applied to the equilibrium state ρ_B, letting the system re-equilibrate at the end to ρ_A. In this backward process, trajectories $\tilde{\gamma}$ are generated obeying a different probability distribution for the work $\tilde{P}(w)$. Crooks fluctuation theorem relates the work probability distributions in forward and backward processes [84]:

$$\frac{P(w)}{\tilde{P}(-w)} = e^{-\beta(w - \Delta F_{AB})}. \tag{3.48}$$

In contrast to the FT in Eq. (3.44), both Eqs. (3.48) and (3.45) are valid for arbitrary times, falling in the class known in the literature as *transient fluctuation theorems*. Furthermore, one usually refers to the fluctuation theorems in the form of Eq. (3.45) as *integral fluctuation theorems*, while Eqs. (3.44) and (3.48) are usually called *detailed fluctuation theorems*. Crooks fluctuation theorem has been experimentally tested in Refs. [110, 112–116]. In Fig. 3.6a we show an example of one of such tests using a colloidal particle trapped by optical tweezers and subjected to a time-dependent nonharmonic potential.

Crooks fluctuation theorem corresponds to a refinement of the second law of thermodynamics, which applied to the sequence of forward and backward processes closing a cycle reads

$$\langle w[\gamma] \rangle + \langle w[\tilde{\gamma}] \rangle \geqslant 0, \tag{3.49}$$

that is, no net (average) work can be extracted from single thermal reservoir by cyclic operation [111]. Indeed Eq. (3.48) implies that the mean of $P(w)$ is always located at the right of the mean of $\tilde{P}(-w)$, while the two distributions intersect at the point ΔF_{AB} (see Fig. 3.6). Furthermore, one can easily derive the Jarzynski equality from Eq. (3.48) by integrating over trajectories

$$\int dw\; P(w)e^{-\beta(w-\Delta F_{AB})} = \int dw\; \tilde{P}(-w) = 1, \tag{3.50}$$

as long as $\tilde{P}(-w)$ is a well defined probability distribution.

Another important result closely related to Crooks fluctuation theorem is the relation between physical and information-theoretical measurements of irreversibility [117, 118]

$$\langle w_{\text{diss}} \rangle = \langle w \rangle - \Delta F_{AB} = k_B T\; D(\rho || \tilde{\rho}), \tag{3.51}$$

where $D(\rho||\tilde{\rho})$ is the Kullback–Leibler divergence [119] (or relative entropy as introduced in Sect. 1.1.6) between the system densities in forward and backward processes, either in path space [117] or phase-space [118]. The densities $\rho(x, p, t)$ and $\tilde{\rho}(x, -p, t)$ have to be measured at the same intermediate but otherwise arbitrary time. Equation (3.51) implies that dissipation results from the difference between the two distributions, being zero if and only if they are equal [118].

We notice that in the processes introduced above, the dissipative work $w_{\text{diss}}[\gamma] = w[\gamma] - \Delta F_{AB}$ is proportional to the entropy production in the forward process connecting the two equilibrium states ρ_A and ρ_B, as $\langle w_{\text{diss}} \rangle = T \Delta_{\text{i}} S$. However, fluctuation theorems can be extended to more arbitrary situations, such as transitions between steady states [120, 121], thermal systems starting in arbitrary initial states [85], or nonequilibrium feedback control [59, 122], among others (see e.g. Ref. [123] and the review [2]), in which the entropy production is expressed in more general terms. This is also the case of the nonequilibrium equality (3.51) which has been shown to hold in a generalized form for a variety of initial conditions [124]. A particularly important generalization of the work fluctuation theorem was derived by Seifert in [85], who proved the integral transient fluctuation theorem for the total entropy production in driven systems governed by the overdamped Langevin equation (3.28) and general stochastic dynamics

$$\langle e^{-\Delta_{\text{i}} s} \rangle = 1. \tag{3.52}$$

The key point in the derivation is the identification of the entropy production over single trajectories

$$\Delta_{\text{i}} s[\gamma] = \Delta s[\gamma] + \Delta s^m[\gamma], \tag{3.53}$$

resulting from the stochastic entropy change in the system $\Delta s[\gamma] = s_\tau - s_{t_0}$ as defined in Eq. (3.38), and from the entropy change in the medium, $\Delta s^m[\gamma] = -q[\gamma]/T$. The fluctuation theorem in Eq. (3.52) remains valid for arbitrary initial and final states, i.e. we do not need to assume an initial equilibrium state with the reservoir at temperature T, nor a final equilibration step at the end of the protocol. Instead, the backward process may be initialized by using the phase space distribution of the system reached at the end of the forward driving protocol, as given by the corresponding Fokker–Planck or master equation [85]. The integral theorem (3.52) has been experimentally checked in Ref. [125] for a two-level system realized as an optically driven defect center in diamond, and more recently for a single-electron box in the presence of different thermal baths [126].

Another kind of detailed fluctuation theorems for the entropy production has been also derived by Esposito and Van den Broeck in Refs. [99, 123, 127]. Under general Markovian [99, 123] or Fokker–Planck [127] dynamics, they obtained a detailed version of theorem (3.52):

$$\frac{P(\Delta_{\rm i}s)}{\tilde{P}(-\Delta_{\rm i}s)} = e^{\Delta_{\rm i}s}. \tag{3.54}$$

In this theorem, the entropy production is identified with the log-ratio of the sampling probabilities of forward trajectories γ in the forward process $P[\gamma]$, and backward trajectories $\tilde{\gamma}$ in the backward (time-reversed) one $\tilde{P}[\tilde{\gamma}]$, that is

$$\Delta_{\rm i}s[\gamma] \equiv \ln \frac{P[\gamma]}{\tilde{P}[\gamma]} = \Delta s[\gamma] + \Delta s^m[\gamma], \tag{3.55}$$

which equals expression (3.53) when considering that the initial condition of system in the backward process is the final state of the forward one at the end of the trajectory γ. The entropy production in the backward process is analogously defined as

$$\Delta_{\rm i}s[\tilde{\gamma}] \equiv \ln \frac{\tilde{P}[\gamma]}{P[\gamma]} = -\Delta_{\rm i}s[\gamma], \tag{3.56}$$

which however does not equals Eq. (3.53) for the backward process, as the stochastic entropy Δs has not definite parity under time-reversal [2]. We also notice that the above FT in Eq. (3.54) differs from the fluctuation theorem in Eq. (3.44). In the former case it relates the probability distribution of forward $P(\Delta_{\rm i}s)$ and backward protocols $\tilde{P}(-\Delta_{\rm i}s)$, while in the later case the fluctuation theorem relates the two tails of the same distribution $P(\Delta_{\rm i}s)$. In general Eq. (3.44) is a much stronger statement than Eq. (3.54), implying infinitely many integral fluctuation theorems for antisymmetric functions of $\Delta_{\rm i}s$ [128]. In Refs. [99, 123, 127] the authors complement the detailed fluctuation theorem in Eq. (3.54) for the total entropy production with the derivation of two further detailed fluctuation theorems for the so-called *non-adiabatic* and *adiabatic* entropy productions

$$\Delta_{\rm i}s[\gamma] = \Delta_{\rm i}s^{\rm na}[\gamma] + \Delta_{\rm i}s^{\rm ad}[\gamma], \tag{3.57}$$

each of them corresponding to the entropy production associated to a different way to bring the system out-of-equilibrium: by means of external driving (non-adiabatic) and by imposing nonequilibrium environmental conditions (adiabatic). Notice that the term adiabatic is employed here in the dynamical sense. The fluctuation theorems for the *non-adiabatic* and *adiabatic* entropy production are obtained by introducing a *dual* dynamics derived from the original one, which can be also time-reversed. Following similar methods as for the total entropy production fluctuation theorem, they obtained

$$\frac{P(\Delta_{\rm i}s^{\rm na})}{\tilde{P}_D(-\Delta_{\rm i}s^{\rm na})} = e^{\Delta_{\rm i}s^{\rm na}}, \qquad \frac{P(\Delta_{\rm i}s^{\rm ad})}{P_D(-\Delta_{\rm i}s^{\rm ad})} = e^{\Delta_{\rm i}s^{\rm ad}}, \tag{3.58}$$

where P_D denotes the probability in the dual dynamics, and $\tilde{P}_D$ the probability in the time-inverted dual dynamics, also called *dual-reverse* dynamics. The fluctuation theorems for the adiabatic and non-adiabatic entropy production constitute a further refinement of the second law for nonequilibrium steady states. They also generalize previous notions introduced in this context as the *house-keeping* heat and *excess* heat [129], corresponding respectively to the heat dissipated in order to maintain a nonequilibrium steady state, and the heat dissipated when the system is driven far from that state.

3.2.3 Quantum Fluctuation Theorems

We have seen some of the most important fluctuation theorems derived for both deterministic and stochastic dynamics in the classical regime. At this point we turn back our view to quantum systems, in which extensions of the above fluctuation theorems are highly desirable. However, a number of difficulties arise when considering the extension of the concepts introduced in stochastic thermodynamics to the quantum regime. The first one is the absence of the same notion of trajectory. As we have seen in Sect. 2.5 quantum trajectories can be defined, but they require the introduction of quantum measurements to monitor the system, which in turn introduces a back-action on the system being measured. Consequently, in most quantum extensions of fluctuation theorems, projective measurements are introduced at the beginning and at the end of the process of interest [3, 89]. We call this framework the *two measurement protocol* (TMP). Within this approach, work fluctuation relations such as the Jarzynski equality and the Crooks fluctuation theorem in both isolated [130–133] and open systems [132, 134–139], as well as various fluctuation theorems for heat and matter exchange in nonequilibrium steady states [105, 136, 140–147], have been derived (see also the reviews [3, 89]). This approach is in contrast with alternative attempts to derive work fluctuation theorems through the expectations of a work operator [13, 87, 148].

We will first discuss the case of work fluctuation theorems for isolated quantum systems and then consider open systems. In this context, a process in the TMP framework consists in assuming at time t_0 the state of the system to be given by some initial density operator ρ_{t_0}, and Hamiltonian $\hat{H}(\lambda)$ depending on the external control parameter λ. A first projective measurement of the energy is hence performed in the system, corresponding to measuring the Hamiltonian

$$\hat{H}(\lambda_A) = \sum_n E_n^A |E_n^A\rangle\langle E_n^A|, \tag{3.59}$$

where $\lambda_A \equiv \lambda(t_0)$. This measurement induces a collapse of the system state to one of the pure states $|E_n^A\rangle$ corresponding to outcome n. In analogy with the classical case, the control parameter is then varied following a prescribed protocol $\Lambda = \{\lambda(t)\}_{t_0}^{\tau}$, the evolution being given by some unitary operator $\hat{U}_\Lambda$ as in Sect. 1.1.4. At time τ the driving is fixed, $\lambda_B \equiv \lambda(\tau)$ and a second measurement of the Hamiltonian is performed

$$\hat{H}(\lambda_B) = \sum_m E_m^B |E_m^B\rangle\langle E_m^B|, \tag{3.60}$$

obtaining some outcome m and the final state $|E_m^B\rangle$. A backward process can be defined as well by inverting the previous sequence (and the driving protocol) while measuring the time-reversed Hamiltonian $\tilde{\hat{H}}(\lambda_B) = \hat{\Theta}\hat{H}(\lambda_B)\hat{\Theta}^\dagger$ and $\tilde{\hat{H}}(\lambda_A) = \hat{\Theta}\hat{H}(\lambda_A)\hat{\Theta}^\dagger$, $\hat{\Theta}$ being the time-reversal anti-unitary operator in quantum mechanics (see Sect. 1.1.4). For the backward process we choose as initial condition some arbitrary initial state denoted $\tilde{\rho}_{t_0}$. In the present setup *trajectories* can be defined by the outcomes of initial and final measurements, $\gamma = \{n, m\}$, together with the driving protocol Λ. According to Born rule, those trajectories are sampled with probability

$$P_\gamma = p_n |\langle E_m^B|\hat{U}_\Lambda|E_n^A\rangle|^2, \tag{3.61}$$

where $p_n \equiv \mathrm{Tr}[\rho_{t_0}|E_n^A\rangle\langle E_n^A|]$ is the probability to obtain outcome n in the initial projective measurement. In the backward process the inverse trajectory $\tilde{\gamma} = \{m, n\}$ associated to the same measurement results and time-reversed driving protocol $\tilde{\Lambda} = \{\lambda(t_0 + \tau - t)\}_{t_0}^{\tau}$ analogously reads

$$\tilde{P}_{\tilde{\gamma}} = \tilde{p}_m |\langle E_n^A|\hat{\Theta}^\dagger \hat{U}_{\tilde{\Lambda}} \hat{\Theta}|E_m^B\rangle|^2, \tag{3.62}$$

with $\tilde{p}_m = \mathrm{Tr}[\tilde{\rho}_{t_0}\hat{\Theta}|E_m^B\rangle\langle E_m^B|\hat{\Theta}^\dagger]$ and $\hat{U}_{\tilde{\Lambda}}$ the unitary governing the time-reversed dynamics. Furthermore, as long as the system is isolated, the work in a trajectory γ can be identified with the change in energy of the system as given by the measurement outcomes:

$$w_\gamma = E_m^B - E_n^A. \tag{3.63}$$

Analogously the work performed in the reverse trajectory generated in the backward process can be computed as

$$w_{\tilde{\gamma}} = E_m^A - E_n^B = -w_\gamma. \tag{3.64}$$

At this point it is important to notice the intimate relation between the probabilities of forward and backward trajectories

$$\frac{P_\gamma}{\tilde{P}_{\tilde{\gamma}}} = \frac{p_n}{\tilde{p}_m} \times \frac{|\langle E_m^B|\hat{U}_\Lambda|E_n^A\rangle|^2}{|\langle E_n^A|\hat{\Theta}^\dagger \hat{U}_{\tilde{\Lambda}} \hat{\Theta}|E_m^B\rangle|^2} = \frac{p_n}{\tilde{p}_m}, \tag{3.65}$$

where the second term above cancels by applying the microreversibility principle for non-autonomous systems derived in Sect. 1.1.4, i.e. using $\hat{U}_\Lambda = \hat{\Theta}^\dagger \hat{U}_{\tilde{\Lambda}}^\dagger \hat{\Theta}$. If we now denote by $P(w) \equiv \sum_{n,m} P_\gamma \, \delta(w - w_\gamma)$ the probability to obtain work w in the forward process, and $\tilde{P}(w) \equiv \sum_{n,m} \tilde{P}_{\tilde{\gamma}} \delta(w - w_{\tilde{\gamma}})$ the probability to obtain work w in the backward one, it follows that

$$\tilde{P}(-w) = \sum_{n,m} \tilde{P}_{\tilde{\gamma}} \, \delta(w - w_{\tilde{\gamma}}) = \sum_{n,m} \frac{\tilde{p}_m}{p_n} P_\gamma \, \delta(w - w_\gamma). \tag{3.66}$$

The Crooks fluctuation theorem, also called Tasaki–Crooks fluctuation theorem in the quantum context [89], follows from this expression by assuming that the initial states of the system in forward and backward processes are thermal equilibrium states at some inverse temperature β:

$$\rho_{t_0} = \frac{e^{-\beta \hat{H}(\lambda_A)}}{Z_A}, \qquad \tilde{\rho}_{t_0} = \hat{\Theta} \frac{e^{-\beta \hat{H}(\lambda_B)}}{Z_B} \hat{\Theta}^\dagger, \tag{3.67}$$

which implies $p_n = e^{-\beta E_n^A}/Z_A$ and $\tilde{p}_m = e^{-\beta E_m^B}/Z_B$. Inserting the expressions for p_n and $\tilde{p}_m$ in Eq. (3.66) we obtain

$$\begin{aligned} \tilde{P}(-w) &= \frac{Z_A}{Z_B} \sum_{n,m} e^{-\beta(E_m^B - E_m^A)} P_\gamma \, \delta(w - w_\gamma) \\ &= \frac{Z_A}{Z_B} e^{-\beta w} \, P(w) = e^{-\beta(w - \Delta F_{AB})} \, P(w), \end{aligned} \tag{3.68}$$

and the Crooks fluctuation theorem in Eq. (3.48) is recovered. In the last step we used the thermodynamic relations linking the Helmholtz free energy with the partition function $F_A = -k_B T \ln Z_A$ and $F_B = -k_B T \ln Z_B$. The derivation of Eq. (3.68) may be complemented with various remarks. First, we have explicitly seen that the Crooks fluctuation theorem follows from two key ingredients: the microreversibility principle for non-autonomous systems and the shape of the Gibbs thermal states. This feature has been widely stressed in the literature [3, 89]. Second, contrary to previous derivations [3, 89], we do not require the Hamiltonian of the system to be invariant under time reversal, $[\hat{\Theta}, \hat{H}(\lambda)] = 0$, at the price that the initial state of the backward process has to be time-reversed, c.f. Eq. (3.67). Third, the assumption of initial thermal equilibrium states implies the access to a thermal reservoir at inverse temperature β, to which the system has to be coupled in order to prepare the initial states. Fourth, integrating the work probability distributions at both sides of Eq. (3.48), Jarzynski equality [Eq. (3.45)] immediately follows.

We notice that in Ref. [89] a complementary derivation of Crooks fluctuation theorem and Jarzynski equality is provided. It is based on the characteristic function of work, defined as the Fourier transform of the work probability distribution

$$G(u) = \int dw\, e^{iuw}\; P(w), \tag{3.69}$$

which contains full information about the statistics of the work w. Talkner et al. showed that in the present case it acquires the form [133, 149]

$$G(u) = \mathrm{Tr}[\hat{U}_\Lambda^\dagger e^{iu\hat{H}(\lambda_B)} \hat{U}_\Lambda e^{-(iu+\beta)\hat{H}(\lambda_A)}]/Z_A. \tag{3.70}$$

For the probability work distribution in the backward process one can analogously define a characteristic function $\tilde{G}(u) = \int dw\; e^{iuw}\;\; \tilde{P}(w)$. These two characteristic functions are related as a consequence of microreversibility, obeying

$$Z_A\; G(u) = Z_B\; \tilde{G}(-u + i\beta). \tag{3.71}$$

Finally, applying the inverse Fourier transform, the Crooks fluctuation theorem is recovered [89]. This approach is of very practical utility as avoiding the reconstruction of the probability distribution of work by projective measurements is of technical importance in the experimental test of fluctuation relations [150, 151]. Instead, different schemes to measure the characteristic function of work $G(u)$ based on interferometric schemes have been proposed [152–154]. It is also worth stressing that the Crooks fluctuation theorem has been experimentally tested in the full quantum regime only recently in a liquid-state NMR platform [155], where the work probability distributions have been assessed indirectly through the characteristic function of work (see Fig. 3.7). The Jarzynski equality has been also tested recently in a single $^{171}\mathrm{Yb}^+$ ion trapped in a harmonic potential using projective measurements [156], following the proposal presented in Ref. [157].

The derivation of the Crooks fluctuation theorem and the Jarzynski equality can also be extended to the case of open quantum systems in contact with thermal reservoirs, both for weak or strong coupling with the environment [89]. In the open system situation the total Hamilton operator reads

$$\hat{H}_{\mathrm{tot}}(\lambda) = \hat{H}_S(\lambda) + \hat{H}_E + \hat{H}_{\mathrm{int}}, \tag{3.72}$$

$\hat{H}_S(\lambda)$ being the Hamilton operator of the system defined as in the isolated case, $\hat{H}_E = \sum_\nu \epsilon_n |\epsilon_\nu\rangle\langle\epsilon_\nu|$ for the reservoir with eigenvalues $\{\epsilon_\nu\}$, and $\hat{H}_{\mathrm{int}}$ for the interaction term. An important assumption here concerns the dependence on the external control parameter λ, which we assumed to enter only in the system Hamiltonian [158]. The global system is considered to be isolated, and hence we may apply the above TMP to the present situation. However, now we are interested in measuring simultaneously the local Hamiltonians of system and environment. A trajectory consists in this case in four outcomes $\gamma = \{n, \nu, m, \mu\}$, where we include the outcomes ν and μ for the

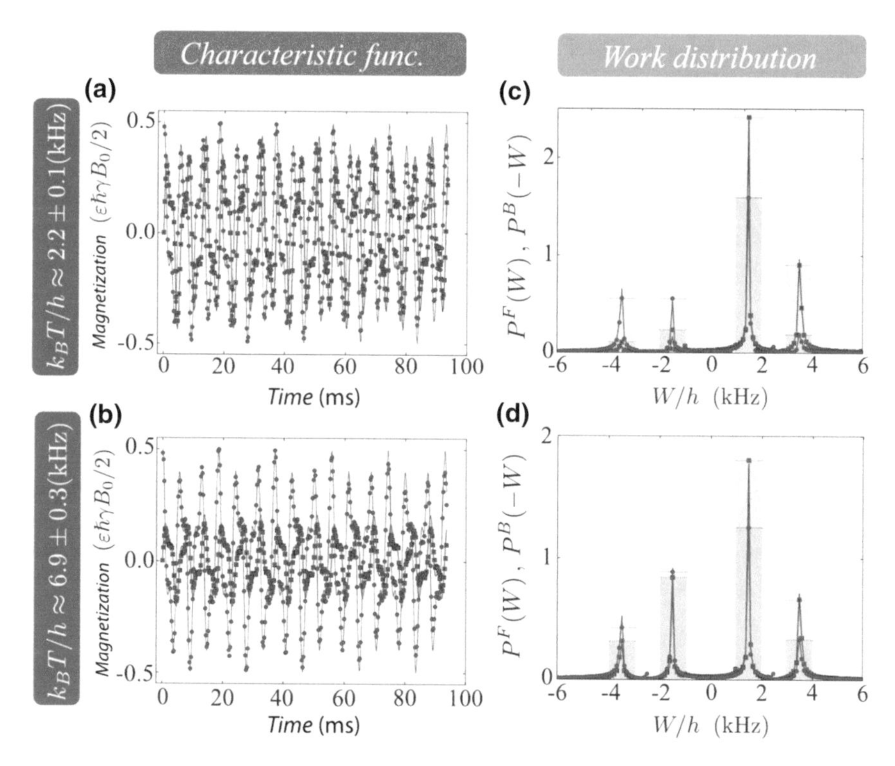

Fig. 3.7 Experimental data for the reconstruction of the characteristic function of work (left column) and the work probability distribution (right column) of forward (red) and backward (blue) processes in the experiment reported in Ref. [155]. There, the ^{13}C and ^{1}H nuclear spins of a chloroform-molecule sample are respectively used as the system driven by a resonant radio-frequency pulse, and ancilla for the reconstruction of the characteristic functions by measuring the x and y components of its transverse magnetization. Picture taken from Ref. [155]

energy measurements in the reservoir at the beginning and at the end of the driving protocol, respectively. The probability to obtain a trajectory γ reads in this case

$$P_\gamma = p_{n,\nu} |\langle E_m^B| \otimes \langle \epsilon_\mu| \, \hat{U}_\Lambda \, |E_n^A\rangle \otimes |\epsilon_\nu\rangle|^2, \tag{3.73}$$

where $p_{n,\nu}$ is the probability to obtain outcomes n and ν in the initial energy measurement, and $\hat{U}_\Lambda$ is generated by the global Hamiltonian $\hat{H}_{\text{tot}}(\lambda)$. The energy changes in the system ΔE, and in the thermal reservoir $\Delta\epsilon$, for the protocol Λ defined as in the previous case, are:

$$\Delta E_\gamma = E_m^B - E_n^A, \qquad \Delta\epsilon_\gamma = \epsilon_\mu - \epsilon_\nu. \tag{3.74}$$

If we assume weak-coupling between system and reservoir the energy associated to the interaction term $\hat{H}_{\text{int}}$ becomes negligible, and the total energy change in the global system (the work externally performed) becomes the sum of system and reservoir changes, $w_\gamma = \Delta E_\gamma + \Delta\epsilon_\gamma$. Furthermore we can identify the heat with the energy changes in the reservoir as $q_\gamma = -\Delta\epsilon_\gamma$, recovering the first law (energy balance) [89]

$$\Delta E_\gamma = w_\gamma + q_\gamma. \tag{3.75}$$

Following the same prescriptions as in the isolated case, one can hence obtain the joint probability distribution for the system energy changes *and* heat $P(\Delta E, q)$, with corresponding characteristic function

$$G(u, v) = \int d(\Delta E)\, dq\; e^{i(u\Delta E + vq)\;P(\Delta E, q)}, \tag{3.76}$$

and analogously for the backward process leading to $\tilde{P}(\Delta E, q)$ and $\tilde{G}(u, v)$. Furthermore, we assume thermal equilibrium initial states of forward and backward process as in Eq. (3.67), where $\hat{H}$ is replaced by the total Hamiltonian $\hat{H}_{\text{tot}}$, and we denote Y_A and Y_B the *total* partition functions at the beginning and end of the protocol Λ respectively. Importantly, the weak coupling assumption implies that the global equilibrium initial state approximately factorizes (see Ref. [138])

$$\rho_{t_0} = \frac{e^{-\beta \hat{H}_{\text{tot}}(\lambda_A)}}{Y_A} \approx \frac{e^{-\beta \hat{H}_S(\lambda_A)}}{Z_A} \otimes \frac{e^{-\beta \hat{H}_E}}{Z_E}, \tag{3.77}$$

where $Z_A = \text{Tr}[e^{-\beta \hat{H}_S(\lambda_A)}]$ and $Z_E = \text{Tr}[e^{-\beta \hat{H}_E}]$. This implies that the probability of outcomes in the initial measurement becomes $p_{n,\nu} = e^{-\beta(E_n^A + \epsilon_\nu)}/Z_A Z_E$, and analogously for the initial state of the backward process, $\tilde{\rho}_{t_0}$, we have $\tilde{p}_{m,\mu} = e^{-\beta(E_m^B + \epsilon_\nu)}/Z_B Z_E$, with $Z_B = \text{Tr}[e^{-\beta \hat{H}_S(\lambda_B)}]$. The above relations, together with the microreversibility principle, are the key points to obtain the following relation, analogous to Eq. (3.71):

$$Z_A\, G(u, v) = Z_B\, \tilde{G}(-u + i\beta, -v + -i\beta), \tag{3.78}$$

which immediately implies [89]

$$\frac{P(\Delta E, q)}{\tilde{P}(-\Delta E, -q)} = e^{\beta(\Delta E - q - \Delta F_{AB})}, \tag{3.79}$$

$\Delta F_{AB} = -k_B T \ln(Z_B/Z_A)$ being the system free energy difference. Finally, performing a change of variable $\Delta E \to w = \Delta E - q$, the right-hand side of Eq. (3.79) becomes $e^{\beta(w - \Delta F_{AB})}$, and integrating both $P(w, q)$ and $\tilde{P}(-w, -q)$ over the heat q, the Crooks fluctuation theorem is finally obtained [89]. We stress that the weak coupling assumption can be overcome by performing global measurements of the total Hamiltonian $\hat{H}_{\text{tot}}(\lambda)$ at the beginning and at the end of the processes, together with the key identification of the system partition function $Z(\lambda) \equiv \text{Tr}[e^{-\beta \hat{H}_{\text{tot}}(\lambda)}]/Z_E \neq \text{Tr}[e^{-\beta \hat{H}_S(\lambda)}]$ in general [139].

Along with the quantum extension of Jarzynski equality and Crooks fluctuation theorems, the Seifert integral fluctuation theorem for the entropy production has been also extended to the quantum regime in Ref. [159]. Assuming an open quantum system weakly coupled to a thermal reservoir at inverse temperature β, the TMP is implemented as in the case of the Crooks fluctuation theorem above. However, in this case the initial state of the open system is arbitrary, while the local measurements performed on the system Hamiltonian are replaced by measurements on the system density operator itself, obtaining eigenvalues $\rho_n^{t_0}$ and ρ_m^{τ} respectively. On the other hand, energetic measurements of the reservoir Hamiltonian are preserved in order to keep trace of the heat dissipated. In this setting, the (information) entropy production for a single realization can be defined as

$$\Sigma_\gamma \equiv \Delta s_\gamma - \beta q_\gamma, \tag{3.80}$$

with $\Delta s_\gamma = s_\tau - s_{t_0} = -\ln \rho_m^\tau + \ln \rho_n^{t_0}$ the stochastic entropy change in the system, and $q_\gamma = -(\epsilon_\mu - \epsilon_\nu)$ the heat entering the system from the thermal reservoir according to Eq. (3.75). In analogy to the classical case, here the term $-\beta q_\gamma \equiv \Delta s_\gamma^{\mathrm{m}}$ can be interpreted as the entropy change in the medium (or minus the entropy flowing into the system). Hence following the same lines as in the Crooks fluctuation theorem for open quantum systems, the authors derive the joint probability distribution for entropy changes in system and reservoir $P(\Delta s, \beta q)$, which performing the variable change $\Delta s \to \Sigma$ and integrating over βq gives

$$P(\Sigma) = \sum_{m,n,\mu,\nu} P_\gamma \, \delta(\Sigma - \Sigma_\gamma). \tag{3.81}$$

The integral fluctuation theorem can be derived from microreversibility and the normalization property of the density operator [159]

$$\langle e^{-\Sigma} \rangle = \int d\Sigma \, e^{-\Sigma} \, P(\Sigma) = \sum_{m,n,\mu,\nu} P_\gamma = 1. \tag{3.82}$$

We finally stress that the classical fluctuation theorems generalizing the Jarzynski equality and the Crooks fluctuation theorem to include feedback control [122], have been extended as well to the quantum regime [160]. Furthermore, a quantum fluctuation theorem capturing the role of classical correlations in heat exchange processes has been also recently derived in Ref. [161].

In Chaps. 7 and 8 we will develop further fluctuation theorems for the entropy production extending the results presented in this section. In particular, we will be interested in obtaining expressions for the entropy production that overcome the limitations imposed by the presence of a thermal environment, allowing more general quantum states for the surroundings leading to nonequilibrium steady states or multiple conserved quantities. In this context our aim is to extend the integral and detailed fluctuation theorems for the entropy production to the quantum case and to more arbitrary nonequilibrium situations. On the other hand, we will also investigate the general split of the entropy production in adiabatic and non-adiabatic contributions and the cases for which the detailed fluctuation theorems presented in the classical case [99, 123, 127] hold for general quantum evolutions described by CPTP maps.

3.3 Quantum Thermal Machines

A second important branch of quantum thermodynamics is the description and analysis of the operation of small quantum thermal machines. A quantum thermal machine is generally intended as a small quantum system operating between different reservoirs and possibly subjected to external driving, which performs a thermodynamic task such as work extraction, refrigeration, pumping heat, or information erasure. When the thermodynamic task operated by the machine is work extraction from a difference of temperatures, we call it a *heat engine*. Analogously, if the operation consists in pumping heat into the hotter body of the configuration we call the machine a *heat pump*, and if the heat is extracted from the coldest body, it will be referred to as a *fridge*. Finally, if the task performed by the machine consists in erasing information in the Landauer's principle sense (see Sect. 3.1.6), we will call it an *eraser*.

The first proposal of a quantum thermal machine is due to Scovil and Schultz-DuBois, who proposed in 1959 to view the three-level maser as a quantum heat engine [8]. This machine may reach Carnot efficiency at the verge of population inversion in the signal transition of the maser. Its operation can be further reversed to obtain a refrigerator, which can also reach maximum efficiency [162]. Inspired by this pioneering work, different models for power-driven heat engines and refrigerators, as well as absorption thermal machines based on lasers and masers have been discussed in the following decades [9, 163–167], while lasing cooling techniques were in parallel developed and implemented in the laboratory [168].

The different models of thermal machines, which can be ultimately reduced to one or few two-level systems, three-level systems or harmonic oscillators, present many common features [169–173]. They provide an ideal platform to study quantum thermodynamics from a theoretical point of view [12], where the role of quantum effects in the performance of the devices can be also addressed in many situations of interest [5]. Furthermore, they may be of practical importance in biological processes [174], quantum state preparation and metrology [175, 176], or for the implementation of future quantum technologies such as quantum computers [177]. Quantum heat engines can operate in cycles where the machine follows several steps, called strokes, in which it can be in contact with different reservoirs or manipulated by means of external driving [1, 178, 179]. On the other hand, they can operate continuously in their steady state regime [171]. They can be further classified as machines with time-dependent fields or external driving, or autonomous machines that function without any external control (see e.g. Refs. [45, 167, 180, 181]).

Recent proposals of quantum heat engines include quantum Otto cycles on single trapped ions or atoms [182, 183], optomechanical setups [184, 185], or driven superconducting qubits [186]. A Stirling cycle in a nanomechanical system controlled by optical fields has been proposed in Ref. [187], and a SWAP engine implemented in solid state platforms in Ref. [188]. Other interesting proposals are thermoelectric devices using ultracold atoms [189], hybrid micro-wave cavities [190], or Josephson junctions [191]. Autonomous realizations of quantum thermal machines have been also recently proposed, e.g. quantum absorption refrigerators using four quantum dots [192], an atom-cavity system [193], or circuit-QED architectures [194, 195]. Other proposals comprise refrigerators driven by sunlight [196, 197], a heat engine generating steady-state entanglement [198], a three-level thermal machine powered by a nonequilibrium electromagnetic field [199], or a rotor heat engine implemented in an optomechanical system [200].

In the following we will introduce one of the most important four-stroke cycles operated in quantum heat engines, namely, the quantum Otto cycle, which has been recently implemented in the laboratory using a single ion in a Paul trap [183]. We will discuss the performance of the cycle and its main properties when the working substance consists of a finite level system or a single bosonic mode. Furthermore, we will also analyze a simple model for an autonomous quantum heat engine consisting of a pair of qubits operating in steady state conditions between two thermal reservoirs at different temperatures and a third quantum system acting as a load [180, 181, 201]. Finally, we will discuss the different configurations leading to genuine quantum features affecting (is some cases dramatically) the performance of quantum thermal machines.

3.3.1 Quantum Otto Cycle

Otto cycles are widely used in most common macroscopic heat engines, such as the internal combustion engine. The quantum version has been discussed e.g. in Refs. [178, 182, 202, 203]. In the quantum Otto cycle the working substance performs four strokes, namely, two isochoric processes and two isentropic processes.

Let us consider two thermal reservoirs at temperatures T_1 and T_2, with $T_2 \geqslant T_1$. We denote the state of the generic quantum working substance performing the cycle at the fourth points between the strokes as ρ_A, ρ_B, ρ_C and ρ_D, respectively. Its Hamiltonian $\hat{H}(\lambda)$ is considered to be externally controlled through the variation of a control parameter λ. The cycle starts with the working substance in thermal equilibrium with the reservoir at the lower temperature T_1 and control parameter at λ_1, that is,

$$\rho_A = \frac{e^{-\beta_1 \hat{H}(\lambda_1)}}{Z_1}, \tag{3.83}$$

with $\beta_1 = 1/k_B T_1$ and Z_1 the partition function. The first stroke $A \to B$ is an isentropic compression with the working substance detached from the reservoirs, during which the control parameter is changed, $\hat{H}(\lambda_1) \to \hat{H}(\lambda_2)$. This operation can be described by means of a unitary operator $\hat{U}_{A\to B}$, and the state of the working substance at point B simply reads $\rho_B = \hat{U}_{A\to B}\rho_A \hat{U}^\dagger_{A\to B}$. In this step there is no heat exchange with the reservoirs, while the external driving performs a work

$$W_{A\to B} = \mathrm{Tr}[\rho_B \hat{H}(\lambda_2)] - \mathrm{Tr}[\rho_A \hat{H}_1]. \tag{3.84}$$

The second stroke $B \to C$ is an isochoric process in which the Hamiltonian of the working substance is kept fixed at $\hat{H}(\lambda_2)$. The working substance is here put in contact with the reservoir at T_2, until it reaches the thermal equilibrium state

$$\rho_C = \frac{e^{-\beta_2 \hat{H}(\lambda_2)}}{Z_2}, \tag{3.85}$$

with $\beta_2 = 1/k_B T_2$ and Z_2 the partition function. In this stroke, all energy changes in the working substance are due to heat absorption from the reservoir involving an entropy flow

$$Q_{B\to C} = \mathrm{Tr}[(\rho_C - \rho_B)\hat{H}(\lambda_2)], \qquad \Delta_{\mathrm{e}} S_{B\to C} = Q_{B\to C}/T_2. \tag{3.86}$$

The third stroke $C \to D$ corresponds to an isentropic expansion, where the working substance is again detached from the reservoirs, and the control parameter is modulated back to its original value, $\hat{H}(\lambda_2) \to \hat{H}(\lambda_1)$. The system state changes to $\rho_D = \hat{U}_{C\to D}\rho_C \hat{U}^\dagger_{C\to D}$ while performing a work

$$W_{C\to D} = \mathrm{Tr}[\rho_D \hat{H}(\lambda_1)] - \mathrm{Tr}[\rho_C \hat{H}_2]. \tag{3.87}$$

Finally, the cycle is closed by means of the forth stroke $D \to A$, corresponding to the second isochoric process occurring at fixed position of the control parameter, and letting the system relax back to ρ_A in contact with the thermal reservoir at temperature T_1. The heat entering the system in this last stroke and its corresponding entropy flow are given by

$$Q_{D\to A} = \mathrm{Tr}[(\rho_A - \rho_D)\hat{H}(\lambda_1)], \qquad \Delta_{\mathrm{e}} S_{D\to A} = Q_{D\to A}/T_1. \tag{3.88}$$

In Fig. 3.8 we show a picture of the recent experiment by Roßnagel et al. [183] implementing a Quantum Otto cycle single ion.

Using the above expressions for work and heat along the cycle, we notice that the first law of thermodynamics adopts the form

$$\Delta U_{\mathrm{cycle}} = W_{A\to B} + Q_{B\to C} + W_{C\to D} + Q_{D\to A} = 0, \tag{3.89}$$

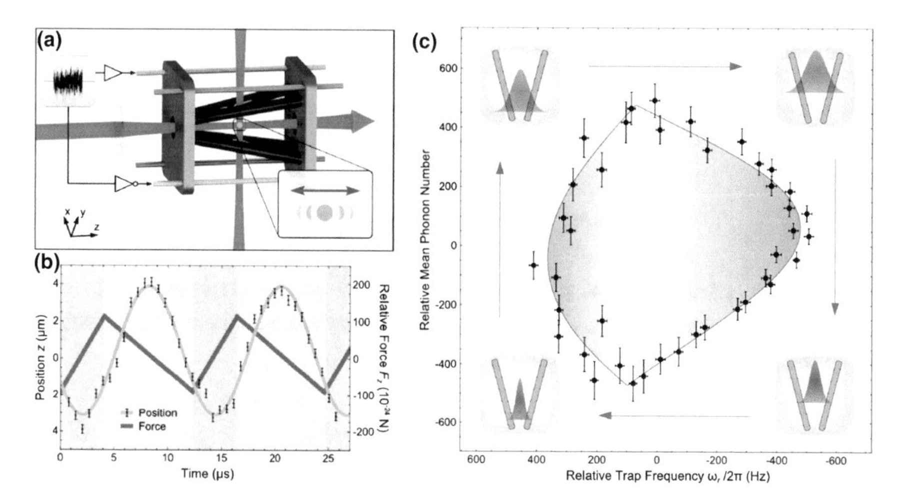

Fig. 3.8 The quantum Otto heat engine recently implemented in the laboratory by Roßnagel et al. [183]. A single ion inside a linear Paul trap with tapered geometry (A, B) generates work to drive harmonic oscillations in the axial position of the ion (C) by performing a quantum Otto cycle (D). Hot and cold reservoirs are engineered by using electric-field noise and laser cooling respectively. In (D) we can identify the pictographs at the corners with the states ρ_A (bottom-right), ρ_B (bottom-left), ρ_C (upper-left) and ρ_D (upper-right) introduced in the text for the ideal case. The points with error bars correspond to the real cycle performed by the ion. Red and blue shaded areas correspond to heating and cooling processes. Picture taken from Ref. [183]

and the total work extracted in the cycle can be hence defined as

$$W_{\text{ext}} \equiv -W_{A\to B} - W_{C\to D} = Q_{B\to C} + Q_{D\to A}. \tag{3.90}$$

A basic condition for the functioning of the heat engine is a positive work extraction $W_{\text{ext}} \geqslant 0$, obtained from the heat absorbed by the hot reservoir $Q_{B\to C} \geqslant 0$. By looking at Eqs. (3.84) and (3.87), we can maximize the work extracted in the isentropic expansion, $A \to B$, and minimize the one wasted in the isentropic compression, $C \to D$. As both processes are isentropic, we notice that maximum work can be extracted in the cycle by ensuring the states ρ_B and ρ_D being of Gibbs form, because these minimizes the energy for fixed entropy. That is

$$\rho_B \equiv \frac{e^{-\beta_1^* \hat{H}(\lambda_2)}}{Z_1^*}, \qquad \rho_D \equiv \frac{e^{-\beta_2^* \hat{H}(\lambda_1)}}{Z_2^*}, \tag{3.91}$$

for some arbitrary parameters β_1^* and β_2^*. The isentropic strokes in such case correspond to quasi-static modulation of the working substance energy eigenstates, such that the quantum adiabatic theorem can be applied (see also Chap. 10). On the other hand, the refrigerator condition implies heat extraction from the cold reservoir $Q_{D\to A} \geqslant 0$, at the price of external input work $W_{\text{in}} \equiv -W_{\text{ext}} \geqslant 0$. Analogous arguments lead to the same conclusion for the quasi-static modulation of the control parameter when maximizing the heat extracted from the cold reservoir.

The energetic efficiency of the cycle operated as a heat engine can be defined as the ratio between the total work output and the heat absorbed from the hot reservoir

$$\eta_{\text{engine}} \equiv \frac{W_{\text{ext}}}{Q_{B\to C}} = 1 - \frac{Q_{D\to A}}{Q_{B\to C}}. \tag{3.92}$$

The efficiency of refrigerators are instead typically measured by the so-called *coefficient of performance* (COP), given by the ratio between the heat extracted from the cold reservoir, divided by the work input in the cycle:

$$\eta_{\text{fridge}} = \frac{Q_{D\to A}}{W_{\text{in}}}. \tag{3.93}$$

We now discuss the performance of the cycle using the second law of thermodynamics. It can be stated as the positivity of the entropy production in the cycle

$$\Delta_{\text{i}} S_{\text{cycle}} = \Delta S_{\text{cycle}} - \Delta_{\text{e}} S_{\text{cycle}} \geqslant 0, \tag{3.94}$$

where $\Delta S_{\text{cycle}} = 0$, and $\Delta_{\text{e}} S_{\text{cycle}} = \Delta_{\text{e}} S_{B\to C} + \Delta_{\text{e}} S_{D\to A}$. Inserting the expressions of the entropy flows during the isochoric strokes, Eqs. (3.86) and (3.88), we obtain

$$-\frac{Q_{D\to A}}{T_1} - \frac{Q_{B\to}}{T_2} \geqslant 0, \tag{3.95}$$

which, by using the first law in Eq. (3.89) can be rewritten in the two following equivalent forms

$$W_{\text{ext}} \leqslant \left(1 - \frac{T_1}{T_2}\right) Q_{B\to C}, \qquad W_{\text{in}} \geqslant \left(1 - \frac{T_1}{T_2}\right) Q_{D\to A}. \tag{3.96}$$

These two relations imply Carnot bounds for the efficiency of the heat engine and fridge configurations for any quantum working substance

$$\eta_{\text{engine}} \leqslant 1 - \frac{T_1}{T_2} = \eta_{\text{carnot}}, \qquad \eta_{\text{fridge}} \leqslant \frac{T_2}{T_2 - T_1} = \eta_{\text{carnot}}^{-1}. \tag{3.97}$$

However, it is important to notice that reaching Carnot efficiency means as well that the work extracted in a cycle vanishes. Indeed, in quantum Otto engines with quasi-static isentropic strokes, the work extracted in a single cycle vanishes at such conditions in all known models [178, 182, 202, 203]. Henceforth a figure of merit for practical applications is the efficiency at maximum power. For the Otto cycle operating as a heat engine with quasi-static driving in the isentropic strokes, and in the high-temperature limit, the efficiency at maximum power is given by the well-known Curzon–Ahlborn formula [204]

$$\eta_{\mathrm{CA}} = 1 - \sqrt{\frac{T_1}{T_2}}. \tag{3.98}$$

This has been checked in Refs. [182, 203] for the case of a quantum harmonic oscillator as working substance. Finally, in the recent proposal of Ref. [205], the quasi-static strokes of the Otto cycle are replaced by finite-time processes generating the same final states, usually called shortcuts to adiabaticity [206], while in Ref. [207] the effects of inner friction due to finite-time transformations and disorder effects in the quantum Otto cycle have been explored.

3.3.2 Autonomous Thermal Machines

Autonomous quantum thermal machines function in steady state conditions via thermal contact to heat baths at different temperatures, powering different thermodynamic operations without the need of any external driving. Some examples of autonomous thermal machines are small quantum absorption refrigerators, which use only two thermal reservoirs, one as a heat source, and the other as a heat sink, in order to cool a system to a temperature lower than that of either of the thermal reservoirs. Models for autonomous thermal machines have been provided for a three-level system [8, 167], two two-level systems [169, 180], or three harmonic oscillators [45], among others [52, 208–210]. The efficiency of these machines has been investigated [8, 211, 212], and quantum effects, such as coherence and entanglement, were shown to enhance their performance [169, 213–216].

The basic quantum absorption fridge composed by a three-level system, whose transitions are weakly coupled to three different heat reservoirs at different temperatures, is analyzed in detail in Chap. 11. There, its fundamental function is analyzed together with possible extensions to multilevel setups. Furthermore, the fluctuations in this basic fridge configuration are studied in Chap. 9 as an application of quantum fluctuation theorems developed in Chaps. 7 and 8. Here we will instead discuss a similar model for an autonomous heat engine introduced in Refs. [181, 217], where the external classical driving field arising in the prototypical three-level amplifier [171] is substituted by a fully quantum system consisting of an infinite ladder of energy levels which acts as a *weight* (see Fig. 3.9).

The heat engine is composed by a pair of two-level systems (or qubits) described by the basis states $\{|0\rangle_1, |1\rangle_1\}$ and $\{|0\rangle_2, |1\rangle_2\}$, and Hamilton operators $\hat{H}_1 = E_1|1\rangle\langle 1|_1$ and $\hat{H}_2 = E_2|1\rangle\langle 1|_2$ respectively. Each qubit is coupled to a different thermal reservoir at temperature T_1 and T_2 respectively as depicted in Fig. 3.9. Taken together, the two qubits form a four-level system with tensor-product basis states

$$\{|0\rangle_1|0\rangle_2, |1\rangle_1|0\rangle_2, |0\rangle_1|1\rangle_2, |1\rangle_1|1\rangle_2\}, \tag{3.99}$$

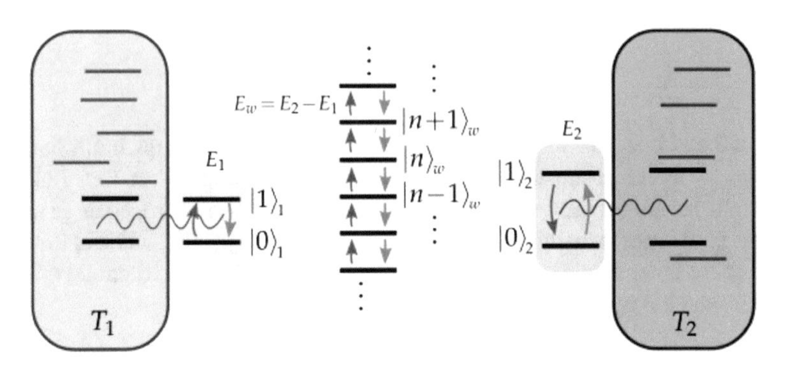

Fig. 3.9 Model of quantum heat engine discussed in the text. The machine is composed by two qubits with energy spacings E_1 and E_2, which are in thermal contact with reservoirs at temperatures T_1 and T_2. Work extraction from an external driving field is substituted by a quantum weight consisting of an infinite energy ladder with spacing $E_{\mathrm{w}} = E_2 - E_1$. Picture taken from Ref. [181]

and corresponding energy eigenvalues $\{0, E_1, E_2, E_1 + E_2\}$. Each qubit is assumed to be in thermal equilibrium with its respective reservoir, $\rho_1 = e^{-\beta_1 \hat{H}_1}/Z_1$ and $\rho_2 = e^{-\beta_2 \hat{H}_2}/Z_2$, with $\beta_1 = 1/k_B T_1$ and $\beta_2 = 1/k_B T_2$. This implies that the populations of the four-level system are given by

$$p_{00} = \frac{1}{Z_1 Z_2}, \quad p_{10} = \frac{e^{-\beta_1 E_1}}{Z_1 Z_2}, \quad p_{01} = \frac{e^{-\beta_2 E_2}}{Z_1 Z_2}, \quad p_{11} = \frac{e^{-\beta_1 E_1 - \beta_2 E_2}}{Z_1 Z_2}.$$

The inner transition of the four level system plays an important role in the model, and will be called the *virtual qubit* of the machine. The ratio of its populations obeys the following Gibbs ratio [181]

$$\frac{p_{01}}{p_{10}} = e^{-\beta_{\mathrm{v}} E_{\mathrm{v}}}, \quad \text{with} \quad \beta_{\mathrm{v}} \equiv \frac{E_2}{E_{\mathrm{v}}}\beta_2 - \frac{E_1}{E_v}\beta_1, \tag{3.100}$$

where $E_{\mathrm{v}} \equiv E_2 - E_1$ and the so-called (inverse) *virtual temperature*, β_{v}, has been introduced. Remarkably β_{v} can take negative values when $E_2\beta_2 \leqslant E_1\beta_1$, i.e. for the ratio between the temperatures of the reservoirs sufficiently large $T_2 \geqslant (E_2/E_1)T_1$. When this condition is met, the virtual qubit levels of the machine show population inversion, a feature which facilitates work extraction from the reservoirs.

As commented before, to achieve work extraction without external manipulation of the machine, a quantum weight is provided, in such a way that lifting the weight corresponds to work extraction. The weight is modeled by an unbounded ladder system with energy levels equally spaced and resonant with the virtual qubit of the machine

$$\hat{H}_{\mathrm{w}} = \sum_{n=-\infty}^{\infty} n E_{\mathrm{v}} |n\rangle\langle n|_{\mathrm{w}}. \tag{3.101}$$

The interaction between machine and weight is given by the following interaction Hamiltonian

$$\hat{H}_{\text{int}} = g \sum_{n=-\infty}^{\infty} |0\rangle\langle 1|_1 \otimes |1\rangle\langle 0|_2 \otimes |n\rangle\langle n+1|_{\text{w}} + \text{h.c.} \tag{3.102}$$

This interaction term allows the conversion between a quantum of energy E_2 into two quanta E_1 and E_{w}, together with the opposite process. Furthermore, it is assumed weak coupling with the weight, such that $g \ll E_{\text{w}}$. The idea of the model is hence to recoil work from the spontaneous heat flow from the hot to the cold reservoirs, promoting transitions $|0\rangle_1|1\rangle_2|n\rangle_{\text{w}} \rightarrow |1\rangle_1|0\rangle_2|n+1\rangle_{\text{w}}$ through the population inversion condition.

In Ref. [169] an open system model is provided to describe the operation of the heat engine in the steady state limit. It is given by the following phenomenological master equation for the global density operator of the machine and the weight

$$\dot{\rho} = -\frac{i}{\hbar}[\hat{H}_1 + \hat{H}_2 + \hat{H}_{\text{w}} + \hat{H}_{\text{int}}, \rho] + \sum_{i=1}^{2} \gamma_0 \left(\rho_i^{\text{th}} \otimes \text{Tr}_i[\rho] - \rho\right), \tag{3.103}$$

where the dissipative terms induce (in absence of the interaction term) an asymptotic decay of the machine's qubits to their respective thermal states ρ_i^{th}, at a constant decay rate γ_0. This kind of master equation has been used to model the dynamics of small thermal machines e.g. in Refs. [169, 180, 211, 215]. In the limit $t \rightarrow \infty$ this model produces a constant raising of the weight, and heat flows from the hot and cold reservoirs given by

$$\dot{W} \equiv \frac{d}{dt}\langle \hat{H}_{\text{w}} \rangle \rightarrow \alpha E_{\text{v}}(p_{01} - p_{10}), \tag{3.104}$$

$$\dot{Q}_1 \equiv \frac{d}{dt} Q_1 \rightarrow -\alpha E_1(p_{01} - p_{10}), \tag{3.105}$$

$$\dot{Q}_2 \equiv \frac{d}{dt} Q_2 \rightarrow \alpha E_2(p_{01} - p_{10}), \tag{3.106}$$

where $\alpha = \frac{g^2\gamma_0}{2g^2+\gamma_0^2}$ is a model-dependent constant. We stress that p_{10} and p_{01} are the populations of the virtual qubit in equilibrium conditions as given above. As long as population inversion implies $p_{01} \geqslant p_{10}$, heat is absorbed from the hot reservoir at temperature T_2, and released to the cold one at temperature T_1 while extracting work which is stored in the weight. The first law of thermodynamics is easily checked to hold in the configuration

$$\dot{Q}_1 + \dot{Q}_2 = \dot{W}. \tag{3.107}$$

Furthermore, we notice that the extracted power and the heat fluxes at steady state fulfill the condition

$$\dot{W} : \dot{Q}_1 : \dot{Q}_2 = E_\mathrm{v} : -E_1 : E_2, \tag{3.108}$$

from which the efficiency of the heat engine can be obtained [181]

$$\eta_\mathrm{engine} = \frac{\dot{W}}{\dot{Q}_2} = 1 - \frac{E_1}{E_2} = \eta_\mathrm{carnot}\left(1 - \frac{-\beta_\mathrm{v}}{\beta_1 - \beta_\mathrm{v}}\right). \tag{3.109}$$

In the last equality we used Eq. (3.100) in order to rewrite the machine efficiency in terms of the inverse virtual temperature β_v. As can be seen from the above expression the heat engine can reach Carnot efficiency when $\beta_\mathrm{v} \to 0^-$, a condition achieved for vanishing population inversion $p_{01} \to p_{10}$, when the transitions

$$|0\rangle_1|1\rangle_2|n\rangle_\mathrm{w} \to |1\rangle_1|0\rangle_2|n+1\rangle_\mathrm{w} \tag{3.110}$$

allowed by the interaction Hamiltonian in Eq. (3.102), becomes only infinitesimally more probable than the opposite transitions

$$|1\rangle_1|0\rangle_2|n+1\rangle_\mathrm{w} \to |0\rangle_1|1\rangle_2|n\rangle_\mathrm{w}. \tag{3.111}$$

We again notice that this implies vanishingly small energy fluxes trough the machine, c.f. Eqs. (3.104)–(3.106). Furthermore, the following equality between the entropy flows from the environment holds

$$-\beta_2\dot{Q}_2 - \beta_1\dot{Q}_1 = -\beta_\mathrm{v}\dot{W} \geqslant 0, \tag{3.112}$$

which follows from Eqs. (3.100) and (3.108) [181]. Henceforth, when Carnot conditions are imposed, $\beta_\mathrm{v} \to 0^-$, we recover the classical reversibility conditions for the entropy flows between the reservoirs

$$\beta_2\dot{Q}_2 \to -\beta_1\dot{Q}_1. \tag{3.113}$$

Finally, we notice that in the steady state operation, the weight continuously increases its energy and spreads at a rate [181]

$$\frac{d}{dt}\sigma^2(\hat{H}_\mathrm{w}) = E_\mathrm{w}^2\left(\alpha(p_{01} + p_{10}) - \alpha'(p_{01} - p_{10})^2\right), \tag{3.114}$$

with $\alpha' = 2g^4\gamma_0(g^2 + 2\gamma_0^2)/(2g^2 + \gamma_0^2)^3$ a second constant depending on the machine couplings and we recall that $\sigma^2(\hat{H}_\mathrm{w})$ stands for the variance of $\hat{H}_\mathrm{w}$. This makes a big difference with models using an external driving field, in which work is defined as the output energy in the classical driving field, which experience no back action in the process. Indeed, the definition of work we use here becomes controversial itself, as the weight may increase its entropy while storing energy, and hence degrades the quality of the energy stored (other related issues concerning work definitions are discussed in the recent Ref. [88]). This happens for instance if the weight starts its evolution in an eigenstate $|n_0\rangle_{\mathrm{w}}$. In such case its motion is similar to a biased random walk [181] and hence the entropy of the weight permanently increases according to Eq. (3.114). This problem may be avoided if the weight starts the evolution in some specific state such that it does not increase its entropy during the machine operation, nor build up correlations with it. In such case the entropy production of the setup in the steady state would be simply given by Eq. (3.112).

The idea of considering a quantum weight to quantify work in thermodynamical tasks has been further explored recently. For instance in Ref. [22] a continuous quantum weight was used to derive a maximum work extraction protocol for individual quantum systems in nonequilibrium states with the help of a thermal reservoir. Further results indicate that implementation of optimal protocols would require a coherent resource, which for infinite-ladder weights can be used catalytically [218] (see also Ref. [219]). Work extraction with a weight has been also considered in the strong-coupling regime in Ref. [220]. Furthermore, in Ref. [221] it has been argued that any unitary can be approximately performed in a quantum system by means of a time-independent global Hamiltonian by using a weight and an ideal *quantum clock*. Nevertheless, the use of an ideal quantum clock has been shown to be problematic due to the fact that it would require an infinite amount of energy for its fabrication [222]. Considering instead non-ideal (finite-sized) clocks introduces non-trivial tradeoffs between the accuracy of the operations that can be implemented and their energetic costs [222].

3.3.3 Quantum Effects in Thermal Machines

So far we have considered the basic functioning of small quantum thermal machines operating between thermal reservoirs at different temperatures. We have seen that the framework of quantum thermodynamics applies, and expressions for the heat fluxes and efficiencies analogous to the classical case are obtained. However, some fundamental questions arise at this point. Is there something specifically quantum, apart from the energy level quantization, in the functioning of the thermal machines? How quantum effects like coherence or quantum correlations affect the performance of the machine? Is there any quantum advantage?

A great effort to answer the above questions is being undertaken nowadays in the current research on quantum thermodynamics. Quantum effects have been shown to enhance the power or efficiency of thermal machines under three main circumstances:

(i) The substitution of the traditional thermal reservoirs by more general nonequilibrium environments with quantum properties. This path started with the pioneering work of Scully et al. on the photo-Carnot engine driven by quantum fuel [223], and proliferated in the last decade with different analysis of thermal machines powered by coherent [224–226], correlated [227] or squeezed thermal reservoirs [213, 228–230].

(ii) The introduction of external control operations inducing coherence or correlations in the working substance of the machine, which henceforth operates in genuinely quantum nonequilibrium conditions. This has been shown to be the case of the prototypical three-level amplifier driven by an external classical field [173], but also feedback engines using quantum measurements in specific basis [76, 216, 231]. In this class we may also include thermal machines operating quantum enhanced finite-time strokes [205, 232].

(iii) The induction of a quantum nonequilibrium state in the working substance by means of degeneracies in the Hamiltonian of the machine. In this third modality we find heat engines profiting from noise-induced coherence [233–235], or the entangled absorption refrigerator studied in Ref. [181].

Circumstances (ii) and (iii) may in general lead to an enhancement in the machines performance in terms of power, that is, work (or heat in the case of fridges) can be extracted at a higher rate by using coherence or correlations in the working substance [173, 181, 216, 233, 235]. This power enhancement appears in some regimes of operation and may consequently lead to greater efficiencies at non-reversible conditions [231, 232, 234], but the (classical) Carnot bound still holds in these situations [236]. A different scenario is provided in the case (i) as the incorporation of nonequilibrium reservoirs may lead to alterations in the bounds imposed by the second law of thermodynamics [237]. In such cases both power and maximum efficiency can be enhanced by using quantum resources provided by the environment. In this thesis we will investigate circumstance (i) for the case of the squeezed thermal reservoir in Chap. 10. There we provide a complete derivation of maximum efficiency bounds and power output, demonstrating that the corresponding enhancements are induced by a genuine entropy exchange between the working substance and the nonequilibrium reservoir.

We finally comment that quantum effects in thermodynamical setups have been also investigated in non-cyclic operations. In this context the amount of work which can be drawn from quantum correlations such as entanglement or quantum discord has attracted a great attention (see e.g. Refs. [69, 73, 75, 238, 239]). Work extraction from coherence has been considered under different scenarios [22, 240], while its role in optimal projection processes has been recently investigated [241]. In addition, enhancements due to the presence of entanglement or coherence in single-shot refrigeration protocols are under current investigation [214, 215].

3.4 Other Topics in Quantum Thermodynamics

In this last section we briefly introduce some other important topics of current research in quantum thermodynamics, namely, equilibration and thermalization in isolated quantum systems, and resource theories. Our aim is to provide a qualitative overview over some important results in the corresponding topics without entering into technical details. Although those topics are not specifically covered in this thesis, their close relation to the field makes them susceptible to profit from the results of the research presented here. For more information we refer the reader to two recent reviews on quantum thermodynamics [1, 5] as well as to two more specific ones focusing on equilibration and thermalization issues [4, 242].

3.4.1 Equilibration and Thermalization

As we have seen in Sect. 3.1, a fundamental assumption in macroscopic phenomenological thermodynamics is the spontaneous evolution to an equilibrium state for any system not exchanging energy nor matter with its surroundings. Furthermore, following statistical mechanics, in such equilibrium state the different allowed configurations, i.e. the possible states compatible with given constraints, occur with equal probabilities (see Sect. 3.1). From the point of view of quantum mechanics, isolated systems are represented by pure states at all times, a fact that seems to be in strong contradiction with statistical mechanics. Therefore it would be highly desirable to reconcile foundational aspects of statistical mechanics with the underlying time-reversible evolution of quantum systems. These issues have recently experimented a renewed attention, motivated by the improvements in experimental techniques to handle quantum systems with many degrees of freedom, the increase of computer power and the introduction of computational techniques for the simulation of quantum systems, and the introduction new mathematical methods from quantum information theory [242].

One first *kinematic* viewpoint has been adopted by using typicality arguments in order to replace the equal a priori probabilities postulate of statistical mechanics. In Refs. [243, 244] it is shown that by looking only at a small subsystem of an isolated quantum system, its state is indistinguishable from the state predicted by the equal a priori probabilities postulates for almost all pure states of the global isolated system. In other words, if the pure states of the global system, say $|\psi\rangle \in \mathcal{H}$, are randomly sampled according to the Haar measure (uniform distribution) in the Hilbert space $\mathcal{H}_R \subseteq \mathcal{H}$ allowed by some restriction R (e.g. a given energy), then the state of the subsystem, $\rho_S = \mathrm{Tr}_E[|\psi\rangle\langle\psi|] \approx \Omega_S$, where Tr_E denotes the partial trace over the subsystem complementary to S (i.e. $\mathcal{H} = \mathcal{H}_S \otimes \mathcal{H}_E$) and $\Omega_S = \mathrm{Tr}_E[\mathbb{1}_R/d_R]$ is the equal a priori probability state given the restriction R [243]. Nonetheless, as pointed in a recent review [5], the use of typicality for sampling states in $\mathcal{H}_R$ is not physically motivated, as most of those states can be never generated from local symmetric Hamiltonians arising in nature in reasonable times, thus the search for extended notions of typicality constitutes an open problem in the field (see Ref. [5] and references therein for a more detailed discussion on this topic).

On the other hand from a *dynamical* viewpoint, a first important question is to consider the *equilibration* dynamics of local observables and subsystems of isolated quantum systems, even if the specific form of the equilibrium state is not approached. As long as an isolated quantum system follows time-reversible unitary evolution

$$|\psi(t)\rangle = e^{-\frac{i}{\hbar}\hat{H}t}|\psi(0)\rangle \in \mathcal{H} \tag{3.115}$$

$\hat{H} = \sum_k E_k \hat{\Pi}_k$ being the spectral decomposition of the Hamiltonian ($\hat{\Pi}_k = |E\rangle_k\langle E|_k$ being projectors on the energy eigenstates $\{|E\rangle_k\}$), recurrences in the state of the system immediately follow. This is in apparent contradiction with the H-theorem, stating that entropy always grows over time and systems spontaneously equilibrate. Nevertheless it results that this contradiction can be solved to a large extend by considering the dynamical properties of local observables and deriving dynamical typicality statements [5]. It is useful to introduce in this context the time-averaged state

$$\omega = \lim_{\tau\to\infty} \frac{1}{\tau}\int_0^\tau dr\rho(t) = \sum_k \hat{\Pi}_k|\psi(0)\rangle\langle\psi(0)|\hat{\Pi}_k, \tag{3.116}$$

which just corresponds to the dephased version of the initial pure state $|\psi(0)\rangle$ in the energy basis. In this context, a subsystem S, where again $\mathcal{H} = \mathcal{H}_S \otimes \mathcal{H}_E$, is said to equilibrate *on average* if its reduced state verifies $\rho_S(t) = \mathrm{Tr}_E[|\psi(t)\rangle\langle\psi(t)|] \approx \omega_S \equiv \mathrm{Tr}_E[\omega]$. More precisely, this statement is fulfilled when [5]

$$\lim_{\tau\to\infty} \frac{1}{\tau}\int_0^\tau T(\rho_S(t), \omega_S) \ll 1, \tag{3.117}$$

where $T(\rho, \sigma)$ is the trace distance as introduced in Sect. 1.1.7. This has been shown to be indeed the case [245] when the Hamiltonian $\hat{H}$ has a small number of degenerate energy gaps [246]. The physical meaning of this condition is that it excludes Hamiltonians with no interaction part between the subsystem S and the environment E, as they contain a high number of degenerate energy gaps. This condition also provides some general bounds on the equilibration times. More recent results provide a stronger statement asserting that any local observable of the subsystem S passes almost all the time very close to the value generated by ω_S, fluctuating around it, provided the initial state does not assign large populations to few energy levels [242]. An alternative form of equilibration that has been found in several models [247, 248] is the *equilibration during intervals*. This kind of equilibration means that local observables of the reduced system are close to the equilibrium state $T(\langle\hat{A}_S\rangle_t, \mathrm{Tr}[\hat{A}_S\omega]) \ll 1$ for all times t inside an interval after a known relaxation time and before the recurrence time. Another important result from the point of view of the foundations of statistical mechanics in the equilibration context is the maximum entropy principle derived in Ref. [249], which states that if the expectation value $\mathrm{Tr}[\rho\hat{A}]$ of an operator $\hat{A}$ equilibrates on average, it equilibrates towards its time average, given by

$$\lim_{\tau\to\infty} \frac{1}{\tau}\int_0^\tau \mathrm{Tr}[\rho(t)\hat{A}] = \mathrm{Tr}[\omega\hat{A}], \tag{3.118}$$

where ω in Eq. (3.116), is the unique state that maximizes the von Neumann entropy given all conserved quantities. This implies that the unitary dynamics of isolated quantum systems alone gives rise to a kind of maximum entropy principle [242].

A second main question investigated from the dynamical point of view is *thermalization* of subsystems in isolated quantum setups. In its stronger form, thermalization is the emergence of Gibbs states as the local equilibrium states of the subsystem. The definitions of thermalization in this context can however exhibit some technical differences. Some general requisites in stronger order have been introduced in Ref. [242]:

- Equilibration is a necessary condition for thermalization.
- The equilibrium state of a small subsystem should be independent of the initial state of that subsystem.
- The equilibrium expectation value of local observables should be almost independent on the initial state of the rest of the system, but only depend on some 'macroscopic properties' such as the energy density.
- The equilibrium state should be approximately diagonal in the energy eigenbasis of a suitably defined 'self-Hamiltonian'.
- Ultimately, one would like to recover that the equilibrium state is in some sense close to the Gibbs thermal state.

In order to introduce the different approaches to thermalization, it is illustrative to explicitly consider the expectation value of a local observable $\hat{A}_S$ of a small subsystem S of the compound $S + E$. Following Ref. [1]:

$$\begin{aligned} \langle\psi(t)|\hat{A}_S|\psi(t)\rangle &= \sum_k |c_k|^2 \langle E_k|\hat{A}_S|E_k\rangle \\ &+ \sum_{k,m} c_m^* c_k e^{i(E_m - E_k)t} \langle E_m|\hat{A}_S|E_k\rangle, \end{aligned} \tag{3.119}$$

where the initial state is $|\psi(0)\rangle = \sum_k c_k|E_k\rangle$, and the coefficients c_k are assumed to be non-zero only in a small band around some given energy E_0 [1]. If energies are non-degenerate, the long time average of this expectation will be given by the first term of Eq. (3.119), which needs to be independent of the coefficients $|c_k|^2$. As explained in Ref. [1] this can happen in three ways. The first one is called the *eigenstate thermalization hypothesis* (ETH), which in its simplest version demands that the quantities $\langle E_k|\hat{A}_S|E_k\rangle$ equal the thermal average of $\hat{A}$ at the mean energy E_k [5]. In such case $\langle E_k|\hat{A}_S|E_k\rangle$ factorizes and the coefficients $|c_k|^2$ will sum up to one. There exist however different variants of the ETH in the literature, in particular not requiring a non-degenerate Hamiltonian [242]. A second approach towards the problem of thermalization independent of the ETH consists in making strong assumptions concerning the energy distribution of the initial state. In this case, thermalization on average has been rigorously proven for both spin and fermionic systems assuming a suitably weak coupling condition between the subsystem and its environment [242]. Turning to Eq. (3.119), this means demanding the coefficients c_k to be constant and non-zero for a subset of indices k, or requiring the coefficients c_k to be uncorrelated with respect to $\langle E_k|\hat{A}_S|E_k\rangle$ [1].

However, if the system has sufficiently many local conserved quantities, that is, if it is integrable, thermalization is not expected. In such case the constants of motion prevent that the subsystem reaches the Gibbs state. On the contrary, one may expect the system will reach a maximally entropy state given the locally conserved quantities [4]. This is a so-called *generalized Gibbs ensemble* (GGE) [250, 251], which has been shown to correctly describe the properties of the equilibrium state of many-body quantum systems after a quantum quench under a variety of circumstances [247, 252–258]. The explicit construction of the GGE for general interacting integrable models remains an open problem, while it has been also shown to fail in non-integrable models [4]. Furthermore, several quantum many-body systems exhibit a so-called *pre-thermalization* [259], that is the apparent equilibration to some meta-stable state in a short time scale, before the system finally relaxes to a state indistinguishable from the thermal state. This has been shown e.g. in almost-integrable systems and continuous models of coupled Bose–Einstein condensates [4].

3.4.2 Resource Theories in Quantum Thermodynamics

Resource theories of quantum thermodynamics are inspired by other resource theories from quantum information, such as the resource theory of entanglement, asymmetry, purity or quantum coherence (see Ref. [260] for a review on quantum resource theories). The general idea of such theories is to quantify quantum resources by introducing a *state space* $\mathcal{S}$, and a set of *allowed operations* $\mathcal{T}$. There will be states that can be obtained using the allowed operations $\mathcal{T}$ from any initial state, which are called *free-resources*, and states which cannot. Those later states are hence considered a *resource* in the theory. The idea is that extra operations can be performed when provided a resource, which is consumed in this process. However, if the resource is not consumed, we may refer to it as a *catalyst*, as it can be used to perform otherwise impossible operations (or to aid in those operations) infinitely many times.

Following Ref. [5], state transformations $\rho \to \sigma$, for $\rho, \sigma \in \mathcal{S}$ are characterized inside the theory by means of some functions $f(\rho, \sigma)$ which determine whether the transformation is possible or not. For example if $\rho \to \sigma \;\Rightarrow\; f(\rho, \sigma) \geqslant 0$, then $f(\rho, \sigma) \geqslant 0$ is a necessary condition for the state transformation $\rho \to \sigma$. Analogously if $f(\rho, \sigma) \geqslant 0 \;\Rightarrow\; \rho \to \sigma$, then $f(\rho, \sigma) \geqslant 0$ is a sufficient condition for the state transformation. A *monotone* of the resource theory is a function m such that

$$f(\rho, \sigma) = m(\rho) - m(\sigma) \geqslant 0, \tag{3.120}$$

is a necessary condition for the state transformation [5].

The adoption of a resource theory perspective in thermodynamics comes back to the ideas of Lieb and Yngvason [261] and was firstly introduced in a quantum setting by Janzing et al. [262]. In this case the set of states are all quantum states of systems equipped with a fixed Hamiltonian $\hat{H}$, and the allowed operations $\mathcal{T}$ are thermalization to a fixed background temperature T, and global unitaries $\hat{U}$ conserving the energy, that is, $[\hat{U}, \hat{H}] = 0$.[1] In this resource theory the *free-resource* states are the Gibbs thermal states $\rho_{\text{th}}(\beta) = e^{-\beta\hat{H}}/Z$, with $\beta = 1/k_B T$ and Z the partition function, and the allowed transformations are called *thermal operations*

$$\mathcal{T}(\rho) = \text{Tr}_p[\hat{U}(\rho \otimes \rho_{\text{th}})\hat{U}^\dagger], \tag{3.121}$$

where here an ancillary system in the free-resource state ρ_{free} is used, and Tr_p denotes the partial trace over any arbitrary subsystem. In Ref. [263] a monotone of the resource theory of thermal operations has been shown to be the (nonequilibrium) free energy introduced in Sect. 3.1, and the optimal rate of a transformation $\rho \rightarrow \sigma$ for a system with Hamiltonian $\hat{H}$ is given by

$$R(\rho \rightarrow \sigma) = \frac{\mathcal{F}(\rho) - \mathcal{F}(e^{-\beta\hat{H}}/Z)}{\mathcal{F}(\sigma) - \mathcal{F}(e^{-\beta\hat{H}}/Z)}. \tag{3.122}$$

Other versions of the free energy apply for the single-shot scenario [263, 264], assuming that a single copy of the state is provided and experiments cannot be repeated.

In order to check that Gibbs thermal states are well-defined *free-resources* in the theory, one may prove that they are useless. This can be done by using the notion of *passive states* [265]. A state is called passive if there is no unitary $\hat{U}$ which decreases the energy $\text{Tr}[\rho\hat{H}] > \text{Tr}[\hat{U}\rho\hat{U}^\dagger\hat{H}]$ [5]. Passive states are all diagonal in the energy basis showing decreasing populations for increasing energy. Nevertheless, a stronger notion of passivity can be also considered in the case in which many ancillary systems in the same state can be used as well

$$\text{Tr}[\rho\hat{H}] > \text{Tr}[\hat{U}\rho^{\otimes n}\hat{U}^\dagger\hat{H}], \tag{3.123}$$

where here $\hat{U}$ is a unitary acting globally on the many copies. If it is still not possible for any $\hat{U}$ to fulfill Eq. (3.123), then the state ρ is called *completely passive*. It turns out that only the Gibbs thermal states $\rho_{\text{th}}(\beta) = e^{-\beta\hat{H}}/Z$ are completely passive. This remains true even if we further allow the many copies to have different Hamiltonians, which justifies the use of arbitrary ancillary systems in Gibbs states as *free-resources* [5].

When the allowed fixed Hamilton operators for the systems of interest are fully degenerate, this framework reduces to the resource theory of noisy operations [266], for which the *free-resource* is the maximally mixed state, the allowed operations reduce to unital CPTP maps, $\mathcal{T}(\mathbb{1}) = \mathbb{1}$, and the monotone reduces to the von Neumann entropy $S(\rho)$ [267]. Analogously to this case, one may think that the set of ther-

[1] We notice that some works allow for unitaries preserving energy only *on average*, i.e. $\text{Tr}[\hat{H}\rho] = \text{Tr}[\hat{H}\hat{U}\rho\hat{U}]$ (see e.g. Ref. [22]), which may require a restriction on the states ρ as noticed in Ref. [218].

mal operations can be substituted by the set of Gibbs-preserving maps, $\mathcal{T}(\rho_{\text{th}}) = \rho_{\text{th}}$. Indeed any thermal operation preserves the Gibbs state. However, it turns out that Gibbs-preserving maps allow transformations which may be not always implemented by just thermal operations [268], as for instance the transformation

$$|1\rangle \rightarrow \frac{1}{\sqrt{2}}(|0\rangle + |1\rangle), \tag{3.124}$$

in a qubit system with Hamiltonian $\hat{H} = E|1\rangle\langle 1|$. Indeed, thermal operations itself cannot create nor manipulate coherence [269], but they need access to an extra source providing it. A coherence reservoir may be hence introduced as the weight of the previous section, i. e. a doubly infinite ladder system with Hamiltonian $\hat{H}_{\text{c}} = \sum_{n=-\infty}^{\infty} nE|n\rangle\langle n|$ in some coherent state such as

$$|\phi\rangle_{\text{c}}(l, L) = \frac{1}{L}\sum_{n=l}^{l+L} |n\rangle. \tag{3.125}$$

This kind of coherent reservoir is affected by back-action when interacts with other systems, which causes the state to spread over the energy ladder. Nonetheless, it has been shown that its coherence properties needed to implement operations are, remarkably, unaltered [218]. This leads to think about coherence as a catalytic resource which can be used to perform arbitrary unitary operations on a system [218, 221]. If the coherence reservoir is instead provided with a ground state, this catalytic property is partially lost and one should require some input energy (but no coherence) in order to use it again.

Another related point of the theory concerns the use of *catalysts*. That is, the use of arbitrary systems in nonequilibrium states σ_C which helps to perform impossible transition between states $\rho_S \rightarrow \sigma_S$ while turning back to its original state at the end of the transformation

$$\rho_S \otimes \omega_C \rightarrow \sigma_S \otimes \omega_C. \tag{3.126}$$

This is called in the literature *exact catalysis*, since the final state of the catalyst exactly coincides with the original one. The study of exact catalysis yielded the derivation of a family of second-laws for the one-shot regime [270]. However, in the prototypical case of many copies, the inclusion of exact catalysis does not modify the usual second law of thermodynamics in terms of the nonequilibrium free energy. If instead of exact catalysis one considers the case in which the catalyst after operation is ϵ-close to its original state

$$\rho \otimes \omega_C \rightarrow \sigma_{SC}, \quad \text{with} \quad T(\text{Tr}_S[\sigma_{SC}], \omega_C) \leqslant \epsilon, \tag{3.127}$$

it results that one can achieve arbitrary transformations for arbitrary small ϵ if all (arbitrary large) catalyst are allowed [270]. On the other hand, by imposing restrictions on the energy and dimension of the catalyst, the free energy constrains can be recovered [271]. Furthermore, limitations imposed by finite-size effects in general transformations has been also recently analyzed in Ref. [27].

Generalizations of the present framework to include the case of time-dependent Hamiltonians may be achieved by introducing quantum *clocks* in the direction pointed in Refs. [221, 222]. This would be a very desirable extension of the resource theory of thermal states because thermodynamics is mostly based on time-dependent Hamiltonians implementing driving protocols, as we have already seen in this chapter. Other generalizations of the resource theory of thermal states are nowadays being proposed in order to include more conserved quantities such as the number of particles or the momentum in the theory, and more general reservoirs in generalized Gibbs ensembles [272–274]. It should be then interesting to consider how those theories are affected when the reservoirs have some quantum property such as squeezing or quantum correlations in comparison with the classical case.

Finally, we point out the interesting case of combining different resource theories such as the resource theory of entanglement and the resource theory of thermal states. Elaborating hybrid theories are important e.g. in determining the costs of information processing tasks in a thermodynamic framework. In the following we mention some examples commented in the review [5], in which the combination of entanglement theory and thermodynamics has lead to fruitful results. It has been for instance shown that when starting in Gibbs thermal states, the generation of correlations has an average input energy cost which depends on the temperature [275, 276], reading for a bipartite state AB

$$W_{\text{corr}} \geqslant k_B T \; I(\rho_{AB}), \tag{3.128}$$

when a thermal bath at temperature T can be used in the process. As mentioned previously, other references studied the work which can be extracted from correlations. In particular, it has been shown that the restriction to the set of LOCC amounts to associate a work value to quantum discord [73]

$$W = k_B T \; \delta(\rho_{AB}), \tag{3.129}$$

obtained in a Maxwell demon configuration in which work extraction from local or global agents are compared. Another interesting result coming from Ref. [239] states that the capacity to store work in the purely form of correlations differs in the quantum and classical cases, and scales with the systems dimension N as

$$\frac{W_{\text{classical}}}{W_{\text{quantum}}} = 1 - \mathcal{O}(N^{-1}). \tag{3.130}$$

Those results can only be obtained when allowing global unitaries to change the total average energy of the systems of interest. Reproducing those results in this framework requires again the introduction a well-behaved weight providing (or recoiling) energy in order to guarantee that the global unitary acting on the system and weight preserves the total energy. A different point of view is taken when investigating the generation of quantum correlations directly from thermal resources, as is the case of the small thermal machine introduced in Ref. [169]. In order to fully incorporate such configurations, more general resource-like theories are needed, including e.g. free access to two different temperatures. This may allow us to operationally characterize quantum information from a thermodynamic perspective in a more general scenario than the isothermal case.

References

1. S. Vinjanampathy, J. Anders, Quantum thermodynamics. Contemp. Phys. **57**, 1–35 (2016)
2. U. Seifert, Stochastic thermodynamics, fluctuation theorems and molecular machines. Rep. Prog. Phys. **75**, 126001 (2012)
3. M. Esposito, U. Harbola, S. Mukamel, Nonequilibrium fluctuations, fluctuation theorems, and counting statistics in quantum systems. Rev. Mod. Phys. **81**, 1665–1702 (2009)
4. J. Eisert, M. Friesdorf, C. Gogolin, Quantum many-body systems out of equilibrium. Nat. Phys. **11**, 124–130 (2015)
5. J. Goold, M. Huber, A. Riera, L. del Rio, P. Skrzypczyk, The role of quantum information in thermodynamics-a topical review. J. Phys. A: Math. Theor. **49**, 143001 (2016)
6. M. Planck, *The Theory of Heat Radiation*, translated by M. Masius (P. Blakiston's Son & Co., Philadelphia, 1914)
7. A. Einstein, Über einen die Erzeugung und Verwandlung des Lichtes betreffenden heuristischen Gesichtspunkt [Concerning an Heuristic Point of View Toward the Emission and Transformation of Light], Ann. Phys. **17**, 132-48 (1905), [English translation in Am. J. Phys. **33**, 367 (1965)]
8. H.E.D. Scovil, E.O. Schulz-DuBois, Three-level masers as heat engines. Phys. Rev. Lett. **2**, 262 (1959)
9. J.E. Geusic, E.O. Schulz-DuBois, H.E.D. Scovil, Quantum equivalent of the carnot cycle. Phys. Rev. **156**, 343 (1967)
10. H. Spohn, J.L. Lebowitz, Irreversible thermodynamics for quantum systems weakly coupled to thermal reservoirs, in *Advances in Chemical Physics: For Ilya Prigogine*, vol. 38, ed. by S.A. Rice (Wiley, Hoboken, 1978)
11. M.H. Partovi, Quantum thermodynamics. Phys. Lett. A **137**, 440–444 (1989)
12. J. Gemmer, M. Michel, G. Mahler, *Quantum Thermodynamics: Emergence of Thermodynamic Behavior Within Composite Quantum Systems*, Lecture Notes in Physics (Springer, Berlin, 2009)
13. P. Talkner, E. Lutz, P. Hänggi, Fluctuation theorems: work is not an observable. Phys. Rev. E **75**, 050102 (2007)
14. A. Engel, R. Nolte, Jarzynski equation for a simple quantum system: comparing two definitions of work. Europhys. Lett. **79**, 10003 (2007)
15. P. Solinas, D.V. Averin, J.P. Pekola, Work and its fluctuations in a driven quantum system. Phys. Rev. B **87**, 060508 (2013)
16. J.M.R. Parrondo, J.M. Horowitz, T. Sagawa, Thermodynamics of information. Nat. Phys. **11**, 131–139 (2015)
17. H.B. Callen, *Thermodynamics and an Introduction to Thermostatistics*, 2nd edn. (Wiley, Singapore, 1985)
18. D. Kondepudi, I. Prigogine, *Modern Thermodynamics* (From Heat Engines to Dissipative Structures (Wiley, Chichester, 1998)
19. L.N.S. Carnot, *Réflexions sur la puissance motrice du feu et sur les machines propres à développer cette puissance [Reflections on the Motive Power of Fire and on Machines Fitted to Develop that Power] (Bachelier, Paris, 1824), [English translation by R* (Manchester University Press, New York, Fox, 1986)

20. R. Clausius, *Mechanical Theory of Heat* (John van Voorst, London, 1867)
21. W. Greiner, L. Neise, H. Stöcker, *Thermodynamics and Statistical Mechanics* (Springer, New York, 1995)
22. P. Skrzypczyk, A.J. Short, S. Popescu, Work extraction and thermodynamics for individual quantum systems. Nat. Commun. **5**, 4185 (2014)
23. M. Esposito, C.V. den Broeck, Second law and Landauer principle far from equilibrium. Europhys. Lett. **95**, 40004 (2011)
24. T. Albash, S. Boixo, D.A. Lidar, P. Zanardi, Quantum adiabatic Markovian master equations. New J. Phys. **14**, 123016 (2012)
25. R. Kosloff, Quantum thermodynamics: a dynamical viewpoint. Entropy **15**, 2100–2128 (2013)
26. K. Jacobs, *Quantum Measurement Theory and its Applications* (Cambridge University Press, Cambridge, 2014)
27. H. Wilming, R. Gallego, J. Eisert, Second law of thermodynamics under control restrictions. Phys. Rev. E **93**, 042126 (2016)
28. M. Hemmo, O. Shenker, Von Neumann's entropy does not correspond to thermodynamic entropy. Philos. Sci. **73**, 153–174 (2006)
29. M. Esposito, K. Lindenberg, C. Van den Broeck, Entropy production as correlation between system and reservoir. New J. Phys. **12**, 013013 (2010)
30. L.F. Santos, A. Polkovnikov, M. Rigol, Entropy of isolated quantum systems after a quench. Phys. Rev. Lett. **107**, 040601 (2011)
31. A. Deville, Y. Deville, Clarifying the link between von Neumann and thermodynamic entropies. Eur. Phys. J. H **38**, 57–81 (2013)
32. J. Gemmer, R. Steinigeweg, Entropy increase in K-step Markovian and consistent dynamics of closed quantum systems. Phys. Rev. E **89**, 042113 (2014)
33. E. Solano-Carrillo, A.J. Millis, Theory of entropy production in quantum many-body systems. Phys. Rev. B **93**, 224305 (2016)
34. W. Nernst, The New Heat Theorem, Its Foundations in Theory and Experiment, translated by G. Barr (Methuen & Co., London, 1926)
35. J. Wilks, *The Thid Law of Thermodynamics* (Oxford University Press, Oxford, 1961)
36. L. Masanes, J. Oppenheim, A derivation (and quantification) of the third law of thermodynamics (2016), arXiv:1412.3828
37. H.S. Leff, Proof of the third law of thermodynamics for ising ferromagnets. Phys. Rev. A **2**, 2368–2370 (1970)
38. M. Aizenman, E.H. Lieb, The third law of thermodynamics and the degeneracy of the ground state for lattice systems. J. Stat. Phys. **24**, 279–297 (1981)
39. G. Ford, R. O'Connell, Entropy of a quantum oscillator coupled to a heat bath and implications for quantum thermodynamics. Physica E **29**, 82–86 (2005)
40. R.F. O'connell, Does the third law of thermodynamics hold in the quantum regime? J. Stat. Phys. **124**, 15–23 (2006)
41. Y. Rezek, P. Salamon, K.H. Hoffmann, R. Kosloff, The quantum refrigerator: the quest for absolute zero. Europhys. Lett. **85**, 30008 (2009)
42. A.E. Allahverdyan, K. Hovhannisyan, G. Mahler, Optimal refrigerator. Phys. Rev. E **81**, 051129 (2010)
43. R. Kosloff, E. Geva, J.M. Gordon, Quantum refrigerators in quest of the absolute zero. J. Appl. Phys. **87**, 8093–8097 (2000)
44. A. Levy, R. Alicki, R. Kosloff, Quantum refrigerators and the third law of thermodynamics. Phys. Rev. E **85**, 061126 (2012)
45. A. Levy, R. Kosloff, Quantum absorption refrigerator. Phys. Rev. Lett. **108**, 070604 (2012)
46. M. Kolář, D. Gelbwaser-Klimovsky, R. Alicki, G. Kurizki, Quantum bath refrigeration towards absolute zero: challenging the unattainability principle. Phys. Rev. Lett. **109**, 090601 (2012)
47. A.E. Allahverdyan, K.V. Hovhannisyan, D. Janzing, G. Mahler, Thermodynamic limits of dynamic cooling. Phys. Rev. E **84**, 041109 (2011)
48. L.-A. Wu, D. Segal, P. Brumer, No-go theorem for ground state cooling given initial system-thermal bath factorization. Sci. Rep. **3**, 1824 (2013)

49. C. Di Franco, M. Paternostro, A no-go result on the purification of quantum states. Sci. Rep. **3**, 1387 (2013)
50. F. Ticozzi, L. Viola, Quantum resources for purification and cooling: fundamental limits and opportunities. Sci. Rep. **4**, 5192 (2014)
51. D. Reeb, M.M. Wolf, An improved Landauer principle with finite-size corrections. New J. Phys. **16**, 103011 (2014)
52. K. Maruyama, F. Nori, V. Vedral, Colloquium: the physics of Maxwell's demon and information. Rev. Mod. Phys. **81**, 1–23 (2009)
53. H.S. Leff, A.F. Rex (eds.), *Maxwell's Demon 2. Entropy, Classical and Quantum Information, Computing* (Taylor & Francis, New York, 2002)
54. L. Szilárd, Ober die Enfropieuerminderung in einem thermodynamischen System bei Eingrifen intelligenter Wesen [On the decrease of entropy in a thermodynamic system by the intervention of intelligent beings], Zeitschrift fur Physik **53**, 840–856 (1929), [English translation in Ref. [53], pp. 110–119]
55. J.M.R. Parrondo, The Szilard engine revisited: entropy, macroscopic randomness, and symmetry breaking phase transitions. Chaos **11**, 725–733 (2001)
56. C.H. Bennett, The thermodynamics of computation-a review. Int. J. Theor. Phys. **21**, 905–940 (1982)
57. R. Landauer, Irreversibility and heat generation in the computing process. IBM J. Res. Dev. **5**, 183–191 (1961)
58. A. Berut, A. Arakelyan, A. Petrosyan, S. Ciliberto, R. Dillenschneider, E. Lutz, Experimental verification of Landauer's principle linking information and thermodynamics. Nature **483**, 187–189 (2012)
59. S. Toyabe, T. Sagawa, M. Ueda, E. Muneyuki, M. Sano, Experimental demonstration of information-to-energy conversion and validation of the generalized Jarzynski equality. Nat. Phys. **6**, 988–992 (2010)
60. E. Roldan, I.A. Martinez, J.M.R. Parrondo, D. Petrov, Universal features in the energetics of symmetry breaking. Nat. Phys. **10**, 457–461 (2014)
61. J.V. Koski, V.F. Maisi, T. Sagawa, J.P. Pekola, Experimental observation of the role of mutual information in the nonequilibrium dynamics of a Maxwell Demon. Phys. Rev. Lett. **113**, 030601 (2014)
62. J.V. Koski, V.F. Maisi, J.P. Pekola, D.V. Averin, Experimental realization of a Szilard engine with a single electron. Proc. Natl. Acad. Sci. **111**, 13786–13789 (2014)
63. M.D. Vidrighin, O. Dahlsten, M. Barbieri, M.S. Kim, V. Vedral, I.A. Walmsley, Photonic Maxwell's Demon. Phys. Rev. Lett. **116**, 050401 (2016)
64. J.V. Koski, A. Kutvonen, I.M. Khaymovich, T. Ala-Nissila, J.P. Pekola, On-chip Maxwell's Demon as an information-powered refrigerator. Phys. Rev. Lett. **115**, 260602 (2015)
65. W.H. Zurek, Maxwell's Demon, Szilard's engine and quantum measurements, in *Frontiers of Nonequilibrium Statistical Physics*, ed. by G.T. Moore, M.O. Scully (Springer, Boston, 1986), pp. 151–161
66. S.W. Kim, T. Sagawa, S. De Liberato, M. Ueda, Quantum Szilard engine. Phys. Rev. Lett. **106**, 070401 (2011)
67. T. Sagawa, M. Ueda, Minimal energy cost for thermodynamic information processing: measurement and information erasure. Phys. Rev. Lett. **102**, 250602 (2009)
68. T. Sagawa, M. Ueda, Erratum: minimal energy cost for thermodynamic information processing: measurement and information erasure [Phys. Rev. Lett. 102, 250602 (2009)]. Phys. Rev. Lett. **106**, 189901 (2011)
69. L. del Rio, J. Åberg, R. Renner, O. Dahlsten, V. Vedral, The thermodynamic meaning of negative entropy. Nature **474**, 61–63 (2011)
70. M. Horodecki, J. Oppenheim, A. Winter, Partial quantum information. Nature **436**, 673–679 (2005)
71. T. Sagawa, M. Ueda, Second law of Thermodynamics with Discrete Quantum Feedback Control. Phys. Rev. Lett. **100**, 080403 (2008)

72. K. Jacobs, Second law of thermodynamics and quantum feedback control: Maxwell's demon with weak measurements. Phys. Rev. A **80**, 012322 (2009)
73. W.H. Zurek, Quantum discord and Maxwell's demons. Phys. Rev. A **67**, 012320 (2003)
74. K. Funo, Y. Watanabe, M. Ueda, Thermodynamic work gain from entanglement. Phys. Rev. A **88**, 052319 (2013)
75. J.J. Park, K.-H. Kim, T. Sagawa, S.W. Kim, Heat engine driven by purely quantum information. Phys. Rev. Lett. **111**, 230402 (2013)
76. J.M. Horowitz, K. Jacobs, Quantum effects improve the energy efficiency of feedback control. Phys. Rev. E **89**, 042134 (2014)
77. J.P.S. Peterson, R.S. Sarthour, A.M. Souza, I.S. Oliveira, J. Goold, K. Modi, D.O. Soares-Pinto, L.C. Céleri, Experimental demonstration of information to energy conversion in a quantum system at the Landauer limit. Proc. Roy. Soc. Lon. A **472**, 20150813 (2016)
78. J.A. Vaccaro, S.M. Barnett, Information erasure without an energy cost. Proc. Roy. Soc. Lon. A **467**, 1770–1778 (2011)
79. M.H. Mohammady, M. Mohseni, Y. Omar, Minimising the heat dissipation of quantum information erasure. New J. Phys. **18**, 015011 (2016)
80. C. Jarzynski, Equalities and inequalities: irreversibility and the second law of thermodynamics at the nanoscale. Annu. Rev. Condens. Matter Phys. **2**, 329–351 (2011)
81. C. V. den Broeck, *Stochastic Thermodynamics: A Brief Introduction, in Course CLXXXIV "Physics of Complex Colloids"*, vol. 184, ed. by C. Bechinger, F. Sciortino, P. Ziherl, Proceedings of the International School of Physics "Enrico Fermi" (2013), pp. 155–193
82. C. Jarzynski, Nonequilibrium equality for free energy differences. Phys. Rev. Lett. **78**, 2690–2693 (1997)
83. C. Jarzynski, Equilibrium free-energy differences from nonequilibrium measurements: a master-equation approach. Phys. Rev. E **56**, 5018–5035 (1997)
84. G.E. Crooks, Entropy production fluctuation theorem and the nonequilibrium work relation for free energy differences. Phys. Rev. E **60**, 2721–2726 (1999)
85. U. Seifert, Entropy production along a stochastic trajectory and an integral fluctuation theorem. Phys. Rev. Lett. **95**, 040602 (2005)
86. D.J. Evans, E.G.D. Cohen, G.P. Morriss, Probability of second law violations in shearing steady states. Phys. Rev. Lett. **71**, 2401–2404 (1993)
87. A.E. Allahverdyan, Nonequilibrium quantum fluctuations of work. Phys. Rev. E **90**, 032137 (2014)
88. R. Gallego, J. Eisert, H. Wilming, Thermodynamic work from operational principles. New J. Phys. **18**, 103017 (2016)
89. M. Campisi, P. Hänggi, P. Talkner, Colloquium: quantum fluctuation relations: foundations and applications. Rev. Mod. Phys. **83**, 771–791 (2011)
90. T. Albash, D.A. Lidar, M. Marvian, P. Zanardi, Fluctuation theorems for quantum processes. Phys. Rev. E **88**, 032146 (2013)
91. K. Funo, Y. Watanabe, M. Ueda, Integral quantum fluctuation theorems under measurement and feedback control. Phys. Rev. E **88**, 052121 (2013)
92. A.E. Rastegin, K. Życzkowski, Jarzynski equality for quantum stochastic maps. Phys. Rev. E **89**, 012127 (2014)
93. J.M. Horowitz, T. Sagawa, Equivalent definitions of the quantum nonadiabatic entropy production. J. Stat. Phys. **156**, 55–65 (2014)
94. B.P. Venkatesh, G. Watanabe, P. Talkner, Transient quantum fluctuation theorems and generalized measurements. New J. Phys. **16**, 015032 (2014)
95. G. Watanabe, B.P. Venkatesh, P. Talkner, Generalized energy measurements and modified transient quantum fluctuation theorems. Phys. Rev. E **89**, 052116 (2014)
96. J. Goold, M. Paternostro, K. Modi, Nonequilibrium quantum Landauer principle. Phys. Rev. Lett. **114**, 060602 (2015)
97. K. Sekimoto, *Stochastic Energetics* (Springer, New York, 2010)
98. U. Seifert, Stochastic thermodynamics: principles and perspectives. Eur. Phys. J. B **64**, 423–431 (2008)

99. M. Esposito, C. Van den Broeck, Three faces of the second law. I. Master equation formulation. Phys. Rev. E **82**, 011143 (2010)
100. D.J. Evans, D.J. Searles, Equilibrium microstates which generate second law violating steady states. Phys. Rev. E **50**, 1645–1648 (1994)
101. N. Merhav, Y. Kafri, Statistical properties of entropy production derived from fluctuation theorems. J. Stat. Mech.: Theor. Exp. P12022 (2010)
102. G. Gallavotti, E.G.D. Cohen, Dynamical ensembles in nonequilibrium statistical mechanics. Phys. Rev. Lett. **74**, 2694–2697 (1995)
103. J. Kurchan, Fluctuation theorem for stochastic dynamics. J. Phys. A: Math. Gen. **31**, 3719 (1998)
104. J.L. Lebowitz, H. Spohn, A Gallavotti-Cohen-type symmetry in the large deviation functional for stochastic dynamics. J. Stat. Phys. **95**, 333–365 (1999)
105. C. Jarzynski, Nonequilibrium work theorem for a system strongly coupled to a thermal environment. J. Stat. Mech.: Theor. Exp. **2004**, P09005 (2004)
106. J. Liphardt, S. Dumont, S.B. Smith, I. Tinoco, C. Bustamante, Equilibrium information from nonequilibrium measurements in an experimental test of Jarzynski's equality. Science **296**, 1832–1835 (2002)
107. G. Hummer, A. Szabo, Free energy reconstruction from nonequilibrium single-molecule pulling experiments. Proc. Natl. Acad. Sci. **98**, 3658–3661 (2001)
108. O. Braun, A. Hanke, U. Seifert, Probing molecular free energy landscapes by periodic loading. Phys. Rev. Lett. **93**, 158105 (2004)
109. N.C. Harris, Y. Song, C.-H. Kiang, Experimental free energy surface reconstruction from single-molecule force spectroscopy using Jarzynski's equality. Phys. Rev. Lett. **99**, 068101 (2007)
110. V. Blickle, T. Speck, L. Helden, U. Seifert, C. Bechinger, Thermodynamics of a colloidal particle in a time-dependent nonharmonic potential. Phys. Rev. Lett. **96**, 070603 (2006)
111. C. Jarzynski, Nonequilibrium work relations: foundations and applications. Eur. Phys. J. B **64**, 331–340 (2008)
112. D. Collin, F. Ritort, C. Jarzynski, S.B. Smith, I. Tinoco, C. Bustamante, Verification of the Crooks fluctuation theorem and recovery of RNA folding free energies. Nature **437**, 231–234 (2005)
113. F. Douarche, S. Ciliberto, A. Petrosyan, Estimate of the free energy difference in mechanical systems from work fluctuations: experiments and models. J. Stat. Mech.: Theor. Exp. P09011 (2005)
114. F. Douarche, S. Ciliberto, A. Petrosyan, I. Rabbiosi, An experimental test of the Jarzynski equality in a mechanical experiment. Europhys. Lett. **70**, 593 (2005)
115. F. Douarche, S. Joubaud, N.B. Garnier, A. Petrosyan, S. Ciliberto, Work fluctuation theorems for harmonic oscillators. Phys. Rev. Lett. **97**, 140603 (2006)
116. O.-P. Saira, Y. Yoon, T. Tanttu, M. Möttönen, D.V. Averin, J.P. Pekola, Test of the Jarzynski and Crooks fluctuation relations in an electronic system. Phys. Rev. Lett. **109**, 180601 (2012)
117. C. Jarzynski, Rare events and the convergence of exponentially averaged work values. Phys. Rev. E **73**, 046105 (2006)
118. R. Kawai, J.M.R. Parrondo, C. Van den Broeck, Dissipation: the phase-space perspective. Phys. Rev. Lett. **98**, 080602 (2007)
119. S. Kullback, R.A. Leibler, On information and sufficiency. Ann. Math. Statist. **22**, 79–86 (1951)
120. T. Hatano, S.-I. Sasa, Steady-state thermodynamics of Langevin systems. Phys. Rev. Lett. **86**, 3463 (2001)
121. T. Speck, U. Seifert, Integral fluctuation theorem for the housekeeping heat. J. Phys. A: Math. Gen. **38**, L581–L588 (2005)
122. T. Sagawa, M. Ueda, Generalized Jarzynski equality under nonequilibrium feedback control. Phys. Rev. Lett. **104**, 090602 (2010)
123. M. Esposito, C. Van den Broeck, Three detailed fluctuation theorems. Phys. Rev. Lett. **104**, 090601 (2010)

124. J.M.R. Parrondo, C. Van den Broeck, R. Kawai, Entropy production and the arrow of time. New J. Phys. **11**, 073008 (2009)
125. C. Tietz, S. Schuler, T. Speck, U. Seifert, J. Wrachtrup, Measurement of stochastic entropy production. Phys. Rev. Lett. **97**, 050602 (2006)
126. J.V. Koski, T. Sagawa, O.-P. Saira, Y. Yoon, A. Kutvonen, P. Solinas, M. Mottonen, T. Ala-Nissila, J.P. Pekola, Distribution of entropy production in a single-electron box. Nat. Phys. **9**, 644–648 (2013)
127. C. Van den Broeck, M. Esposito, Three faces of the second law. II. Fokker-Planck formulation. Phys. Rev. E **82**, 011144 (2010)
128. D. Luposchainsky, A.C. Barato, H. Hinrichsen, Strong fluctuation theorem for nonstationary nonequilibrium systems. Phys. Rev. E **87**, 042108 (2013)
129. Y. Oono, M. Paniconi, Steady state thermodynamics. Prog. Theor. Phys. Supplement **130**, 29–44 (1998)
130. J. Kurchan, A Quantum Fluctuation Theorem (2000), arXiv:cond-mat/0007360
131. H. Tasaki, Jarzynski relations for quantum systems and some applications (2000), arXiv:cond-mat/0009244
132. S. Mukamel, Quantum extension of the Jarzynski relation: analogy with stochastic dephasing. Phys. Rev. Lett. **90**, 170604 (2003)
133. P. Talkner, P. Hänggi, The Tasaki-Crooks quantum fluctuation theorem. J. Phys. A: Math. Theor. **40**, F569–F571 (2007)
134. W. De Roeck, C. Maes, Quantum version of free-energy-irreversiblework relations. Phys. Rev. E **69**, 026115 (2004)
135. T. Monnai, Unified treatment of the quantum fluctuation theorem and the Jarzynski equality in terms of microscopic reversibility. Phys. Rev. E **72**, 027102 (2005)
136. M. Esposito, S. Mukamel, Fluctuation theorems for quantum master equations. Phys. Rev. E **73**, 046129 (2006)
137. G. E. Crooks, On the Jarzynski relation for dissipative quantum dynamics. J. Stat. Mech.: Theor. Exp. **10**, P10023 (2008)
138. P. Talkner, M. Campisi, P. Hänggi, Fluctuation theorems in driven open quantum systems. J. Stat. Mech.: Theor. Exp. P02025 (2009)
139. M. Campisi, P. Talkner, P. Hänggi, Fluctuation theorem for arbitrary open quantum systems. Phys. Rev. Lett. **102**, 210401 (2009)
140. D. Andrieux, P. Gaspard, Fluctuation theorem for transport in mesoscopic systems. J. Stat. Mech.: Theor. Exp. **2006**, P01011 (2006)
141. W. de Roeck, C. Maes, Steady state fluctuations of the dissipated heat for a quantum stochastic model. Rev. Math. Phys. **18**, 619–653 (2006)
142. K. Saito, A. Dhar, Fluctuation theorem in quantum heat conduction. Phys. Rev. Lett. **99**, 180601 (2007)
143. M. Esposito, U. Harbola, S. Mukamel, Fluctuation theorem for counting statistics in electron transport through quantum junctions. Phys. Rev. B **75**, 155316 (2007)
144. K. Saito, Y. Utsumi, Symmetry in full counting statistics, fluctuation theorem, and relations among nonlinear transport coefficients in the presence of a magnetic field. Phys. Rev. B **78**, 115429 (2008)
145. J. Dereziński, W. De Roeck, C. Maes, Fluctuations of quantum currents and unravelings of master equations. J. Stat. Phys. **131**, 341–356 (2008)
146. D. Andrieux, P. Gaspard, T. Monnai, S. Tasaki, The fluctuation theorem for currents in open quantum systems. New J. Phys. **11**, 043014 (2009)
147. M. Campisi, P. Talkner, P. Hänggi, Fluctuation theorems for continuously monitored quantum fluxes. Phys. Rev. Lett. **105**, 140601 (2010)
148. A.E. Allahverdyan, T.M. Nieuwenhuizen, Fluctuations of work from quantum subensembles: the case against quantum workfluctuation theorems. Phys. Rev. E **71**, 066102 (2005)
149. P. Talkner, P. Hänggi, M. Morillo, Microcanonical quantum fluctuation theorems. Phys. Rev. E **77**, 051131 (2008)

150. M. Heyl, S. Kehrein, Crooks Relation in optical spectra: universality in work distributions for weak local quenches. Phys. Rev. Lett. **108**, 190601 (2012)
151. J.P. Pekola, P. Solinas, A. Shnirman, D.V. Averin, Calorimetric measurement of work in a quantum system. New J. Phys. **15**, 115006 (2013)
152. L. Mazzola, G. De Chiara, M. Paternostro, Measuring the characteristic function of the work distribution. Phys. Rev. Lett. **110**, 230602 (2013)
153. R. Dorner, S.R. Clark, L. Heaney, R. Fazio, J. Goold, V. Vedral, Extracting quantum work statistics and fluctuation theorems by single-qubit interferometry. Phys. Rev. Lett. **110**, 230601 (2013)
154. M. Campisi, R. Blattmann, S. Kohler, D. Zueco, P. Hänggi, Employing circuit QED to measure non-equilibrium work fluctuations. New J. Phys. **15**, 105028 (2013)
155. T.B. Batalhão, A.M. Souza, L. Mazzola, R. Auccaise, R.S. Sarthour, I.S. Oliveira, J. Goold, G. De Chiara, M. Paternostro, R.M. Serra, Experimental reconstruction of work distribution and study of fluctuation relations in a closed quantum system. Phys. Rev. Lett. **113**, 140601 (2014)
156. S. An, J.-N. Zhang, M. Um, D. Lv, Y. Lu, J. Zhang, Z.-Q. Yin, H.T. Quan, K. Kim, Experimental test of the quantum Jarzynski equality with a trapped-ion system. Nat. Phys. **11**, 193–199 (2015)
157. G. Huber, F. Schmidt-Kaler, S. Deffner, E. Lutz, Employing trapped cold ions to verify the quantum Jarzynski equality. Phys. Rev. Lett. **101**, 070403 (2008)
158. P. Hänggi, P. Talkner, The other QFT. Nat. Phys. **11**, 108–110 (2015)
159. S. Deffner, E. Lutz, Nonequilibrium entropy production for open quantum systems. Phys. Rev. Lett. **107**, 140404 (2011)
160. Y. Morikuni, H. Tasaki, Quantum Jarzynski-Sagawa-Ueda relations. J. Stat. Phys. **143**, 1–10 (2011)
161. S. Jevtic, T. Rudolph, D. Jennings, Y. Hirono, S. Nakayama, M. Murao, Exchange fluctuation theorem for correlated quantum systems. Phys. Rev. E **92**, 042113 (2015)
162. J.E. Geusic, E.O.S.-D. Bois, R.W. De Grasse, H.E.D. Scovil, Three level spin refrigeration and maser action at 1500 mc/sec. J. Appl. Phys. **30**, 1113–1114 (1959)
163. R. Levine, O. Kafri, Thermodynamic analysis of chemical laser systems. Chem. Phys. Lett. **27**, 175–179 (1974)
164. R. Kosloff, A quantum mechanical open system as a model of a heat engine. J. Chem. Phys. **80**, 1625–1631 (1984)
165. E. Geva, R. Kosloff, A quantum-mechanical heat engine operating in finite time. A model consisting of spin-1/2 systems as the working fluid. J. Chem. Phys. **96**, 3054–3067 (1992)
166. E. Geva, R. Kosloff, Three-level quantum amplifier as a heat engine: a study in finite-time thermodynamics. Phys. Rev. E **49**, 3903–3918 (1994)
167. J.P. Palao, R. Kosloff, Quantum thermodynamic cooling cycle. Phys. Rev. E **64**, 056130 (2001)
168. W.D. Phillips, Nobel lecture: laser cooling and trapping of neutral atoms. Rev. Mod. Phys. **70**, 721–741 (1998)
169. N. Brunner, M. Huber, N. Linden, S. Popescu, R. Silva, P. Skrzypczyk, Entanglement enhances cooling in microscopic quantum refrigerators. Phys. Rev. E **89**, 032115 (2014)
170. L.A. Correa, J.P. Palao, G. Adesso, D. Alonso, Optimal performance of endoreversible quantum refrigerators. Phys. Rev. E **90**, 062124 (2014)
171. R. Kosloff, A. Levy, Quantum heat engines and refrigerators: continuous devices. Annu. Rev. Phys. Chem. **65**, 365–393 (2014)
172. D. Gelbwaser-Klimovsky, W. Niedenzu, G. Kurizki, Chapter Twelve - Thermodynamics of quantum systems under dynamical control, in *Advances in Atomic, Molecular, and Optical Physics*, vol. 64, ed. by C. C. Lin, E. Arimondo, S. F. Yelin (Academic Press, 2015), pp. 329–407
173. R. Uzdin, A. Levy, R. Kosloff, Equivalence of quantum heat machines, and quantum-thermodynamic signatures. Phys. Rev. X **5**, 031044 (2015)
174. H. J. Briegel, S. Popescu, Entanglement and intra-molecular cooling in biological systems? - A quantum thermodynamic perspective (2008), arXiv:0806.4552

175. V. Giovannetti, S. Lloyd, L. Maccone, Quantum-enhanced measurements: beating the standard quantum limit. Science **306**, 1330 (2004)
176. V. Giovannetti, S. Lloyd, L. Maccone, Quantum metrology. Phys. Rev. Lett. **96**, 010401 (2006)
177. M.A. Nielsen, I.L. Chuang, *Quantum Computation and Quantum Information* (Cambridge University Press, Cambridge, 2000)
178. H.T. Quan, Y.-X. Liu, C.P. Sun, F. Nori, Quantum thermodynamic cycles and quantum heat engines. Phys. Rev. E **76**, 031105 (2007)
179. H.T. Quan, Quantum thermodynamic cycles and quantum heat engines. II. Phys. Rev. E **79**, 041129 (2009)
180. N. Linden, S. Popescu, P. Skrzypczyk, How small can thermal machines be? The smallest possible refrigerator. Phys. Rev. Lett. **105**, 130401 (2010)
181. N. Brunner, N. Linden, S. Popescu, P. Skrzypczyk, Virtual qubits, virtual temperatures, and the foundations of thermodynamics. Phys. Rev. E **85**, 051117 (2012)
182. O. Abah, J. Roßnagel, G. Jacob, S. Deffner, F. Schmidt-Kaler, K. Singer, E. Lutz, Single-Ion heat engine at maximum power. Phys. Rev. Lett. **109**, 203006 (2012)
183. J. Roßnagel, S.T. Dawkins, K.N. Tolazzi, O. Abah, E. Lutz, F. Schmidt-Kaler, K. Singer, A single-atom heat engine. Science **352**, 325–329 (2016)
184. K. Zhang, F. Bariani, P. Meystre, Quantum optomechanical heat engine. Phys. Rev. Lett. **112**, 150602 (2014)
185. C. Elouard, M. Richard, A. Auffèves, Reversible work extraction in a hybrid opto-mechanical system. New J. Phys. **17**, 055018 (2015)
186. A.O. Niskanen, Y. Nakamura, J.P. Pekola, Information entropic superconducting microcooler. Phys. Rev. B **76**, 174523 (2007)
187. A. Dechant, N. Kiesel, E. Lutz, All-optical nanomechanical heat engine. Phys. Rev. Lett. **114**, 183602 (2015)
188. M. Campisi, J. Pekola, R. Fazio, Nonequilibrium fluctuations in quantum heat engines: theory, example, and possible solid state experiments. New J. Phys. **17**, 035012 (2015)
189. J.-P. Brantut, C. Grenier, J. Meineke, D. Stadler, S. Krinner, C. Kollath, T. Esslinger, A. Georges, A thermoelectric heat engine with ultracold atoms. Science **342**, 713–715 (2013)
190. C. Bergenfeldt, P. Samuelsson, B. Sothmann, C. Flindt, M. Büttiker, Hybrid microwave-cavity heat engine. Phys. Rev. Lett. **112**, 076803 (2014)
191. P.P. Hofer, J.-R. Souquet, A.A. Clerk, Quantum heat engine based on photon-assisted Cooper pair tunneling. Phys. Rev. B **93**, 041418 (2016)
192. D. Venturelli, R. Fazio, V. Giovannetti, Minimal self-contained quantum refrigeration machine based on four quantum dots. Phys. Rev. Lett. **110**, 256801 (2013)
193. M.T. Mitchison, M. Huber, J. Prior, M.P. Woods, M.B. Plenio, Realising a quantum absorption refrigerator with an atomcavity system. Quantum Sci. Technol. **1**, 015001 (2016)
194. Y.-X. Chen, S.-W. Li, Quantum refrigerator driven by current noise. Europhys. Lett. **97**, 40003 (2012)
195. P. P. Hofer, M. Perarnau-Llobet, J. B. Brask, R. Silva, M. Huber, N. Brunner, Autonomous Quantum Refrigerator in a Circuit- QED Architecture Based on a Josephson Junction. Phys. Rev. B **94**, 235420 (2016), arXiv:1607.05218
196. A. Mari, J. Eisert, Cooling by heating: very hot thermal light can significantly cool quantum systems. Phys. Rev. Lett. **108**, 120602 (2012)
197. B. Cleuren, B. Rutten, C. Van den Broeck, Cooling by heating: refrigeration powered by photons. Phys. Rev. Lett. **108**, 120603 (2012)
198. J.B. Brask, G. Haack, N. Brunner, M. Huber, Autonomous quantum thermal machine for generating steady-state entanglement. New J. Phys. **17**, 113029 (2015)
199. B. Leggio, B. Bellomo, M. Antezza, Quantum thermal machines with single nonequilibrium environments. Phys. Rev. A **91**, 012117 (2015)
200. A. Roulet, S. Nimmrichter, J. M. Arrazola, V. Scarani, Autonomous Rotor Heat Engine. Phys. Rev. E **95**, 062131 (2017), arXiv:1609.06011
201. M. Youssef, G. Mahler, A.-S.F. Obada, Quantum optical thermodynamic machines: lasing as relaxation. Phys. Rev. E **80**, 061129 (2009)

202. T.D. Kieu, The second law, Maxwell's Demon, and work derivable from quantum heat engines. Phys. Rev. Lett. **93**, 140403 (2004)
203. Y. Rezek, R. Kosloff, Irreversible performance of a quantum harmonic heat engine. New J. Phys. **8**, 83 (2006)
204. F. Curzon, B. Ahlborn, Efficiency of a carnot engine at maximum power input. Am. J. Phys. **43**, 22–24 (1975)
205. A. del Campo, J. Goold, M. Paternostro, More bang for your buck: super-adiabatic quantum engines. Sci. Rep. **4**, 6208 (2014)
206. E. Torrontegui, S. Ibáñez, S. Martínez-Garaot, M. Modugno, A. del Campo, D. Guéry-Odelin, A. Ruschhaupt, X. Chen, J. G. Muga, Chapter 2 - Shortcuts to adiabaticity, in advances in atomic, molecular, and optical physics, in *Advances In Atomic, Molecular, and Optical Physics*, vol. 62, ed. by P. R. B. Ennio Arimondo, C. C. Lin (Academic Press, 2013), pp. 117–169
207. A. Alecce, F. Galve, N. Lo Gullo, L. Dell'Anna, F. Plastina, R. Zambrini, Quantum Otto cycle with inner friction: finite-time and disorder effects. New J. Phys. **17**, 075007 (2015)
208. D. Gelbwaser-Klimovsky, G. Kurizki, Heat-machine control by quantum-state preparation: from quantum engines to refrigerators. Phys. Rev. E **90**, 022102 (2014)
209. L. A. Correa, Multistage quantum absorption heat pumps. Phys. Rev. E, 042128 (2014)
210. R. Silva, P. Skrzypczyk, N. Brunner, Small quantum absorption refrigerator with reversed couplings. Phys. Rev. E **92**, 012136 (2015)
211. P. Skrzypczyk, N. Brunner, N. Linden, S. Popescu, The smallest refrigerators can reach maximal efficiency. J. Phys. A: Math. Theor. **44**, 492002 (2011)
212. L. A. Correa, J. P. Palao, G. Adesso, D. Alonso, Performance bound for quantum absorption refrigerators. Phys. Rev. E 042131 (2013)
213. L. A. Correa, J. P. Palao, D. Alonso, G. Adesso, Quantum enhanced absorption refrigerators. Sci. Rep. 3949 (2014)
214. M.T. Mitchison, M.P. Woods, J. Prior, M. Huber, Coherenceassisted single-shot cooling by quantum absorption refrigerators. New J. Phys. **17**, 115013 (2015)
215. J.B. Brask, N. Brunner, Small quantum absorption refrigerator in the transient regime: time scales, enhanced cooling, and entanglement. Phys. Rev. E **92**, 062101 (2015)
216. M.F. Frenzel, D. Jennings, T. Rudolph, Quasi-autonomous quantum thermal machines and quantum to classical energy flow. New J. Phys. **18**, 023037 (2016)
217. N. Linden, S. Popescu, P. Skrzypczyk, The smallest possible heat engines (2010), arXiv:1010.6029
218. J. Åberg, Catalytic coherence. Phys. Rev. Lett. **113**, 150402 (2014)
219. M. P. Woods, N. Ng, S. Wehner, The maximum efficiency of nano heat engines depends on more than temperature (2016), arXiv:1506.02322
220. R. Gallego, A. Riera, J. Eisert, Thermal machines beyond the weak coupling regime. New J. Phys. **16**, 125009 (2014)
221. A.S.L. Malabarba, A.J. Short, P. Kammerlander, Clockdriven quantum thermal engines. New J. Phys. **17**, 045027 (2015)
222. M. P. Woods, R. Silva, J. Oppenheim, Autonomous quantum machines and finite sized clocks (2017), arXiv:1607.04591
223. M.O. Scully, M.S. Zubairy, G.S. Agarwal, H. Walther, Extracting work from a single heat bath via vanishing quantum coherence. Science **299**, 862–864 (2003)
224. H.T. Quan, P. Zhang, C.P. Sun, Quantum-classical transition of photon-Carnot engine induced by quantum decoherence. Phys. Rev. E **73**, 036122 (2006)
225. H. Li, J. Zou, W.-L. Yu, B.-M. Xu, J.-G. Li, B. Shao, Quantum coherence rather than quantum correlations reflect the effects of a reservoir on a system's work capability. Phys. Rev. E **89**, 052132 (2014)
226. A.Ü.C. Hardal, Ö.E. Müstecaplığlu, Superradiant quantum heat engine. Sci. Rep. **5**, 12953 (2015)
227. R. Dillenschneider, E. Lutz, Energetics of quantum correlations. Europhys. Lett. **88**, 50003 (2009)

228. X.L. Huan, T. Wang, X.X. Yi, Effects of reservoir squeezing on quantum systems and work extraction. Phys. Rev. E **86**, 051105 (2012)
229. J. Roßnagel, O. Abah, F. Schmidt-Kaler, K. Singer, E. Lutz, Nanoscale heat engine beyond the carnot limit. Phys. Rev. Lett. **112**, 030602 (2014)
230. R. Long, W. Liu, Performance of quantum Otto refrigerators with squeezing. Phys. Rev. E **91**, 062137 (2015)
231. K. Brandner, M. Bauer, M.T. Schmid, U. Seifert, Coherenceenhanced efficiency of feedback-driven quantum engines. New J. Phys. **17**, 065006 (2015)
232. J. Jaramillo, M. Beau, A. del Campo, Quantum supremacy of many-particle thermal machines. New J. Phys. **18**, 075019 (2016)
233. M.O. Scully, K.R. Chapin, K.E. Dorfman, M.B. Kim, A. Svidzinsky, Quantum heat engine power can be increased by noise-induced coherence. Proc. Natl. Acad. Sci. **108**, 15097–15100 (2011)
234. U. Harbola, S. Rahav, S. Mukamel, Quantum heat engines: a thermodynamic analysis of power and efficiency. Europhys. Lett. **99**, 50005 (2012)
235. G.-K. David, N. Wolfgang, B. Paul, K. Gershon, Power enhancement of heat engines via correlated thermalization in a three-level "working fluid". Sci. Rep. **5**, 14413 (2015)
236. K. Brandner, U. Seifert, Periodic thermodynamics of open quantum systems. Phys. Rev. E **93**, 062134 (2016)
237. O. Abah, E. Lutz, Efficiency of heat engines coupled to nonequilibrium reservoirs. Europhys. Lett. **106**, 20001 (2014)
238. J. Oppenheim, M. Horodecki, P. Horodecki, R. Horodecki, Thermodynamical approach to quantifying quantum correlations. Phys. Rev. Lett. **89**, 180402 (2002)
239. M. Perarnau-Llobet, K.V. Hovhannisyan, M. Huber, P. Skrzypczyk, N. Brunner, A. Acín, Extractable work from correlations. Phys. Rev. X **5**, 041011 (2015)
240. K. Korzekwa, M. Lostaglio, J. Oppenheim, D. Jennings, The extraction of work from quantum coherence. New J. Phys. **18**, 023045 (2016)
241. P. Kammerlander, J. Anders, Coherence and measurement in quantum thermodynamics. Sci. Rep. **6**, 22174 (2016)
242. C. Gogolin, J. Eisert, Equilibration, thermalisation, and the emergence of statistical mechanics in closed quantum systems. Rep. Prog. Phys. **79**, 056001 (2016)
243. S. Popescu, A.J. Short, A. Winter, Entanglement and the foundations of statistical mechanics. Nat. Phys. **2**, 754–758 (2006)
244. S. Goldstein, J.L. Lebowitz, R. Tumulka, N. Zanghì, Canonical typicality. Phys. Rev. Lett. **96**, 050403 (2006)
245. N. Linden, S. Popescu, A.J. Short, A. Winter, Quantum mechanical evolution towards thermal equilibrium. Phys. Rev. E **79**, 061103 (2009)
246. A.J. Short, T.C. Farrelly, Quantum equilibration in finite time. New J. Phys. **14**, 013063 (2012)
247. M. Cramer, C.M. Dawson, J. Eisert, T.J. Osborne, Exact relaxation in a class of nonequilibrium quantum lattice systems. Phys. Rev. Lett. **100**, 030602 (2008)
248. M. Cramer, J. Eisert, A quantum central limit theorem for nonequilibrium systems: exact local relaxation of correlated states. New J. Phys. **12**, 055020 (2010)
249. C. Gogolin, M.P. Müller, J. Eisert, Absence of thermalization in nonintegrable systems. Phys. Rev. Lett. **106**, 040401 (2011)
250. M. Rigol, V. Dunjko, V. Yurovsky, M. Olshanii, Relaxation in a completely integrable many-body quantum system: an ab initio study of the dynamics of the highly excited states of 1d lattice hard-core bosons. Phys. Rev. Lett. **98**, 050405 (2007)
251. M.A. Cazalilla, Effect of suddenly turning on interactions in the luttinger model. Phys. Rev. Lett. **97**, 156403 (2006)
252. M. Rigol, V. Dunjko, M. Olshanii, Thermalization and its mechanism for generic isolated quantum systems. Nature **452**, 854–858 (2008)
253. P. Calabrese, F.H.L. Essler, M. Fagotti, Quantum quench in the transverse-field ising chain. Phys. Rev. Lett. **106**, 227203 (2011)

254. A.C. Cassidy, C.W. Clark, M. Rigol, Generalized thermalization in an integrable lattice system. Phys. Rev. Lett. **106**, 140405 (2011)
255. J.-S. Caux, R.M. Konik, Constructing the generalized gibbs ensemble after a quantum quench. Phys. Rev. Lett. **109**, 175301 (2012)
256. M. Fagotti, F.H.L. Essler, Reduced density matrix after a quantum quench. Phys. Rev. B **87**, 245107 (2013)
257. T. Langen, S. Erne, R. Geiger, B. Rauer, T. Schweigler, M. Kuhnert, W. Rohringer, I.E. Mazets, T. Gasenzer, J. Schmiedmayer, Experimental observation of a generalized Gibbs ensemble. Science **348**, 207–211 (2015)
258. R. Hamazaki, T.N. Ikeda, M. Ueda, Generalized Gibbs ensemble in a nonintegrable system with an extensive number of local symmetries. Phys. Rev. E **93**, 032116 (2016)
259. J. Berges, S. Borsányi, C. Wetterich, Prethermalization. Phys. Rev. Lett. **93**, 142002 (2004)
260. M. Horodecki, J. Oppenheim, (Quantumness in the context of) resource theories. Int. J. Mod. Phys. B **27**, 1345019 (2013)
261. E.H. Lieb, J. Yngvason, The physics and mathematics of the second law of thermodynamics. Phys. Rep. **310**, 1–96 (1999)
262. D. Janzing, P. Wocjan, R. Zeier, R. Geiss, T. Beth, Thermodynamic cost of reliability and low temperatures: tightening Landauer's principle and the second law. Int. J. Theor. Phys. **39**, 2717–2753 (2000)
263. F.G.S.L. Brandão, M. Horodecki, J. Oppenheim, J.M. Renes, R.W. Spekkens, Resource theory of quantum states out of thermal equilibrium. Phys. Rev. Lett. **111**, 250404 (2013)
264. M. Horodecki, J. Oppenheim, Fundamental limitations for quantum and nanoscale thermodynamics. Nat. Commun. **4**, 2059 (2013)
265. W. Pusz, S.L. Woronowicz, Passive states and KMS states for general quantum systems. Comm. Math. Phys. **58**, 273–290 (1978)
266. M. Horodecki, P. Horodecki, J. Oppenheim, About Reversible transformations from pure to mixed states and the unique measure of information. Phys. Rev. A **67**, 062104 (2003)
267. G. Gour, M.P. Müller, V. Narasimhachar, R.W. Spekkens, N.Y. Halpern, The resource theory of informational nonequilibrium in thermodynamics. Phys. Rep. **583**, 1–58 (2015)
268. P. Faist, J. Oppenheim, R. Renner, Gibbs-preserving maps outperform thermal operations in the quantum regime. New J. Phys. **17**, 043003 (2015)
269. M. Lostaglio, K. Korzekwa, D. Jennings, T. Rudolph, Quantum coherence, time-translation symmetry, and thermodynamics. Phys. Rev. X **5**, 021001 (2015)
270. F. Brandão, M. Horodecki, N. Ng, J. Oppenheim, S. Wehner, The second laws of quantum thermodynamics. Proc. Natl. Acad. Sci. **112**, 3275–3279 (2015)
271. N.H.Y. Ng, L. Manĉinska, C. Cirstoiu, J. Eisert, S. Wehner, Limits to catalysis in quantum thermodynamics. New J. Phys. **17**, 085004 (2015)
272. M. Weilenmann, L. Krämer, P. Faist, R. Renner, Axiomatic relation between thermodynamic and information-theoretic entropies. Phys. Rev. Lett. **117**, 260601 (2016), arXiv:1501.06920
273. N.Y. Halpern, J.M. Renes, Beyond heat baths: generalized resource theories for small-scale thermodynamics. Phys. Rev. E **93**, 022126 (2016)
274. Y. Guryanova, S. Popescu, A.J. Short, R. Silva, P. Skrzypczyk, Thermodynamics of quantum systems with multiple conserved quantities. Nat. Commun. **7**, 12049 (2016)
275. M. Huber, M. Perarnau-Llobet, K.V. Hovhannisyan, P. Skrzypczyk, C. Klöckl, N. Brunner, A. Acín, Thermodynamic cost of creating correlations. New J. Phys. **17**, 065008 (2015)
276. D.E. Bruschi, M. Perarnau-Llobet, N. Friis, K.V. Hovhannisyan, M. Huber, Thermodynamics of creating correlations: limitations and optimal protocols. Phys. Rev. E **91**, 032118 (2015)

Part II
Quantum Synchronization Induced by Dissipation in Many-Body Systems

Chapter 4
Transient Synchronization and Quantum Correlations

In Chaps. 1 and 2 we provided an introduction to the most important concepts and general methods employed in the description of open quantum systems. In particular, we introduced the central concept of quantum correlations as a characteristic trait of quantum mechanics, responsible of a rich and striking phenomenology which has fascinated scientists from over almost one century ago (see Sect. 1.4). Nowadays quantum correlations are considered the basic resource in modern applications of quantum information and quantum computation, while still being the central subject of a wide range of fundamental research. The detailed dynamical study of quantum correlations in open systems plays an important role, as the understanding of the mechanisms creating, preserving, or destroying quantum correlations becomes a topic of major importance when exploring the quantum-to-classical boundary [1]. The dynamics of quantum correlations in few-body open systems such as entanglement has been investigated during decades (for a review see [2]), while quantum discord has been only more recently considered [3–9].

Moreover, the generation of large quantum correlations have been recognized as an indicator of the presence of other interesting phenomena, such as quantum phase transitions [10–12]. The other way around has been also explored: for example the presence of entanglement may be revealed by internal energy [13] or by deviations in the scaling of a solid heat capacity [14]. In this chapter we report our results in establishing a connection between the phenomenon of *mutual synchronization* and the presence of robust quantum correlations.[1] In order to do it, we consider a fundamental quantum system, two detuned interacting quantum harmonic oscillators dissipating into the environment. We will carry out a dynamical analysis using the methods developed in Sect. 2.4. In the following, we show how the emergence of synchronous dynamics in the system due to the presence of common dissipation is accompanied by the robust, slow decay of quantum discord. On the other hand,

[1]The results in this chapter have been published in Ref. [15].

G. Manzano Paule, *Thermodynamics and Synchronization in Open Quantum Systems*, Springer Theses, https://doi.org/10.1007/978-3-319-93964-3_4

in the case of independent dissipation for the system components, quantum discord and synchronization quickly disappear.

The chapter is structured as follows. We start in Sect. 4.1 by introducing the phenomenon of synchronization and discussing previous works considering it in quantum systems. In Sect. 4.2 we characterize our system study, two detuned quantum harmonic oscillators dissipating into the environment, together with their dynamical description depending on the form of the dissipation. We identify in Sect. 4.3 the conditions leading to this spontaneous phenomenon, showing that the ability of the system to synchronize is related to the existence of disparate decay rates. In Sect. 4.4 we further show that the existence of such disparate decay rates is accompanied by robust quantum discord and mutual information between the oscillators, preventing the leak of information from the system into the environment. Further, we dedicate Sect. 4.5 to analyze the dependence on initial conditions, showing that they do not play a significant role. Some conclusions about this first study are presented in Sect. 4.6. Further technical details on the master equation and the equations of motion employed for describing the dynamics can be found in Appendix.

4.1 Synchronization Phenomena and Previous Works

Synchronization phenomena, from its first scientific description in the 17th century by Christiaan Huygens [16], have been observed in a broad range of physical, chemical and biological systems under a variety of circumstances [17]. The development of a general framework for the description of the phenomenon beyond the specific details of each system, allow us to distinguish between different types of synchronization (complete, phase, lag, ...). In some instances, synchronization is induced by the presence of an external forcing or driving which acts as a pacemaker. This is usually called *entrainment*, as it typically occurs when the influence of one among several oscillatory objects is unidirectional, and can be considered as an external periodic driving producing synchronization to its own frequency. Some examples of entrainment can be found in circadian rhythms, radio-controlled clocks or artificial pacemakers [18]. In other circumstances, synchronization appears spontaneously as a mutual cooperative behavior of different elements, which when coupled start to oscillate at a common frequency. This last case is the most relevant from the complex systems point of view since it appears as an emergent phenomenon that takes place as a consequence of the mutual influence between the elements, despite their natural differences. We will refer to this type of synchronization as *spontaneous*, *mutual* or *collective synchronization*. The simplest description of collective synchronization can be given in terms of coupled self-sustained oscillators, while it has been found in many different systems such as relaxation oscillator circuits, networks of neurons, hearth cardiac pacemaker cells or fireflies that flash in unison [19]. A key ingredient for it to appear is dissipation, which is the responsible for collapsing any trajectory of the system in phase space into a lower dimensional manifold.

Synchronization has also been studied in the quantum world in the case of entrainment induced by an external driving. Some interesting examples are dissipative driven two-level systems [20], a kicked particle falling in a static field [21], or nanomechanical beam oscillators [22]. Difficulties in addressing quantum spontaneous synchronization come from the fact that in linear oscillators, amenable to analytical treatment, dissipation will lead to the death of the oscillations after a transient. On the other hand, nonlinear oscillators can be considered, but then one needs to invoke different approximations limiting an insightful treatment. Here we take a first step towards the understanding of quantum spontaneous synchronization, showing that it is possible to fully characterize synchronization during the transient dynamics in an harmonic, i.e. linear, system. We find indeed that synchronization can arise even in absence of nonlinear dynamics depending on the dissipation. Different groups have recently approached this subject considering the synchronization of nano/microscopic systems susceptible of having quantum behavior, such as optomechanical cells [23], or micro [24] and nanomechanical oscillators [25]. In such cases, synchronization is studied by focusing on first-order momenta (mean values of position and momentum). However, this provides just a classical description of synchronization, while in order to go beyond, we must take into account higher order moments characterizing the quantum fluctuations of the system and the full correlations between the oscillatory objects.

4.2 Two Dissipative Harmonic Oscillators

In this chapter we consider two coupled quantum harmonic oscillators dissipating into the environment [26–28] with different frequencies [29], which is arguably one of the most fundamental prototypical models. Current experimental realizations in the quantum regime include nanoelectromechanical structures (NEMS) [30], optomechanical devices [31–33], or separately trapped ions whose direct coupling has been recently reported [34, 35]. The system Hamiltonian for unit masses is

$$\hat{H}_S = \frac{\hat{p}_1^2}{2} + \frac{\hat{p}_2^2}{2} + \frac{1}{2}(\omega_1^2\hat{x}_1^2 + \omega_2^2\hat{x}_2^2) + \lambda\hat{x}_1\hat{x}_2, \tag{4.1}$$

where $|\lambda| < \omega_1\omega_2$ (as required for an attractive potential) and we allow for frequencies diversity, i.e. $\omega_1 \neq \omega_2$. The free Hamiltonian is diagonalized by a rotation in the $\hat{x}_1$ and $\hat{x}_2$ plane

$$\begin{aligned} \hat{X}_- &= \cos\theta\,\hat{x}_1 - \sin\theta\,\hat{x}_2, \\ \hat{X}_+ &= \sin\theta\,\hat{x}_1 + \cos\theta\,\hat{x}_2, \end{aligned} \tag{4.2}$$

θ being the angle that gives the eigenvectors (or normal modes) $\{\hat{X}_\pm\}$ as a function of the coupling: $\tan 2\theta = 2\lambda/\omega_2^2 - \omega_1^2$. Applying the same rotation in Eq. (4.2) to $\hat{p}_1$ and $\hat{p}_2$, we can rewrite the system Hamiltonian in the normal modes basis

$$\hat{H}_S = \frac{1}{2}\sum_{i=\pm} \hat{P}_i^2 + \Omega_i \hat{X}_i^2, \quad \text{where} \quad [\hat{X}_j, \hat{P}_i] = i\hbar\delta_{i,j} \tag{4.3}$$

which corresponds to a pair of uncoupled harmonic oscillators with frequencies

$$\Omega_\pm^2 = \frac{1}{2}\left(\omega_1^2 + \omega_2^2 \pm \sqrt{4\lambda^2 + (\omega_2^2 - \omega_1^2)^2}\right). \tag{4.4}$$

We consider the two different dissipation scenarios introduced in Sect. 2.4: in the first one, the oscillators couple to independent (but equivalent) separate baths (SB) [36, 37]. In the second one, the two oscillators equally couple to the same common bath (CB). As we have discussed in Sect. 2.4, these two scenarios emerge in extended environments. In the simplest case of dissipation into isotropic surrounding media (e.g. electromagnetic radiation in free space), the transition occurs for some distance ξ_E depending on the system frequency and on the environment dispersion [38, 39]. If the ξ_E is smaller than the distance between the two oscillators, they would feel independent (uncorrelated) environmental noise (SB scenario). In the opposite case, both oscillators would feel the same noise fluctuations and hence the environment can be considered to be common (CB scenario).

We model the first case, SB, by considering two equivalent bosonic thermal baths independent of each other:

$$\begin{aligned}\hat{H}_B^{(1)} &= \sum_\alpha \hbar\tilde{\Omega}_\alpha \left(\hat{b}_\alpha^{(1)\dagger}\hat{b}_\alpha^{(1)} + \frac{1}{2}\right),\\ \hat{H}_B^{(2)} &= \sum_\alpha \hbar\tilde{\Omega}_\alpha \left(\hat{b}_\alpha^{(2)\dagger}\hat{b}_\alpha^{(2)} + \frac{1}{2}\right).\end{aligned} \tag{4.5}$$

The operators $\hat{b}_\alpha^i$ $(\hat{b}_\alpha^{i\dagger})$ annihilate (create) an excitation with energy $\hbar\tilde{\Omega}_\alpha$ over the αth mode of the ith thermal bath. The interaction Hamiltonian between the oscillators and the environments is

$$\hat{H}_I^{SB} = \sum_{i=1}^{2}\sum_\alpha g_\alpha\, \hat{x}_i \otimes \hat{Q}_\alpha^{(i)}, \tag{4.6}$$

where $\hat{Q}_\alpha^{(i)} = \sqrt{\hbar/2\tilde{\Omega}_\alpha}(\hat{b}_\alpha^{(i)} + \hat{b}_\alpha^{(i)\dagger})$, and the coupling coefficients g_α are related to the spectral density $J(\Omega)$ of the baths through $J(\Omega) \equiv \sum_\alpha \delta(\Omega - \Omega_\alpha)g_\alpha^2/\tilde{\Omega}_\alpha$. We assume Ohmic environments (see Sect. 2.3) with a Lorentz–Drude cut-off function, whose spectral density is given by

$$J(\Omega) = \frac{2\gamma_0}{\pi}\Omega\frac{\Lambda^2}{\Lambda^2 + \Omega^2}, \tag{4.7}$$

where γ_0 controls the coupling strength between oscillators and bath, and we assume for the cutoff frequency $\Lambda = 50\omega_1$. On the other hand, in the case of CB, we have a single bosonic reservoir with Hamiltonian given by Eq. (4.5). For the interaction Hamiltonian between the system and the environment we assume the form

$$\hat{H}_I^{CB} = \sum_\alpha g_\alpha \hat{x}_+ \otimes \hat{Q}_\alpha, \tag{4.8}$$

where $\hat{x}_+ = \hat{x}_1 + \hat{x}_2$. The spectral density of the bath is that of Eq. (4.7), with the same parameters γ_0 and Λ introduced before.

Master equations for both SB and CB have been compared also analyzing entanglement decay time in Ref. [29] where both the similarity of the frequencies of the oscillators and the coupling strength were shown to contribute to preserve entanglement for CB, leading to asymptotic entanglement in the case of identical frequencies [26, 27, 40–42]. The transition from SB to one CB underlies the capability of entanglement generation discussed in Ref. [43], and a physical implementation of the latter has been proposed in Ref. [44]. Following Ref. [29], the system dynamics is described by a master equation valid in the weak coupling limit between system and environment, without rotating wave approximation [45]. Even if the obtained master equation has the same form as the exact one [28], the coefficients are approximated for weak coupling. This equation for strong coupling can lead to unphysical values for the reduced density, and violation of positivity can appear at low temperatures and for certain initial states [45]. In the following we restrict our analysis to weak coupling, $\gamma_0 = 0.01\omega_1^2$, where we never encounter any unphysical dynamics. This is consistent with the fact that actually deviations of this master equation from one in the Lindblad form (preserving positivity) are small for high temperatures (here $T = 10k_B^{-1}\hbar\omega_1$).

Particularly useful to understand the physical behavior of the oscillators dissipation is the master equation in the basis of the normal modes of the system Hamiltonian (see Sect. 2.4), valid for both SB and CB:

$$\begin{aligned}\frac{d\rho(t)}{dt} = &-\frac{i}{\hbar}[\hat{H}_S, \rho(t)] - \frac{i}{2\hbar^2}\sum_{i,j}\tilde{\Gamma}_{ij}[\hat{X}_i, \{\hat{P}_j, \rho(t)\}] \\ &- \frac{1}{2\hbar^2}\sum_{i,j}\tilde{D}_{ij}[\hat{X}_i, [\hat{X}_j, \rho(t)]] + \frac{1}{2\hbar^2}\sum_{i,j}\tilde{F}_{ij}[\hat{X}_i, [\hat{P}_j, \rho(t)]],\end{aligned} \tag{4.9}$$

for $i, j = \{+, -\}$. Here the damping, diffusion and anomalous diffusion coefficients, $\{\tilde{\Gamma}_{ij}, \tilde{D}_{ij}, \tilde{F}_{ij}\}$ respectively, are different for CB and SB (see Appendix A.1). The terms $i = j$ are related to the dissipation of each normal mode by direct contact with the bath(s) while the terms $i \neq j$ are related to indirect channels of dissipation.

In the following we will first focus in the second order moments to gain insight into the noise dynamics and assume initial vacuum states. On the other hand, we will consider non-vacuum initial states conditions and the relevant first-order dynamics later when discussing classical synchronization. From the master equation (4.9) we obtain the following equations of motion for the second-order moments of the normal modes:

$$\frac{d\langle \hat{X}_i \hat{X}_j\rangle}{dt} = \frac{1}{2}\left(\{\hat{X}_i, \hat{P}_j\} + \{\hat{X}_j, \hat{P}_i\}\right), \tag{4.10}$$

$$\begin{aligned}\frac{d\langle \hat{P}_i \hat{P}_j\rangle}{dt} &= -\frac{1}{2}(\Omega_i^2\langle\{\hat{X}_i, \hat{P}_j\}\rangle + \Omega_j^2\langle\{\hat{X}_j, \hat{P}_i\}\rangle) \\ &\quad - (\tilde{\Gamma}_{i,i} + \tilde{\Gamma}_{j,j})/\hbar\langle \hat{P}_i \hat{P}_j\rangle - \tilde{\Gamma}_{i,-i}/\hbar\langle \hat{P}_j \hat{P}_{-i}\rangle \\ &\quad - \tilde{\Gamma}_{j,-j}/\hbar\langle \hat{P}_i \hat{P}_{-j}\rangle + \tilde{D}_{i,j},\end{aligned} \tag{4.11}$$

$$\begin{aligned}\frac{d\langle\{\hat{X}_i, \hat{P}_j\}\rangle}{dt} &= 2\langle \hat{P}_i \hat{P}_j\rangle - 2\Omega_j^2\langle \hat{X}_i \hat{X}_j\rangle + \tilde{F}_{i,j} \\ &\quad - \tilde{\Gamma}_{j,j}/\hbar\langle\{\hat{X}_i, \hat{P}_j\}\rangle - \tilde{\Gamma}_{j,-j}/\hbar\langle \hat{X}_i \hat{P}_{-j}\rangle.\end{aligned} \tag{4.12}$$

An important observation is that actually the results shown in the following do not depend on the specific choice of this master equation. In particular, in Appendix A.2 we compare our results with those from a master equation in the Lindblad form, obtained by a rotating wave approximation. Within this approximation the master equation is known to be in the Lindblad form [45, 46] and we find almost exactly the same results as with the master equation (4.9). Therefore, the phenomena predicted in the following do not depend on the specific details of the master equation.

4.3 Synchronization

The dynamical behavior of the oscillators can be analyzed through their average positions, variances and correlations, as we deal here with Gaussian states. The presence of a CB or of two (even if identical) SB leads to different friction terms in the dynamical equations of both first-order and second-order moments with profound consequences. We remind that for CB only the positions sum $\hat{x}_+ = \hat{x}_1 + \hat{x}_2$ is actually dissipating and this does not coincide with $\hat{X}_+$ unless the oscillators are identical.

Figure 4.1a shows the variance dynamics of two oscillators starting from two vacuum squeezed states. To quantify synchronization between two functions $f(t)$ and $g(t)$, we adopted a commonly used indicator, namely the *Pearson* indicator of synchronization

$$C_{f,g}(t, \Delta t) = \frac{\overline{\delta f\ \delta g}}{\sqrt{\overline{\delta f^2}\ \overline{\delta g^2}}}, \tag{4.13}$$

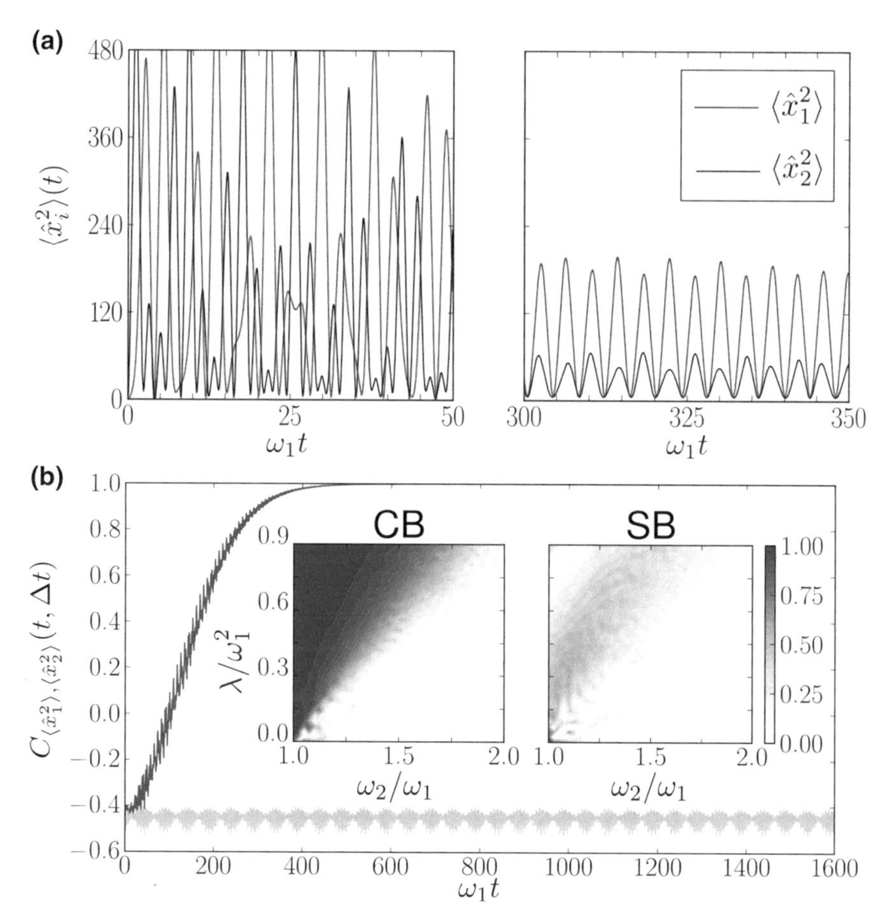

Fig. 4.1 a Temporal evolution of second-order moments $\langle \hat{x}_1^2(t) \rangle$ (red line) and $\langle \hat{x}_2^2(t) \rangle$ (black line) for $\omega_2 = 1.4\omega_1$ and $\lambda = 0.7\omega_1^2$ starting from squeezed states for CB and **b** synchronization $C_{\langle \hat{x}_1^2 \rangle, \langle \hat{x}_2^2 \rangle}(t, \Delta t)$ (being $\Delta t = 15\omega_1^{-1}$) for CB (blue) and SB (green) for temperature $T = 10\hbar\omega_1/k_B$. The insets show synchronization values $|C_{\langle \hat{x}_1^2 \rangle \langle \hat{x}_2^2 \rangle}|$ varying ω_2/ω_1 and λ/ω_1^2 at $t = 300\omega_1^{-1}$. Here and for the rest of the chapter we set $\gamma = 0.01\omega_1$. The initial state is separable with squeezing parameter $r = 2$ and $r = 4$, respectively, in the two oscillators

where the bar stands for a time average $\overline{f} = \int_t^{t+\Delta t} dt' f(t')$ with time window Δt and $\delta f = f - \overline{f}$. For 'similar' evolutions $|C| \sim 1$, while it decreases to zero for different dynamics. The position variances for CB [Fig. 4.1a] show a transient dynamics without any similarity, also in anti-phase ($C_{\langle \hat{x}_1^2 \rangle \langle \hat{x}_2^2 \rangle} < 0$), before reaching full synchronization [Fig. 4.1b].

A comprehensive analysis shows that this behavior is actually robust considering (i) different initial conditions, (ii) any second-order moments of the two oscillators (either of positions $\hat{x}_{1,2}$ or momenta $\hat{p}_{1,2}$, or any arbitrary quadrature) and (iii) a whole range of couplings and detunings. Regarding (i), an important observation is that, while in an isolated system the dynamics is strongly determined by the initial conditions, this is not the case in presence of an environment. After a transient (in which the initial conditions have an important role), we actually find synchronization independently on the initial state, implying that detuning and oscillators coupling are the only relevant parameters. The full analysis (iii) for CB allows us to conclude that synchronization arises faster for nearly resonant oscillators and that the deteriorating effect of detuning can be *proportionally* compensated by strong coupling, as represented in the CB inset of Fig. 4.1b.

Moving now to the case of separate baths, a completely different scenario appears. The quality of the synchronization is generally poor (small $|C|$), not improving in time and dependent on the initial condition. The full parameters map for $|C|$ is shown in the SB inset of Fig. 4.1b. In this case the oscillators do not synchronize in spite of their coupling even considering long times when, finally, the system thermalizes.

The appearance of a synchronous dynamics only for CB can be understood considering the time evolution of the second moments. The time evolution of the vector $\mathbf{R}$ of all the 10 second moments can be written in a compact matrix form as

$$\dot{\mathbf{R}} = \mathcal{M}\mathbf{R} + \mathbf{N}, \tag{4.14}$$

where the matrix $\mathcal{M}$ governing the time evolution [29] (see also Appendix A.1) has complex eigenvalues $\{\mu_i\}$ $(i = 1, \ldots, 10)$, named in the following *dynamical eigenvalues*. Their real and imaginary parts determine the decays and oscillatory dynamics of all second-order moments and variances. As shown in Fig. 4.2a when $\lambda = 0$, all the eigenvalues are along the line -0.01 and for increasing coupling — in the case of one CB — they move in the complex plane taking on three different real values. On the other hand, for SB all dynamical eigenvalues have similar real parts that remain almost unchanged when varying parameters. Hence for SB the ratio between the maximum and the minimum eigenvalues $\text{Re}(\mu_M)/\text{Re}(\mu_m) \approx 1$ is almost constant for all parameters while for CB *and* for parameters for which synchronization is found [CB inset in Fig. 4.1b], $\text{Re}(\mu_M)/\text{Re}(\mu_m) << 1$ as shown in Fig. 4.2b. In this parameters regime, after a transient time, only the least damped eigenvector survives then fixing the frequency of the whole dynamics of the moments. As a consequence of this mechanism, synchronization is observed considering the expectation values of any quadrature of the oscillators as well as higher order moments.

We obtain an approximated analytical estimation of time scales by considering the master equation in the eigenbasis [Eq. (4.9)] of the free Hamiltonian, Eq. (4.1). As we commented previously, both master equations for common and separate baths have the same expression in the case of detuned oscillators and the nature of dissipation (CB or SB) only appears in the form of the damping coefficients. By eliminating the oscillating terms in the dynamics in the interaction picture, one obtains that, within this approximation, the decay rates of $\langle \hat{P}_\pm^2 \rangle$ are given by

$$\begin{aligned} \tilde{\Gamma}_{--}^{SB} &= \cos^2\theta\, \Gamma_{11} + \sin^2\theta\, \Gamma_{22} - \cos\theta \sin\theta\, \Gamma_{12}, \\ \tilde{\Gamma}_{++}^{SB} &= \cos^2\theta\, \Gamma_{22} + \sin^2\theta\, \Gamma_{11} + \cos\theta \sin\theta\, \Gamma_{12}, \end{aligned} \tag{4.15}$$

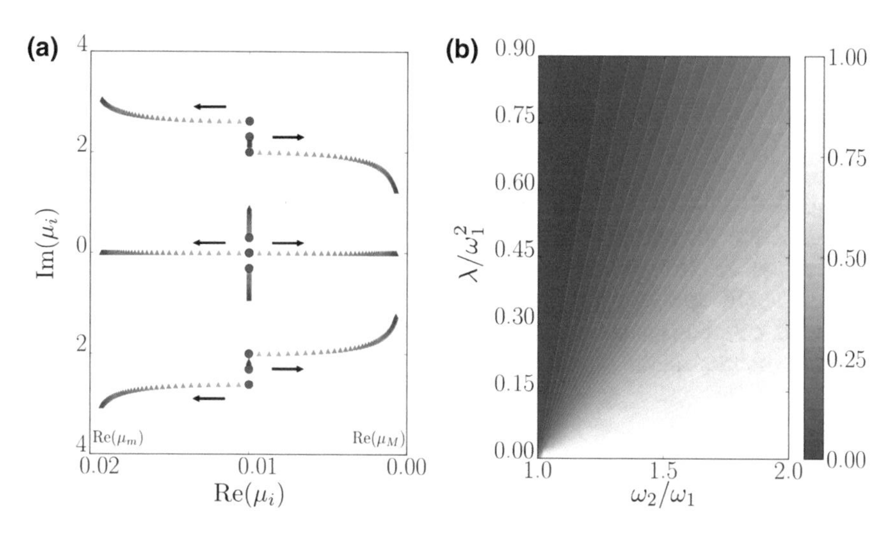

Fig. 4.2 **a** The 10 dynamical eigenvalues μ_i in the complex plane for CB. We used $\omega_2/\omega_1 = 1.31$ and then increase the coupling from $\lambda = 0$ (blue dots) to $\lambda = 0.9\omega_1^2$ (dark triangles). Notice that for $\lambda = 0$ the eigenvalue with $\text{Re}(\mu_i) = -0.01$ and $\text{Im}(\mu_i) = 0$ is degenerated. The arrows highlight the emergence of disparate real components for the eigenvalues leading to a separation of time scales for the decay of the normal modes. **b** Ratio between minimum and maximum eigenvalue $\text{Re}(\mu_m)/\text{Re}(\mu_M)$ for CB as a function of ω_2/ω_1 and λ/ω_1^2

for SB, while for CB:

$$\tilde{\Gamma}^{CB}_{\pm\pm} = (\cos\theta \pm \sin\theta)(\cos\theta\,\Gamma_{11} \pm \sin\theta\,\Gamma_{22}) + \\ + (1 \pm 2\sin\theta\cos\theta)\,\Gamma_{12}, \tag{4.16}$$

where θ is the previously defined diagonalization angle of $\hat{H}_S$, and $\Gamma_{11,22,12}$ appear in the original master equation (see Appendix A.1). These approximated decays for the variances, together with their average $(\tilde{\Gamma}_{--} + \tilde{\Gamma}_{++})/2$ (for $\langle \hat{P}_+ \hat{P}_- \rangle$), for a CB and SB do agree very well with the real parts of the dynamical eigenvalues (see Appendix A.2).

As mentioned before, synchronization (for CB) is observed looking at both the dynamics of first-order and second-order moments and, as a matter of fact, the ratio between minimum and maximum dynamical eigenvalues is the same in both cases. Still our interest is in the second-order moments due to their relevance for the quantum information shared by the oscillators. As a further remark, inspection of first-order moments dynamics allows us to establish connections with what is known in classical systems [17]

$$\frac{d\langle\hat{p}_1\rangle}{dt} = -\omega_1^2\langle\hat{x}_1\rangle - \lambda\langle\hat{x}_2\rangle - (\Gamma_{11}+\Gamma_{12})\langle\hat{p}_1\rangle - (\Gamma_{22}+\Gamma_{12})\langle\hat{p}_2\rangle,$$
$$\frac{d\langle\hat{p}_2\rangle}{dt} = -\omega_2^2\langle\hat{x}_2\rangle - \lambda\langle\hat{x}_1\rangle - (\Gamma_{11}+\Gamma_{12})\langle\hat{p}_1\rangle - (\Gamma_{22}+\Gamma_{12})\langle\hat{p}_2\rangle.$$

Two studied scenarios for classical synchronization are the 'diffusive' coupling where both oscillators dampings depend on the difference of the velocity and the 'direct' coupling where each one depends on the velocity of the other [17]. The quantum harmonic oscillators here considered for CB display in their first-order moments a 'diffusive' coupling up to a change of sign and this explains the *anti-phase* character of their synchronization.

4.4 Quantum Correlations

Once established the conditions for the emergence of synchronization, we explore this phenomenon focusing on information aspects, through mutual information shared by the oscillators and their quantum correlations. In particular, the total correlations between the oscillators are measured by the mutual information $I(1:2) = S(\varrho_1) + S(\varrho_2) - S(\varrho)$ where S stands for the Von Neumann entropy, and $\varrho_{1(2)}$ is the reduced density matrix of each harmonic oscillator (see Sect. 1.4.2). As we explained in Sect. 1.4.3, a possible partition of correlations into quantum and classical parts that has lately received great attention is given by the quantum discord [47–49]. It reads $\delta(1:2) = \min_{\{\hat{\Pi}_i\}}\left[S(\varrho_2) - S(\varrho) + S(\varrho_1|\{\hat{\Pi}_i\})\right]$ with the conditional entropy defined as $S(\varrho_1|\{\hat{\Pi}_j\}) = \sum_i p_i S(\varrho_{1|\hat{\Pi}_i}), \varrho_{1|\hat{\Pi}_i} = \hat{\Pi}_i\varrho\hat{\Pi}_i/p_i$ being the density matrix after a complete measurement $\{\hat{\Pi}_j\}$ on the second oscillator and $p_i = \mathrm{Tr}_{12}(\hat{\Pi}_i\varrho)$. Here we use quantum discord as a measure of the quantum correlations between the two oscillators by numerical computation of the *Gaussian discord* [50, 51] through the covariance matrix V_{12}, and minimizing over single mode generalized Gaussian measurements (see details in Sect. 1.4.3).

Dissipation degrades all quantum and classical correlations [52]. Nevertheless, important differences are found when comparing CB and SB for the same parameters choice. In Fig. 4.3 we show a fast decay of the total (a) as well as quantum (b) correlations for SB. On the other hand, for CB we find that, after a short transient, both mutual information and discord oscillate around an almost constant value and their decay is nearly frozen. For these parameters and a common environment, the oscillators synchronize and $C_{\langle\hat{x}_1^2\rangle\langle\hat{x}_2^2\rangle} = 0.95$ at $t \sim 270\omega_1^{-1}$. The robustness of quantum correlations in long times for synchronizing oscillators in a CB and the deep differences with the case of SB is surprising also because their respective asymptotic values are really similar for detuned oscillators. In other words, the upper CB curve in Fig. 4.3a or b will eventually thermalize converging to a value very similar the one obtained for SB, while strong differences in the asymptotic values actually appear only in the case of identical oscillators [26, 40]. As a further result, the effect of

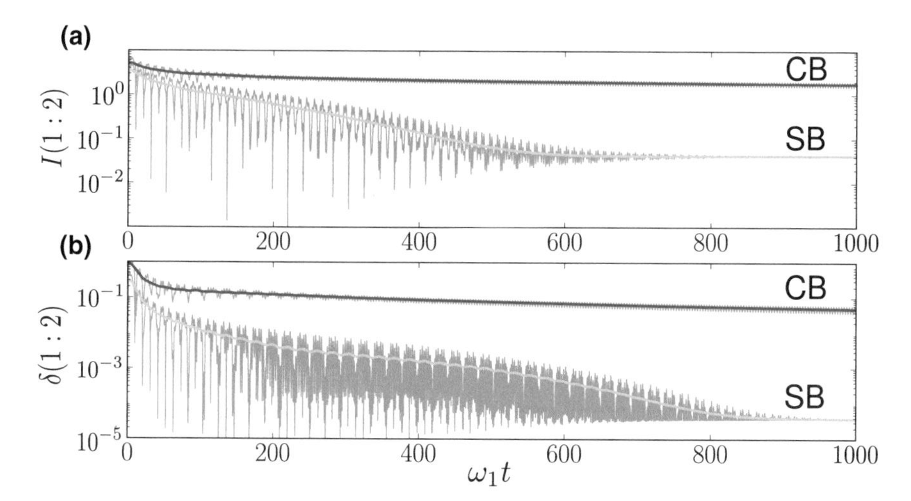

Fig. 4.3 Logarithmic plot of mutual information (**a**) and discord (**b**) for common (CB) and separate baths (SB). The exact time evolutions of both quantities are shown in gray while the thick blue and thick yellow lines correspond to their filtered versions (using a Gaussian filter) which eliminates rapid oscillations. Here we used $\omega_2 = 1.05\omega_1$ and $\lambda = 0.3\omega_1^2$ in both plots

increasing the temperature is mostly on the asymptotic state while the main features of the dynamics described here are still observed.

We now focus on the case of CB to look for specific quantum features of the synchronization in different parameters regimes. The comparison of mutual information and discord in cases in which there is synchronization or the system dissipates without having time to synchronize is given in Fig. 4.4 (upper and lower curves, respectively) where we filter out the fast oscillations to highlight the decay dynamics. For small coupling and large detuning, discord (shown in Fig. 4.4 for $\lambda/\omega_1^2 = 0.3$, $\omega_2/\omega_1 = 1.4$) and mutual information are rapidly degraded. In this case, when $t = 200\omega_1^{-1}$ there is not synchronous dynamics and $C_{\langle\hat{x}_1^2\rangle\langle\hat{x}_2^2\rangle} \sim 0$. On the other hand, for strong coupling or for small detuning, synchronization occurs fast: for $\lambda/\omega_1^2 = 0.8$, $\omega_2/\omega_1 = 1.05$ $C_{\langle\hat{x}_1^2\rangle\langle\hat{x}_2^2\rangle}(t = 200\omega_1^{-1}) \sim 1$. In this case, after a short transient, the dynamics of discord is almost frozen and it remains *robust* against decoherence. Exploring different parameter regimes we conclude that fast decay of classical and quantum correlations is found in cases in which there is no synchronization while the emergence of synchronization accompanies robust correlations against dissipation (frozen decay). The inset in Fig. 4.4 represents the value of the discord after the fast decay (here for $t = 300\omega_1 - 1$) where it is expected to be already in the plateau. There is a rather suggestive similarity with the synchronization 'map' for CB, shown in the inset of Fig. 4.1b. Considering that also entropy shows in this regime a slow growth, we conclude that synchronized oscillators are characterized by a reduced leakage of information into the environment.

One might wonder if the presence of a synchronous dynamics has any effect on entanglement as, in contrast to pure states, mixed states with large quantum correlations can even have vanishing entanglement [53, 54]. The presence of the

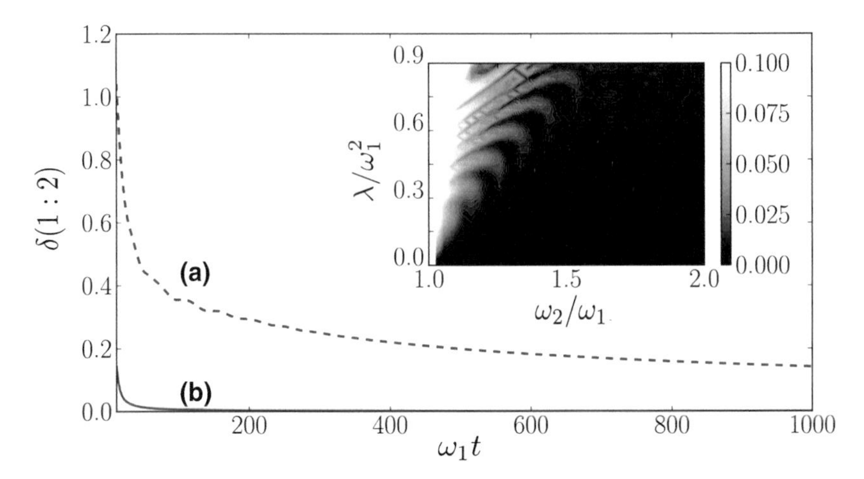

Fig. 4.4 Evolution of the discord for common bath and parameters $\omega_2/\omega_1 = 1.05$, $\lambda = 0.8\omega_1^2$ (blue dashed line A) and $\omega_2/\omega_1 = 1.4$, $\lambda = 0.3\omega_1^2$ (green solid line B). The inset represents the quantum discord at time $t = 300\omega_1^{-1}$ for CB as a function of ω_2/ω_1 and λ/ω_1^2

environment for oscillators with different frequencies leads to a complete loss of entanglement in finite short times unless the couplings to the CB are 'balanced' [29]. In the general case of detuned oscillators, even for large coupling, entanglement decay is typically faster than the time scales at which the system reaches synchronous dynamics both for CB and SB, mostly at this temperature ($T = 10k_B^{-1}\hbar\omega_1$). Still, longer survival times for entanglement in CB are found for small detunings and strong couplings [29].

4.5 Dependence on Initial Conditions

We mentioned before that initial conditions do not play any important role in the appearance of synchronization. Indeed synchronous dynamics of the moments appears when an eigenmode dominates because of its slow dissipation rate and this goes beyond the specificity of the choice of the initial state. However the details of the dynamics do depend on the latter as we illustrate for the following initial conditions for the two oscillators

1. Separable vacuum state:
$$\rho = |0\rangle\langle 0| \otimes |0\rangle\langle 0|. \tag{4.17}$$

2. Two-mode squeezed states:
$$\rho = \hat{U}_{12}(r)\left(|0\rangle\langle 0| \otimes |0\rangle\langle 0|\right)\hat{U}_{12}^{\dagger}(r), \tag{4.18}$$

   where $\hat{U}_{12}(r) = \exp\left[-r(\hat{a}_1^{\dagger}\hat{a}_2^{\dagger} - \hat{a}_1\hat{a}_2)/2\right]$ and $\hat{a}_i(\hat{a}_i^{\dagger})$ are the usual annihilation (creation) operators.

3. Separable squeezed state:

$$\rho = \hat{U}_1(r_1)|0\rangle\langle 0|\hat{U}_1^\dagger(r_1) \otimes \hat{U}_2(r_2)|0\rangle\langle 0|\hat{U}_2^\dagger(r_2), \tag{4.19}$$

with $\hat{U}_i(r_i) = \exp\left[-r(\hat{a}_i^{\dagger 2} - \hat{a}_i^2)/2\right]$.

Quantum correlations ($\delta(1:2)$) depend on the initial condition in the sense that more or less of the latter will be present. However, after the short transient, they always reach a plateau where information leakage to the environment is hugely reduced. Both information leakage reduction and synchronization are part of the same underlying phenomenon: that of a dissipation channel being much slower than the other. This behavior is seen in Fig. 4.5 where quantum correlations and the synchronization indicator are displayed for different initial conditions. We must further stress here that, since the asymptotic thermal state has $\delta(1:2) \sim 10^{-4}$, the plateau is expected to be very long.

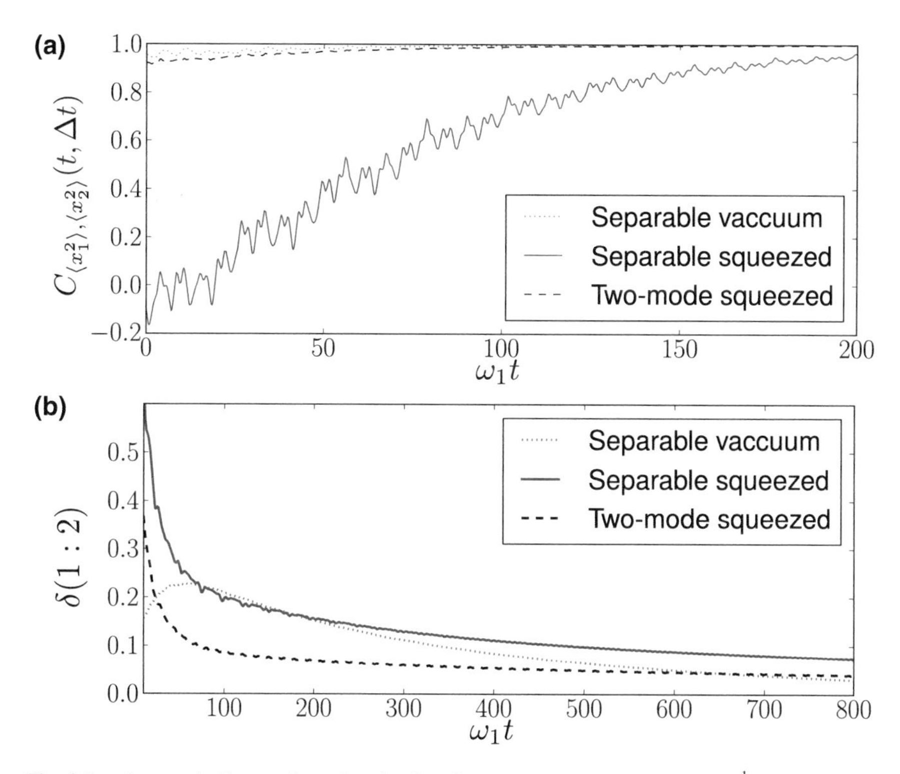

Fig. 4.5 **a** Pearson indicator of synchronization $C_{\langle \hat{x}_1^2\rangle,\langle \hat{x}_2^2\rangle}(t, \Delta t)$ with $\Delta t = 15\omega_1^{-1}$ and **b** decay of quantum correlations for different initial conditions in the case of common bath: Separable squeezed state with squeezing parameters $r_1 = 2$ and $r_2 = 4$ (green solid lines), separable vacuum state (red dotted lines), and an entangled two-mode squeezed state (black dashed lines) with squeezing $r = 2$. Here we employed $\omega_2/\omega_1 = 1.1$ and $\lambda = 0.8\omega_1^2$

4.6 Conclusions

Our analysis of the dynamics of dissipative quantum harmonic oscillators allows us to establish under which conditions synchronization appears. This phenomenon can appear in rather different forms but, to the best of our knowledge, it is the first time that it is reported during the transient dynamics of a (quantum or classical) system coupled to an environment and relaxing towards equilibrium. The emergence of synchronization is explained in terms of different temporal decays governing the system evolution and related to a separation between the eigenvalues of the matrix generating the dynamics. Indeed, we trace synchronization in second-order moments from the existence of a slowly decaying eigenmode and find approximated expressions for the variances decay coefficients in very good agreement with the real parts of the dynamical eigenvalues. We find that synchronization arises in presence of a common bath but not for separate ones, while it could be of interest to study the transition between so different scenarios [38, 43, 55–58]. An extensive analysis considering different parameters shows that a synchronous dynamics for common bath is degraded when increasing the detuning or weakening their direct coupling with the ratio shown in Fig. 4.1b (CB inset). The relevant parameters are λ/ω_1^2 and ω_2/ω_1 being the dependence on initial conditions actually weak.

We have then characterized mutual synchronization from a quantum information perspective. In order to do that, the dynamics of mutual information and quantum discord have been explored for different regimes of parameters. Our results indicate a signature of transient synchronization in the information shared by the oscillators: discord and mutual information are more robust when the oscillators synchronize. In spite of the fact that after thermalizing the asymptotic discord is negligible both for CB and SB, the decay towards this equilibrium value is clearly frozen in presence of synchronization. In this case, total and quantum correlations display a very slow decay (*plateau*) and the leak of information into the bath is reduced.

The identification of the conditions for the occurrence of synchronization and its connection with quantum correlations reported here provide the path towards extensions such as the study of arrays and networks, in which the presence of many normal modes in the dynamics opens a broader landscape with a richer phenomenology for synchronization phenomena. This is partially addressed in Chap. 5 for a system of three harmonic oscillators dissipating into a common environment, and more systematically explored in the general case of an arbitrary network of harmonic oscillators in Chap. 6, together with the analysis of the role of different environments. Another interesting direction for future investigation would be the exploration of eventual connections with biological systems, in which synchronization is a widespread phenomenon.

Different perspectives arise when turning to the definition of synchronization in the quantum realm. For instance in Ref. [59] phase and complete synchronization are discussed, and the ultimate bounds on the complete synchronization of quantum systems imposed by the uncertainty principle has been reported. An extension of such concepts has been also very recently proposed in Ref. [60]. For a recent review on the different measurements employed to measure quantum synchronization in different systems see Ref. [61].

Finally, from the experimental point of view, we stress that there have been few recent experiments showing the emergence of mutual synchronization in the average position and momenta (first order moments) in nanomechanical resonators coupled via a photonic resonator [62], and arrays of silicon nitride micromechanical oscillators coupled through an optical radiation field [63]. Autonomous coupled microelectromechanical oscillators [64], and anharmonic nanoelectromechanical resonators [65] have been also shown to display mutual synchronization. Finally, a phase-coherent regime in the oscillations of snowflake optomechanical crystals conforming an array has been predicted [66]. However, experiments testing the quantum signatures of synchronization and its relations to classical and quantum correlations have not been reported yet to date.

Appendix

A.1 Master Equation Details

The master equation describing the evolution of the reduced density matrix of the system of two different oscillators, up to the second order in the coupling strength, has been derived in Sect. 2.4 of Chap. 2 both for common and separate baths in the normal modes basis. Here we show that this is equivalent to the master equation in the original basis reported in Ref. [29]. We also stress that the exact master equation at all coupling orders has the same structure as the one reported here, the difference being in the form of its coefficients [28, 45]. For weak coupling this equation is a very good approximation to the exact one. Furthermore, in the following appendix we show that the full evolution almost perfectly matches that of a master equation obtained by a rotating-wave approximation, the latter having Lindblad form.

A.1.1 Separate Baths

As stated in Sect. 4.2, the normal modes of $\hat{H}_S$ are expressed in terms of the original position and momentum operators of the oscillators as

$$\hat{X}_- = \cos\theta\, \hat{x}_1 - \sin\theta\, \hat{x}_2, \qquad \hat{P}_- = \cos\theta\, \hat{p}_1 - \sin\theta\, \hat{p}_2, \tag{A.1}$$

$$\hat{X}_+ = \sin\theta\, \hat{x}_1 + \cos\theta\, \hat{x}_2, \qquad \hat{P}_+ = \sin\theta\, \hat{p}_1 + \cos\theta\, \hat{p}_2, \tag{A.2}$$

with $\tan 2\theta = 2\lambda/\omega_2^2 - \omega_1^2$. The system eigenfrequencies $\Omega_\pm$ are always different due to the coupling

$$2\Omega_{\pm}^2 = \omega_1^2 + \omega_2^2 \pm \sqrt{4\lambda^2 + (\omega_2^2 - \omega_1^2)^2}. \tag{A.3}$$

Neglecting energy renormalization, the master equation in the original basis reads [29]

$$\begin{aligned} \frac{d\rho(t)}{dt} &= -\frac{i}{\hbar}[\hat{H}_S, \rho(t)] - \frac{i}{2\hbar^2}\sum_{i,j=1}^{2}\Gamma_{ij}[\hat{x}_i, \{\hat{p}_j, \rho(t)\}] \\ &-\frac{1}{2\hbar^2}\sum_{i,j=1}^{2} D_{ij}[\hat{x}_i, [\hat{p}_j, \rho(t)]] + \frac{1}{2\hbar}\sum_{i,j=1}^{2} F_{ij}[\hat{x}_i, [\hat{p}_j, \rho(t)]], \end{aligned} \tag{A.4}$$

where the damping and diffusion coefficients are

$$\Gamma_{ij} = 2\hbar \int_0^{\tau} d\tau\, \beta_{ij}(\theta, \tau) \int_0^{\infty} d\Omega J(\Omega) \sin(\Omega\tau), \tag{A.5}$$

$$D_{ij} = 2\hbar \int_0^{\tau} d\tau\, \alpha_{ij}(\theta, \tau) \int_0^{\infty} d\Omega J(\Omega) \cos(\Omega\tau) \coth\left(\frac{\hbar\Omega}{2k_B T}\right), \tag{A.6}$$

$$F_{ij} = 2\hbar \int_0^{\tau} d\tau\, \beta_{ij}(\theta, \tau) \int_0^{\infty} d\Omega J(\Omega) \cos(\Omega\tau) \coth\left(\frac{\hbar\Omega}{2k_B T}\right), \tag{A.7}$$

with T the temperature of the bath, and

$$\alpha_{11}(\theta, \tau) = \cos^2\theta \cos(\Omega_-\tau) + \sin^2\theta \cos(\Omega_+\tau), \tag{A.8}$$

$$\alpha_{22}(\theta, \tau) = \alpha_{11}(\pi/2 - \theta, \tau), \tag{A.9}$$

$$\alpha_{12}(\theta, \tau) = \frac{\sin(2\theta)}{2}(\cos(\Omega_+\tau) - \cos(\Omega_-\tau)), \tag{A.10}$$

$$\alpha_{21}(\theta, \tau) = \alpha_{12}(\theta, \tau), \tag{A.11}$$

$$\beta_{11}(\theta, \tau) = \cos^2\theta \frac{\sin(\Omega_-\tau)}{\Omega_-} + \sin^2\theta \frac{\sin(\Omega_+\tau)}{\Omega_+}, \tag{A.12}$$

$$\beta_{22}(\theta, \tau) = \beta_{11}(\pi/2 - \theta, \tau), \tag{A.13}$$

$$\beta_{12}(\theta, \tau) = \frac{\sin(2\theta)}{2}\left(\frac{\sin(\Omega_+\tau)}{\Omega_+} - \frac{\sin(\Omega_-\tau)}{\Omega_-}\right), \tag{A.14}$$

$$\beta_{21}(\theta, \tau) = \beta_{12}(\theta, \tau). \tag{A.15}$$

Now introducing the normal modes, Eq. (A.1), into the master equation (A.4), we obtain Eq. (4.9) in Sect. 4.2, with the redefined coefficients:

$$
\begin{aligned}
\tilde{D}^{SB}_{--} &= c^2 D_{11} + s^2 D_{22} - 2cs D_{12}, \\
\tilde{D}^{SB}_{++} &= s^2 D_{11} + c^2 D_{22} + 2cs D_{12}, \\
\tilde{D}^{SB}_{+-} &= cs(D_{11} - D_{22}) + (c^2 - s^2) D_{12} = 0, \\
\tilde{F}^{SB}_{--} &= c^2 F_{11} + s^2 F_{22} - 2cs F_{12}, \\
\tilde{F}^{SB}_{++} &= s^2 F_{11} + c^2 F_{22} + 2cs F_{12}, \\
\tilde{F}^{SB}_{-+} &= \tilde{F}^{SB}_{+-} = cs(F_{11} - F_{22}) + (c^2 - s^2) F_{12} = 0, \\
\tilde{\Gamma}^{SB}_{--} &= c^2 \Gamma_{11} + s^2 \Gamma_{22} - 2cs \Gamma_{12}, \\
\tilde{\Gamma}^{SB}_{++} &= s^2 \Gamma_{11} + c^2 \Gamma_{22} + 2cs \Gamma_{12}, \\
\tilde{\Gamma}^{SB}_{-+} &= \tilde{\Gamma}^{SB}_{+-} = cs(\Gamma_{11} - \Gamma_{22}) + (c^2 - s^2) \Gamma_{12} = 0,
\end{aligned} \tag{A.16}
$$

being $c = \cos\theta$ and $s = \sin\theta$. Once these coefficients are used instead of those coming from a common bath, the equations of motion are formally identical to those in Eqs. (4.10)–(4.12). Notice that these expressions reproduce the coefficients reported in Eq. (2.126) for the derivation of the master equation in the normal modes basis when we identify the following basis-change matrix

$$
f = \begin{pmatrix} \cos\theta & \sin\theta \\ -\sin\theta & \cos\theta \end{pmatrix}, \quad \text{where} \begin{pmatrix} \hat{x}_1 \\ \hat{x}_2 \end{pmatrix} = f \begin{pmatrix} \hat{X}_- \\ \hat{X}_+ \end{pmatrix}. \tag{A.17}
$$

A.1.2 Common Bath

Neglecting again energy renormalization, the master equation in the original basis is now [29]

$$
\begin{aligned}
\frac{d\rho(t)}{dt} = &-\frac{i}{\hbar}[\hat{H}_S, \rho(t)] - \frac{i}{2\hbar^2} \sum_{i=1}^{2} \gamma_i [\hat{x}_+, \{\hat{p}_i, \rho(t)\}] \\
&- \frac{1}{2\hbar^2} \sum_{i=1}^{2} d_i [\hat{x}_+, [\hat{p}_i, \rho(t)]] + \frac{1}{2\hbar^2} \sum_{i=1}^{2} f_i [\hat{x}_+, [\hat{p}_i, \rho(t)]],
\end{aligned} \tag{A.18}
$$

with coefficients $\gamma_i = \Gamma_{ii} + \Gamma_{12}$, $d_i = D_{ii} + D_{12}$ and $f_i = F_{ii} + F_{12}$, being Γ_{ij}, D_{ij} and F_{ij} defined before. When implementing the change into the normal modes basis, Eq. (A.18) transforms into Eq. (4.9), but the coefficients are modified as follows

$$
\begin{aligned}
\tilde{D}_{--}^{CB} &= (c-s)(cD_{11} - sD_{22}) + (1-2sc)D_{12}, \\
\tilde{D}_{++}^{CB} &= (c+s)(cD_{11} + sD_{22}) + (1+2sc)D_{12}, \\
\tilde{D}_{+-}^{CB} &= \frac{1}{2}(c^2 - s^2)(D_{11} + D_{22} + 2D_{12}) + sc(D_{11} - D_{22}), \\
\tilde{F}_{--}^{CB} &= (c-s)(cF_{11} - sF_{22}) + (1-2sc)F_{12}, \\
\tilde{F}_{++}^{CB} &= (c+s)(cF_{11} + sF_{22}) + (1+2sc)F_{12}, \\
\tilde{F}_{-+}^{CB} &= (c-s)(cF_{22} + sF_{11}) + (c^2 - s^2)F_{12}, \\
\tilde{F}_{+-}^{CB} &= (c+s)(cF_{11} - sF_{22}) + (c^2 - s^2)F_{12}, \\
\tilde{\Gamma}_{--}^{CB} &= (c-s)(c\Gamma_{11} - s\Gamma_{22}) + (1-2sc)\Gamma_{12}, \\
\tilde{\Gamma}_{++}^{CB} &= (c+s)(c\Gamma_{11} + s\Gamma_{22}) + (1+2sc)\Gamma_{12}, \\
\tilde{\Gamma}_{-+}^{CB} &= (c-s)(c\Gamma_{22} + s\Gamma_{11}) + (c^2 - s^2)\Gamma_{12}, \\
\tilde{\Gamma}_{+-}^{CB} &= (c+s)(c\Gamma_{11} - s\Gamma_{22}) + (c^2 - s^2)\Gamma_{12},
\end{aligned}
\tag{A.19}
$$

being again $c = \cos\theta$ and $s = \sin\theta$. We finally recall that the above expressions can be obtained as well by direct derivation of the master equation in the normal mode basis as we reported in Sect. 2.4.

A.2 Independent Decay Rates

The normal modes (A.1)–(A.2) diagonalize the Hamiltonian of the system $\hat{H}_S$ but are still indirectly coupled through the heat bath(s) as can be seen from Eqs. (4.10–4.12). This means that the normal modes cannot be considered as independent channels for dissipation. Yet, if we rewrite their master equation in interaction picture, we can neglect fast oscillating terms, as usual in the rotating wave approximation. That is, elimination of exponents like $\exp(\pm i(\Omega_+ + \Omega_-)t)$ due to their highly oscillatory behavior in comparison with the overall slower dynamics [46]. If we take this approach to the extreme, we can also eliminate terms which also rotate more slowly, those like $\exp(\pm i(\Omega_+ - \Omega_-)t)$, and only keep non-rotating terms. Finally, this procedure leads to an effective total decoupling of the normal modes, which then dissipate independently to the heat bath(s) with the decay rates $\tilde{\Gamma}_{\pm\pm}^{CB}$ and $\tilde{\Gamma}_{\pm\pm}^{SB}$ [Eqs. (4.15) and (4.16)]. In some sense this time averaging approximation can be seen as renormalizing all dissipation coefficients having mixed indices $+-$ (and $-+$) to zero, hence rendering the master equation as a tensor product of two independent evolutions.

This could seem a bit too far fetched, but comparison of the full dynamics and this approximation seems to be quite accurate as clear form Fig. 4.6, where we compare $\tilde{\Gamma}_{\pm\pm}^{CB}$ and their average with the three different values of $\mathrm{Re}(\mu_i)$.

Inspection of these analytical expressions when varying system parameters, confirms that the normal modes decays for SB have all similar real parts (being in general Γ_{12} small and $\Gamma_{11} \simeq \Gamma_{22}$), while for a CB the decays can be significantly different

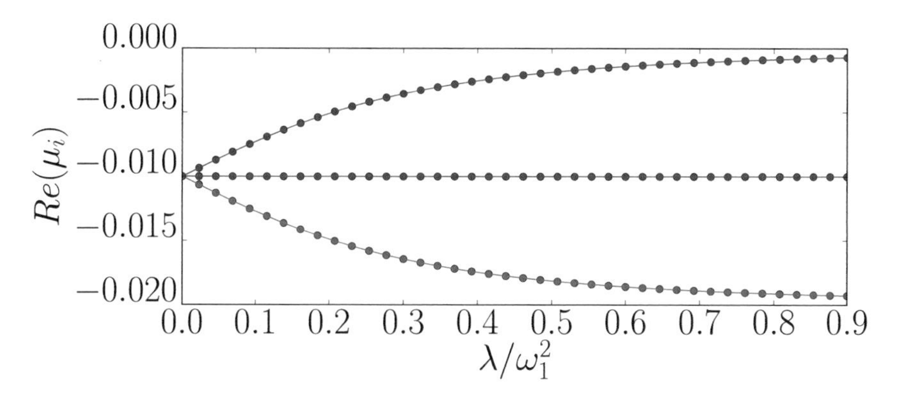

Fig. 4.6 The rates $\tilde{\Gamma}^{CB}_{++}$, $\tilde{\Gamma}^{CB}_{--}$, from Eq. (4.16) in Sect. 4.2, and $(\tilde{\Gamma}^{CB}_{++} + \tilde{\Gamma}^{CB}_{--})/2$ (dots) are compared with the (three different) real parts of the dynamical eigenvalues $\mathrm{Re}(\mu_i)$ (continuous line) in the case of common bath. Here $\omega_2/\omega_1 = 1.31$

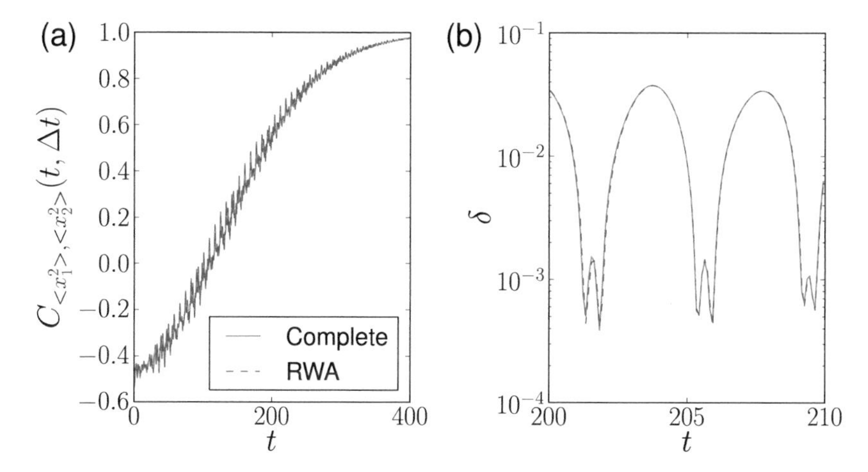

Fig. 4.7 Synchronization (**a**) and discord (**b**) obtained from the complete master equation (4.9) in Sect. 4.2, are compared with the values obtained after the rotating wave approximation as described in the text in the case of common bath. Here $\omega_2/\omega_1 = 1.4$, and $\lambda = 0.7\omega_1^2$

(a factor 20 in Fig. 4.6). This difference between decay rates can be up to several orders of magnitude for parameters where synchronization appears faster. Synchronization is therefore linked to imbalanced dissipation rates of the normal modes, allowing the mode which survives longer to govern the dynamics. Within the discussed approximation, its frequency is found to be $2\Omega_-$ with Ω_- the previously defined frequency of the normal mode $\hat{X}_-$. This is independent of bath coefficients, and we find very good agreement with the exact frequency.

It can be easily seen that the rotating wave approximation described in this Appendix leads to (CB and SB) master equations in Linblad form. In spite of formal differences with Eq. (4.9), we actually find a very good agreement between their dynamical evolutions. In Fig. 4.7 we show that, in the limit of weak coupling here considered, predictions for synchronization and discord are actually almost indistinguishable. Maximum deviations in this case are at least two order of magnitude smaller than the represented quantities. As expected, larger deviations are found for strong coupling.

References

1. T. Yu, J.H. Eberly, Sudden death of entanglement. Science **323**, 598–601 (2009)
2. L. Aolita, F. de Melo, L. Davidovich, Open-system dynamics of entanglement:a key issues review. Rep. Prog. Phys. **78**, 042001 (2015)
3. T. Werlang, S. Souza, F.F. Fanchini, C.J.V. Boas, Robustness of quantum discord to sudden death. Phys. Rev. A **80**, 024103 (2009)
4. J. Maziero, L.C. Céleri, R.M. Serra, V. Vedral, Classical and quantum correlations under decoherence. Phys. Rev. A **80**, 044102 (2009)
5. L. Mazzola, J. Piilo, S. Maniscalco, Sudden transition between classical and quantum decoherence. Phys. Rev. Lett. **104**, 200401 (2010)
6. F.F. Fanchini, L.K. Castelano, A.O. Caldeira, Entanglement versus quantum discord in two coupled double quantum dots. New J. Phys. **12**, 073009 (2010)
7. J.-S. Xu, X.-Y. Xu, C.-F. Li, C.-J. Zhang, X.-B. Zou, G.-C. Guo, Experimental investigation of classical and quantum correlations under decoherence. Nat. Commun. **1**, 7 (2010)
8. A. Streltsov, H. Kampermann, D. Bruß, Behavior of quantum correlations under local noise. Phys. Rev. Lett. **107**, 170502 (2011)
9. L.A. Correa, A.A. Valido, D. Alonso, Asymptotic discord and entanglement of nonresonant harmonic oscillators under weak and strong dissipation. Phys. Rev. A **86**, 012110 (2012)
10. A. Osterloh, L. Amico, G. Falci, R. Fazio, Scaling of entanglement close to a quantum phase transition. Nature **416**, 608–610 (2002)
11. L.-A. Wu, M.S. Sarandy, D.A. Lidar, Quantum phase transitions and bipartite entanglement. Phys. Rev. Lett. **93**, 250404 (2004)
12. T. Werlang, C. Trippe, G.A.P. Ribeiro, G. Rigolin, Quantum correlations in spin chains at finite temperatures and quantum phase transitions. Phys. Rev. Lett. **105**, 095702 (2010)
13. G. Tóth, Entanglement witnesses in spin models. Phys. Rev. A **71**, 010301(R) (2005)
14. M. Wieśniak, V. Vedral, Č. Brukner, Heat capacity as an indicator of entanglement. Phys. Rev. B **78**, 064108 (2008)
15. G.L. Giorgi, F. Galve, G. Manzano, P. Colet, R. Zambrini, Quantum correlations and mutual synchronization. Phys. Rev. A **85**, 052101 (2012)
16. C. Huygens, Horologium oscillatorium [The Pendulum Clock] (A.F. Muguet, Paris, France, 1673), [English translation by R.J. Blackwell (Iowa State University Press, Ames, USA, 1986)]
17. A. Pikovsky, M. Rosenblum, J. Kurths, *Synchronization. A Universal Concept in Nonlinear Sciences* (Cambridge University Press, Cambridge, 2001)
18. M. Rosenblum, A. Pikovsky, Synchronization: from pendulum clocks to chaotic lasers and chemical oscillators. Contemp. Phys. **44**, 4011–416 (2003)
19. S.H. Strogatz, *Nonlinear Dynamics and Chaos: With Applications to Physics, Biology, Chemistry, and Engineering* (Westview Press, Colorado, 2001)
20. I. Goychuk, J.C. Pascual, M. Morillo, J. Lehmann, P. Hänggi, Quantum stochastic synchronization. Phys. Rev. Lett. **97**, 210601 (2006)
21. O.V. Zhirov, D.L. Shepelyansky, Synchronization and bistability of a qubit coupled to a driven dissipative oscillator. Phys. Rev. Lett. **100**, 014101 (2008)
22. S.-B. Shim, M. Imboden, P. Mohanty, Synchronized oscillation in coupled nanomechanical oscillators. Science **316**, 95 (2007)
23. G. Heinrich, M. Ludwig, J. Qian, B. Kubala, F. Marquardt, Collective dynamics in optomechanical arrays. Phys. Rev. Lett. **107**, 043603 (2011)

24. M. Zhang, G.S. Wiederhecker, S. Manipatruni, A. Barnard, P. McEuen, M. Lipson, Synchronization of micromechanical oscillators using light. Phys. Rev. Lett. **109**, 233906 (2012)
25. C.A. Holmes, C.P. Meaney, G.J. Milburn, Synchronization of many nanomechanical resonators coupled via a common cavity field. Phys. Rev. E **85**, 066203 (2012)
26. J.P. Paz, A.J. Roncaglia, Dynamics of the entanglement between two oscillators in the same environment. Phys. Rev. Lett. **100**, 220401 (2008)
27. K.-L. Liu, H.-S. Goan, Non-Markovian entanglement dynamics of quantum continuous variable systems in thermal environments. Phys. Rev. A **76**, 022312 (2007)
28. B.L. Hu, J.P. Paz, Y. Zhang, Quantum Brownian motion in a general environment: exact master equation with nonlocal dissipation and colored noise. Phys. Rev. D **45**, 2843 (1992)
29. F. Galve, G.L. Giorgi, R. Zambrini, Entanglement dynamics of nonidentical oscillators under decohering environments. Phys. Rev. A **81**, 062117 (2010)
30. T. Rocheleau, T. Ndukum, C. Macklin, J.B. Hertzberg, A.A. Clerk, K.C. Schwab, Preparation and detection of a mechanical resonator near the ground state of motion. Nature **463**, 72–75 (2010)
31. D.V. Thourhout, J. Roels, Optomechanical device actuation through the optical gradient force. Nat. Photonics **4**, 211–217 (2010)
32. F. Marino, F.S. Cataliotti, A. Farsi, M.S. de Cumis, F. Marin, Classical signature of ponderomotive squeezing in a suspended mirror resonator. Phys. Rev. Lett. **104**, 073601 (2010)
33. P. Verlot, A. Tavernarakis, T. Briant, P.-F. Cohadon, A. Heidmann, Backaction amplification and quantum limits in optomechanical measurements. Phys. Rev. Lett. **104**, 133602 (2010)
34. K.R. Brown, C. Ospelkaus, Y. Colombe, A.C. Wilson, D. Leibfried, D.J. Wineland, Coupled quantized mechanical oscillators. Nature **471**, 196–199 (2011)
35. M. Harlander, R. Lechner, M. Brownnutt, R. Blatt, W. Hänsel, Trapped-ion antennae for the transmission of quantum information. Nature **471**, 200–203 (2011)
36. A.K. Rajagopal, R.W. Rendell, Decoherence, correlation, and entanglement in a pair of coupled quantum dissipative oscillators. Phys. Rev. A **63**, 022116 (2001)
37. A. Serafini, F. Illuminati, M.G.A. Paris, S. De Siena, Entanglement and purity of two-mode Gaussian states in noisy channels. Phys. Rev. A **69**, 022318 (2004)
38. G.M. Palma, K.-A. Suominen, A.K. Ekert, Quantum computers and dissipation. Proc. R. Soc. Lond. A **452**, 567 (1996)
39. F. Galve, A. Mandarino, M.G.A. Paris, C. Benedetti, R. Zambrini, Microscopic description for the emergence of collective dissipation in extended quantum systems. Sci. Rep. **7**, 42050 (2017), arXiv:1606.03390
40. J.P. Paz, A.J. Roncaglia, Dynamical phases for the evolution of the entanglement between two oscillators coupled to the same environment. Phys. Rev. A **79**, 032102 (2009)
41. J.S. Prauzner-Bechcicki, Two-mode squeezed vacuum state coupled to the common thermal reservoir. J. Phys. A Math. Gen. **37**, L173 (2004)
42. F. Benatti, R. Floreanini, Entangling oscillators through environment noise. J. Phys. A Math. Theor. **39**, 2689 (2006)
43. T. Zell, F. Queisser, R. Klesse, Distance dependence of entanglement generation via a Bosonic heat bath. Phys. Rev. Lett. **102**, 160501 (2009)
44. C. Cormick, J.P. Paz, Observing different phases for the dynamics of entanglement in an ion trap. Phys. Rev. A **81**, 022306 (2010)
45. H.-P. Breuer, F. Petruccione, *The Theory of Open Quantum Systems* (Oxford University Press, New York, 2002)
46. A. Rivas, S.F. Huelga, *Open Quantum Systems: An Introduction* (Springer, Berlin, 2012)
47. W.H. Zurek, Einselection and decoherence from information theory perspective. Ann. Phys. **9**, 855–864 (2000)
48. H. Ollivier, W.H. Zurek, Quantum discord: a measure of the quantumness of correlations. Phys. Rev. Lett. **88**, 017901 (2001)
49. L. Henderson, V. Vedral, Classical, quantum and total correlations. J. Phys. A Math. Gen. **34**, 6899 (2001)
50. P. Giorda, M.G.A. Paris, Gaussian quantum discord. Phys. Rev. Lett. **105**, 020503 (2010)

51. G. Adesso, A. Datta, Quantum versus classical correlations in Gaussian states. Phys. Rev. Lett. **105**, 030501 (2010)
52. G.L. Giorgi, F. Galve, R. Zambrini, Robustness of different indicators of quantumness in the presence of dissipation. Int. J. Quantum Inform. **09**, 1825 (2011)
53. F. Galve, G.L. Giorgi, R. Zambrini, Maximally discordant mixed states of two qubits. Phys. Rev. A **83**, 012102 (2011)
54. F. Galve, G.L. Giorgi, R. Zambrini, Erratum: maximally discordant mixed states of two qubits. Phys. Rev. A **83**, 069905 (2011)
55. P. Zanardi, Dissipation and decoherence in a quantum register. Phys. Rev. A **57**, 3276–3284 (1998)
56. J.H. Reina, L. Quiroga, N.F. Johnson, Decoherence of quantum registers. Phys. Rev. A **65**, 032326 (2002)
57. R. Doll, M. Wubs, P. Hänggi, S. Kohler, Limitation of entanglement due to spatial qubit separation. Europhys. Lett. **74**, 547–553 (2006)
58. D.P.S. McCutcheon, A. Nazir, S. Bose, A.J. Fisher, Longlived spin entanglement induced by a spatially correlated thermal bath. Phys. Rev. A **80**, 022337 (2009)
59. A. Mari, A. Farace, N. Didier, V. Giovannetti, R. Fazio, Measures of quantum synchronization in continuous variable systems. Phys. Rev. Lett. **111**, 103605 (2013)
60. W. Li, C. Li, H. Song, Quantum synchronization in an optomechanical system based on Lyapunov control. Phys. Rev. E **93**, 062221 (2016)
61. F. Galve, G.-L. Giorgi, R. Zambrini, Quantum correlations and synchronization measures. Lectures on General Quantum Correlations and their Applications. Quantum Science and Technology (Springer, 2017), arXiv:1610.05060
62. M. Bagheri, M. Poot, L. Fan, F. Marquardt, H.X. Tang, Photonic cavity synchronization of nanomechanical oscillators. Phys. Rev. Lett. **111**, 213902 (2013)
63. M. Zhang, S. Shah, J. Cardenas, M. Lipson, Synchronization and phase noise reduction in micromechanical oscillator arrays coupled through light. Phys. Rev. Lett. **115**, 163902 (2015)
64. D.K. Agrawal, J. Woodhouse, A.A. Seshia, Observation of locked phase dynamics and enhanced frequency stability in synchronized micromechanical oscillators. Phys. Rev. Lett. **111**, 084101 (2013)
65. M.H. Matheny, M. Grau, L.G. Villanueva, R.B. Karabalin, M.C. Cross, M.L. Roukes, Phase synchronization of two anharmonic nanomechanical oscillators. Phys. Rev. Lett. **112**, 014101 (2014)
66. M. Ludwig, F. Marquardt, Quantum many-body dynamics in optomechanical arrays. Phys. Rev. Lett. **111**, 073603 (2013)

Chapter 5
Noiseless Subsystems and Synchronization

In the previous chapter we have seen that common dissipation leads to the emergence of mutual synchronization between two oscillators, and we have also shown its relation with the slow decay of quantum correlations. Furthermore, this phenomenon is stronger the closer the natural frequencies of the oscillators are, as in the limiting case of equal frequencies only the normal mode corresponding to the center of mass position, $\hat{X}_+ = (\hat{x}_1 + \hat{x}_2)/2$, couples to the environment, while the other normal mode, the relative position $\hat{X}_- = (\hat{x}_1 - \hat{x}_2)/2$, is effectively uncoupled from any environmental action, leading to asymptotic entanglement between the two oscillators [1–5]. When a subsystem of a larger many-body system is effectively uncoupled from the environment due to symmetries in its interaction, we call it a noiseless subsystem (NS) [6]. Other authors refer to it as a decoherence-free subsystem or subspace (DFS) [7], as its dynamics is unitary and preserves any initial coherence.

In this chapter we show how to obtain NSs in a system of three harmonic oscillators and bypass decoherence independently of the bath properties.[1] We consider different frequencies, couplings and boundary conditions for the harmonic oscillators, in the presence of a common bath. We also analyze how, by using these NSs, quantum correlations like entanglement can persist (by two different mechanisms) in the asymptotic limit of the dynamical evolution. In this three-body scenario, we explore synchronization phenomenon and its connection to the presence of robust quantum correlations, extending the analysis performed in Chap. 4. We find that even for three different oscillators, a variety of regimes emerge for different parameters.

The chapter is organized as follows. We start in Sect. 5.1 motivating our work in the context of prevention of decoherence and dissipation in open quantum systems. In Sect. 5.2 we introduce the model for the system of three harmonic oscillators dissipating into a common bath, in terms of the normal modes of the system. For certain particular values of the system's parameters (and independently of the bath

[1]The results presented in this chapter have been published in Ref. [8].

G. Manzano Paule, *Thermodynamics and Synchronization in Open Quantum Systems*, Springer Theses, https://doi.org/10.1007/978-3-319-93964-3_5

characteristics), one of several normal modes can be protected from decoherence. We find analytically these conditions in Sect. 5.3, introducing some specific cases that are analyzed in more detail. In Sect. 5.4 deviations of the NSs conditions are considered. This leads to dynamical relaxation of the system that converges towards a thermal state. We conclude with Sect. 5.5 summarizing our main results. In Appendix A.1 we provide further technical details of the calculation of the asymptotic entanglement and the equations of motion for the Markovian transient dynamics.

5.1 Prevention of Decoherence and Dissipation

Prevention of decoherence and dissipation in open quantum systems is a fundamental condition for the presence of quantum phenomena in warm macroscopic everyday world. Decoherence has been extensively studied from the early 1980s [9, 10] to the present, providing a natural explanation to the quantum-to-classical transition induced by monitoring the environment (for a list of reviews see for instance [11–14]). Indeed decoherence and the leak of information to the environment have been identified as major obstacles in quantum processing of information and construction of quantum memories [15]. Different mechanisms to avoid decoherence have been discussed in recent years, including strategies to engineer it for applications [16–18]. Furthermore, some macroscopic systems from photosyntetic marine algae [19] to metal carboxylates [20], which can present quantum correlations at high temperatures, suggest that avoiding a complete quantum-to-classical transition can also occur inherently in natural phenomena. The mechanisms that produce such survival or even construction of coherence and correlations at large time scales remain almost unclear, but different theoretical strategies have been proposed in order to predict it, mostly motivated in the context of quantum computation [21–26].

In this context, one of the strategies to bypass decoherence is exploiting dynamical symmetries in the system-environment interaction. In order to generate unitary evolution in a certain subspace of the Hilbert space of an open system, a common dissipation where several units equally couple to the same environment (see Sect. 2.4) has been first used in a two-qubit system [23, 27–31] and later extended to multiple qubits [32–34] and continuous variable systems [1–3, 5, 35–41]. A general framework has been developed with several contributions (see, for example, [7], and references therein) agglutinating the main concepts of decoherence-free subspaces and subsystems (DFSs) [24], noiseless subsystems (NSs) [6], or more recently, information-preserving structures (IPSs) [26]. DFSs and NSs have been experimentally tested and realized in the lab [42–45], and reservoir engineering techniques [46] has been proposed to obtain them [17, 47].

In this chapter we extend previous studies in the context of continuous variable systems, exploring the vaster landscape offered by three coupled harmonic oscillators in the search for NSs, in comparison with the simpler case of two oscillators already considered in Chap. 4. Previous works on dissipative harmonic oscillators reported that in presence of identical frequencies and couplings between oscillators, the symmetry of the collective motion can lead to the effective decoupling from the bath of some normal modes [1–3, 36, 38]. In addition, the consideration of different frequencies [5, 40] or couplings [39] opens a huge field of possibilities which is instead less studied and understood. The natural step of considering three harmonic oscillators beyond the symmetric configuration of identical oscillators already provides much more phenomenological richness, while at the same time allows for analytic treatment and gives valuable intuition when pursuing a further extension to the case of N oscillators.

5.2 Three Oscillators in a Common Environment

We start with a Hamiltonian describing three coupled quantum harmonic oscillators with arbitrary frequencies and coupling constants. For simplicity we suppose unit masses:

$$\hat{H}_S = \frac{1}{2}\sum_{i=1}^{3}\left(\hat{p}_i^2 + \omega_i^2\hat{q}_i^2\right) + \sum_{i<j}\lambda_{ij}\hat{q}_i\hat{q}_j, \tag{5.1}$$

where $\hat{p}_i$ and $\hat{q}_i$ represent, respectively, the momentum and position operators of each harmonic oscillator ($[\hat{q}_i, \hat{p}_j] = i\hbar\delta_{ij}$). This equation is conveniently expressed in quadratic matrix form as

$$\hat{H}_S = \frac{1}{2}\left(\mathbf{p}^T\mathbb{1}\,\mathbf{p} + \mathbf{q}^T\mathcal{H}\,\mathbf{q}\right), \tag{5.2}$$

where $\mathbb{I}$ is the identity (3×3) matrix, we have introduced the vectors $\mathbf{p}^T = (\hat{p}_1, \hat{p}_2, \hat{p}_3)$, $\mathbf{q}^T = (\hat{q}_1, \hat{q}_2, \hat{q}_3)$, and $\mathcal{H}$ contains all the parameters of the system, i.e. the squared frequencies and couplings between oscillators. We will only consider $\mathcal{H}$ with positive eigenvalues, so as to have bounded states (attractive potential).

The environment is introduced by equally coupling each oscillator of the system to the same thermal bath (see Sect. 2.4), which is described by an infinite collection of independent bosonic modes:

$$\hat{H}_B = \frac{1}{2}\sum_{\alpha=1}^{\infty}\left(\hat{\Pi}_\alpha^2 + \tilde{\Omega}_\alpha^2\hat{X}_\alpha^2\right), \tag{5.3}$$

where $[\hat{X}_\alpha, \hat{\Pi}_\beta] = i\hbar\delta_{\alpha\beta}\hat{\mathbb{1}}$. We will use throughout the paper Greek subscripts to refer to bath modes, while Latin ones are reserved for system oscillators (i, j) and normal modes (k, n). The system-bath interaction reads

$$\hat{H}_I = \sum_{i=1}^{N}\hat{q}_i\sum_{\alpha=1}^{\infty}\lambda_\alpha\hat{X}_\alpha, \tag{5.4}$$

with a factorized form $\hat{H}_I = \hat{S} \otimes \hat{B}$ of an operator $\hat{S}$ acting only on the system's Hilbert space, and $\hat{B}$ acting on the environment one. As usual, this type of interaction yields a renormalization of the frequencies that we may include directly in our model by performing the change [48]

$$\omega_i^2 \to \omega_i^2 + \sum_\alpha \frac{\lambda_\alpha^2}{2\tilde{\Omega}_\alpha^2}. \tag{5.5}$$

The normal modes basis of the system, Eq. (5.2), is obtained after a canonical transformation of the system operators through the orthogonal basis-change matrix $\mathcal{F}$:

$$\hat{q}_i = \sum_{k=1}^{N} \mathcal{F}_{ik} \hat{Q}_k \quad , \quad \hat{p}_i = \sum_{k=1}^{N} \mathcal{F}_{ik} \hat{P}_k, \tag{5.6}$$

which diagonalizes $\mathcal{H}$ ($\mathbf{q}^T \mathcal{H} \mathbf{q} = \mathbf{Q}^T \Omega \mathbf{Q}$). Here $\Omega = \mathcal{F}^T \mathcal{H} \mathcal{F}$ is a diagonal matrix containing the squared frequencies of the normal modes, Ω_n with $n = 1, 2, \ldots, N$. In this basis H_S now represents the Hamiltonian for a $N = 3$ uncoupled harmonic oscillators, or normal modes, related with the original (natural) modes by $\mathcal{F}$. Henceforth we can rewrite the system Hamiltonian as

$$\hat{H}_S = \frac{1}{2} \sum_{n=1}^{N} \left(\hat{P}_n^2 + \Omega_n^2 \, \hat{Q}_n^2 \right), \tag{5.7}$$

and the interaction Hamiltonian of Eq. (5.4) as

$$\hat{H}_I = \sum_{n=1}^{3} \kappa_n \hat{Q}_n \sum_{\alpha=1}^{\infty} \lambda_\alpha \hat{X}_\alpha. \tag{5.8}$$

Notice from comparison between Eqs. (5.4) and (5.8), that even if the oscillators are coupled with the same strength to the bath center of mass, $\sum_\alpha \lambda_\alpha \hat{X}_\alpha$, the couplings of the normal modes positions, $\hat{Q}_n$, to the bath center of mass, are not homogeneous, but given by

$$\kappa_n \equiv \sum_{i=1}^{N} \mathcal{F}_{in}. \tag{5.9}$$

These *effective couplings* κ_n only depend on the canonical transformation, i.e. on the system's parameters and arrangement defined by $\mathcal{H}$. This suggests a strategy to protect one or more normal modes from the environment action based on proper tuning of the system parameters $\{\omega_1, \omega_2, \omega_3\}$ and $\{\lambda_{12}, \lambda_{23}, \lambda_{13}\}$. Our analysis in Sect. 5.3 addresses this point while deviations form the condition of vanishing effective coupling of a system normal mode and the environment are explored in Sect. 5.4.

We mention that while here we focus on the case of three coupled harmonic oscillators, the description in terms of effective couplings is rather general and applies for arbitrary networks of N harmonic oscillators, as we show in Chap. 6. The case of a common bath for all oscillators in the system corresponds to situations where the correlation length in the environment is larger than the system size. This assumption is not crucial for our discussion, although any other choice would produce different specific analytic expressions. Finally, we have seen in Chap. 4 that the case of a separate bath for each oscillator yields equal decoherence for all normal modes and therefore neither NSs nor synchronization.

Furthermore, the equal coupling of each system oscillator to the bath might seem an arbitrary restriction. Imagine for example that each oscillator is at a different distance from the common heat bath, leading to an interaction

$$\hat{H}_I = \sum_{i=1}^{3} \gamma_i \hat{q}_i \sum_{\alpha=1}^{\infty} \lambda_\alpha \hat{X}_\alpha, \tag{5.10}$$

where the different oscillators feel a coupling of strength $0 \leq \gamma_i \leq 1$, with $\sum_i \gamma_i = 1$. The immediate consequence is that the effective couplings become

$$\kappa_n = \sum_{i=1}^{3} \gamma_i \mathcal{F}_{in}. \tag{5.11}$$

Here we will consider $\gamma_i = 1$, but the unbalanced case would be solved following exactly the same procedure as we outline in the next section.

5.3 Noiseless Subsystems and Asymptotic Properties

In this section we discuss the conditions to achieve noiseless subsystems with dissipation avoided in one or two of the system's normal modes. The properties of our system are specified completely by the matrix $\mathcal{H}$ appearing in Eq. (5.2):

$$\mathcal{H} = \begin{pmatrix} \omega_1^2 & \lambda_{12} & \lambda_{13} \\ \lambda_{12} & \omega_2^2 & \lambda_{23} \\ \lambda_{13} & \lambda_{23} & \omega_3^2 \end{pmatrix} \tag{5.12}$$

and we aim to derive the set of conditions for the system parameters leading to one or two normal modes decoupled from the environment, i.e. whose effective coupling κ_n is zero.

Let us consider a normal mode δ with normal frequency Ω_δ. The eigenvalue problem is expressed adequately by

$$(\mathcal{H} - \Omega_\delta^2 \mathbb{1})\, \mathbf{C}_\delta = 0, \tag{5.13}$$

involving three equations, one for each of the components of the vector $\mathbf{C}_\delta = (\mathcal{F}_{1\delta}, \mathcal{F}_{2\delta}, \mathcal{F}_{3\delta})^T$ with $\mathcal{F}_{ij}$ defined in Eq. (5.6). The condition for normal mode δ to be non-dissipative (out of the bath influence) leads to a constraint as follows

$$\kappa_\delta = 0 \;\Leftrightarrow\; \mathcal{F}_{1\delta} + \mathcal{F}_{2\delta} + \mathcal{F}_{3\delta} = 0. \tag{5.14}$$

From Eqs. (5.13), (5.14) and the normalization condition, we can obtain analytically $\mathbf{C}_\delta$, Ω_δ with a further constraint for the system parameters. In other words, not all parameter choices lead to NSs, but it is possible for some configurations of frequencies and couplings of the set of three oscillators (satisfying some constraint).

The normal mode δ in terms of the system parameters reads

$$\mathbf{C}_\delta = c \begin{pmatrix} \lambda_{13}\lambda_{12} + \lambda_{23}(\Omega_\delta^2 - \omega_1^2) \\ (\Omega_\delta^2 - \omega_2^2)(\Omega_\delta^2 - \omega_1^2) - \lambda_{12}^2 \\ \lambda_{13}\lambda_{23} + \lambda_{12}(\Omega_\delta^2 - \omega_3^2) \end{pmatrix} \tag{5.15}$$

where c is the normalization constant. Applying Eq. (5.14) we can obtain its eigenfrequency Ω_δ^2 as:

$$\begin{aligned} \Omega_\delta^2 = &\left(\frac{\omega_1^2 + \omega_3^2}{2}\right) - \left(\frac{\lambda_{12} + \lambda_{23}}{2}\right) \\ &\pm \sqrt{\Delta^2 + \left(\frac{\lambda_{12} + \lambda_{23}}{2}\right)^2 + \Delta(\lambda_{23} - \lambda_{12}) + \lambda_{13}(\lambda_{13} - \lambda_{12} - \lambda_{23})}, \end{aligned} \tag{5.16}$$

where $\Delta \equiv (\omega_1^2 - \omega_3^2)/2$. Therefore, by defining the quantities:

$$\Sigma \equiv (\omega_1^2 + \omega_3^2)/2 - \omega_2^2, \tag{5.17}$$

$$\begin{aligned} \mathcal{R} \equiv &- (\lambda_{12} + \lambda_{23})/2 \\ &\pm \sqrt{(\Delta + (\lambda_{12} + \lambda_{23})/2 - \lambda_{13})^2 + 2\Delta(\lambda_{13} - \lambda_{12})}, \end{aligned} \tag{5.18}$$

the constraint relation ($\kappa_\delta = 0$) ensuring a one-mode NS reads:

$$\begin{aligned} &2\lambda_{12}\lambda_{23}\mathcal{R} + \lambda_{13}(\lambda_{12}^2 + \lambda_{23}^2) + \lambda_{13}^2(\mathcal{R} + \Sigma) \\ &\qquad - (\mathcal{R} + \Delta)(\mathcal{R} - \Delta)(\mathcal{R} + \Sigma) = 0. \end{aligned} \tag{5.19}$$

The above Eq. (5.19) is one of our main results, and represents a hypersurface in the d-dimensional parameters space [being $d = (N + 1)N/2 = 6$ for $N = 3$ oscillators] whereby a normal mode is allowed to evolve freely and without dissipation. Such manifold is restricted to regions in which the normal mode frequency Ω_δ is real and positive, and the normalization constant, c in Eq. (5.15), is well defined. It is worth noting that looking only to non-dissipative modes (imposing the condition $\kappa_\delta = 0$) has been crucial in order to solve analytically the above equations. Otherwise we have to deal with complicated expressions involving third order equations corresponding to the general expression for a normal mode in Eq. (5.13).

Therefore, when Eq. (5.19) is fulfilled, we obtain a NS composed by (at least) a single normal mode that is effectively uncoupled to the reservoir. This could be performed artificially by tuning one of the $d = 6$ parameters of $\mathcal{H}$, such as, for instance, the natural frequency of one oscillator. In experiments where it is possible to control the local potentials, such as ions confined to individual traps, this modification should be rather straightforward (see e.g. [49]). It should be stressed that noise models for ion traps typically favor a SB interpretation in terms of fluctuating uncorrelated surface dipoles [50, 51], though other microscopic models based on charge diffusion [52] in the electrode surface question whether the bath's correlation length could in fact be larger than the distance of the ion to the electrode. For the moment, this is an open problem.

Configurations in which a NS consisting of two normal modes is produced can also be obtained analytically in the specific case of three oscillators. Indeed, we find that, when two normal modes uncouple from the environmental action, the third one must necessarily coincide with the center of mass (c. m.) of the system. Explicitly, the condition for the center of mass being a normal mode is:

$$(\mathcal{H} - \Omega^2_{\text{c.m.}} \mathbb{1})\, \mathbf{C}_{\text{c.m.}} = 0 \;\Leftrightarrow\; \Omega^2_{\text{c.m.}} = \omega_i^2 + \sum_{j \neq i} \lambda_{ij} \tag{5.20}$$

$\forall i = 1, 2, 3$, and where $\mathbf{C}_{\text{c.m.}} = (1, 1, 1)^T/\sqrt{3}$. The latter constraint can be captured in the next two relations that have to be fulfilled simultaneously by the system parameters:

$$\omega_1^2 = \omega_2^2 + \lambda_{23} - \lambda_{13}, \tag{5.21}$$

$$\omega_3^2 = \omega_2^2 + \lambda_{12} - \lambda_{13}. \tag{5.22}$$

Furthermore, since we want to remain in the domain of attractive potentials, we have to restrict ourselves to regions of the parameter space where $\Omega^2_{\text{c.m.}} = \omega_1^2 + \omega_3^2 - \omega_2^2 + 2\lambda_{13} > 0$.

In order to see the scope of the conditions (5.19), (5.21), and (5.22) we give in the following some examples of configurations in which a NS of one or two modes is produced. We consider simple situations in which the six-dimensional parameter space is reduced, first by assuming two of the three natural frequencies to be equal ($\omega \equiv \omega_1 = \omega_3 \neq \omega_2$), and, second, considering two of the three couplings equal ($\lambda \equiv \lambda_{12} = \lambda_{23} \neq \lambda_{13}$). This is sufficient to obtain some different scenarios appearing in open an closed chains configurations, as is schematically shown in Fig. 5.1.

Let us start from the case of two equal frequencies ($\omega \equiv \omega_1 = \omega_3 \neq \omega_2$). Then the quantities defined in Eq. (5.19) are simply $\Delta = 0$, $\Sigma = \omega^2 - \omega_2^2$ and $\mathcal{R} =$

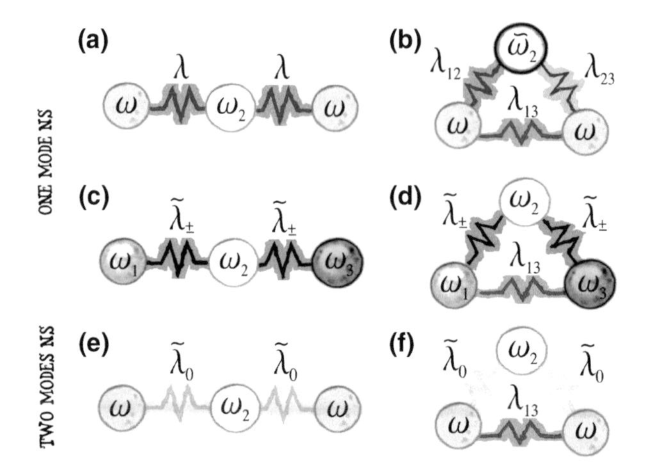

Fig. 5.1 Different configurations for a chain of three coupled oscillators in which a NS of one (**a**–**d**) or two (**e**–**f**) normal modes is predicted. The tilde on parameters indicates a fixed value depending on the other non-tilted ones as described in the text

$\{-\lambda_{13}, \lambda_{13} - \lambda_{12} - \lambda_{23}\}$, the latter implying two different consistent solutions to Eq. (5.19). The first one is $\lambda_{12} = \lambda_{23}$, and the second one $\omega_2 = \tilde{\omega}_2$, with

$$\tilde{\omega}_2^2 = \omega^2 + \frac{2\lambda_{13}(\lambda_{12} + \lambda_{23} - \lambda_{13}) - 2\lambda_{12}\lambda_{23}}{\lambda_{12} + \lambda_{23} - 2\lambda_{13}}. \tag{5.23}$$

Both conditions can be simultaneously fulfilled as well. In this case we would have

$$\lambda_{12} = \lambda_{23} \equiv \lambda, \qquad \lambda = \omega^2 - \omega_2^2 - \lambda_{13} \equiv \tilde{\lambda}_0, \tag{5.24}$$

which satisfies Eqs. (5.21) and (5.22). Therefore this defines a two-mode NS. These three situations correspond respectively to configurations in Fig. 5.1a ($\lambda_{12} = \lambda_{23}$), Fig. 5.1b ($\omega_2^2 = \tilde{\omega}_2{}^2$) and Fig. 5.1f ($\lambda = \tilde{\lambda}_0$). It is worth noting that the configuration in Fig. 5.1a is valid also for the closed chain ($\lambda_{13} \neq 0$), as well as the one in Fig. 5.1b for the open chain (when $\lambda_{13} = 0$).

On the other hand, by assuming two equal couplings we have three different solutions: $\lambda = 0$, and $\lambda = \tilde{\lambda}_{\pm}$. The first one is trivial, accounting for a separated pair of coupled oscillators, together with an uncoupled one. The second solution allows for the situations in Fig. 5.1c, d, defined by:

$$\tilde{\lambda}_{\pm} = \lambda_{13} \pm \sqrt{(\omega_2^2 - \omega_1^2)(\omega_2^2 - \omega_3^2)}. \tag{5.25}$$

Finally, when $\omega_1 = \omega_3$, $\lambda_{13} = 0$, and $\lambda = \omega^2 - \omega_2^2 = \tilde{\lambda}_0$, we have a two-mode NS solution [Fig. 5.1e].

The presence of one or two non dissipating normal modes prevents the full thermalization of the system in the long time run. On the contrary, it leads to an asymptotic state whose features are analyzed in the following, focusing on entanglement and on quantum synchronization between the oscillators.

5.3.1 Asymptotic Entanglement

When a NS is enabled, decoherence can be prevented in the system leading to asymptotic entanglement that would be absent in the thermal state. As a measure of entanglement between a pair of oscillators, we will use the well known logarithmic negativity which is computable for bipartite Gaussian states [53, 54] as is our case (see Sect. 1.4.1 in Chap. 1)

$$E_{\mathcal{N}} = \max\{0, -\log \nu_-\}, \tag{5.26}$$

where ν_- is the minimum symplectic eigenvalue of the partial transposed covariance matrix $\tilde{V}_{AB}$, corresponding to time reflection of one party. With the help of the general expressions, we can calculate analytically the asymptotic entanglement when a NS is produced.

Here we present our results for the external pair of oscillators in open chain with equal couplings to the inner one, $\lambda_{12} = \lambda_{23} \equiv \lambda$, and frequencies $\omega_1 = \omega_3 \equiv \omega \neq \omega_2$. We consider first the case when only one of the three normal modes is not subjected to dissipation [Fig. 5.1a], and second the case when only one of them is dissipating [Fig. 5.1e], by imposing $\lambda = \tilde{\lambda}_0$. The details of the calculations are reported in Appendix A.1. As initial condition for the natural oscillators we choose a squeezed separable vacuum state given by

$$\langle \hat{q}_i^2(0) \rangle = \frac{\hbar}{2\,\omega_i}\, e^{-2r_i}, \qquad \langle \hat{p}_i^2(0) \rangle = \frac{\hbar\omega_i}{2}\, e^{2r_i}, \tag{5.27}$$

where any other first-order or second-order moments are zero.

5.3.1.1 One-Mode NS

As a paradigmatic example of the case in which there is one frozen normal mode, let us consider the configuration given in Fig. 5.1a. As for the initial condition, $\omega_1 = \omega_3 \equiv \omega$ in Eq. (5.27), and we will assume the same squeezing factor for the external pair, i.e. $r_1 = r_3 \equiv r$, while the squeezing in the central oscillator r_2 will be irrelevant.

Normal modes coupled to the environment will reach a thermal equilibrium state asymptotically, whose variances are given by

$$\langle \hat{Q}_k^2 \rangle_{\text{th}} = \frac{\hbar}{2\Omega_k} \coth\left(\frac{\hbar\Omega_k}{2k_B T}\right),$$
$$\langle \hat{P}_k^2 \rangle_{\text{th}} = \frac{\hbar\Omega_k}{2} \coth\left(\frac{\hbar\Omega_k}{2k_B T}\right), \tag{5.28}$$

while the uncoupled one evolves freely. The asymptotic covariance matrix of the external oscillators can be obtained by expressing the second-order moments of the natural oscillators in terms of the normal modes. Then we substitute respectively the asymptotic expressions corresponding to the frozen mode (not coupled to the bath) or the thermalized ones. This yields the following analytical expression for the entanglement:

$$E_{\mathcal{N}} = \max\{0, E_0 + \Delta E(1 + \cos(2\omega t))\}, \tag{5.29}$$

that is defined by a minimum value E_0 and an oscillatory term with amplitude ΔE and frequency 2ω

$$E_0 \equiv \begin{Bmatrix} r - r_0^+ & \text{for} \;\; r \geq 2r_c \\ r_0^- - r & \text{for} \;\; r < 2r_c \end{Bmatrix} \tag{5.30}$$

$$\Delta E \equiv \begin{Bmatrix} 2r_c & \text{for} \;\; r \geq 2r_c \\ 2r & \text{for} \;\; r < 2r_c \end{Bmatrix} \tag{5.31}$$

where $r_c \equiv (r_0^+ + r_0^-)/4$ and the critical values are defined by the following expressions

$$r_0^+ \equiv \frac{1}{2}\log(4\lambda^2\sigma_Q), \qquad r_0^- \equiv -\frac{1}{2}\log(4\lambda^2\sigma_P). \tag{5.32}$$

Coefficients σ_P and σ_Q depend both on the bath's temperature and on the system parameters via the shapes and frequencies of the dissipative normal modes as can be seen in their definition in Appendix A.1 [Eq. (A.10)]. Note that while decoupling of normal modes from the bath is a temperature independent feature, the amount of entanglement generated depends on it via the thermalized degrees of freedom.

The presence of asymptotic entanglement between the external pair of oscillators in a symmetric chain (independent of the frequency of the central one, but depending on the temperature and initial squeezing) is illustrated in Fig. 5.2. The minimum entanglement E_0 is plotted both for low (left panel) and high temperatures (right panel) in the relevant squeezing ranges. Different regions are distinguished in the map and are labeled following Paz and Roncaglia notation in Ref. [3]: sudden death is reached (SD), the asymptotic state consisting of an infinite sequence of sudden death and revivals (SDR) and finally, when non-zero entanglement is present at all times [no sudden death (NSD)].

An asymptotic entangled state with strictly $E_{\mathcal{N}} > 0$, can be generated both when $r > r_0^+$ ($> 2r_c$) or equivalently when $r < r_0^-$ ($< 2r_c$) with different origins. In the first case ($r > r_0^+$) the entanglement oscillates between $r - r_0^+$ and $r + r_0^-$ and the initial squeezing in the natural oscillators is employed as a resource to generate an entangled state, while the bath contribution r_0^+ acts as a source for its degradation.

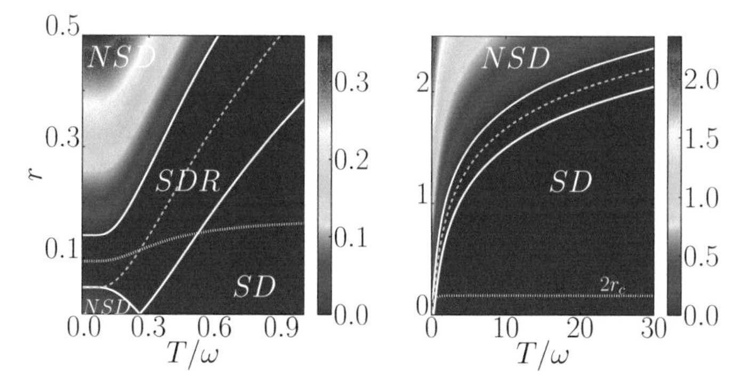

Fig. 5.2 Minimum entanglement E_0 generated in the asymptotic limit between external oscillators of the chain in configuration in Fig. 5.1a for low temperatures (left panel) and high temperatures (right panel). The different phases (SD, SDR and NSD) are bounded by the two critical values $r_0^{\pm}$ (separating NSD phase from SDR phase) and $-r_0^{-}$ (separating SDR from SD) that are represented by continuous white lines. The dotted line corresponds to $2r_c$ and the dashed colored one to $(r_0^+ - r_0^-)/2$. We have set $\omega_2 = 1.2\omega$ and $\lambda = 0.6\omega^2$

It is interesting to see that r_0^+ is strongly dependent on the bath's temperature while the system parameters play a secondary role, only important at low temperatures. Indeed when temperature increases ($T \gg \omega$) sudden death of entanglement can be only avoided by increasing r to be greater than

$$r_0^+ \to \frac{1}{2}\log\left(4\lambda^2 T\omega\left(\frac{c_+^2}{\Omega_+^2} + \frac{c_-^2}{\Omega_-^2}\right)\right) = \frac{1}{2}\log(T) + ct. \tag{5.33}$$

On the other hand, the amplitude of the oscillations in this case is $\Delta E = 2r_c$, that has a very weak dependence on temperature, quickly reaching a constant value when increasing the temperature

$$\Delta E \to \frac{1}{4}\log\left(\omega^2\left(\frac{c_+^2}{\Omega_+^2} + \frac{c_-^2}{\Omega_-^2}\right)(c_+^2 + c_-^2)^{-1}\right). \tag{5.34}$$

The second case ($r < r_0^-$) only appears at low temperatures (of order $0.1k_B^{-1}\hbar\omega$). Here entanglement oscillates around $r_0^- + r$ with amplitude $2r$. This means that introducing no squeezing in the initial state leads to a constant entanglement at r_0^-, while adding squeezing (a resource in the former case) makes entanglement tend to a SDR phase by widening its oscillatory amplitude. We stress that the fact that thermalization can lead to entanglement at low temperatures is well known [55].

Finally, we can relate critical values $r_0^{\pm}$ with the uncertainty induced by the environment in the virtual oscillator $\hat{\tilde{q}} = (\hat{q}_1 + \hat{q}_3)/\sqrt{2}$ position and momentum, which corresponds to the center of mass of the external oscillators of the chain:

$$\langle \hat{\tilde{q}}^2 \rangle_{\text{th}} = e^{2r_0^+}/2\omega, \qquad \langle \hat{\tilde{p}}^2 \rangle_{\text{th}} = \omega e^{-2r_0^-}/2\,. \tag{5.35}$$

This reveals that when $r_0^- > 0$, a squeezing in momentum is generated ($\Delta\tilde{p} < \omega/2$), yielding entanglement as we have commented above. However, note that we have never a minimum uncertainty state with $r_0^+ > r_0^-$ for all temperatures and physical regimes of the system parameters. Indeed, the uncertainty relation can be expressed for the virtual oscillator, $\hat{\tilde{q}}$, as

$$\Delta\tilde{q}\;\Delta\tilde{p} \;=\; \frac{e^{r_0^+ - r_0^-}}{2} > \frac{1}{2}\,. \tag{5.36}$$

The quantity r_c can be also related with virtual oscillator uncertainties in position and momentum as $e^{-r_c} = \Delta\tilde{p}/\omega\Delta\tilde{q} < 1$.

In the left panel of Fig. 5.2, we can see the two regions in which $E_0 > 0$ (NSD phases): the big one at the left top corner corresponding to entanglement generation by using the initial squeezing in the external oscillators as a resource (once $r > r_0^+$), and the small left bottom island, that represents the environment yielding entanglement via the squeezing generated in $\hat{\tilde{q}}$ when $r < r_0^-$. The SDR phase is centered around $2r_c$ (white dotted line) for low temperatures, and their amplitude is given by the separation of the dashed colored line $(r_0^+ - r_0^-)/2$ from the zero squeezing axis. For temperatures greater than that for which $r_0^- = 0$ (cross point between the dotted and dashed lines), they interchange their roles acting now $(r_0^+ - r_0^-)/2$ (dashed colored line) as the center of the SDR region, and $2r_c$ as the amplitude. The SD phase is bounded by the quantity $r_0^+ - 4r_c = -r_0^-$, corresponding to the case in which $E_0 + 2\Delta E < 0$, and thus no entanglement is present in the asymptotic limit. For high temperatures (right panel) we can see how $2r_c$ reaches a constant value while the pronounced curvature in the SDR region reveals that we can always obtain robust entanglement by increasing the squeezing parameter r logarithmically with temperature.

We have to point out that our results resemble those obtained in Ref. [3] for two resonant harmonic oscillators. There, a similar entanglement phases diagram has been found, and the same two different mechanism for entanglement generation appear. In that context, both oscillations and the appearance of the low temperatures NSD phase were attributed to non-Markovian effects, while here follow by simply considering a final asymptotic (Gibbsian) state for the normal modes coupled to the bath (that can be reproduced by a Markovian Lindblad dynamics as is pointed in Sect. 5.4). Moreover, the presence of a third oscillator in the system, allows for manipulation of the width of entanglement phases at low temperatures (specially the low squeezings NSD one) by tuning the free system parameters ω_2 and λ.

5.3.1.2 Two-Modes NS

Let us now consider the case in which two modes become decoupled from the bath. In particular, we focus on the configuration in Fig. 5.1e. This is indeed a symmetric open chain configuration as before, but now we have a special value of the couplings, $\lambda = \tilde{\lambda}_0$, leading to a larger NS. The calculation is similar to the previous one, while now only one of the normal modes thermalizes, and the other two have a free evolution decoupled from the bath. This leads to a less compact expression for the asymptotic entanglement between the external oscillators of the open chain (more details are reported in Appendix A.1). Still, a similar phase diagram can be found in this case by numerical evaluation of logarithmic negativity from Eq. (A.12) in Appendix A.1. Our results are shown in Fig. 5.3 in the same range of squeezing and temperatures as in the previous (one mode NS) case. For low temperatures (left panel) the low temperature low squeezing NSD island of Fig. 5.2, that corresponds to the environment acting as a resource for entanglement generation, disappears, since the bigger one expands to low squeezing. Degradation of resources by the environmental action here is not sufficient to prevent entanglement production even in the non-squeezed ($r = 0$) case for $T < T_c$, since actually the mode ε is also contributing to entanglement generation. On the other hand, the entanglement phases shows the same structure for high temperatures (right panel), where the only difference resides in the attenuated growth for entanglement when r increases (see color bars).

Notice that all the expressions have been calculated in the limit of weak coupling, assuming a final Gibbs state for decohered eigenmodes and free evolution for nondecohered ones. Of course this situation can only be perturbative, since for stronger coupling to the bath the eigenmodes become increasingly coupled among them, through second-order processes mediated by the bath. This necessarily leads to decoherence of all eigenmodes, at a low rate though.

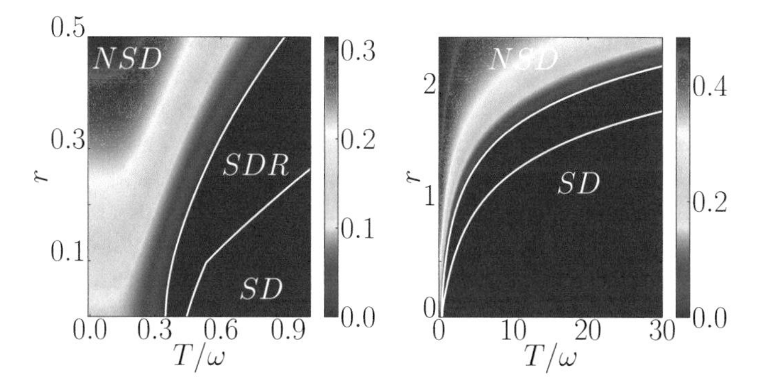

Fig. 5.3 Minimum entanglement in the asymptotic limit between external oscillators of the chain in the two-mode NS configuration in Fig. 5.1e for low temperatures (left panel) and high temperatures (right panel). The different phases (SD, SDR and NSD) are bounded by the continuous white lines obtained by numerical evaluation. We have set $\omega_2 = 1.2\omega$

5.3.2 *Quantum Synchronization*

In this section we analyze the dynamics of the system showing the existence of a parameter manifold where the harmonic oscillators oscillate in phase, synchronously, in spite of having different natural frequencies. The possibility to have synchronization in this system is important for two reasons: (i) this phenomenon has been largely studied in classical non-linear systems but we show that, for dissipation in a common bath, it can arise even among harmonic oscillators; (ii) few attempts have been done to extend it to the quantum regime, and we show here that one can have robust quantum correlations in a synchronous steady state accompanied by asymptotic entanglement.

In Chap. 4, we actually considered the phenomenon of mutual synchronization extended to the quantum regime, where two coupled harmonic oscillators with different frequencies, were studied during their relaxation towards a thermal equilibrium state. Synchronization was reported in first-order and second-order moments, characterizing the full dynamics for Gaussian states, during a long transient and accompanied by the robust preservation of quantum correlations (as measured by quantum discord) between oscillators. Two oscillators dissipating in a common bath are actually preserving asymptotically their entanglement and retaining a larger energy than in the thermal equilibrium state only if they are identical [5]. In this (symmetric) case they also evolve towards a synchronous asymptotic state.

When three elements are considered, we have shown above that the symmetric chain can reach an asymptotic regime with entanglement between the external oscillators, independently on the frequency of the central one. Then asymptotic synchronization between the external pair is also expected. Beyond this symmetric case, more interesting is the possibility offered by a chain to freeze all the oscillators out of the thermal equilibrium state when their frequencies are all different, as discussed below.

As we commented previously, the long time dynamics of our system can be straightforwardly calculated by assuming that normal modes which are coupled to the bath get thermalized, while uncoupled ones have a free evolution. This is sufficient to analyze the presence of synchronization in the asymptotic state. Quantum mutual synchronization appears always in one-mode NSs among natural oscillators linked by the non-dissipative mode, as long as they have an asymptotic dynamics with only one oscillatory contribution. Phase or anti-phase synchronization at the non-dissipating normal frequency is possible in first-order moments depending on the sign of their $\mathcal{F}$ matrix coefficients [Eq. (5.6)], while only in-phase synchronization occurs for second-order moments at twice the frequency of first-order ones. Let us illustrate it in some situations and compare with the time evolution of $\langle q_i^2 \rangle$ $\forall i = 1, 2, 3$ (Fig. 5.4) when considering simple Markovian dynamics in the weak coupling limit (see Sect. 5.4 below).

Consider first the specific case of an open chain with equal couplings and frequencies in the external oscillators [corresponding to situation in Fig. 5.1a]. The form of the non-dissipative normal mode is $\mathbf{C}_\delta = (1, 0, -1)^T/\sqrt{2}$, and hence synchronization will emerge only between external oscillators in anti-phase for position and

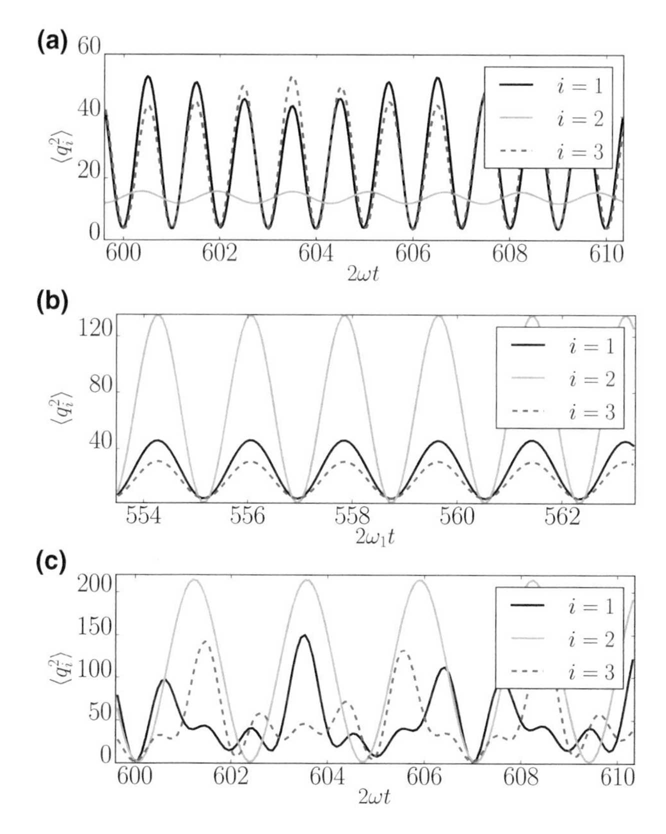

Fig. 5.4 Evolution of position variances for each oscillator in the open chain (see legend) for **a** configuration in Fig. 5.1a where a one-mode NS is generated ($\omega = 1.3\omega_2$, $\lambda = 0.4\omega_2^2$) synchronizing the external oscillators at 2ω; **b** configuration in Fig. 5.1c where a different one-mode NS is generated ($\omega_1 = 1.2\omega_2$, $\omega_3 = 1.3\omega_2$, $\lambda = 0.4\omega_2^2$) producing synchronization in all pairs of oscillators at $2\Omega_\varepsilon$; and **c** configuration in Fig. 5.1e where a two-mode NS is generated ($\omega = 1.3\omega_2$, $\lambda = \tilde{\lambda}_0$) and synchronization is lost. Bath parameters for the simulation are in all cases $T = 10k_B^{-1}\hbar\omega_2$, $\gamma_0 = 0.07\omega_2^2$ and $\Lambda = 50\omega_2$

momentum at frequency $\Omega_\delta = \omega$ (the normal mode frequency) and for the second-order moments (necessarily in-phase and at 2ω). The central oscillator instead decays into the thermal equilibrium state, its initial oscillations being suppressed in the long time dynamics. This case is shown in Fig. 5.4a, where synchronization appears after a transient only for the external oscillators of the open chain, while the central oscillator looses oscillation amplitude.

In the latter case synchronization appears between identical unlinked (λ_{13}) oscillators in a symmetric chain [Fig. 5.1a]. More peculiar is the case in which all oscillators have different frequencies and eventually couplings. In the case of Fig. 5.1c, we actually have that the non-dissipative mode involves all the three oscillators

$$\mathbf{C}_\varepsilon = c_\varepsilon(\omega_3^2 - \omega_2^2 \,,\ \omega_3^2 - \omega_2^2 - \tilde{\lambda}_\pm \,,\ \tilde{\lambda}_\pm)^T, \tag{5.37}$$

with $\Omega_\varepsilon = \sqrt{\omega_2^2 + \tilde{\lambda}_\pm}$. This can actually give rise to synchronous dynamics of all the oscillators, in spite of the difference in their natural frequencies. Since one of the components has different sign than the other two in $\mathbf{C}_\varepsilon$, two of the oscillators first moments will synchronize in-phase between them, and in anti-phase with the third one. In Fig. 5.4b a total synchronization is produced involving all three (different) oscillators, consistently with the fact that the non-dissipative normal mode, ε, involves all three oscillators.

A different situation is produced when we have a two-mode NS, since two oscillatory contributions are present in the asymptotic limit of the natural oscillators. Here synchronization is only possible when the two normal modes frequencies are the same. An example is the open chain with a two-mode NS [see Fig. 5.1e], where apart from the previous non-dissipative mode $\mathbf{C}_\delta$, actually the collective mode $\mathbf{C}_\varepsilon = (1, -2, 1)^T/\sqrt{6}$ with frequency $\Omega_\varepsilon = \sqrt{2\omega_2^2 - \omega^2}$ does not dissipate either. In this case, synchronization is destroyed by the presence of the mode ε, and it can be only recovered when Ω_ε equals Ω_δ, i.e. in the trivial case of independent ($\lambda = 0$) identical oscillators ($\omega_2 = \omega$). Lack of synchronization as well as a multimode oscillation are shown in Fig. 5.4c.

The initial state employed for simulations is a squeezed separable vacuum state, where the squeezing parameters have been chosen to be different ($r_1 = 2, r_2 = 2.5$ and $r_3 = 3$). In general, we have tried to avoid special initial conditions that could have filtered just one normal mode into the dynamics. What we discussed is therefore the emergence of synchronization as a dynamical process when considering more general initial states, leading to robust conclusions.

The scenarios here discussed allow to establish the effect of having a NS with one or two modes in the configurations of open chains [Fig. 5.1a, e]. The same analysis can be extended to other cases, where a different normal mode is uncoupled from the environment. For instance, the configurations in Fig. 5.1c, d, admit only one non-dissipative normal mode that involves the three oscillators, producing then a collective synchronization of the chain.

5.4 Thermalization and Robustness of Quantum Correlations

Creation of NSs is a powerful tool to avoid decoherence and produce synchronized dynamics as we have seen in the previous sections. However, the conditions leading to NSs are satisfied only in some parameter manifolds. It is relevant analyze the effect of deviations from these couplings and frequencies, that could also arise from the difficulty of experimental tuning. In this case, dissipation is present in all normal modes, and the effective couplings of Eq. (5.8) are all different from zero. Henceforth a thermal equilibrium state is finally reached in all the degrees of freedom in the long time run of the dynamical evolution.

In absence of NS, entanglement is lost after a finite time. Although the asymptotic state is simply the Gibbs state, damping dynamics of the normal modes with different decoherence and relaxation time scales is present, producing a rich behavior in which synchronization or high quantum correlations may emerge during a large transient before the final thermalization of the system. These effects have been reported in Chap. 4 in the case of two harmonic oscillators, where disparate decay rates between the two normal modes is produced for small deviations from the resonant case.

A dynamical description of the system weakly interacting with the environment reveals the central influence of the effective couplings in the relaxation time scales of the different normal modes. By using the general Born and Markov approximations as well as an initial product state we may easily obtain a Markovian master equation for the reduced density matrix of the open system in the normal modes basis (see Sect. 2.4). The resulting equation is not of the Lindblad form, thus complete positivity (CP) is not guaranteed [56]. This issue can be solved in two different ways, either considering a rotating wave approximation (RWA)

$$\hat{x}_i \hat{x}_j \quad \rightarrow \quad \hat{a}_i \hat{a}_j^\dagger + \hat{a}_i^\dagger \hat{a}_j, \tag{5.38}$$

in the interaction Hamiltonian (5.4), or performing a strong-type RWA in the non-Lindbladian master equation by eliminating oscillatory terms of the form $\exp(\pm i(\Omega_i \pm \Omega_j)t)$ that appear in the interaction picture. The latter is the one we will pursue. The advantages of this method not only reside in obtaining a master equation in Lindblad form (thus CP), but also in that dynamical evolution can be solved analytically. However, an exhaustive analysis in the case of two harmonic oscillators shows a very well agreement between results using the original non-Lindbladian master equation and the strong RWA here used (see Appendix A.2 in Chap. 4). The Markovian master equation for the evolution of the reduced density matrix for a common bath in the strong RWA is then

$$\begin{aligned}\frac{d\rho(t)}{dt} = & -\frac{i}{\hbar}[\hat{H}_S, \rho(t)] \\ & - \frac{1}{4\hbar^2}\sum_{n=1}^{3} i\Gamma_n \left([\hat{Q}_n, \{\hat{P}_n, \rho(t)\}] - [\hat{P}_n, \{\hat{Q}_n, \rho(t)\}]\right) \\ & + D_n \left([\hat{Q}_n, [\hat{Q}_n, \rho(t)]] - \frac{1}{\Omega_n^2}[\hat{P}_n, [\hat{P}_n, \rho(t)]]\right).\end{aligned} \tag{5.39}$$

Here Γ_n and D_n are constant coefficients (by virtue of the Markov approximation) accounting for the damping and diffusion effects respectively. Note that under this approximation, each normal mode is dissipating separately to the bath, i.e. they have independent decay rates. The bath has been considered to be in thermal equilibrium at temperature T, and to be composed by a continuum of frequencies characterized by a spectral density $J(\Omega) \equiv \sum_\alpha \delta(\Omega - \tilde{\Omega}_\alpha)\lambda_\alpha^2/\tilde{\Omega}_\alpha$. For simplicity it has been considered to be Ohmic with a sharp cutoff $J(\Omega) = \frac{2\gamma_0}{\pi}\Omega\,\Theta(\Lambda - \Omega)$, where $\Theta(x)$ is the Heaviside step function, Λ is the largest frequency present in the environment (cutoff frequency) and γ_0 is a constant quantifying the strength of system-environment interaction (thus in the weak-coupling limit we have always $\gamma_0 \ll \Omega_i$ $\forall i = 1, 2, 3$). This assumptions leads to the following definitions of the master equation coefficients

$$\Gamma_n = \kappa_n^2 \frac{\hbar\pi}{2} \frac{J(\Omega_n)}{\Omega_n} = \hbar\gamma_0\,\kappa_n^2, \tag{5.40}$$

$$D_n = \kappa_n^2 \frac{\hbar\pi}{2\hbar} J(\Omega_n) \coth\left(\frac{\hbar\Omega_n}{2k_B T}\right) = \hbar\gamma_0\,\kappa_n^2\Omega_n \coth\left(\frac{\hbar\Omega_n}{2k_B T}\right),$$

where we also assume $\Omega_i < \Lambda\ \forall i = 1, 2, 3$. The equations governing the second-order moments of the normal modes from the master equation (5.39) are reported in Appendix A.2.

In this context, the ratio between the two smallest decay rates, defined as

$$R \equiv \frac{\Gamma_0}{\Gamma_1} = \frac{\kappa_0^2}{\kappa_1^2}, \tag{5.41}$$

provides important information about the dynamics of the system. This is in fact one of the central figures (but not the only one) in order to predict the robustness of correlations between oscillators or the emergence of synchronization as we will see in the following. In presence of disparate decay rates ($R << 1$), a large time interval appears between thermalization of the two modes with largest damping coefficients (*strongly-damped* modes) and thermalization of the mode with the smallest one (*weakly-damped* mode). This produces the emergence, after a transient, of a long interval in which the *weakly-damped* mode is effectively the only one present in the dynamics, hence producing the synchronization between pairs of oscillators linked by this normal mode, and the slow decay of quantum discord between these pairs. On the other hand, when the decay rates are similar ($R \sim 1$), the different modes are present for all times inhibiting synchronization, and the survival of correlations associated to one of the modes for long times is lost. These phenomena will be next exemplified in the scenario of an open chain with equal couplings ($\lambda_{12} = \lambda_{23} \equiv \lambda$).

In Fig. 5.5 we represent R showing broad regions in which a *weakly-damped* mode exists (white regions) near the NSs manifolds corresponding to the configurations in Fig. 5.1a (dashed line), c (dashed-dotted hyperbola), e (the crossing point). Out of these regions, there is no separation of scales for the decay rates (blue regions), and all rates become progressively similar. We point out that the blue region wrapping the diagonal in Fig. 5.5, acts as boundary for the two white ones, since a different mode (with radically different shape) is *weakly-damped* in each white region. The coupling λ is related to the position of the dashed-dotted hyperbola by Eq. (5.25) and the width of white regions, making them broader as λ increases, and tighter when it decreases.

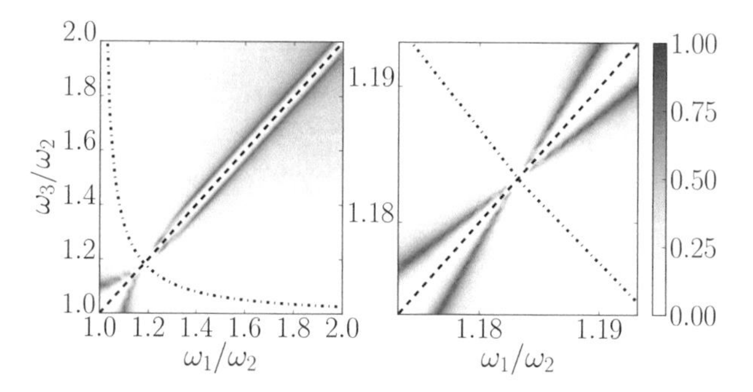

Fig. 5.5 Map of R in Eq. (5.41) for $\lambda = 0.4\omega_2^2$ as a function of the open chain frequencies. Dashed and dashed-dotted lines represent the non-dissipative parameters manifolds of Fig. 5.1a, c, respectively. The right panel is a zoom of the vicinities of the two-mode NS cross point in the left panel

5.4.1 *Quantum Correlations*

Even if out of the NS conditions entanglement suffers a sudden death, other indicators such as quantum discord can remain robust in regions with disparate decay rates ($R << 1$) (as in the case of two detuned oscillators in Chap. 4). By using an adapted measure of discord for Gaussian bipartite states [57, 58] (see Sect. 1.4.3 in Chap. 1), we observe the existence of a *plateau* in the dynamical evolution of discord between single pairs of oscillators, which are linked by a *weakly-damped* normal mode. More precisely, in the white region of Fig. 5.5, close to the dashed-dotted hyperbola, the *weakly-damped* mode links the three natural oscillators, producing a *plateau* in the evolution of discord for all pairs. Moving to the tighter white region, close to the dashed diagonal line, the *weakly-damped* mode only involves the external oscillators pair of the open chain, leading to a slowly decaying discord only for this pair of oscillators. On the contrary, in blue regions, no *plateau* is observed for discord, reaching in shorter times the value corresponding to the thermal equilibrium state for each pair of oscillators.[2]

Figure 5.6 shows time evolution of discord in logarithmic scale for the three pairs of oscillators (see colors) for a selection of parameters close and far away from the dashed-dotted hyperbola [Fig. 5.6a, b respectively]. The initial condition has been taken to be a squeezed separable vacuum state with same squeezing parameters as in Fig. 5.4 and will be kept for further simulations. A Gaussian filter has been employed to eliminate fast oscillations (original quantities are plotted in gray), in order to make it easier to identify the *plateau* characterizing discord robustness, as in Chap. 4.

As already seen for asymptotic entanglement, the effect of increasing the bath's temperature is, in general, a degradation of quantum effects. It is therefore important to see how robustness of discord is a feature also present in hotter environments. The main effect when increasing T is that the final thermal state displays lower

[2]Notice that quantum discord between pairs of oscillators in a global Gibbs state is non-zero due to the coupling between them.

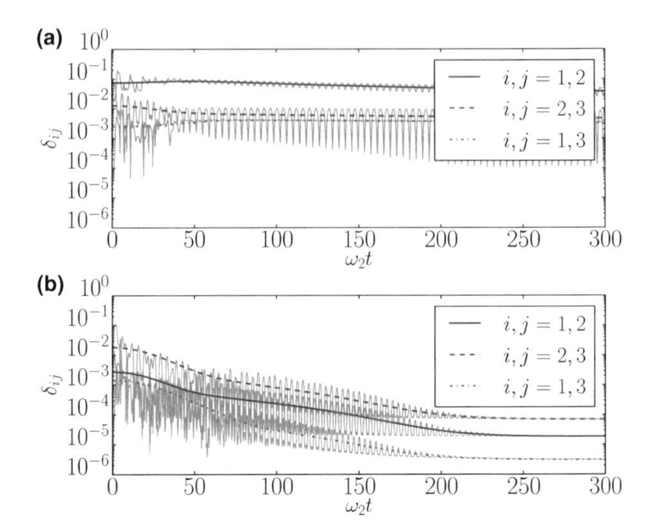

Fig. 5.6 Time evolution of discord between pairs of oscillators for the three pairs in the open chain (see legend) in two different regions of Fig. 5.5. We set $\lambda = 0.4\omega_2^2$, $\omega_3 = 1.6\omega_2$ and change ω_1. **a** Near the dashed-dotted hyperbola ($\omega_1 = 1.1\omega_2$) and **b** far away from it ($\omega_1 = 1.9\omega_2$). The exact time evolutions are shown in grey while the thick color lines (solid, dashed and dashed-dotted) represent the filtered ones (see text). Bath parameters for the simulations are $T = 10k_B^{-1}\hbar\omega_2$, $\gamma_0 = 0.07\omega_2^2$ and cutoff frequency $\Lambda = 50\omega_2$

correlations, implying that the amount of discord that can be maintained in a robust way diminishes. In Fig. 5.7 we show the evolution of discord for a pair of linked oscillators (1, 2) when T is increased by factors 3 and 6 [the other parameters are as in Fig. 5.6a]. While the *plateau* is present for all temperatures and their (negative) slope is very similar, a lower amount of discord is now generated in the short initial

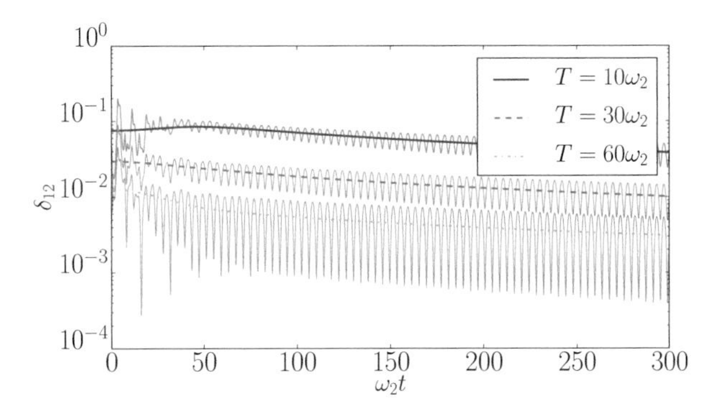

Fig. 5.7 Time evolution of discord for a pair of linked oscillators (1, 2) of the open chain for different bath temperatures (see legend). We have set $\lambda = 0.4\omega_2^2$, $\omega_3 = 1.6\omega_2$ and $\omega_1 = 1.1\omega_2$. The rest of bath parameters has been kept $\gamma_0 = 0.07\omega_2^2$ and $\Lambda = 50\omega_2$

transient, producing a shift of the curves to lower values (notice that oscillations are increased by the logarithmic scale of the plot). This degradation by temperature effects can be avoided by increasing the squeezing in the initial separable vacuum state, similarly to the case of entanglement presented in Sect. 5.3.1.

5.4.2 *Synchronous Thermalization*

With respect to the emergence of synchronization for pairs of oscillators when the NS is lost, we have to point out that, when thermalizing, the system reaches a stationary state where oscillations are suppressed. We therefore restrict our analysis to a transient (which becomes longer the more we approach one of the NS conditions) where oscillations in the first-order and second-order moments are still present in the dynamics. In this situation synchronization of first-order and second-order moments can be estimated quantitatively by using the Pearson indicator C introduced in the previous chapter [Eq. 4.13]. When evolutions are phase or anti-phase synchronized we will obtain $|C| \sim 1$, while for very different dynamics we will obtain a value of C near to zero.

Figure 5.8 shows the synchronization indicator $C_{\langle q_i^2 \rangle, \langle q_j^2 \rangle}$ with position variances of (a) the external pair of oscillators $i, j = 1, 3$ and (b) for $i, j = 1, 2$ of the open chain with identical couplings and varying the external oscillators frequencies (in the same range as in Fig. 5.5). We see immediately the high resemblance with the R map of Fig. 5.5 and some interesting differences induced by the shape of the normal modes. Effectively the external pair of oscillators $(1, 3)$ synchronizes $(C \sim 1)$ in all regions where disparate decay rates $(R \ll 1)$ are predicted since these two oscillators are linked by the *weakly-damped* normal mode in these regions [Fig. 5.8a]. As for the

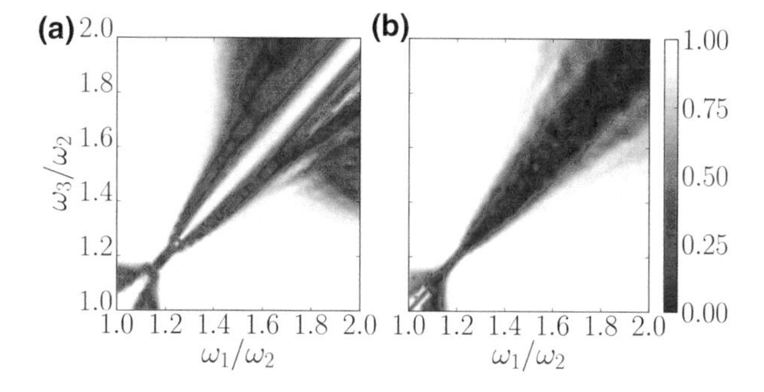

Fig. 5.8 Absolute value of the synchronization indicator $|C(t, \Delta t)|$ for position variances $(\langle \hat{q}_i^2 \rangle)$ for **a** the external pair of oscillators $(1, 3)$, and **b** a linked pair $(1, 2)$. The synchronization factor is plotted at time $t = \min\{t_{\max}, \Gamma_0^{-1}\}$ where $t_{\max} = 5000\omega_2^{-1}$ (the maximum time used in the simulations) in order to obtain a map in which oscillations were not yet suppressed. We have used $\Delta t = 15\omega_2^{-1}$ and the same bath parameters as in former figures

internal pair $(1, 2)$, it does depend on the *weakly-damped* mode in the vicinities of the dashed-dotted hyperbola where synchronization is actually found. On the other hand, near the diagonal the *weakly-damped* mode approximates to $\mathbf{C}_\delta = (1, 0, -1)/\sqrt{2}$, excluding the central oscillator from the induced collective motion. Consistently, the $(1, 2)$ pair [Fig. 5.8b] does not synchronize for $\omega_1 \sim \omega_2$ (near the diagonal).

We finally point out that, as expected, the synchronization frequency is that of the *weakly-damped* mode, Ω_0, for the first-order moments (position and momenta), and $2\Omega_0$ for the second-order momenta.

5.5 Conclusions

Decoherence in an open quantum system can be avoided or reduced by tuning the system parameters in a common environment context. The shape of the interaction Hamiltonian between system and bath can be used in order to engineer the protection of some degrees of freedom from the environmental action. In this chapter we solved the case of three coupled harmonic oscillators in contact with a bosonic bath in thermal equilibrium, developing the necessary general relations so as to obtain a NS composed by one or two non-dissipative normal modes.

Different open and close chain configurations have been explored, highlighting the richer variety of NS configurations available when the dissipative system is extended from two to three harmonic oscillators. For a symmetric open chain with equal frequencies of the external pair of oscillators and same coupling, a closed analytical expression for the asymptotic entanglement between the external pair (as given by the logarithmic negativity) has been derived observing the appearance of three different phases depending on temperature and squeezing of the initial state (sudden death of entanglement, a infinite series of sudden death events and revivals and asymptotic robust entanglement). Sudden death of entanglement can be avoided for arbitrarily high bath temperatures by increasing the squeezing in the initial state for both cases of one-mode or two-mode NS. Remarkably this critical squeezing in order to avoid sudden death depends logarithmically on temperature. Asymptotic robust entanglement is also reached for a small region of parameters corresponding to low temperatures of the bath ($T \sim 0.1 k_B \hbar\omega$) even in the absence of initial squeezing. This small island of asymptotic entanglement has been previously reported in the case of two identical oscillators [3] where it has been attributed to non-Markovian dynamical effects. From our analysis it becomes clear that this is not the case, being just produced by the thermalization in a subspace of the system. On the other hand, multipartite entanglement in the strong local dissipation regime for the open chain configuration has been recently explored in Ref. [59].

Other dynamical effects such as the emergence of synchronization of mean values and variances have been analyzed in different situations by simply assuming relaxation to a thermal equilibrium state of the normal modes coupled to the environment. Coherent oscillations appear when only a surviving normal mode is present in the dynamics, inducing synchronization in the natural oscillators that depend on it.

Interestingly, the parameter manifold leading to NSs include several not symmetric configurations: for instance an hyperbolic relation among frequencies can be satisfied for identical couplings in an open chain; in this case both asymptotic entanglement and synchronization are predicted even if all the oscillators natural frequencies are different, a possibility offered by a chain of three oscillators and absent in the case of two.

Furthermore an analysis of situations in which the NS conditions are not accomplished at all has been performed. Indeed, important properties can be present although, when deviating from NS conditions, entanglement does not survive: robust conservation of discord during a long transient dynamics and the emergence of synchronous oscillations are found before thermalization. These effects are interpreted in relation to disparate decay rates for the normal modes, clarifying and extending our previous results for the case of two dissipative oscillators in Chap. 4. As long as there is a *weakly-damped* mode surviving among several *strongly-damped* modes, effects such as robust discord and synchronization arise among the oscillators following this normal mode. On the other hand, if this separation of damping time scales for the normal modes does not exists, synchronization is lost and any initial discord quickly decays to its asymptotic (small) thermal equilibrium value.

Our results for the system of three oscillators in presence of common dissipation may be implemented with ions in linear Paul traps by following the proposal in Ref. [60]. Experimental realization of coupled harmonic oscillators appear in optical [61, 62] and superconducting [63, 64] cavities as well as trapped ions [65, 66] or nanoelectromechanical resonators [67]. Three coupled elements architectures are also known to allow for isochronous synchronization of semiconductor lasers with delayed coupling or neuronal models [68]. While in this chapter we have focused in specific cases of three oscillator configurations (in which calculations are greatly simplified) the strategy provided here is rather general and applies straightforwardly to other choices of system parameters that produce the decoupling of one or several normal modes from the environment. The methods presented here may be extended to more complicated systems such as disordered harmonic lattices or complex networks. This opens the possibility of an engineering of the normal modes of complex quantum many-body systems in order to induce noiseless subsystems for its use e.g. in quantum information or quantum computational tasks. The case of arbitrary complex networks of dissipative harmonic oscillators is analyzed in Chap. 6, where a wide range of possibilities are shown, including different forms of dissipation taking place across the system, selective protection against decoherence and dissipation of clusters of oscillators in the network, or the induction of synchronized states in the whole network to a common frequency by tuning a single parameter.

Appendix

A.1 Analytical Derivation of Asymptotic Entanglement

As we pointed in Sect. 1.4.1, all the information about bipartite quantum correlations for a Gaussian continuous-variable state is condensed in its covariance matrix defined through the ten second-order moments of $\hat{q}_{(A,B)}$ and $\hat{p}_{(A,B)}$ (in our case first-order moments are initially zero). This bipartite covariance matrix defined for a system of two oscillators A and B, can be written as

$$V_{AB} = \begin{pmatrix} \alpha & \gamma \\ \gamma^t & \beta \end{pmatrix}, \tag{A.1}$$

where α, β and γ are (2×2) blocks: $\alpha(\beta)$ contains the second-order moments of oscillator subsystem A (B), and γ contains correlations of both subsystems. The minimum symplectic eigenvalue (of the covariance matrix corresponding to the partially transposed density matrix), necessary to calculate the logarithmic negativity, is given by

$$\nu_- = \sqrt{\frac{1}{2}(a + b - 2g - \sqrt{(a + b - 2g)^2 - 4s}}, \tag{A.2}$$

with $a = 4\det(\alpha)/\hbar^2$, $b = 4\det(\beta)/\hbar^2$, $g = 4\det(\gamma)/\hbar^2$ and $s = 16\det V_{AB}/\hbar^4$. Normal modes coupled to the environment will reach in the asymptotic limit a thermal state, given by the Gibbs distribution. For a normal mode (k), that is

$$\rho_{\text{th}}^{(k)} = \frac{e^{-\frac{\hat{H}_k}{k_B T}}}{\text{Tr}[e^{-\frac{\hat{H}_k}{k_B T}}]}, \tag{A.3}$$

with $\hat{H}_k = \frac{1}{2}\left(\hat{P}_k^2 + \Omega_k^2 \hat{Q}_k^2\right)$, yielding the second-order moments

$$\begin{aligned} \langle \hat{Q}_k^2 \rangle_{\text{th}} &= \frac{\hbar}{2\Omega_k} \coth\left(\frac{\hbar\Omega_k}{2k_B T}\right), \\ \langle \hat{P}_k^2 \rangle_{\text{th}} &= \frac{\hbar\Omega_k}{2} \coth\left(\frac{\hbar\Omega_k}{2k_B T}\right), \end{aligned} \tag{A.4}$$

where Ω_k is the corresponding frequency of the normal mode, and T the reservoir temperature. On the other hand, the uncoupled modes evolve freely. This means that the asymptotic covariance matrix can be calculated by expressing all second-order moments of natural oscillators in terms of the normal modes, and then substituting the asymptotic expressions corresponding to free modes or thermalized ones.

The covariance matrix in the asymptotic limit can be separated into three parts corresponding to the contributions of each normal mode, which we call V_i, for

$i = 1, 2, 3$. In terms of the blocks we have

$$\alpha = \sum_{i=1}^{3} \mathcal{F}_{Ai}^2 V_i, \quad \beta = \sum_{i=1}^{2} \mathcal{F}_{Bi}^2 V_i, \quad \gamma = \gamma^T = \sum_{i=1}^{2} \mathcal{F}_{Ai}\mathcal{F}_{Bi} V_i, \tag{A.5}$$

where V_i can correspond either to a non-dissipative, or to a dissipative mode. For a non-dissipative normal mode, say n, we have:

$$V_{\text{no-diss}} = \begin{pmatrix} \langle \hat{Q}_n^2 \rangle & \frac{\langle \{\hat{Q}_n, \hat{P}_n\} \rangle}{2} \\ \frac{\langle \{\hat{Q}_n, \hat{P}_n\} \rangle}{2} & \langle \hat{P}_n^2 \rangle \end{pmatrix}, \tag{A.6}$$

and for a dissipative one, say k, we get

$$V_{\text{diss}} = \begin{pmatrix} \langle \hat{Q}_k^2 \rangle_{\text{th}} & 0 \\ 0 & \langle \hat{P}_k^2 \rangle_{\text{th}} \end{pmatrix}. \tag{A.7}$$

While elements in V_{diss} are given by the expressions (A.4), those of $V_{\text{no-diss}}$ are the ones corresponding to a free evolution of an harmonic oscillator. This analysis gives all the necessary elements in order to calculate the asymptotic entanglement for pairs of oscillators in every particular situation, in which one or two of the normal modes are uncoupled from the environmental action.

A.1.1 One-Mode NS

Consider the specific case of an open chain ($\lambda_{13} = 0$) in which we have two equal frequencies ($\omega_1 = \omega_3 \equiv \omega$) and two equal couplings ($\lambda_{12} = \lambda_{23} \equiv \lambda \neq 0$) [Fig. 5.1a in Sect. 5.3]. In this case we get only one normal mode decoupled from the bath. In order to calculate the expression of the minimum symplectic eigenvalue, we have to first calculate the elements of the three normal modes, that are shown here as vector columns

$$\mathbf{C}_\delta = \frac{1}{\sqrt{2}} \begin{pmatrix} 1 \\ 0 \\ -1 \end{pmatrix}, \qquad \mathbf{C}_\pm = c_\pm \begin{pmatrix} \lambda \\ \Omega_\pm^2 - \omega^2 \\ \lambda \end{pmatrix}.$$

Here we have labeled the non-dissipative mode as δ and the other two modes as $\{\pm\}$. Their corresponding frequencies are $\Omega_\delta = \omega$ and $\Omega_\pm = \sqrt{(\omega_2^2 + \omega^2)/2 \pm \sqrt{\Delta}}$, defining $\Delta \equiv (\frac{\omega_2^2 - \omega^2}{2})^2 + 2\lambda^2$, and $c_\pm$ being nothing but a normalization constant. We can now obtain all the terms appearing in $V_\pm$.

The initial condition given in Eq. (5.27), can be now rewritten in terms of the non-dissipative normal mode as

$$\langle \hat{Q}_\delta^2(0)\rangle = \frac{\hbar}{2\omega}e^{-2r}, \quad \langle \hat{P}_\delta^2(0)\rangle = \frac{\hbar\omega}{2}e^{2r}, \quad \langle\{\hat{Q}_\delta, \hat{P}_\delta\}(0)\rangle = 0,$$

and then their free evolution is given by

$$\begin{aligned}
\langle Q_\delta^2\rangle &= \frac{\hbar}{2\omega}(e^{2r}\sin^2(\omega t) + e^{-2r}\cos^2(\omega t)),\\
\langle P_\delta^2\rangle &= \frac{\hbar\omega}{2}(e^{-2r}\sin^2(\omega t) + e^{2r}\cos^2(\omega t)),\\
\langle\{Q_\delta, P_\delta\}\rangle &= 2\hbar\sinh(2r)\cos(\omega t)\sin(\omega t),
\end{aligned} \tag{A.8}$$

where we have already used that $\Omega_\delta = \omega$. By substituting the above expressions in $V_{\text{no-diss}}$ [Eq. (A.6)] we can now obtain the expressions of the determinants a, b, g and s. This yields for the minimum symplectic eigenvalue [Eq. (A.2)]:

$$\frac{\nu_-(t)^2}{2\lambda^2} = \mathcal{G}_0 + \mathcal{G}_1\cos(2\omega t) - \sqrt{(\mathcal{G}_0 + \mathcal{G}_1\cos(2\omega t))^2 - 4\sigma_P\sigma_Q}, \tag{A.9}$$

which is an oscillatory function with frequency 2ω. Here

$$\mathcal{G}_0 = (\sigma_Q + \sigma_P)\cosh(2r), \qquad \mathcal{G}_1 = (\sigma_Q - \sigma_P)\sinh(2r),$$

and the dependence on the bath temperature and on the shape of the dissipative normal modes is given by

$$\begin{aligned}
\sigma_P &= \frac{\Omega_+}{2\omega}c_+^2\coth\left(\frac{\Omega_+}{2T}\right) + \frac{\Omega_-}{2\omega}c_-^2\coth\left(\frac{\Omega_-}{2T}\right),\\
\sigma_Q &= \frac{\omega}{2\Omega_+}c_+^2\coth\left(\frac{\Omega_+}{2T}\right) + \frac{\omega}{2\Omega_-}c_-^2\coth\left(\frac{\Omega_-}{2T}\right).
\end{aligned} \tag{A.10}$$

From Eq. (A.9), we can obtain the minimum entanglement (obtained for $t = (2n+1)\frac{\pi}{2\omega}$; $n = 1, 2, 3, \ldots$) and the maximum one (for $t = (n+1)\frac{\pi}{\omega}$; $n = 1, 2, 3, \ldots$) in order to recover Eq. (5.29) with the proper definitions specified there.

A.1.2 Two-Mode NS

On the other hand, if we move to situation represented in Fig. 5.1e of Sect. 5.3 by fixing $\lambda = \tilde{\lambda}_0$ [see Eq. (5.25)], we have that the normal modes transform into

$$\mathbf{C}_\delta = \frac{1}{\sqrt{2}}\begin{pmatrix}1\\0\\-1\end{pmatrix}, \quad \mathbf{C}_\varepsilon = \frac{1}{\sqrt{6}}\begin{pmatrix}1\\-2\\1\end{pmatrix}, \quad \mathbf{C}_{\text{c.m.}} = \frac{1}{\sqrt{3}}\begin{pmatrix}1\\1\\1\end{pmatrix},$$

being the center of mass, $\mathbf{C}_{\text{c.m.}}$, the only dissipative mode. Their corresponding frequencies are respectively

$$\Omega_\delta = \omega, \quad \Omega_\varepsilon = \sqrt{2\omega_2^2 - \omega^2}, \quad \Omega_{\text{c.m.}} = \sqrt{2\omega^2 - \omega_2^2}. \tag{A.11}$$

Naturally, we have to restrict ourselves to the regime $2\omega_3^2 > \omega > \omega_3^2/2$ in order for these quantities to be real and positive.

Keeping the same initial condition as in the previous case, we have that nothing changes in the expression of the free evolution of mode δ [Eq. (A.8)], while the free evolution of mode ε is given by

$$\begin{aligned}
\langle \hat{Q}_\varepsilon^2 \rangle &= \frac{2\omega_2 + \omega}{6\Omega_\varepsilon^2} \hbar e^{2r} \sin^2(\Omega_\varepsilon t) + \frac{2\omega + \omega_2}{6\omega\omega_2} \hbar e^{-2r} \cos^2(\Omega_\varepsilon t), \\
\langle \hat{P}_\varepsilon^2 \rangle &= \frac{2\omega + \omega_2 \Omega_\varepsilon^2}{6\omega\omega_2} \hbar e^{-2r} \sin^2(\Omega_\varepsilon t) + \frac{2\omega_2 + \omega}{6} \hbar e^{2r} \cos^2(\Omega_\varepsilon t), \\
\langle \{\hat{Q}_\varepsilon, \hat{P}_\varepsilon\} \rangle &= \left(\frac{2\omega_2 + \omega}{3\Omega_\varepsilon} e^{2r} - \frac{(2\omega + \omega_2)\Omega_\varepsilon}{3\omega\omega_2} e^{-2r} \right) \times \\
&\quad \times \hbar \cos(\Omega_\varepsilon t) \sin(\Omega_\varepsilon t).
\end{aligned}$$

We have assumed the same squeezing parameter r in the central oscillator of the chain (notice that in the previous case the initial state of the central oscillator is not relevant and then we did not specify it). Following the same procedure as above, we calculate the expression for the minimum symplectic eigenvalue. It is worth noticing that in this case we have two contributions to the determinants of the free type $V_{\text{no-diss}}$ [Eq. (A.6)], corresponding to the two non dissipative modes, and a single dissipative one V_{diss} [Eq. (A.7)], corresponding to the center of mass mode.

The minimum symplectic eigenvalue yields:

$$2\nu_-(t)^2 = \mathcal{A}_0 + \mathcal{A}_1(t) - \sqrt{(\mathcal{A}_0 + \mathcal{A}_1(t))^2 - \mathcal{B}_0 - \mathcal{B}_1(t)} \tag{A.12}$$

where we have defined the following quantities in order to simplify the expression. The constant terms

$$\begin{aligned}
\mathcal{A}_0 &\equiv \cosh(2r) \left(4(\sigma_Q + \sigma_P) + \mathcal{J}_+(\Omega_\varepsilon^2 + \omega^2)\right), \\
\mathcal{B}_0 &\equiv 64\sigma_P\sigma_Q + \frac{4(\omega + \omega_2)^2}{81\omega\omega_2} + \frac{32\Omega_\varepsilon\omega\mathcal{J}_+}{3} \left(\frac{\omega\sigma_P}{\Omega_\varepsilon} + \frac{\Omega_\varepsilon\sigma_Q}{\omega} \right),
\end{aligned}$$

and the oscillating terms

$$\begin{aligned}\mathcal{A}_1(t) \equiv\ & 4\cos(2\omega t)\sinh(2r)(\sigma_Q - \sigma_P) \\ & + \mathcal{J}_+ \cos(2\omega t)\sinh(2r)(\Omega_\varepsilon^2 + \omega^2) \\ & + \mathcal{J}_- \cos(2\Omega_\varepsilon t)\cosh(2r)(\Omega_\varepsilon^2 - \omega^2) \\ & - \mathcal{J}_- \cos(2(\Omega_\varepsilon - \omega)t)\sinh(2r)\frac{(\Omega_\varepsilon + \omega)^2}{2} \\ & - \mathcal{J}_- \cos(2(\Omega_\varepsilon + \omega)t)\sinh(2r)\frac{(\Omega_\varepsilon - \omega)^2}{2}, \\ \mathcal{B}_1(t) \equiv\ & \cos(2\Omega_\varepsilon t)\frac{32\Omega_\varepsilon \omega \mathcal{J}_+}{3}\left(\frac{\Omega_\varepsilon \sigma_Q}{\omega} - \frac{\omega \sigma_P}{\Omega_\varepsilon}\right),\end{aligned}$$

where different frequencies coming from the two non-dissipative modes are present. We have used $\mathcal{J}_\pm \equiv \frac{1}{12\omega}\left(e^{2r}\frac{2\omega_2+\omega}{\Omega_\varepsilon^2} \pm e^{-2r}\frac{2\omega+\omega_2}{\omega\omega_2}\right)$ and the two bath-dependent functions are now given by the contribution of the c.m. mode:

$$\begin{aligned}\sigma_P &= \frac{\Omega_{\text{c.m.}}}{6\omega}\coth\left(\frac{\hbar\Omega_{\text{c.m.}}}{2k_B T}\right), \\ \sigma_Q &= \frac{\omega}{6\Omega_{\text{c.m.}}}\coth\left(\frac{\hbar\Omega_{\text{c.m.}}}{2k_B T}\right).\end{aligned} \tag{A.13}$$

A.2 Equations of Motion for the Second-Order Moments

As we are interested in classical and quantum correlations of the system oscillators, a description for the evolution of the first-order and second-order moments is necessary. The equations of motion for position, momenta, and the variances can be obtained from the Markovian master equation governing the dissipative dynamics, (5.39). In analogy to the case of two oscillators (Chap. 4) they can be indeed written in a simple form as $\dot{\mathbf{R}} = \mathcal{M}\mathbf{R} + \mathbf{N}$, where $\mathbf{R}$ is a column vector, now containing the $M = (2N+1)N$ for $N = 3$ independent second-order moments of the normal modes. The matrix $\mathcal{M}$ condenses all the information about their dynamical evolution and $\mathbf{N}$ determines the stationary values for long times (when $\dot{\mathbf{R}} = 0$). The dynamics of $\mathbf{R}$ can be solved in terms of the eigenvalues of $\mathcal{M}$:

$$\{\mu_{ij}\} = \{-\frac{\Gamma_i + \Gamma_j}{2} \pm i\left|\Omega_i \pm \Omega_j\right|\}, \quad i \le j \tag{A.14}$$

where the $i = j$ eigenvalues determine the evolution of $\langle \hat{Q}_i^2 \rangle$, $\langle \hat{P}_i^2 \rangle$ and $\langle \{\hat{Q}_i, \hat{P}_i\} \rangle$, while the ones with $i \neq j$ determine that of $\langle \hat{Q}_i \hat{Q}_j \rangle$, $\langle \hat{P}_i \hat{P}_j \rangle$ and $\langle \{\hat{Q}_i, \hat{P}_j\} \rangle$. Note that by virtue of Eqs. (5.40) and (A.14) the decay of the normal modes is entirely governed by the effective couplings mentioned above, thus differences in their magnitude produce disparate temporal scales for the dissipation and diffusion of normal modes.

We further stress that the stationary state of the dynamics is found to be ($\mathbf{R}_\infty = \mathcal{M}^{-1}\mathbf{N}$):

$$\langle \hat{Q}_i^2 \rangle_\infty = \frac{D_i}{2\Gamma_i \Omega_i^2} = \frac{\hbar}{2\Omega_i} \coth\left(\frac{\Omega_i}{2k_B T}\right),$$

$$\langle \hat{P}_i^2 \rangle_\infty = \frac{D_i}{2\Gamma_i} = \frac{\hbar \Omega_i}{2} \coth\left(\frac{\Omega_i}{2k_B T}\right),$$

being all the other second-order moments equal to zero. Note that these expressions for the asymptotic limit recover the thermal state of the system at the bath temperature T given by the Gibbs distribution in Eq. (A.4).

References

1. J.S. Prauzner-Bechcicki, Two-mode squeezed vacuum state coupled to the common thermal reservoir. J. Phys. A: Math. Gen. **37**, L173 (2004)
2. K.-L. Liu, H.-S. Goan, Non-Markovian entanglement dynamics of quantum continuous variable systems in thermal environments. Phys. Rev. A **76**, 022312 (2007)
3. J.P. Paz, A.J. Roncaglia, Dynamics of the Entanglement between Two Oscillators in the Same Environment. Phys. Rev. Lett. **100**, 220401 (2008)
4. J.P. Paz, A.J. Roncaglia, Dynamical phases for the evolution of the entanglement between two oscillators coupled to the same environment. Phys. Rev. A **79**, 032102 (2009)
5. F. Galve, G.L. Giorgi, R. Zambrini, Entanglement dynamics of nonidentical oscillators under decohering environments. Phys. Rev. A **81**, 062117 (2010)
6. E. Knill, R. Laflamme, L. Viola, Theory of Quantum Error Correction for General Noise. Phys. Rev. Lett. **84**, 2525–2528 (2000)
7. D.A. Lidar, K.B. Whaley, Decoherence-Free Subspaces and Subsystems, in *Irreversible Quantum Dynamics*, vol. 622, ed. by F. Benatti, R. Floreanini (Springer, Berlin Heidelberg, Germany, 2003), pp. 83–120
8. G. Manzano, F. Galve, R. Zambrini, Avoiding dissipation in a system of three quantum harmonic oscillators. Phys. Rev. A **87**, 032114 (2013)
9. W.H. Zurek, Pointer basis of quantum apparatus: Into what mixture does the wave packet collapse? Phys. Rev. D **24**, 1516–1525 (1981)
10. W.H. Zurek, Environment-induced superselection rules. Phys. Rev. D **26**, 1862 (1982)
11. W.H. Zurek, Decoherence, einselection, and the quantum origins of the classical. Rev. Mod. Phys. **75**, 715–775 (2003)
12. E. Joos, H.D. Zeh, C. Kiefer, D.J.W. Giulini, J. Kupsch, I.-O. Stamatescu, *Decoherence and the Appearance of a Classical World in Quantum Theory* (Springer, Berlin, Germany, 2003)
13. M. Schlosshauer, Decoherence, the measurement problem, and interpretations of quantum mechanics. Rev. Mod. Phys. **76**, 1267 (2005)
14. M. Schlosshauer, *Decoherence and the Quantum-to-Classical Transition* (Springer, Berlin Heidelberg, Germany, 2008)
15. M.A. Nielsen, I.L. Chuang, *Quantum Computation and Quantum Information* (Cambridge University Press, Cambridge, UK, 2000)
16. S. Diehl, A. Micheli, A. Kantian, B. Kraus, H.P. Bchler, P. Zoller, Quantum states and phases in driven open quantum systems with cold atoms. Nat. Phys. **4**, 878–883 (2008)
17. F. Verstraete, M.M. Wolf, J.I. Cirac, Quantum computation and quantum-state engineering driven by dissipation. Nat. Phys. **5**, 633–636 (2009)

18. J.T. Barreiro, M. Maller, P. Schindler, D. Nigg, T. Monz, M. Chwalla, M. Hennrich, C.F. Roos, P. Zoller, R. Blatt, An open-system quantum simulator with trapped ions. Nature **470**, 486–491 (2011)
19. E. Collini, C.Y. Wong, K.E. Wilk, P.M.G. Curmi, P. Brumer, G.D. Scholes, Coherently wired light-harvesting in photosynthetic marine algae at ambient temperature. Nature **463**, 644–647 (2010)
20. A.M. Souza, D.O. Soares-Pinto, R.S. Sarthour, I.S. Oliveira, M.S. Reis, P. Brando, A.M. dos Santos, Entanglement and Bell's inequality violation above room temperature in metal carboxylates. Phys. Rev. B **79**, 054408 (2009)
21. R. Alicki, Limited thermalization for the Markov mean-field model of N atoms in thermal field. Physica A **150**, 455–461 (1988)
22. G.M. Palma, K.-A. Suominen, A.K. Ekert, Quantum Computers and Dissipation. Proc. Roy. Soc. Lon. A **452**, 567 (1996)
23. P. Zanardi, Dissipation and decoherence in a quantum register. Phys. Rev. A **57**, 3276–3284 (1998)
24. D.A. Lidar, I.L. Chuang, K.B. Whaley, Decoherence-Free Subspaces for Quantum Computation. Phys. Rev. Lett. **81**, 2594 (1998)
25. F. Galve, L.A. Pachn, D. Zueco, Bringing Entanglement to the High Temperature Limit. Phys. Rev. Lett. **105**, 180501 (2010)
26. R. Blume-Kohout, H.K. Ng, D. Poulin, L. Viola, Informationpreserving structures: A general framework for quantum zero-error information. Phys. Rev. A **82**, 062306 (2010)
27. L.-M. Duan, G.-C. Guo, Preserving Coherence in Quantum Computation by Pairing Quantum Bits. Phys. Rev. Lett. **79**, 1953 (1997)
28. D. Braun, Creation of Entanglement by Interaction with a Common Heat Bath. Phys. Rev. Lett. **89**, 277901 (2002)
29. M.S. Kim, J. Lee, D. Ahn, P.L. Knight, Entanglement induced by a single-mode heat environment. Phys. Rev. A **65**, 040101 (2002)
30. J.H. Reina, L. Quiroga, N.F. Johnson, Decoherence of quantum registers. Phys. Rev. A **65**, 032326 (2002)
31. F. Benatti, R. Floreanini, M. Piani, Environment Induced Entanglement in Markovian Dissipative Dynamics. Phys. Rev. Lett. **91**, 070402 (2003)
32. S. Schneider, G.J. Milburn, Entanglement in the steady state of a collective-angular-momentum (Dicke) model. Phys. Rev. A **65**, 042107 (2002)
33. N.B. An, J. Kim, K. Kim, Nonperturbative analysis of entanglement dynamics and control for three qubits in a common lossy cavity. Phys. Rev. A **82**, 032316 (2010)
34. F. Benatti, A. Nagy, Three qubits in a symmetric environment: Dissipatively generated asymptotic entanglement. Ann. Phys. **326**, 740–753 (2011)
35. F. Benatti, R. Floreanini, Entangling oscillators through environment noise. J. Phys. A: Math. Theor. **39**, 2689 (2006)
36. C.-H. Chou, T. Yu, B.L. Hu, Exact master equation and quantum decoherence of two coupled harmonic oscillators in a general environment. Phys. Rev. E **77**, 011112 (2008)
37. T. Zell, F. Queisser, R. Klesse, Distance Dependence of Entanglement Generation via a Bosonic Heat Bath. Phys. Rev. Lett. **102**, 160501 (2009)
38. G.-X. Li, L.-H. Sun, Z. Ficek, Multi-mode entanglement of N harmonic oscillators coupled to a non-Markovian reservoir. J. Phys. B: At. Mol. Opt. Phys. **43**, 135501 (2010)
39. M.A. de Ponte, S.S. Mizrahi, M.H.Y. Moussa, State protection under collective damping and diffusion. Phys. Rev. A **84**, 012331 (2011)
40. L.A. Correa, A.A. Valido, D. Alonso, Asymptotic discord and entanglement of nonresonant harmonic oscillators under weak and strong dissipation. Phys. Rev. A **86**, 012110 (2012)
41. G.L. Giorgi, F. Galve, G. Manzano, P. Colet, R. Zambrini, Quantum correlations and mutual synchronization. Phys. Rev. A **85**, 052101 (2012)
42. P.G. Kwiat, A.J. Berglund, J.B. Altepeter, A.G. White, Experimental Verification of Decoherence-Free Subspaces. Science **290**, 498–501 (2000)

43. L. Viola, E.M. Fortunato, M.A. Pravia, E. Knill, R. Laflamme, D.G. Cory, Experimental Realization of Noiseless Subsystems for Quantum Information Processing. Science **293**, 2059–2063 (2001)
44. D. Kielpinski, V. Meyer, M.A. Rowe, C.A. Sackett, W.M. Itano, C. Monroe, D.J. Wineland, A Decoherence-Free Quantum Memory Using Trapped Ions. Science **291**, 1013–1015 (2001)
45. T. Monz et al., Realization of Universal Ion-Trap Quantum Computation with Decoherence-Free Qubits. Phys. Rev. Lett. **103**, 200503 (2009)
46. J.F. Poyatos, J.I. Cirac, P. Zoller, Quantum Reservoir Engineering with Laser Cooled Trapped Ions. Phys. Rev. Lett. **77**, 4728–4731 (1996)
47. A.R.R. Carvalho, P. Milman, R. L. d. M. Filho, and L. Davidovich, Decoherence, Pointer Engineering, and Quantum State Protection. Phys. Rev. Lett. **86**, 4988–4991 (2001)
48. H.-P. Breuer, F. Petruccione, *The theory of open quantum systems* (Oxford University Press, New York, USA, 2002)
49. R. Bowler, J. Gaebler, Y. Lin, T.R. Tan, D. Hanneke, J.D. Jost, J.P. Home, D. Leibfried, D.J. Wineland, Coherent Diabatic Ion Transport and Separation in a Multizone Trap Array. Phys. Rev. Lett. **109**, 080502 (2012)
50. P.F. Weck, A. Safavi-Naini, P. Rabl, H.R. Sadeghpour, Microscopic model of electric-field-noise heating in ion traps. Phys. Rev. A **84**, 023412 (2011)
51. P.F. Weck, A. Safavi-Naini, P. Rabl, H.R. Sadeghpour, Erratum: Microscopic model of electric-field-noise heating in ion traps. Phys. Rev. A **84**, 069901 (2011)
52. C. Henkel, B. Horovitz, Noise from metallic surfaces: Effects of charge diffusion. Phys. Rev. A **78**, 042902 (2008)
53. G. Vidal, R.F. Werner, Computable measure of entanglement. Phys. Rev. A **65**, 032314 (2002)
54. G. Adesso, A. Serafini, F. Illuminati, Quantification and Scaling of Multipartite Entanglement in Continuous Variable Systems. Phys. Rev. Lett. **93**, 220504 (2004)
55. J. Anders, Thermal state entanglement in harmonic lattices. Phys. Rev. A **77**, 062102 (2008)
56. A. Rivas, S.F. Huelga, *Open Quantum Systems : An Introduction* (Springer, Berlin Heidelberg, Germany, 2012)
57. P. Giorda, M.G.A. Paris, Gaussian Quantum Discord. Phys. Rev. Lett. **105**, 020503 (2010)
58. G. Adesso, A. Datta, Quantum versus Classical Correlations in Gaussian States. Phys. Rev. Lett. **105**, 030501 (2010)
59. A.A. Valido, L.A. Correa, D. Alonso, Gaussian tripartite entanglement out of equilibrium. Phys. Rev. A **88**, 012309 (2013)
60. A. Serafini, A. Retzker, M.B. Plenio, Manipulating the quantum information of the radial modes of trapped ions: linear phononics, entanglement generation, quantum state transmission and nonlocality tests. New J. Phys. **11**, 023007 (2009)
61. J.M. Raimond, M. Brune, S. Haroche, Reversible Decoherence of a Mesoscopic Superposition of Field States. Phys. Rev. Lett. **79**, 1964 (1997)
62. M. Bayindir, B. Temelkuran, E. Ozbay, Tight-Binding Description of the Coupled Defect Modes in Three-Dimensional Photonic Crystals. Phys. Rev. Lett. **84**, 2140 (2000)
63. M. Mariantoni, F. Deppe, M.R. Gross, F.K. Wilhelm, E. Solano, Two-resonator circuit quantum electrodynamics: A superconducting quantum switch. Phys. Rev. B **78**, 104508 (2008)
64. M. Mariantoni et al., Photon shell game in three-resonator circuit quantum electrodynamics. Nat. Phys. **7**, 287–293 (2011)
65. A. Naik, O. Buu, M.D. LaHaye, A.D. Armour, A.A. Clerk, M.P. Blencowe, K.C. Schwab, Cooling a nanomechanical resonator with quantum back-action. Nature **443**, 193–196 (2006)
66. K.R. Brown, C. Ospelkaus, Y. Colombe, A.C. Wilson, D. Leibfried, D.J. Wineland, Coupled quantized mechanical oscillators. Nature **471**, 196–199 (2011)
67. J. Eisert, M.B. Plenio, S. Bose, J. Hartley, Towards Quantum Entanglement in Nanoelectromechanical Devices. Phys. Rev. Lett. **93**, 190402 (2004)
68. I. Fischer, R. Vicente, J.M. Buld, M. Peil, C.R. Mirasso, M.C. Torrent, J. Garca-Ojalvo, Zero-Lag Long-Range Synchronization via Dynamical Relaying. Phys. Rev. Lett. **97**, 123902 (2006)

Chapter 6
Dissipative Complex Quantum Networks

Most of the classical literature about synchronization phenomena in networks deals with self-sustained phase oscillators in Kuramoto-type models, or with identical nonlinear oscillators studied through the master stability formalism [1]. We instead continue focusing on synchronization during the relaxation dynamics of different linear oscillators driven out of equilibrium and exploring the key role of dissipation. A first step to characterize quantum spontaneous synchronization, considering quantum fluctuations and correlations beyond the classical limit, has been considered in Chap. 4 where synchronization between one pair of damped quantum harmonic oscillators has been reported. We have already seen that, depending on the damping, a pair of oscillators with different frequencies can exhibit synchronous evolution emerging after a transient, as well as robust (slowly decaying) non-classical correlations [2]. This connection has been extended in Chap. 5, where we showed that synchronization may occur between three oscillators or in a single pair depending on the symmetries of the system [3], discussing both transient and relaxation effects.

In this chapter we extend our analysis to the dynamical properties of arbitrary networks of quantum harmonic oscillators dissipating into the environment.[1] In particular, we focus in the spontaneous synchronization phenomena occurring in the network and the possibility of preserving or even generating quantum correlations between some of its components by engineering of its normal modes. This is a relevant feature, as long as a first approximation to a great variety of controllable quantum systems, such as electromagnetic modes [5–7], trapped ions [8, 9] or nanoelectromechanical resonators [10], is given by a set of coupled quantum harmonic oscillators, susceptible to experience spontaneous synchronization. Moreover, beyond physical systems, there is an increasing awareness that quantum phenomena might play an important role in terms of efficiency of biological processes [11–14].

As we pointed in Sect. 2.4, the form in which dissipation occurs in a spatially extended system has deep consequences. We stress that the importance of symmetries present in the system-bath coupling has been recognized in many contexts. In

[1]The results presented in this chapter have been published in Ref. [4].

G. Manzano Paule, *Thermodynamics and Synchronization in Open Quantum Systems*, Springer Theses, https://doi.org/10.1007/978-3-319-93964-3_6

classical systems, this fundamental issue was already discussed in the seminal work of Lord Rayleigh analyzing damping effects on normal modes in vibrating systems [15]. Indeed, the role of dissipation to reduce detrimental effects of vibrations is fundamental in many areas of mechanical, civil and aerospace engineering [16, 17]. On the other hand, in the context of quantum systems, symmetries in the coupling between qubits and the environment allow for decoherence-free subspaces [18], entangled states preparation [19, 20] and dissipative quantum computing [21, 22]. When several dissipative quantum oscillators coupled in a network are considered dissipation can act globally or locally (in a node) and, depending on the correlation length in the bath with respect to the size of the system, a variety of surprising phenomena are observed.

In the following sections, we show that the distribution and form of losses through the network amounts to synchronous dynamics in spite of the nodes diversity, witnessing the presence of robust quantum correlations as measured by quantum discord and entanglement (see Sects. 1.4.1 and 1.4.3). More importantly, we find that synchronization can actually be induced by local tuning of one (even newly attached) oscillator of a generic (regular or random) network and derive precise conditions for its emergence in both the whole network or in an arbitrary part of it. Stemming both from the structure of the system and from the form of system-bath coupling we further show the possibility to tune the system to configurations in which nodes do not thermalize and relax into a synchronous and non-classical asymptotic state.

The chapter is organized as follows. In Sect. 6.1 we present our approach for modeling dissipative networks of quantum harmonic oscillators, discussing the characterization of synchronization phenomena occurring on it. We then consider the possibility of collective synchronization of all the oscillators in the network by tuning one of them in Sect. 6.2. In Sect. 6.3 we instead characterize the conditions for obtaining synchronization only in a cluster of nodes embedded in the network. Furthermore, we show in Sect. 6.4 how steady entanglement can be generated between unlinked nodes by properly coupling them to a network. We finalize in Sect. 6.5 by presenting our general conclusions and discussing the implications of our results. Some further details can be found in Appendix.

6.1 Dissipation Mechanisms and Synchronization

We consider generic networks of N non-resonant, coupled quantum harmonic oscillators, given by the Hamiltonian

$$\hat{\mathrm{H}}_S = \frac{1}{2}\left(\mathbf{p}^T \mathbb{1} \mathbf{p} + \mathbf{q}^T \mathcal{H} \mathbf{q}\right) \tag{6.1}$$

where $\mathbf{q}^T = (\hat{q}_1, \ldots, \hat{q}_N)$ is the vector of canonical position operators and $\mathbf{p}$ are momenta, satisfying $[\hat{q}_j, \hat{p}_k] = i\hbar\delta_{jk}\hat{\mathbb{1}}$, and $\mathcal{H}_{m,n} = \omega_m^2\delta_{mn} + \lambda_{mn}(1 - \delta_{mn})$ is the matrix containing the topological properties of the network (frequencies ω_m and couplings λ_{mn}). The eigenmodes (or normal modes) of the system, $\mathbf{Q}$, result from diagonalization of this Hamiltonian through the transformation matrix $\mathcal{F}$, this is, $\mathbf{Q} = \mathcal{F}^T \mathbf{q}$, which defines the diagonal matrix $\Omega = \mathcal{F}^T \mathcal{H} \mathcal{F}$ containing the (squared) normal modes frequencies.

Any realistic model of the network needs to include also environment effects [23–25] and different forms of dissipation can be envisaged for an extended network. For example we may consider that all units dissipate into separate identical baths (SB), Fig. 6.1a, as early decoherence models of quantum registers [26] or cavity optical modes [23, 25]. Otherwise, we may assume that all the nodes feel a "similar" dissipation (see Sect. 2.4). This common bath (CB) scenario, Fig. 6.1b, is known to create decoherence-free subspaces [18], noiseless subsystems [27, 28], and asymptotic entanglement [29–34], as we have analyzed in detail for a system of three harmonic oscillators in Chap. 5. A third, limiting, case of a local bath (LB) in which a specific oscillator d dissipates much faster than any other node [Fig. 6.1b] is also considered here.

In a microscopic description with independent oscillators modeling the environment, the system-bath interaction Hamiltonian for SB takes the form

$$\hat{H}_I^{SB} = -\gamma_0 \sum_{m=1}^{N} \hat{q}_m \hat{B}^{(m)}, \text{ with } \hat{B}^{(m)} = \sum_{\alpha=1}^{\infty} \lambda_\alpha \hat{X}_\alpha^{(m)}, \tag{6.2}$$

being γ_0 the system-bath coupling strength (explicitly shown for the ease of understanding), $\hat{X}_\alpha^{(m)}$ the position operators for each environment oscillator α (representing for instance a vibrational mode, or an optical one, etc…) of the bath $\hat{B}^{(m)}$ in which the network unit (m) is dissipating. As mentioned before, this situation occurs when the coherence length of the environment is smaller than the spatial extension of the

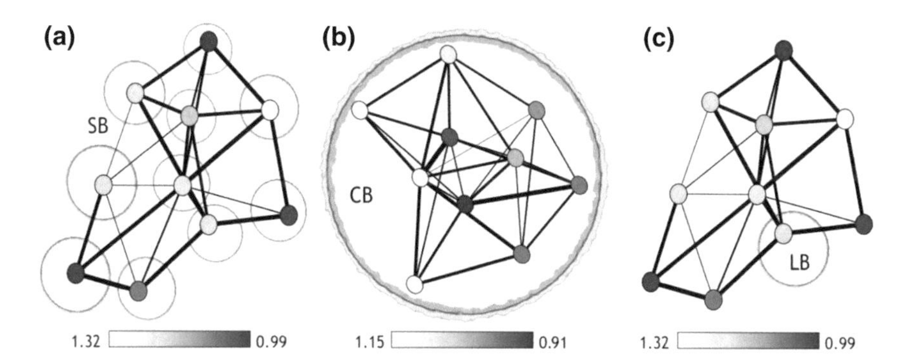

Fig. 6.1 **a** Network of oscillators (represented by the network nodes) dissipating into separate baths (SB, represented by the gray circles surrounding the nodes). Links, representing couplings, have different strengths (lines thickness) and nodes have different natural frequencies (corresponding to different colors as given in the color bar). **b** Network of oscillators dissipating into a common bath (CB). **c** Network of oscillators with dissipation restricted to one node, local bath (LB)

system. In the opposite case, a common bath is seen by all oscillators, resulting in an interaction Hamiltonian

$$\hat{H}_I^{CB} = -\gamma_0 \sum_{m=1}^{N} \hat{q}_m \hat{B}, \tag{6.3}$$

and actually involving only the average position (here the center of mass) of the network, Fig. 6.1b. Notice that in the normal modes basis

$$\hat{H}_I^{CB} = -\gamma_0 \sum_{m} \kappa_m \hat{Q}_m \hat{B} \quad \text{, with } \kappa_m = \sum_{n} \mathfrak{F}_{nm}. \tag{6.4}$$

The *effective couplings* κ_m are different and determined by characteristics of the network such as topology, coupling strengths, and frequencies, encoded in the diagonalization matrix $\mathfrak{F}$. This is in stark contrast to the case of identical SB (6.2) where all normal modes have equal effective couplings to the baths. To see this, notice that we can transform the bath operators $\hat{X}_\alpha^{(m)}$ to a new basis which exactly cancels the transformation $\mathfrak{F}$; these new "oscillators" can be shown to have the same statistical properties as the others, thus resulting in equivalent heat baths. Finally, the case of a given node d dissipating much faster than any other is modeled by

$$\hat{H}_I^{LB} = -\gamma_0 \hat{q}_d \hat{B}. \tag{6.5}$$

This local bath (LB) situation does also lead to non-uniform environment interaction in some of the normal modes with effective couplings κ_m:

$$\hat{H}_I^{LB} = -\gamma_0 \sum_{m} \kappa_m \hat{Q}_m \hat{B} \quad \text{, with } \kappa_m = \mathfrak{F}_{dm}. \tag{6.6}$$

One of the key insights of our work comes from noting that the coupling of real oscillators to the bath (taken here to be equal, γ_0) differs from those of eigenmodes. The latter are found to be $\gamma_0\kappa_m$ (with $\kappa_m^{CB} = \sum_n \mathfrak{F}_{nm}$, $\kappa_m^{SB} = 1$ and $\kappa_m^{LB} = \mathfrak{F}_{dm}$) meaning that, except the SB situation, the eigenmodes have different decay rates. Then for CB and LB only the least dissipative eigenmode will survive to thermalization, thus governing the motion of all oscillators overlapping with it. It is then useful identifying the less dissipating normal mode with smallest *effective coupling*, κ_σ, and also the next one, κ_η, such that $|\kappa_\sigma| \leq |\kappa_\eta|$.

A standard procedure allows us to obtain the evolution of the reduced density matrix for the state of the system, this is, the network of different oscillators (see Sect. 2.4). After a (post-trace) rotating wave approximation, the master equations in the weak coupling limit for separate, common, and local baths are in the Lindblad form, guarantying a well-behaved system dynamics. These equations are obtained by generalization of the problem of two and three dissipative coupled oscillators addressed in Chaps. 4 and 5. For the purpose of our analysis it is interesting to consider the master equation in the normal modes basis

$$\begin{aligned}\frac{d\rho(t)}{dt} &= -\frac{i}{\hbar}[\hat{H}_S, \rho(t)] \\ &-\frac{1}{4\hbar^2}\sum_n i\Gamma_n\Big([\hat{Q}_n, \{\hat{P}_n, \rho(t)\}] - [\hat{P}_n, \{\hat{Q}_n, \rho(t)\}]\Big) \\ &+ D_n\left([\hat{Q}_n, [\hat{Q}_n, \rho(t)]] - \frac{1}{\Omega_n^2}[\hat{P}_n, [\hat{P}_n, \rho(t)]]\right)\end{aligned} \tag{6.7}$$

where Ω_n are the normal modes frequencies of $\hat{H}_S$. The damping and diffusion coefficients read

$$\Gamma_n = \kappa_n^2\hbar\gamma_0, \quad D_n = \kappa_n^2\hbar\gamma_0\Omega_n \coth\left(\frac{\hbar\Omega_n}{2k_BT}\right), \tag{6.8}$$

for an Ohmic bath at temperature T with spectral density $J(\omega) = (2\gamma_0/\pi)\omega\Theta(\Lambda - \omega)$, $\Theta(x)$ being the Heaviside step function, $\Lambda \gg \Omega_n\ \forall n$ the frequency cutoff, and κ_n the effective couplings [see Eqs. (6.4) and (6.6)]. With the appropriate definition of the couplings, this equation is valid both for common and local bath, while for SB we have: $\Gamma_n = \gamma_0$ and $D_n = \gamma_0\Omega_n \coth(\frac{\Omega_n}{2T})$ i.e. we obtain the same damping coefficient for all normal modes. The main differences between our three models of dissipation reside in these expressions for the master equation coefficients that will produce different friction terms in the equations of motion, determining collective or individuals behaviors (the complete set of equations of motion is given in Appendix). We stress that the choice of this master-equation representation is not critical for our main conclusions, as in the case of two oscillators (see Appendix A.2 in Chap. 4).

Knowledge of the normal modes of a complex network and of their dissipation rates (or effective couplings) allows to fully characterize a large variety of phenomena. Indeed this is a simple but powerful approach, even if diagonalization of the problem needs to be performed numerically except in a few (highly symmetric) configurations. By diagonalizing matrix $\mathcal{F}$ and system-bath interaction Hamiltonian we obtain the conditions to have a dominating mode during a transient. This mode dissipating most slowly, $|\kappa_\sigma| < |\kappa_j|\ \forall j \neq \sigma$, is found either for CB or LB. Even more important, it is possible to identify normal modes completely protected against dissipation, which do not thermalize. Indeed, a normal mode σ is protected against decoherence if $\kappa_\sigma = 0$. For a pair of oscillators interacting with a CB, this condition is accomplished only in the trivial case of identical frequencies [31, 34] but this is not the case when more than two nodes are considered [3]. We find that asymptotic synchronized quantum states can then be observed even in random networks where all nodes have different natural frequencies.

Full characterization of the quantum state evolution of the network comes from moments of all orders of the oscillator operators $\hat{q}_j$ and $\hat{p}_j$ [25]. We start considering the classical limit given by the expectation values of positions and momenta, in virtue of Ehrenfest theorem [35]. For SB, average positions (and momenta) are characterized by irregular oscillations before thermalization [Fig. 6.2a]. On the other hand, for dissipation in CB and after a transient, regular phase locked oscillations

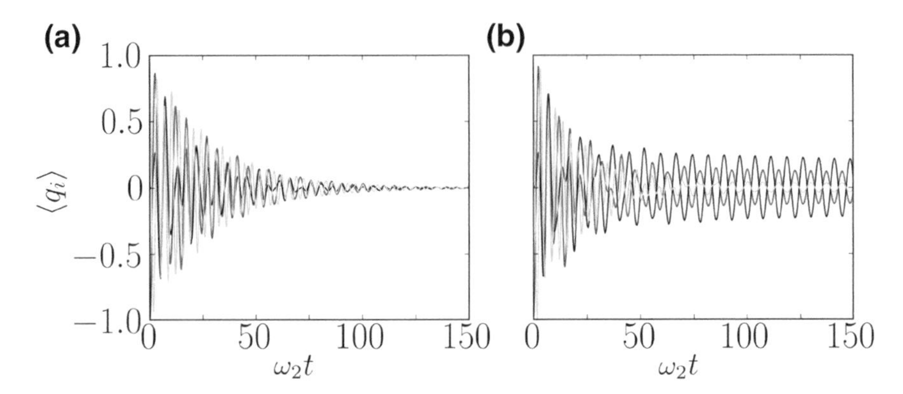

Fig. 6.2 First order moments for initial conditions $-\langle \hat{q}_1 \rangle = \langle \hat{q}_3 \rangle = 1.0$, $\langle \hat{q}_2 \rangle = 0.0$, and vanishing momenta in the case of an open chain of three oscillators with $\omega_1 = 1.2\omega_2$, $\omega_3 = 1.8\omega_2$, non-vanishing couplings $\lambda_{12} = \lambda_{23} = 0.4\ \omega_2^2$, temperature $T = 10k_B^{-1}\hbar\omega_2$, $\gamma_0 = 0.07\omega_2^2$, bath cutoff $50\omega_2$, for SB (**a**) and for CB (**b**)

can arise, as shown in Fig. 6.2b. Synchronization between detuned nodes can be found during a rather long and slow relaxation, like in the case of just one pair [2] (see Chap. 4). Further, the oscillations can remain robust even asymptotically if the condition $\kappa_\sigma = 0$ is satisfied.

Beyond the classical limit given by average positions and momenta, let us now consider the full quantum dynamics stemming from the evolution of higher moments (see details in Appendix). At the microscopic level, quantum fluctuations also oscillate in time (even for initial vacuum states for which first order moments vanish at any time). This collective periodic motion is associated to a slow energy decay and witnesses the presence of robust quantum correlations against decoherence [24]. Our approach points to a wide range of appealing possibilities in quantum networks. In the following we show how a whole random network (or a part of it) can be brought to a synchronized state retaining quantum correlations via local tuning of just one of the nodes, or how two external oscillators can be linked to a random network leading to their entanglement and locked oscillations.

Synchronization between two time series $f(t)$ and $g(t)$ can be characterized by using the Pearson indicator $C_{f,g}$ already employed in Chaps. 4 and 5 [c.f. Eq. (4.13)]. For "similar" and in-phase (anti-phase) evolutions $C \sim 1$ (-1), while it tends to vanish otherwise. As a figure of merit for global synchronization in the whole network we look at the product (neglecting the sign) of this indicator for all pairs of oscillators in the system. When the time series correspond to positions second moments we have

$$\mathcal{S} = \Pi_{i<j} |C_{\langle \hat{q}_i^2 \rangle \langle \hat{q}_j^2 \rangle}|. \tag{6.9}$$

This collective synchronization factor $\mathcal{S}$ can reach unit value only in presence of synchronous dynamics between all the pairs of oscillators in the network.

6.2 Collective Synchronization by Tuning One Oscillator

Let us consider an Erdös–Rényi dissipative network [1] of oscillators with different node frequencies, links and weights (Fig. 6.1). In the Erdös–Rényi random graph model the network is constructed from a given number of nodes by choosing each of the possible links between any pair of nodes with a fixed probability [36, 37]. Here we further choose frequencies and coupling strengths randomly from a suitable range of values. We focus on the relaxation dynamics of energy and quantum correlations. The node dynamics is mostly incoherent and even if initializing the network in a non-classical state, quantum correlations generally disappear due to decoherence [24]. Independently on the form of the network, for dissipation in SB, all nodes thermalize on a time scale γ_0^{-1} [see Eq. (6.2)]. As anticipated before, this is not the case in the presence of a dissipation acting not-uniformly within the network.

6.2.1 Common Dissipation Bath

In presence of CB, an arbitrary network of N nodes can reach a synchronized state before thermalization if there is a weak effective coupling κ_σ. As a matter of fact just by tuning one of the node frequencies, ω_v, even maintaining fixed the rest of the network frequencies $\{\omega_{l\neq v}\}$ and its topology (λ_{ij} couplings) it is possible to decrease the weakest coupling κ_σ. One may wonder from the important consequences of this feature, as it means that an extra oscillator of properly selected frequency $\{\omega_{l\neq v}\}$ (like a synchronizer) can be added to a random network, even if weakly coupled, and it will lead to a collective synchronization of the whole system at some frequency (Ω_σ), generally different from ω_v. Fig. 6.3 displays the average global synchronization and quantum correlations established in the network. Synchronization arises after a transient across the whole network by tuning one of the frequencies ω_v to a particular value $\bar{\omega}_v$, while it is not present when moving a few percent away from this value. Equivalently one could have tuned one of the couplings $\lambda_{vv'}$. In the following we consider separately the case in which κ_σ is significantly smaller than the other effective couplings and the case in which it vanishes.

Conditions for global synchronization follow from small ratio between the damping rates of the two slowest normal modes of the network $R = \kappa_\sigma/\kappa_\eta \to 0$. Interestingly, this is a necessary but not sufficient condition for *collective* synchronization. This is due to the fact that the presence of a slowly dissipating normal mode needs to be accompanied by a significant overlap between this mode ($\hat{Q}_\sigma$), or virtual oscillator, and all the real ones ($\hat{q}_1, \ldots, \hat{q}_N$). An analytical estimation of the synchronization time must hence take into account both the importance (overlap with individual oscillator) and decay of few normal modes in the system. The contributions of the different normal modes to the motion of second moments can be schematically written (for the position) as

$$\langle q_n^2\rangle(t) = \frac{\mathcal{F}_{n0}}{2}^2 e^{-\Gamma_0 t} g_0(2\Omega_0 t) + \frac{\mathcal{F}_{n1}}{2}^2 e^{-\Gamma_1 t} g_1(2\Omega_1 t) \\ + \frac{\mathcal{F}_{n0}\mathcal{F}_{n1}}{2} e^{-\frac{\Gamma_0+\Gamma_1}{2}t} g_{01}(|\Omega_0 + \Omega_1|t, |\Omega_0 - \Omega_1|t) + \ \ldots \tag{6.10}$$

and the damping coefficients Γ_i are labeled in increasing order 0, 1, 2, ..., N, from the minimum to the maximum (positive) value, while the g_i functions represent oscillating terms whose amplitude are determined by the initial conditions. Assuming that $\Gamma_0 >> \Gamma_1$, that is $R \ll 1$, synchronization is achieved when the contributions other than the first one can be neglected in this dynamical evolution. By equating the maximum amplitudes of each normal mode contribution $k = 1, 2, 3, \ldots$ to the first one (labeled by 0) we obtain the time for oscillator j to start oscillating at the less damped frequency Ω_σ

$$t^{(j)} \equiv \max_{\{k\neq\sigma\}} 2\frac{\log \mathcal{F}_{jk} - \log \mathcal{F}_{j\sigma}}{\Gamma_k - \Gamma_\sigma}\,, \tag{6.11}$$

where maximization is over all normal modes k different from the slowest one σ. This expression corresponds to the minimum time for which the network oscillator (j) starts oscillating at the synchronization frequency Ω_0, the eigenfrequency of the less-damped mode. Collective synchronization time hence corresponds to

$$t_{\text{sync}} = \max_{\{j\}}\{t^{(j)}\}, \tag{6.12}$$

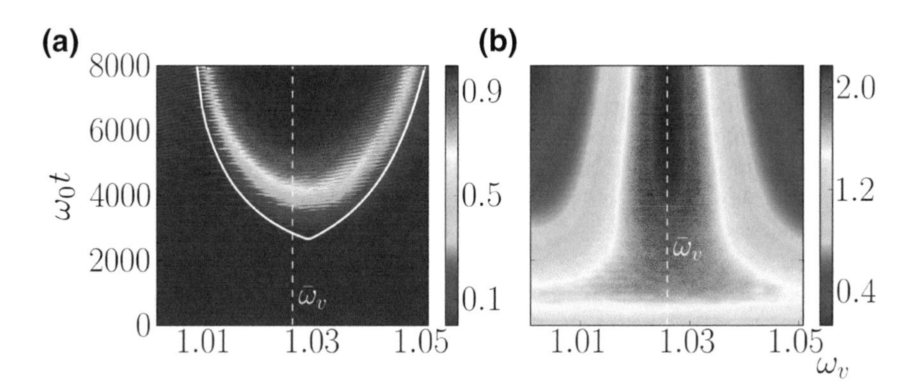

Fig. 6.3 **a** Time evolution of the collective synchronization factor $\mathcal{S}$, and **b** quantum correlations quantified by the discord between pairs of oscillators $\langle\delta\rangle \times 10^3$, when varying one node frequency ω_v. Results are shown for a random network (connection probability $p = 0.6$) of 10 oscillators. Frequencies of nodes are sampled from a uniform distribution from $0.9\omega_0$ to $1.2\omega_0$ and couplings from a Gaussian distribution around $-0.1\omega_0^2$ with standard deviation $0.05\omega_0^2$. Collective synchronization $\mathcal{S}$ and (averaged and filtered) discord δ are obtained considering all oscillator pairs of the network. Dashed line identifies the frequency $\bar{\omega}_v$ for which $\kappa_\sigma = 0$. Continuous line in **a** corresponds to the estimated synchronization time t_{sync}

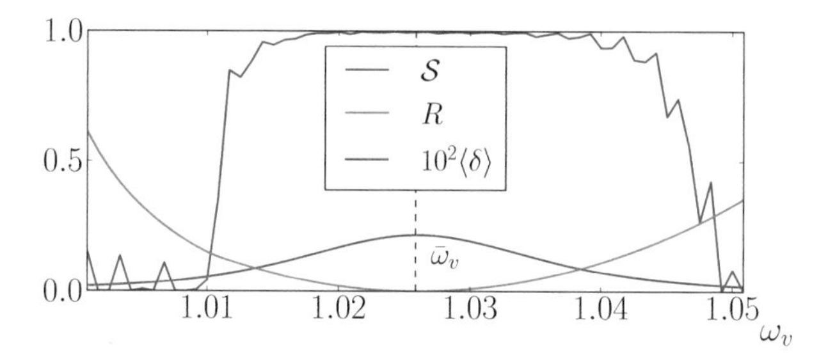

Fig. 6.4 Collective synchronization $\mathcal{S}$, ratio between the two smallest damping rates R and mean discord between pair of oscillators $\langle\delta\rangle(\times 10^2)$ at long times $(8000\omega_0^{-1})$ for the same random network of 10 oscillators and probability connection $p = 0.6$. The dashed line indicates the tuning value $\omega_v = \bar{\omega}_v$ for which the mode σ decouples from the bath

that is, when even the last oscillator joins the synchronous dynamics dominated by the less damped mode. Then phase-locking in the evolution of all oscillators, namely in their (all order) moments, can arise before thermalization, when there is significant separation between largest time decays Γ_η, Γ_σ, and overlap between slowest normal modes and each system node. Global network synchronization obtained from the full dynamical evolution and the estimated synchronization time t_{sync} are in good agreement, as seen in Fig. 6.3a.

We now look at the quantumness of the state in presence of collective synchronization. Generally decoherence is independent of specific features such as the oscillation frequency in a system [38]. Still, synchronization is a consequence of a reduced dissipation in some system mode and indeed witnesses the robustness of quantum correlations, as evident form the average quantum discord between pairs of oscillators in the network, $\langle\delta\rangle$, represented in Fig. 6.3b. For the same network, the ratio between the two smallest damping rates R, the collective synchronization $\mathcal{S}$ in Eq. (6.9), and average discord between pairs of oscillators $\langle\delta\rangle$ at long times $(t = 8000\omega_0^{-1})$ can be seen in Fig. 6.4. After a transient dynamics in which the couplings in the network create quantum correlations [10], even when starting from separable states, discord does actually decay to small values for ω_v different from $\bar{\omega}_v$ (non-synchronized network) while it maintains large values for the case of a properly tuned node $(\omega_v \sim \bar{\omega}_v)$ for which synchronization $\mathcal{S}$ reaches its maximum.

The case $\omega_v = \bar{\omega}_v$, leading to $\kappa_\sigma = 0$, needs special attention. After a transient all the nodes will oscillate at a locked common frequency, the one of the undamped normal mode Ω_σ, which we call a *frozen* mode. As before [Eq. (6.11)], the possibility to synchronize the whole network also requires a second condition, namely that the undamped mode involves all the network nodes. The case in which the latter condition applies only to some nodes is discussed below. When both the conditions

$$\kappa_\sigma = \sum_{k=1}^{N} \mathcal{F}_{k\sigma} = 0, \quad \text{and} \quad \mathcal{F}_{k\sigma} \neq 0 \ \forall k, \tag{6.13}$$

are met, there is a frozen normal mode linking all oscillators. This leads to collective synchronization in the whole network and allows for mutual information and quantum correlations remaining strong even asymptotically, being orders of magnitude larger than for the fully thermalized state, when synchronization is not present (Fig. 6.3). The undamped mode gives actually rise to a decoherence-free dynamics for the whole system of oscillators where quantum correlations and mutual information survive.

The phenomena above are found for nodes dissipating at equal rates into a CB, while in the presence of N independent environments (SB) all oscillators thermalize incoherently, synchronization is not found, and decoherence times for all oscillators are of the same order. As a final observation we mention the special case in which the center of mass of the system is one of the normal modes; then there will be a large decoherence-free subspace (corresponding to the other $N-1$ modes) but no synchronization will appear for a CB.

6.2.2 Local Dissipation Bath

Common and separate baths correspond to two extreme situations in which all oscillators have equivalent interactions with the environment(s). We now consider the case of a local bath, as a limit case in which one oscillator is dissipating stronger, Fig. 6.1c. Here collective synchronization requires that the frozen normal mode σ must not overlap with the dissipative oscillator (labeled by d) while involving all the other nodes ($\mathcal{F}_{i\sigma} \neq 0 \; \forall \; i \neq d$). Then, synchronization of the whole network (except for the dissipative oscillator) arises. This occurs when

$$\mathcal{F}_{d\sigma} = 0, \quad \text{with} \quad \mathcal{F}_{dj} \neq 0 \quad \forall j \neq \sigma, \tag{6.14}$$

meaning that the undamped mode σ involves a cluster of oscillators not including the lossy one. We find synchronization and robust quantum effects across the network as for CB, with the difference that for LB the dissipating node is now excluded.

In Fig. 6.5a we show the ratio R between the two weakest effective couplings when varying the frequency in one node (ω_ν, $\nu \neq 1$) comparing the cases of LB and CB. We see that, depending on the frequency of the tuned node, the necessary condition for synchronization $R \sim 0$ can be satisfied in presence of both dissipation mechanisms. Still, even for identical networks, the "tuned" node produces synchronization for different frequencies (ω_ν) depending on whether dissipation takes place in a CB or in a LB. We also find that for a largely detuned oscillator ($|\omega_\nu - \omega_j| \gg 0 \, \forall j$), the rest of the network becomes rather insensitive to its frequency [see the behavior of R at small and large frequencies in Fig. 6.5a]. This can be expected as for strong detuning the respective dynamics of the $N-1$ network and of the ν oscillator tends to be effectively decoupled, the latter becoming one of the normal modes. We notice that for LB, in this large detuning limit, there is a normal mode ν orthogonal to the dissipating one (here node d) so that it will not dissipate ($\kappa_\nu = 0$) leading to a vanishing ratio

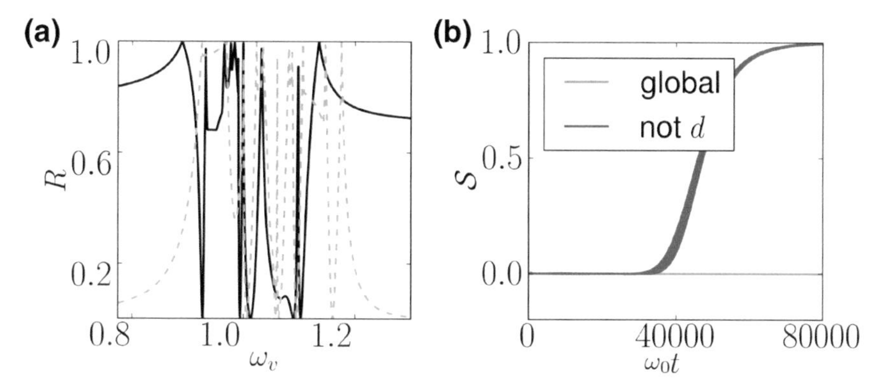

Fig. 6.5 **a** Ratio R between the less-damped modes for a random network (connection probability $p = 0.6$) of 10 nodes coupled to a common bath (black continuous line) and coupled to a local bath (orange dashed line) when the tuning frequency ω_v is varied. **b** Collective synchronization factor $\mathcal{S}$ for a local bath when the dissipative oscillator is included into the factor $\mathcal{S}$ (line labeled "global") or it is not included (line labeled "not d"). Other parameters are $\gamma_0 = 0.01\omega_0^2$, $T = 10k_b^{-1}\hbar\omega_0$ and $\Lambda = 50\omega_0$

$R = \kappa_\sigma/\kappa_\eta$. Still, this node will be frozen but there is no synchronization of a whole cluster, as conditions $\mathcal{F}_{d\sigma} = 0$ with $\mathcal{F}_{dj} \neq 0 \, \forall \, j \neq \sigma$ are not satisfied.

Figure 6.5b shows that the measure $\mathcal{S}$ indicates synchronization for $\omega_v = \bar{\omega}_v$ when it includes contributions from all nodes excluding the lossy one, while synchronization is not evidenced when also this node is taken into account in the calculation of $\mathcal{S}$. Figure 6.5b also shows that the time required for emergence of collective synchronization is larger for local dissipation than for dissipation through the center of mass (see CB in Fig. 6.3) by a factor of N, as expected as here we have one (instead of N) dissipation channels.

6.3 Synchronization of Clusters and Linear Motifs

The possibility to synchronize a whole network, in presence of different dissipation mechanisms, just by tuning one local parameter opens-up the perspective of control that can be explored considering the dynamical variation of a control-node frequency. In particular we find similar qualitative results both for random networks and for disordered lattices consisting of regular networks with inhomogeneous frequencies and couplings, being the latter largely studied in ultracold atomic gases [39]. Local tuning to collective synchronization is not only a general feature of different kind of networks but can also be established in motifs within the network. As we show in the following, the system can be tuned to a partial synchronization, involving some nodes of a network independently of the rest of it. Indeed, even if the whole system is coupled to a CB, we can identify the conditions for having a synchronized cluster, like the 3-node linear motif considered in Fig. 6.6.

The main conditions for emergence of synchronization in a localized cluster of a network can be derived as follows: first consider the existence of a normal mode $\hat{Q}_\sigma$ that involves only the cluster oscillators, i.e.

$$\hat{Q}_\sigma = \sum_{k \in \mathcal{C}_M} \mathcal{F}_{k\sigma} \hat{q}_k \,, \qquad \mathcal{F}_{k\sigma} \neq 0, \tag{6.15}$$

where the cluster is denoted by the ensemble $\mathcal{C}_M$ of M oscillators in the network. We impose that this mode is a normal mode of the cluster by writing

$$(\Omega_\sigma^2 - \mathcal{H}_M)\mathbf{C}_\sigma = 0, \tag{6.16}$$

where $\mathcal{H}_M$ is the corresponding Hamiltonian matrix of the cluster $\mathcal{C}_M$, and $\mathbf{C}_\sigma$ is a column vector containing the $\mathcal{F}_{k\sigma}$ coefficients of Eq. (6.15).

Now we can calculate (numerically in general and analytically for some particular situations) the coefficients of $\mathbf{C}_\sigma$ and the frequency Ω_σ in terms of the cluster parameters. This is done straightforwardly by using Eq. (6.16) and the orthonormality condition for $\mathbf{C}_M$. In order to have (asymptotically surviving) synchronization for both common and local baths we have to impose $\kappa_\sigma = 0$, that yields to a relation among the cluster coefficients. Then by tuning only one parameter of the cluster we can obtain the desired synchronization. However note that in the case of a local bath this later condition is nothing but stating that the oscillator which is locally coupled to the bath cannot pertain to the cluster ensemble.

Finally we have to check that the normal mode which links the cluster involves only the cluster oscillators and induces a collective motion in the cluster (when $\kappa_\sigma \sim 0$) different to the rest of the network, that is:

$$(\Omega_\sigma^2 - \mathcal{H})\mathbf{C}_\sigma = 0, \tag{6.17}$$

where we complete $\mathbf{C}_\sigma$ with zeros in the positions of oscillators others than the cluster ones. Note that the last equation is equivalent to Eq. (6.16) when the next condition is fulfilled:

$$\sum_{k \in \mathcal{C}_M} \mathcal{F}_{k\sigma} \lambda_{kj} = 0 \,, \qquad \forall j \notin \mathcal{C}_M, \tag{6.18}$$

that fixes a relation for the couplings between the cluster and any other oscillator outside. We can point out from the last equation that the cluster must be coupled to any other oscillator at least by a pair of coupling terms. This analysis is valid in general even when the cluster is considered to be the whole network. The only variation in this case is that we do not have to ensure any more the condition (6.18) as long as we do not consider anything outside the cluster.

In Fig. 6.6 we consider the synchronization of a three-oscillator linear motif, $\mathcal{C}_1$, for which analytical expressions can be derived (see the details in Appendix A.3). Here

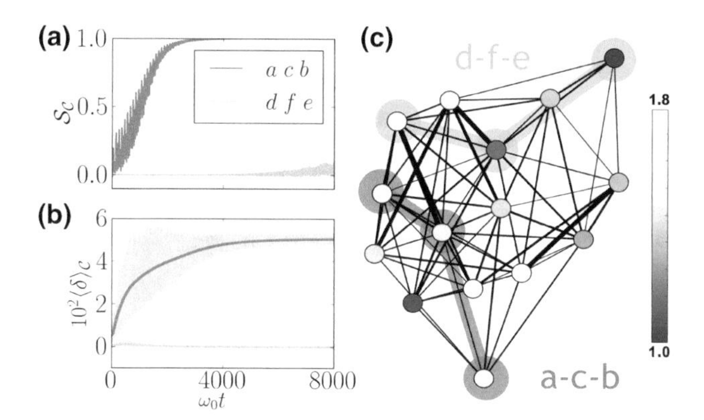

Fig. 6.6 **a** Synchronization factors $\mathcal{S}_\mathcal{C}$ and **b** average discord $\delta_\mathcal{C} \times 10^2$ evaluated for linear 3-node motifs (hence the subindex $\mathcal{C}$) in a random network (connection probability $p = 0.6$) of 15 oscillators (**c**). A tuned non-dissipative motif $\mathcal{C}_1$ composed by the three nodes $a - c - b$ is compared with another equivalent non-tuned motif $\mathcal{C}_2$ composed by nodes $d - f - e$. Frequencies in the network are sampled from a uniform distribution from ω_0 to $1.8\omega_0$, and couplings with a Gaussian distribution around $-0.1\omega_0^2$ with standard deviation $0.05\ \omega_0^2$. In order to avoid dissipation in the $a - c - b$ motif we have set $\omega_c = 1.51\ \omega_0$, being $\lambda_{ac} = -0.09\ \omega_0^2$ and $\lambda_{bc} = -0.11\ \omega_0^2$

two non-directly linked nodes a and b of the motif are asymptotically synchronized through another one, here c, and this leads to a common oscillation dynamics along the whole motif *a-c-b*. The condition for synchronization of the motif, namely its dependence on a frozen normal mode with frequency Ω_σ, reads from the above expressions

$$\frac{\lambda_{ac}}{\Omega_\sigma^2 - \omega_a^2} + \frac{\lambda_{bc}}{\Omega_\sigma^2 - \omega_b^2} = -1. \tag{6.19}$$

This case is an example of the general result stating that given any network, a part of it (in our case a linear motif, $\mathcal{C}_1$) can be synchronized by tuning one of its components, for instance a frequency or coupling of the motif. A key point is that this is independent of the frequencies and links of the rest of the network, provided the motif is properly embedded in the network. The links between $\mathcal{C}_1$ and the rest of the network should satisfy

$$\left(\frac{\lambda_{ac}}{\Omega_s^2 - \omega_a^2}\right)\lambda_{aj} + \left(\frac{\lambda_{bc}}{\Omega_s^2 - \omega_b^2}\right)\lambda_{bj} + \lambda_{cj} = 0\ ,\ \forall j\ . \tag{6.20}$$

This is equivalent to saying that a synchronized motif with robust quantum correlations can preserve these features when linked to an *arbitrary* network, if some constraints on the reciprocal links are satisfied. For instance, each node of the synchronized motif needs to share with the rest of the network more than one link. In Fig. 6.6 we compare the behavior of two linear motifs of a large network, where a first motif $\mathcal{C}_1$ is synchronized, satisfying Eqs. (6.19)–(6.20) while the second one $\mathcal{C}_2$

is not. After a transient a frozen mode tames the dynamics of $\mathcal{C}_1$, which then shows a synchronous evolution and robust correlations. It can also be shown that quantum purity and energy reach higher values of a stationary non-thermal state (see also Chap. 5). This is compared with the non-synchronized motif $\mathcal{C}_2$ whose dynamics quickly relaxes to a thermal state. The case of a three oscillator chain is an example showing the possibility to tune synchronization and quantum effects in a motif within the network when the proper link conditions are satisfied.

6.4 Entangling Two Oscillators Through a Network

The same technique discussed in the previous section can be applied to the case in which we aim to synchronize few, even if not directly connected, elements of a network. But this does not mean that any set of nodes can be synchronized asymptotically. In fact, we find that for a CB we can only synchronize two different and not directly linked ($\lambda_{ab} = 0$) oscillators if we synchronize also other intermediate linked elements (like in the linear motif example, Fig. 6.6) or when these two oscillators are identical, which is the case that we discuss in the present section.

We consider the case of two identical oscillators (i.e. with $\omega_a = \omega_b$) prepared in a separable state, with some local squeezing. They are not directly coupled ($\lambda_{ab} = 0$) but are connected through an arbitrary network. In general they will dissipate reaching the thermal state, but with the proper conditions we find an important result: because their frequencies are identical it is possible to construct a frozen normal mode involving only these two nodes, given by $\hat{Q}_\sigma = \mathcal{F}_{a\sigma}\ \hat{q}_a + \mathcal{F}_{b\sigma}\ \hat{q}_b$, with $\mathcal{F}_{a\sigma}, \mathcal{F}_{b\sigma} \neq 0$ and this can be obtained, for instance, by a proper choice of their coupling to the network. In other words, it is possible to have both oscillators relaxing onto a frozen mode, so that they will be synchronized and will keep a higher energy than otherwise. Most importantly, in this case entanglement can actually be generated between oscillators initially in a separable state and remains high asymptotically.

In order to entangle the oscillators, their coupling to the rest of the network needs to fulfill the following condition [similar to Eq. (6.20)]

$$\sum_{k=a,b} \mathcal{F}_{k\sigma} \lambda_{kj} = 0, \tag{6.21}$$

which is achieved by proper tuning of coupling strengths of the active links (j) with the rest of the network $\lambda_{aj}, \lambda_{bj}$. As in Chap. 5 we quantify entanglement throughout the logarithmic negativity $E_{\mathcal{N}} = \max(0, -\ln \nu_-)$, with ν_- the smallest symplectic eigenvalue of the partially transposed density matrix [40] (see also Sect. 1.4 in Chap. 1). In Fig. 6.7a and b we show the evolution of energy and entanglement of the oscillators a and b when linked to a random network. As we see in Fig. 6.7c there is not direct link between a and b ($\lambda_{ab} = 0$) and the whole system dissipates in a common environment. The case where the oscillators a and b are coupled to the network following the prescription (6.21) is compared to another case in which their links are

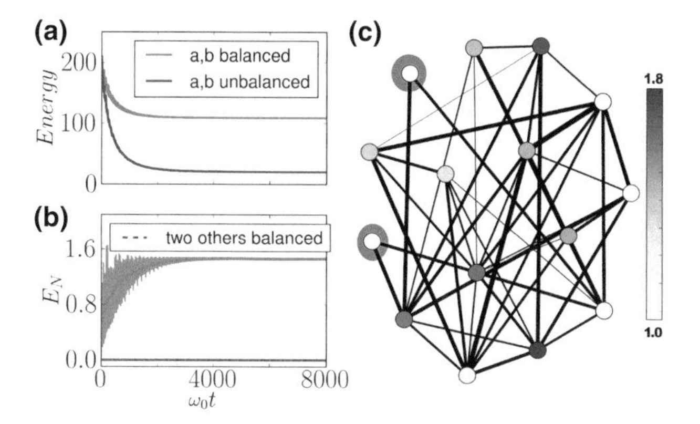

Fig. 6.7 **a** Energy evolution and **b** entanglement (logarithmic negativity) between two nodes with identical frequency ω_0 [we call these nodes a and b and are plotted in red in the network displayed in **c**]. The network is a random one (connection probability $p = 0.6$ of 15 oscillators and same frequency and couplings distribution as in Fig. 6.6. We compare the situations in which the couplings from the red nodes to the rest of the network (they are directly connected to other nodes called c and d) are properly balanced in order to avoid dissipation ($\lambda_{ac} = \lambda_{bc} = -0.15\omega_0^2$ and $\lambda_{ad} = \lambda_{bd} = -0.12\omega_0^2$) with the case when this balance is perturbed ($\lambda_{ac} + 0.04\omega_0^2$ and $\lambda_{ad} + 0.04\omega_0^2$) .The third line in **b** shows the entanglement between other two arbitrary oscillators in the situation in which a and b are balanced

not properly balanced (we slightly change the coupling strengths). Both energy and entanglement are shown to be sensitive to the structure of the reciprocal links and the possibility to actually bring the added nodes into an entangled state that will survive asymptotically is guaranteed by Eq. (6.21). The importance of this result is twofold: in terms of applications it shows that it is possible to dynamically generate entanglement between two non-linked nodes embedded in a random network by tuning their connections to it, and on the other hand it enlarges the scenario for asymptotic entanglement generation through the environment. It is known that large entanglement can be generated between far oscillators during a transient due to a sudden-switch [10] or through parametric driving [41]. On the other hand, a common environment leads to entanglement between a pair of spins [18] or oscillators [31]. In our case, the network may be seen as a structured part of the reservoir, whose engineering allows to dynamically produce entanglement between the otherwise uncoupled oscillators.

6.5 Conclusions

Our results on synchronization in dissipative harmonic networks and its optimization give a flavor of all the possibilities that show up once the mechanism behind the phenomenon is understood. At difference from widely considered self-sustained non-linear oscillators, here we focused on a linear system showing synchronization after a transient for dissipation processes introducing inhomogeneous decay rates among the system normal modes. A synchronous oscillation is predicted either in a long transient during relaxation to the equilibrium state or in a stationary non-thermal state, extending the results in Chaps. 4 and 5. We considered the most significant examples of systems exposed to collective dissipation (CB) and of a node of the network more strongly exposed to dissipation (LB), displaying synchronous dynamics. On the other hand, for independent environments (SB) on different nodes the resulting dynamics remains incoherent even when increasing the strength of the reciprocal couplings in the network.

The presence of synchronization in the whole or a part of the network witnesses the survival of quantum correlations and entanglement between the involved nodes. This connection between a coherent oscillation in the network and its non-classical state is a powerful result in the context of complex quantum systems, considering the abundance of this phenomenon. Indeed, the condition underlying synchronization provides a strategy to protect a system subspace from decoherence. Our discussion and methodological approach are general, but we show specific consequences of our analysis, such as global or partial synchronization in a network through local tuning in one node (synchronizer) as well as the possibility of connecting two nodes (not linked between them) to a network and synchronize and entangle them, even starting from separable states. Even if the reported results refer to random networks, our analysis applies to generic ones, also including homogeneous and disordered lattices and do not require all-to-all connectivity.

In some sense, tuning part of a network so that the rest of it reaches a synchronous, highly correlated state can be seen as a kind of reservoir engineering, where here the tuned part of the network would be a part of the reservoir. This is to be compared with recent proposals of dissipative engineering for quantum information, where special actions are performed to target a desired non-classical state [19–22]. In the context of quantum communications and considering recent results on quantum Internet [42, 43], our study can offer some insight in designing a network with coherent information transport properties. Furthermore, implications of our approach can be explored in the context of efficient transport in biological systems [12, 13]. An interesting methodological connection is also with transport through (classical) random networks [44]. On the other hand, our analysis, when restricted to the classical limit, also gives some insight about vibrations in an engineering context, providing the conditions for undamped normal modes and their effect [15–17]. This is at the basis of the recent analysis of synchronization in closed systems reported in Ref. [45].

The formalism presented above could be applied in principle in more complex settings such as the production of independently synchronized parts of the system beating at different frequencies, or in the presence of more complicated dissipation situations (for example several local baths of different strengths). Though probably more difficult to analyze mathematically, the conceptual structure and the methodology to be followed in such cases is equivalent to that presented here.

Appendix

A.1 Master Equation for Nodes

From the master equation in the basis of normal modes after the (post-trace) rotating wave approximation, given in Sect. 6.1 [Eq. (6.7)] (see also Appendix A.2 in Chap. 5), we can derive an equivalent expression turning to the basis of the original oscillators by simply applying the change of basis matrix $\mathcal{F}$, defined by diagonalization of $\mathcal{H}$. Rearranging terms one obtains

$$\begin{aligned}\frac{d\rho(t)}{dt} &= -i[\hat{H}_S, \rho(t)] - \\ &- \frac{1}{4\hbar^2}\sum_{jk} i\tilde{\Gamma}_{jk}\left([\hat{q}_j, \{\hat{p}_k, \rho(t)\}] - [\hat{p}_k, \{\hat{q}_j, \rho(t)\}]\right) + \\ &+ \tilde{D}^a_{jk}[\hat{q}_j, [\hat{q}_k, \rho(t)]] - \tilde{D}^b_{jk}[\hat{p}_j, [\hat{p}_k, \rho(t)]]. \end{aligned} \tag{A.1}$$

Here we have introduced new master equation coefficients denoted by a tilde and defined from the previous ones as

$$\tilde{\Gamma}_{jk} = \sum_n \mathcal{F}_{jn}\mathcal{F}_{kn}\Gamma_n, \tag{A.2}$$

$$\tilde{D}^a_{jk} = \sum_n \mathcal{F}_{jn}\mathcal{F}_{kn}D_n, \tag{A.3}$$

$$\tilde{D}^b_{jk} = \sum_n \mathcal{F}_{jn}\mathcal{F}_{kn}\frac{D_n}{\Omega_n^2}. \tag{A.4}$$

Those are valid for all the cases considered in the paper, namely, common bath, local bath and separate baths, with the proper definitions of the untilded coefficients for each case (see Sect. 6.1). Note however that for the case of separate baths (assuming an Ohmic frequency spectral distribution with sharp cutoff in the bath) the damping coefficients in the master equation reduce simply to $\tilde{\Gamma}_{ij} = \gamma\delta_{ij}$, i.e. all the nodes in the network dissipate through their own bath at the same rate, determined by the equivalence of the separate baths. This further simplification in the case of separate baths marks its difference from the common or local bath cases, producing a different structure for the friction terms in the equations of motion, as we will see in the next sections of this Appendix.

A.2 Equations for the First- and Second-Order Moments

For Gaussian states, the full dynamics of the oscillators is embedded in the first- and second-order moments [25] and the former give the classical limit of this quantum system, obtained neglecting quantum fluctuations. From the master equation we obtain the evolution of the first-order moments

$$\frac{d}{dt}\langle\hat{Q}_n\rangle = \langle\hat{P}_n\rangle - \frac{\Gamma_n}{2\hbar}\langle\hat{Q}_n\rangle, \tag{A.5}$$

$$\frac{d}{dt}\langle\hat{P}_n\rangle = -\Omega_n^2\langle\hat{Q}_n\rangle - \frac{\Gamma_n}{2\hbar}\langle\hat{P}_n\rangle, \tag{A.6}$$

where the first term corresponds to the free evolution of uncoupled oscillators and the second one is a damping term stemming from the influence of the bath. For the second order moments we obtain the more complicated expressions:

$$\begin{aligned}\frac{d}{dt}\langle\hat{Q}_n\hat{Q}_m\rangle &= \frac{1}{2}\langle\{\hat{Q}_n,\hat{P}_m\} + \{\hat{P}_n,\hat{Q}_m\}\rangle \\ &\quad - \left(\frac{\Gamma_n+\Gamma_m}{2\hbar}\right)\langle\hat{Q}_n\hat{Q}_m\rangle + D_n\frac{\delta_{nm}}{2\Omega_n^2},\end{aligned} \tag{A.7}$$

$$\begin{aligned}\frac{d}{dt}\langle\hat{P}_n\hat{P}_m\rangle &= -\frac{\Omega_n^2}{2}\langle\{\hat{Q}_n,\hat{P}_m\}\rangle - \frac{\Omega_m^2}{2}\langle\{\hat{Q}_m,\hat{P}_n\}\rangle \\ &\quad - \left(\frac{\Gamma_n+\Gamma_m}{2\hbar}\right)\langle\hat{P}_n\hat{P}_m\rangle + D_n\frac{\delta_{nm}}{2},\end{aligned} \tag{A.8}$$

$$\begin{aligned}\frac{d}{dt}\langle\{\hat{Q}_n,\hat{P}_m\}\rangle &= 2\langle\hat{P}_n\hat{P}_m\rangle - 2\Omega_m^2\langle\hat{Q}_n\hat{Q}_m\rangle \\ &\quad - \left(\frac{\Gamma_n+\Gamma_m}{2\hbar}\right)\langle\{\hat{Q}_n,\hat{P}_m\}\rangle,\end{aligned} \tag{A.9}$$

where the first two terms arise from the reduced motion of the free normal modes, and the last ones are induced by the environmental action, which combines damping and diffusion effects.

We also notice that a common environment gives rise to a rather symmetric damping, also known as diffusive coupling (apart from an irrelevant change of sign) [2]. This kind of diffusive coupling is a typical phenomenological assumption when synchronization is modeled in classical systems [46]. This can be seen by looking at the first order moments, for which we obtain different expressions in the case of common, local and separate baths. In the first two cases we have

$$\frac{d}{dt}\langle\hat{q}_n\rangle = \langle\hat{p}_n\rangle - \frac{1}{2\hbar}\sum_k\tilde{\Gamma}_{nk}\langle\hat{q}_k\rangle, \tag{A.10}$$

$$\frac{d}{dt}\langle\hat{p}_n\rangle = -\omega_n^2\langle\hat{q}_n\rangle - \sum_k\lambda_{nk}\langle\hat{q}_k\rangle - \frac{1}{2\hbar}\sum_k\tilde{\Gamma}_{nk}\langle\hat{p}_k\rangle. \tag{A.11}$$

while for the separate baths case the expression transforms into:

$$\frac{d}{dt}\langle q_n \rangle = \langle \hat{p}_n \rangle - \frac{1}{2\hbar}\tilde{\Gamma}\langle \hat{q}_n \rangle, \tag{A.12}$$

$$\frac{d}{dt}\langle \hat{p}_n \rangle = -\omega_n^2 \langle \hat{q}_n \rangle - \sum_k \lambda_{nk} \langle \hat{q}_k \rangle - \frac{1}{2\hbar}\tilde{\Gamma}\langle \hat{p}_n \rangle. \tag{A.13}$$

It is immediately seen that the presence of a common bath, a local bath or N separate (even if identical) baths, leads to different friction terms in the dynamical equations. While the damping of oscillators in the common and local bath cases depends on all the network oscillators weighed by the effective couplings (κ_n^2) through the tilded damping coefficients of Eq. (A.2), in the separate bath case each oscillator decays independently from the rest of the network, being coupled only through the Hamiltonian part of the dynamical evolution.

A.3 Three-Oscillator Motif Details

Here we give the analytical expressions for the synchronization of the three-oscillators linear motif, i.e. an open chain of three oscillators embedded in a bigger network. We are able to give the specific parameter relations that have to be fulfilled in order to obtain a non-dissipative mode, that is, to make the effective coupling for a motif mode $\kappa_\sigma = 0$.

By solving Eq. (6.16) for this particular case, we obtain:

$$\mathcal{F}_{a\sigma} = C\left(\frac{\lambda_{ac}}{\Omega_\sigma^2 - \omega_a^2}\right), \tag{A.1}$$

$$\mathcal{F}_{b\sigma} = C\left(\frac{\lambda_{bc}}{\Omega_\sigma^2 - \omega_b^2}\right), \tag{A.2}$$

$$\mathcal{F}_{c\sigma} = C, \tag{A.3}$$

where $C^2 = 1/\left(1 + \left(\frac{\lambda_{ac}}{\Omega_\sigma^2 - \omega_a^2}\right)^2 + \left(\frac{\lambda_{bc}}{\Omega_\sigma^2 - \omega_b^2}\right)^2\right)$.

Now we can obtain a explicit expression for the effective coupling of the normal mode Q_σ to the heat bath:

$$\kappa_\sigma = C\left(1 + \frac{\lambda_{ac}}{\Omega_\sigma^2 - \omega_a^2} + \frac{\lambda_{bc}}{\Omega_\sigma^2 - \omega_b^2}\right), \tag{A.4}$$

that enables a dissipation-free channel, i.e. no coupling with the bath ($\kappa_\sigma = 0$) when

$$\frac{\lambda_{ac}}{\Omega_\sigma^2 - \omega_a^2} + \frac{\lambda_{bc}}{\Omega_\sigma^2 - \omega_b^2} = -1. \tag{A.5}$$

This last condition gives another different expression for the synchronization frequency in this regime:

$$\Omega_\sigma^2 = \frac{\omega_a^2+\omega_b^2}{2} - \frac{\lambda_{ac}+\lambda_{bc}}{2} \pm \sqrt{\left(\frac{\omega_a^2-\omega_b^2}{2}\right)^2 + \left(\frac{\lambda_{ac}+\lambda_{bc}}{2}\right)^2 - \frac{(\omega_a^2-\omega_b^2)(\lambda_{ac}-\lambda_{bc})}{2}}, \tag{A.6}$$

where we have to check that Ω_σ^2 is real and positive, i.e. that $(\omega_a^2-\omega_b^2)^2+(\lambda_{ac}+\lambda_{bc})^2 > 2(\omega_a^2-\omega_b^2)(\lambda_{ac}-\lambda_{bc})$.

From the explicit expression of Ω_σ and the previous equations, a consistency relation for the selected natural frequencies and coupling of the a, b and c oscillators follows by substituting the expression of Ω_σ^2 into the equation

$$\Omega_\sigma^2 - \omega_c^2 = \frac{\lambda_{ac}^2}{\Omega_\sigma^2-\omega_a^2} + \frac{\lambda_{bc}^2}{\Omega_\sigma^2-\omega_b^2}, \tag{A.7}$$

whose solution for λ_{ac}, is

$$\lambda_{ac} = \frac{\lambda_{bc}^2-\lambda_{bc}(\omega_a^2-\omega_b^2)}{2\lambda_{bc}-\omega_a^2+\omega_c^2} \pm \frac{(\lambda_{bc}-\omega_a^2+\omega_c^2)\sqrt{\lambda_{bc}^2-(\omega_a^2-\omega_b^2)(\omega_b^2-\omega_c^2)}}{2\lambda_{bc}-\omega_a^2+\omega_c^2}, \tag{A.8}$$

corresponding to two different branches of solutions. These two bran-ches intersect when we have that $\lambda_{bc}=\omega_a^2-\omega_c^2$ or equivalently $\lambda_{ac}=\omega_b^2-\omega_c^2$, in this case we have the simpler relation for the couplings $\lambda_{ac}-\lambda_{bc}=\omega_b^2-\omega_a^2$ and here the mode $\hat{Q}_\sigma$ is degenerated, i.e. there are two non-dissipative normal modes with different frequencies. It is worth noticing that when we have different branches it is necessary to impose the condition $\lambda_{bc}^2 > (\omega_a^2-\omega_b^2)(\omega_b^2-\omega_c^2)$ in order to obtain λ_{ac} real.

References

1. A. Arenas, A. Diaz-Guilera, J. Kurths, Y. Moreno, C. Zhou, Synchronization in complex networks. Phys. Rep. **469**, 93–153 (2008)
2. G.L. Giorgi, F. Galve, G. Manzano, P. Colet, R. Zambrini, Quantum correlations and mutual synchronization. Phys. Rev. A **85**, 052101 (2012)
3. G. Manzano, F. Galve, R. Zambrini, Avoiding dissipation in a system of three quantum harmonic oscillators. Phys. Rev. A **87**, 032114 (2013)
4. G. Manzano, F. Galve, G.-L. Giorgi, E. Hernndez-Garcia, R. Zambrini, Synchronization, quantum correlations and entanglement in oscillator networks. Sci. Rep. **3**, 1439 (2013)
5. J.M. Raimond, M. Brune, S. Haroche, Reversible decoherence of a mesoscopic superposition of field states. Phys. Rev. Lett. **79**, 1964 (1997)

6. M. Bayindir, B. Temelkuran, E. Ozbay, Tight-binding description of the coupled defect modes in three-dimensional photonic crystals. Phys. Rev. Lett. **84**, 2140 (2000)
7. M. Mariantoni et al., Photon shell game in three-resonator circuit quantum electrodynamics. Nat. Phys. **7**, 287–293 (2011)
8. K.R. Brown, C. Ospelkaus, Y. Colombe, A.C. Wilson, D. Leibfried, D.J. Wineland, Coupled quantized mechanical oscillators. Nature **471**, 196–199 (2011)
9. M. Harlander, R. Lechner, M. Brownnutt, R. Blatt, W. Hansel, Trapped-ion antennae for the transmission of quantum information. Nature **471**, 200–203 (2011)
10. J. Eisert, M.B. Plenio, S. Bose, J. Hartley, Towards quantum entanglement in nanoelectromechanical devices. Phys. Rev. Lett. **93**, 190402 (2004)
11. E.M. Gauger, E. Rieper, J.J.L. Morton, S.C. Benjamin, V. Vedral, Sustained quantum coherence and entanglement in the avian compass. Phys. Rev. Lett. **106**, 040503 (2011)
12. G. Panitchayangkoon, D.V. Voronine, D. Abramavicius, J.R. Caram, N.H.C. Lewis, S. Mukamel, G.S. Engel, Direct evidence of quantum transport in photosynthetic light-harvesting complexes. Proc. Natl. Acad. Sci. **108**, 20908–20912 (2011)
13. G.S. Engel, T.R. Calhoun, E.L. Read, T.-K. Ahn, T. Mančal, Y.-C. Cheng, R.E. Blankenship, G.R. Fleming, Evidence for wavelike energy transfer through quantum coherence in photosynthetic systems. Nature **446**, 782–786 (2007)
14. N. Lambert, Y.-N. Chen, Y.-C. Cheng, C.-M. Li, G.-Y. Chen, F. Nori, Quantum biology. Nat. Phys. **9**, 10–18 (2013)
15. J.W.S. Rayleigh, *The Theory of Sound* (Dover Publishers, New York, 1945)
16. S. Adhikari, Damping models for structural vibration, Ph.D. thesis, University of Cambridge, UK (2000)
17. W.J. Bottega, *Engineering Vibrations* (CRC Taylor & Francis, New York, 2006)
18. D.A. Lidar, K.B. Whaley, Decoherence-free subspaces and subsystems, in *Irreversible Quantum Dynamics*, vol. 622, ed. by F. Benatti, R. Floreanini (Springer, Berlin, 2003), pp. 83–120
19. S. Diehl, A. Micheli, A. Kantian, B. Kraus, H.P. Buchler, P. Zoller, Quantum states and phases in driven open quantum systems with cold atoms. Nat. Phys. **4**, 878–883 (2008)
20. J.T. Barreiro, P. Schindler, O. Guhne, T. Monz, M. Chwalla, C.F. Roos, M. Hennrich, R. Blatt, Experimental multiparticle entanglement dynamics induced by decoherence. Nat. Phys. **6**, 943–946 (2010)
21. F. Verstraete, M.M. Wolf, J.I. Cirac, Quantum computation and quantum-state engineering driven by dissipation. Nat. Phys. **5**, 633–636 (2009)
22. J.T. Barreiro, M. Muller, P. Schindler, D. Nigg, T. Monz, M. Chwalla, M. Hennrich, C.F. Roos, P. Zoller, R. Blatt, An open-system quantum simulator with trapped ions. Nature **470**, 486–491 (2011)
23. U. Weiss, *Quantum Dissipative Systems* (World Scientific, Singapore, 2008)
24. M. Schlosshauer, *Decoherence and the Quantum-to-Classical Transition* (Springer, Berlin, 2008)
25. C. Gardiner, P. Zoller, *Quantum Noise*, 3rd edn. (Springer, Berlin, 2004)
26. W.G. Unruh, Maintaining coherence in quantum computers. Phys. Rev. A **51**, 992–997 (1995)
27. E. Knill, R. Laflamme, L. Viola, Theory of quantum error correction for general noise. Phys. Rev. Lett. **84**, 2525–2528 (2000)
28. L. Viola, E.M. Fortunato, M.A. Pravia, E. Knill, R. Laflamme, D.G. Cory, Experimental realization of noiseless subsystems for quantum information processing. Science **293**, 2059–2063 (2001)
29. J.S. Prauzner-Bechcicki, Two-mode squeezed vacuum state coupled to the common thermal reservoir. J. Phys. A Math. Gen. **37**, L173 (2004)
30. K.-L. Liu, H.-S. Goan, Non-Markovian entanglement dynamics of quantum continuous variable systems in thermal environments. Phys. Rev. A **76**, 022312 (2007)
31. J.P. Paz, A.J. Roncaglia, Dynamics of the entanglement between two oscillators in the same environment. Phys. Rev. Lett. **100**, 220401 (2008)
32. C.-H. Chou, T. Yu, B.L. Hu, Exact master equation and quantum decoherence of two coupled harmonic oscillators in a general environment. Phys. Rev. E **77**, 011112 (2008)

33. T. Zell, F. Queisser, R. Klesse, Distance dependence of entanglement generation via a bosonic heat bath. Phys. Rev. Lett. **102**, 160501 (2009)
34. F. Galve, G.L. Giorgi, R. Zambrini, Entanglement dynamics of nonidentical oscillators under decohering environments. Phys. Rev. A **81**, 062117 (2010)
35. G. Auletta, M. Fortunato, G. Parisi, *Quantum Mechanics into a Modern Perspective* (Cambridge University Press, New York, 2009)
36. P. Erdos, A. Renyi, On random graphs I. Publ. Math. Debr. **6**, 290–297 (1959)
37. E.N. Gilbert, Random graphs. Ann. Math. Stat. **30**, 1141–1144 (1959)
38. A.O. Caldeira, A.J. Leggett, Influence of damping on quantum interference: an exactly soluble model. Phys. Rev. A **31**, 1059 (1985)
39. L. Sanchez-Palencia, M. Lewenstein, Disordered quantum gases under control. Nat. Phys. **6**, 87–95 (2010)
40. R. Horodecki, P. Horodecki, M. Horodecki, K. Horodecki, Quantum entanglement. Rev. Mod. Phys. **81**, 865–942 (2009)
41. F. Galve, L.A. Pachon, D. Zueco, Bringing entanglement to the high temperature limit. Phys. Rev. Lett. **105**, 180501 (2010)
42. H.J. Kimble, The quantum internet. Nature **453**, 1023–1030 (2008)
43. S. Ritter, C. Nlleke, C. Hahn, A. Reiserer, A. Neuzner, M. Uphoff, M. Mucke, E. Figueroa, J. Bochmann, G. Rempe, An elementary quantum network of single atoms in optical cavities. Nature **484**, 195–200 (2012)
44. M. Kim, Y. Choi, C. Yoon, W. Choi, J. Kim, Q.-H. Park, W. Choi, Maximal energy transport through disordered media with the implementation of transmission eigenchannels. Nat. Photonics **6**, 581–585 (2012)
45. C. Benedetti, F. Galve, A. Mandarino, M.G.A. Paris, R. Zambrini, Minimal model for spontaneous quantum synchronization. Phys. Rev. A **94**, 052118 (2016)
46. A. Pikovsky, M. Rosenblum, J. Kurths, *in Synchronization* (A Universal Concept in Nonlinear Sciences (Cambridge University Press, Cambridge, 2001)

Part III
Quantum Fluctuation Theorems and Entropy Production

Chapter 7
Fluctuation Theorems for Quantum Maps

Part II of the thesis has been devoted to the study of dynamical properties of dissipative quantum many-body systems, where we employed open quantum system theory and its related quantum information tools in order to predict the emergence of synchronization phenomena and unveil the behavior of quantum correlations. In the remaining of the thesis we will turn our view to the nonequilibrium thermodynamic properties of open quantum systems, using the full methods introduced in Chaps. 1 and 2.

In Chap. 3 we introduced quantum thermodynamics as an emergent field at the intersection of quantum information theory and small-scale thermodynamics. There we stated that one of the major achievements of the latter is the introduction of a thermodynamic description of single particles subjected to fluctuations in out of equilibrium situations. In this context we have seen that one of the main fruitful tools are a set of universal relations known as *fluctuation theorems* (see Sect. 3.2). Some of them have been recently extended to the quantum realm, such as the Crooks fluctuation theorem, the Jarzynski equality, or various fluctuation theorems for the exchange of heat and particles between equilibrium reservoirs at different temperatures and chemical potentials [1, 2]. Along the present Part III of the thesis, we present and apply a systematic approach for the development of quantum fluctuation theorems in open systems, which allows the reproduction of many previous results in a generalized framework, while opening new possibilities for the study and understanding of genuine quantum effects in thermodynamics.

Most of the fluctuation theorems developed for open quantum systems are based on Hamiltonian dynamics, which require a microscopic model for the system, the environment and their interaction. The most common framework assumes a *two measurement protocol* (TMP) in which the whole system starting in equilibrium conditions is measured before and after some relevant process occurs, the environment is composed by one of more equilibrium thermal reservoirs, the interaction between system and environment is fixed and usually weak, and the driving operates only on the open system [1] (see Sect. 3.2). Although this kind of framework has

G. Manzano Paule, *Thermodynamics and Synchronization in Open Quantum Systems*, Springer Theses, https://doi.org/10.1007/978-3-319-93964-3_7

produced important results, it turns out that abandoning one or few of the above assumptions in order to obtain generalized results constitutes in general a complicated problem, which sometimes may lead to alterations in the results or directly the breaking of the fluctuation theorems [3, 4] (see also Ref. [5]). Furthermore, the inclusion of projective measurements on the whole thermal reservoir representing the environment are unpractical. A promising alternative route is the development of quantum fluctuation theorems for open system dynamics described by completely-positive and trace-preserving (CPTP) maps, as they provide a compact description of physical processes condensing the main effects of the environmental action in a set of few relevant variables. As we have seen in Sect. 2.1, CPTP maps capture a vast diversity of quantum dynamical evolutions, including arbitrary open system dynamics such as decoherence, measurement, and thermal relaxation [6–9]. Consequently, the thermodynamic analysis of processes described by CPTP maps can be considered as a major issue in the development of quantum thermodynamics [10–14].

In this chapter we present a new fluctuation theorem valid for a broad class of quantum CPTP maps.[1] It is based in the concept of a nonequilibrium potential, an intrinsic fluctuating property of the map which allows the thermodynamic description at the single trajectory level in most situations of interest. The chapter is organized as follows. In Sect. 7.1 we review previous work in fluctuation theorems in the context of CPTP maps, highlighting the limitations of previous approaches and the novel features introduced here. Next, in Sect. 7.2 after a brief review of CPTP maps and the Kraus representation, we introduce the nonequilibrium potential and the dual-reverse map, necessary to state the fluctuation theorem. In Sect. 7.3 we prove the general theorem for single maps and for a series of concatenated maps. Some applications are discussed in Sect. 7.4. Finally, in Sect. 7.5 we summarize our results and present the main conclusions of the chapter.

7.1 Fluctuation Theorems, Unital Maps and Beyond

In recent years, there have been several derivations of fluctuations theorems for specific classes of CPTP maps falling into two broad categories: detailed fluctuation theorems for quantum trajectories and fluctuation theorems for thermodynamic variables, such as work and entropy. Campisi et al. obtained a detailed fluctuation theorem for a unitary, driven evolution punctuated by *unital* maps —maps for which the identity matrix is invariant— [1, 16]. This work was followed and extended by Watanabe et al. [17]. Quantum Markov semigroups were explored by Crooks using a time-reversed or dual map [18], which was then applied by Horowitz et al. to nonequilibrium quantum jump trajectories [13, 19]. An alternative, operator formulation for driven Lindblad master equations was independently developed by Chetrite and Mallick [20], and its equivalence to the quantum jump approach was investigated by Liu [21, 22].

[1] Most of the results in the chapter have been published in Ref. [15].

Fluctuation theorems for thermodynamic quantities in the TMP —where the dynamics are described by the specific set of unital (or bistochastic) CPTP maps— have also appeared in some recent works [23, 24] where unitality plays an equivalent role as microreversibility do in closed evolutions. However, when non-unital CPTP maps are considered the usual form of the fluctuation theorem is broken by the appearance of a so-called *correction* [25]. In such case fluctuation theorems in integral from have been derived

$$\langle e^{-\sigma} \rangle = a, \tag{7.1}$$

where σ is some quantity of interest such as energy change, heat, or information-theoretic entropy, and a is a process dependent correction factor [23, 25–27]. In the following sections we will see that this correction factor can be abandoned by means of a generalized detailed balance relation involving the quantum operations in which the map can always be decomposed, i.e. by means of its Kraus operator-sum representation. This allows us to derive a general fluctuation theorem in both detailed and integral forms that includes and extends many of the previous results. The quantity σ can then be given a clear interpretation as a trajectory version of the entropy production [28] in most setups of physical interest. As a consequence, our result also clarifies the minimal hypotheses needed to derive a fluctuation theorem for quantum maps. Furthermore, our theorem is independent of the physical nature of the process that induces the CPTP map. This is a relevant feature as it makes the fluctuation theorem general enough to be applied to situations far from equilibrium, like systems in contact with coherent or squeezed reservoirs [29–32]. Moreover, such a general result could be useful to analyze the thermodynamics of quantum processes whose physical details are not completely known, such as decoherence or quantum collapse.

In order to derive this general fluctuation theorem for CPTP maps, we introduce the concept of a nonequilibrium potential, which is proportional to the logarithm of the invariant density matrix associated to the map. This potential is the analogue of the one used by Hatano and Sasa in a classical context [33], and it has been implicitly used by Sagawa in Ref. [11] for quantum maps as well as Yukawa [10] and Spohn [34] for continuous-time quantum dynamics. It vanishes for unital maps and coincides with the heat flow between the system and the reservoir in the case of thermalization maps. For classical systems subjected to nonequilibrium constraints, this potential allows one to split the entropy production into adiabatic and non-adiabatic contributions [see Eq. (3.57)], and is also a key ingredient to characterize the response of a system to external time-dependent perturbations in the linear regime [35].

7.2 Quantum Operations and Dual-Reverse Dynamics

Consider a generic CPTP quantum map $\rho \to \rho' \equiv \mathcal{E}(\rho)$ acting on the density matrix ρ of a quantum system. As we have seen in Sect. 2.1, any CPTP map admits a (non-unique) Kraus representation in terms of a collection of linear operators $\{\hat{M}_k\}$ as [7]

$$\mathcal{E}(\rho) = \sum_k \mathcal{E}_k(\rho) = \sum_k \hat{M}_k \rho \hat{M}_k^\dagger, \tag{7.2}$$

with $\sum_k \hat{M}_k^\dagger \hat{M}_k = \hat{\mathbb{1}}$ ensuring the trace-preserving property of the map $\mathcal{E}$. If the Hilbert space of the system has finite dimension N, there exists a Kraus representation for any map with at most N^2 operators. However, using more than N^2 operators is sometimes necessary for a complete description of the physical process associated to the map (as we will see below).

7.2.1 Quantum Trajectories and Unconditional States

The Kraus representation (7.2) is not just a mathematical way of writing the map; it also provides a physical picture of the map as a set of random transformation of pure states. A specific representation decomposes the map into a number of operations $\mathcal{E}_k(\rho) = \hat{M}_k \rho \hat{M}_k^\dagger$. Each operation transforms a pure state $|\psi\rangle$ into a new pure state

$$|\psi_k'\rangle = \frac{\hat{M}_k|\psi\rangle}{||\hat{M}_k|\psi\rangle||}, \tag{7.3}$$

with probability $p_k(|\psi\rangle) \equiv ||\hat{M}_k|\psi\rangle||^2$ ($\sum_k p_k(|\psi\rangle) = 1$). This picture extends to mixed states of the form $\rho = \sum_i p_i |\psi_i\rangle\langle\psi_i|$, which represents a classical ensemble of pure states $|\psi_i\rangle$ each sampled with probability p_i (see Sect. 1.1). Thus, the probability that operation k occurs is in general given by $p_k(\rho) = \text{Tr}[\mathcal{E}_k(\rho)]$ and the final state conditioned on this operation is $\rho_k' = \mathcal{E}_k(\rho)/p_k(\rho)$, c.f. Eq. (2.14) in Sect. 2.1.2. If we know which operation $\mathcal{E}_k$ has occurred, then k can be seen as the outcome of a generalized measurement and ρ_k' as the conditional post-measurement state of the system. If we do not know which operation took place (or we decide not to incorporate that information into our description), then the state after the transformation is $\rho' = \mathcal{E}(\rho) = \sum_k p_k(\rho)\rho_k'$, usually referred to as the unconditional post-measurement state, although the transformation given by the map $\rho' = \mathcal{E}(\rho)$ does not necessarily imply any measurement and not even a specific Kraus representation. This setup defines an *efficient generalized measurement* in quantum mechanics (see Sect. 1.3.3), more restrictive than generalized measurements where the observer has access only to a function $f(k)$ of the operation index k, which may not be one-to-one [9].

A generic Markovian quantum evolution can hence be described by a concatenation of maps $\mathcal{E}_r$ with Kraus operators $\hat{M}_k^{(r)}$. For the initial state $\rho(0)$, the unconditional state evolves as

$$\rho(r) = \mathcal{E}_r \mathcal{E}_{r-1} \dots \mathcal{E}_1 \rho(0). \tag{7.4}$$

This density matrix $\rho(r)$ can be interpreted as the average of the stochastic evolution. If the initial state is pure $\rho(0) = |\psi(0)\rangle\langle\psi(0)|$, then a stochastic trajectory $\gamma \equiv (k_1, k_2, \ldots, k_r)$ is given by the operations k_r that occurred in the application of map $\mathcal{E}_r$ and determines the evolution of the pure state

$$|\psi(r)\rangle = \hat{M}_{k_r}^{(r)} \hat{M}_{k_{r-1}}^{(r-1)} \ldots \hat{M}_{k_1}^{(1)} |\psi(0)\rangle. \tag{7.5}$$

Notice that this is an equivalent way of introducing quantum trajectories based on the sequence of discrete processes generated by the operations $\mathcal{E}_k$, analogue to one introduced in the derivation of the stochastic Scrhödinger equation in Sect. 2.5.

7.2.2 *Dual-Reverse Dynamics*

Now consider a particular Kraus representation of a map $\mathcal{E} = \sum_k \mathcal{E}_k$, and suppose that the map has a positive-definite invariant state π (not necessarily unique), i.e.,

$$\mathcal{E}(\pi) = \pi. \tag{7.6}$$

Notice that, as mentioned in Sect. 2.1.1, any CPTP map has always at least one invariant state (or fixed point), but it is in general not guaranteed to be positive-definite. For such maps, we introduce an auxiliary or *dual-reverse* map $\tilde{\mathcal{E}}$ with respect to π and to a fixed, arbitrary unitary or anti-unitary operator $\hat{\mathcal{A}}$. Inspired by Crooks, we define this dual-reverse map through the equality [18, 19][2]

$$\mathrm{Tr}\left[\mathcal{E}_{k_2}\mathcal{E}_{k_1}(\pi)\right] = \mathrm{Tr}\left[\tilde{\mathcal{E}}_{k_1}\tilde{\mathcal{E}}_{k_2}(\tilde{\pi})\right] \tag{7.7}$$

where $\tilde{\pi} \equiv \hat{\mathcal{A}}\,\pi\,\hat{\mathcal{A}}^\dagger$ is the invariant state transformed by $\hat{\mathcal{A}}$. Equation (7.7) states that the probability of observing the outcome k_1 followed by k_2 when we apply the map twice to the invariant state π equals the probability of observing the reverse outcome —k_2 followed by k_1— when the dual-reverse map is applied twice to $\tilde{\pi}$. In this way, the dual-reverse map induces a dynamics in the invariant state that is the reverse of the original one. Following the derivation introduced by Crooks in Ref. [18], one can prove that the Kraus operators of the dual-reverse map are given by

$$\hat{\tilde{M}}_k \equiv \hat{\mathcal{A}}\,\pi^{\frac{1}{2}}\hat{M}_k^\dagger\pi^{-\frac{1}{2}}\,\hat{\mathcal{A}}^\dagger. \tag{7.8}$$

Trace preservation ($\sum_k \hat{\tilde{M}}_k^\dagger \hat{\tilde{M}}_k = \mathbb{1}$) follows immediately from $\mathcal{E}(\pi) = \pi$, and one can verify that the dual-reverse map preserves the invariant state $\tilde{\mathcal{E}}(\tilde{\pi}) = \tilde{\pi}$.

[2] Notice however that we are changing the nomenclature with respect to Ref. [18], in which the map $\tilde{\mathcal{E}}_k$ is called the dual or time-reversed map. The reasons for this change will be specified in the next chapter of this thesis.

Henceforth, following Sect. 2.1, as the dual-reverse map can be written in Kraus form and it preserves the trace, it is also completely positive (CP).

The inclusion of the operator $\hat{\mathcal{A}}$ in the definition of the dual-reverse map is not mathematically necessary to derive the fluctuation theorem. In fact, $\hat{\mathcal{A}}$ does not appear in the original definition by Crooks [18]. However, in some situations an appropriate choice of the operator $\hat{\mathcal{A}}$ is needed to find a dual-reverse dynamics with a precise physical interpretation or that is suitable of being implemented in the laboratory. The customary choice is the time-reversal operator $\hat{\mathcal{A}} = \hat{\Theta}$ that changes the sign of odd variables, like linear and angular momenta. The time-reversal operator $\hat{\Theta}$ is an anti-linear, anti-unitary operator, satisfying $\hat{\Theta}^2 = \hat{\Theta}^\dagger\hat{\Theta} = \hat{\Theta}\hat{\Theta}^\dagger = \mathbb{1}$ [36, 37]. For instance, $\hat{\Theta}$ acts on a spinless particle by complex conjugation of the wave function in the position representation. The need of $\hat{\Theta}$ in the definition of the dual-reverse process is clear, for example, if the map is a unitary evolution, i.e., a map given by a unique Kraus operator $\hat{U}$ with $\hat{U}^\dagger = \hat{U}^{-1}$. In that case the invariant state is proportional to the identity matrix and the dual-reverse dynamics reads

$$\hat{\tilde{U}} = \hat{\Theta}\,\hat{U}^\dagger\,\hat{\Theta}^\dagger. \tag{7.9}$$

The dual-reverse map is again a unitary evolution given by the unitary operator $\hat{\tilde{U}}$ and corresponds to the *operational* time reversal of the original unitary evolution given by $\hat{U}$ [38] (see Sect. 1.1.4). For instance, if $\hat{U}$ is the evolution of a system under a constant Hamiltonian $\hat{H}, \hat{U} = e^{-i\hat{H}t/\hbar}$, and $\hat{H}$ is time-reversal invariant, $[\hat{H}, \hat{\Theta}] = 0$, then $\hat{\tilde{U}} = \hat{U}$, i.e., the dual-reverse map is identical to the original one. On the other hand, if the Hamiltonian depends on time according to some protocol, and $\hat{U}$ is the evolution between $t = 0$ and $t = \tau$, then $\hat{\tilde{U}}$ is the evolution that results when the protocol is reversed (which is, in general, different from $\hat{U}^\dagger$).

The operator $\hat{\mathcal{A}}$ can also account for other transformations of the system state that are necessary to exploit dynamical and static symmetries. In fact, this freedom has a classical counterpart in fluctuation theorems that incorporate various symmetry transformations [39–41].

7.3 Fluctuation Theorems

7.3.1 Nonequilibrium Potential and Detailed Balance

We now prove a general fluctuation theorem for a large family of CPTP maps. To begin our introduction of these maps, let us focus on an important class of maps that admit the following Kraus representation

$$\hat{M}_{ji} = \alpha_{ji}|\pi_j\rangle\langle\pi_i|, \tag{7.10}$$

in terms of the eigenstates $\{|\pi_i\rangle\}$ of the invariant density π, that is $\pi|\pi_i\rangle = \pi_i|\pi_i\rangle$. Here the Kraus operators are labelled by two indices (i, j) that identify jumps or transitions between eigenstates of π, i.e. $|\pi_i\rangle \to |\pi_j\rangle$, each occurring with probability $||\hat{M}_{ji}|\pi_i\rangle||^2 = |\alpha_{ji}|^2$. These maps are special in that a single application of $\mathcal{E}$ destroys any coherences between eigenstates of π in the initial state ρ, reducing the subsequent action of the map to a classical Markov chain on the eigenstates $\{|\pi_i\rangle\}$. Therefore, the dynamics induced by CPTP maps of the form (7.10) is essentially classical. On the other hand, quantum effects arise if the Kraus operators are linear combinations of the transition operators $|\pi_j\rangle\langle\pi_i|$, preserving coherences between eigenstates of the invariant density matrix.

The family of maps that obey our fluctuation theorem go beyond the 'classical' case outlined above [Eq. (7.10)]. To be precise, we assign to each eigenstate $|\pi_i\rangle$, whose strictly positive eigenvalue is denoted by π_i, a *nonequilibrium potential*, similar to the one used in the classical Hatano-Sasa theorem [33],

$$\phi_i \equiv -\ln \pi_i. \tag{7.11}$$

Then the maps that obey our fluctuation theorem are those where each Kraus operator $\hat{M}_k$ is a superposition of jump operators, all of them inducing the same change in nonequilibrium potential $\Delta\phi_k$:

$$\hat{M}_k = \sum_{i,j} m^k_{ji}|\pi_j\rangle\langle\pi_i|, \tag{7.12}$$

with $m^k_{ji} = 0$ if $\phi_j - \phi_i \neq \Delta\phi_k$. That is, by measuring the operation $\hat{M}_k$ we know without uncertainty the change in the nonequilibrium potential, even though that change could have occurred through a superposition of jumps.

One simple example of this construction is given in the context of a harmonic oscillator coupled to an equilibrium reservoir of resonant two-level atoms at temperature T [42] (see also the bosonic collisional model developed in Sect. 2.3.2). In absence of external forces, the maps governing the harmonic oscillator evolution for an infinitesimal time-step dt are Gibbs-preserving maps [43], for which the invariant state is the equilibrium thermal state $\pi = \exp[-\hat{H}/kT]/Z$ being $\hat{H} = \hbar\omega\hat{a}^\dagger\hat{a}$ the Hamilton operator of the system. Furthermore, by measuring the reservoir, we are able to detect jumps in the oscillator energy ladder as specified by the Kraus operators

$$\hat{M}_\downarrow = \sqrt{dt\gamma_\downarrow}\hat{a}, \qquad \hat{M}_\uparrow = \sqrt{dt\gamma_\uparrow}\hat{a}^\dagger, \tag{7.13}$$

with jump rates fulfilling $\gamma_\uparrow = e^{-\hbar\omega/kT}\gamma_\downarrow$. The absence of jumps is associated to the operator $\hat{M}_0 = \mathbb{1} - dt[i\hat{H} + \frac{1}{2}(\gamma_\downarrow\hat{a}^\dagger\hat{a} + \gamma_\uparrow\hat{a}\hat{a}^\dagger)]$ (see Sect. 2.5). In this case, following definition (7.11), the nonequilibrium potential is the energy of each $\hat{H}$ eigenstate divided by kT, and the changes in the nonequilibrium potential $\Delta\phi_0 = 0$ and $\Delta\phi_{\downarrow\uparrow} = \pm\hbar\omega/kT$, correspond to the energy transferred to the reservoir as heat divided by temperature (entropy transferred to the reservoir). Notice also that the jump operators are not of the 'classical' form (7.10) but (7.12), and in general do not decohere superposition states in the energy basis, $\hat{M}_{\downarrow\uparrow}(|n\rangle + |m\rangle) \propto |n \mp 1\rangle + |m \mp 1\rangle$.

Turning to the general case, it is straightforward to check that condition (7.12) is equivalent to

$$\begin{aligned}[\hat{M}_k, \ln \pi] &= \Delta\phi_k \hat{M}_k \\ [\hat{M}_k^\dagger, \ln \pi] &= -\Delta\phi_k \hat{M}_k^\dagger\end{aligned} \tag{7.14}$$

and, consequently $[\hat{M}_k^\dagger \hat{M}_k, \ln \pi] = [\hat{M}_k^\dagger \hat{M}_k, \pi] = 0$. These commutation relations are similar to those satisfied by the Lindblad operators that appear in Davies' theory of systems weakly coupled to thermal baths (see Refs. [44–47] and Sect. 2.2). They indicate that the pair $\hat{M}_k$, $\hat{M}_k^\dagger$ act as ladder operators, inducing jumps between the eigenstates $|\pi_i\rangle$ of π with a fixed change $\Delta\phi_k$ in the nonequilibrium potential ϕ. Finally, (7.12) ensures that the dual-reverse Kraus operators obey a generalized detailed balance condition

$$\hat{\tilde{M}}_k = e^{\Delta\phi_k/2}\, \hat{\mathcal{A}} \hat{M}_k^\dagger \hat{\mathcal{A}}^\dagger \tag{7.15}$$

that can be obtained by plugging (7.12) into (7.8). One can also prove that the form (7.12) is the only one for which the dual-reverse operators $\hat{\tilde{M}}_k$ in Eq. (7.8) are proportional to $\hat{\mathcal{A}}\ \hat{M}_k^\dagger\ \hat{\mathcal{A}}^\dagger$. Remarkably, for maps with multiple invariant states the quantities $\Delta\phi_k$ do not depend on the specific invariant state π chosen to define the nonequilibrium potential and the dual-reverse dynamics.[3] In other words, the set of values $\Delta\phi_k$ associated to the Kraus representation $\mathcal{E}_k$ is a property of the map $\mathcal{E}$.

7.3.2 *Fluctuation Theorem for a Single CPTP Map*

The basis of our fluctuation theorem is the proportionality between Kraus operators and their dual-reverse counterpart in Eq. (7.15). This generalized detailed balance condition connects the probability to observe a given jump, say k, with the probability to observe the same jump in the dual-reverse dynamics. Specifically, suppose that we initially prepare the system in the pure state $|\psi_n\rangle$, and then apply the map $\mathcal{E}$, registering the occurrence of the operation k. We then perform a quantum yes/no measurement of the projector $|\varphi_m\rangle\langle\varphi_m|$. The subscripts n and m are added to the initial and final states so that later on we can consider measurements of arbitrary observables with eigenstates $|\psi_n\rangle$ and $|\varphi_m\rangle$.

Now, let $p_{m,k|n}$ be the probability that given an initial state $|\psi_n\rangle$ we observe operation k *and* the final state $|\varphi_m\rangle$, that is, the probability to observe the jump $|\psi_n\rangle \to |\varphi_m\rangle$ under the action of $\hat{M}_k$. Let also $\tilde{p}_{n,k|m}$ be the probability to observe the

[3] F. Fagnola, private communication.

inverse jump $|\tilde{\varphi}_m\rangle \to |\tilde{\psi}_n\rangle$, with $|\tilde{\psi}\rangle = \hat{\mathcal{A}}|\psi\rangle$, under the action of the dual-reverse operator $\hat{\tilde{M}}_k$. Using Eq. (7.15), the ratio of these two conditional probabilities is

$$\begin{aligned}\frac{p_{m,k|n}}{\tilde{p}_{n,k|m}} &= \frac{|\langle\varphi_m|\hat{M}_k|\psi_n\rangle|^2}{|\langle\tilde{\psi}_n|\hat{\tilde{M}}_k|\tilde{\varphi}_m\rangle|^2} = \frac{|\langle\varphi_m|\hat{M}_k|\psi_n\rangle|^2}{|\langle\psi_n|\hat{\mathcal{A}}^\dagger\hat{\tilde{M}}_k\,\hat{\mathcal{A}}|\varphi_m\rangle|^2} \\ &= \frac{|\langle\varphi_m|M_k|\psi_n\rangle|^2}{|\langle\psi_n|\hat{M}_k^\dagger|\varphi_m\rangle|^2}\frac{1}{e^{\Delta\phi_k}} = e^{-\Delta\phi_k}. \end{aligned} \tag{7.16}$$

Equation (7.16) can be considered as a modified detailed balance relation for the operation $\mathcal{E}_k$ and its dual-reverse $\tilde{\mathcal{E}}_k$, which remarkably is independent of the initial and final states. Notice that when the map is unital, i.e. $\pi = \hat{\mathbb{1}}$ and hence $\Delta\phi_k = 0\ \forall k$, the right-hand side of above equation equals unity and hence reduces to the one obtained in Ref. [23], which is independent of the operation $\mathcal{E}_k$.

Suppose now that we prepare the system in the initial mixture $\rho_\mathrm{i} = \sum_n p_n^\mathrm{i}|\psi_n\rangle\langle\psi_n|$ and apply the map $\mathcal{E}$. By measuring the initial state $|\psi_n\rangle$, the operation $\mathcal{E}_k$ and a final state $|\varphi_m\rangle$ we obtain a *trajectory* (m, k, n) that is observed with a probability $p_{m,k,n} = p_{m,k|n}\,p_n^\mathrm{i}$. We compare this to a dual-reverse process induced by the map $\tilde{\mathcal{E}}$ applied to the initial state $\tilde{\rho}_\mathrm{f} = \sum_m \tilde{p}_m^\mathrm{f}|\tilde{\varphi}_m\rangle\langle\tilde{\phi}_m|$. The dual-reverse trajectory (n, k, m) is given as well by the initial state $|\tilde{\varphi}_m\rangle$, the dual-reverse operation $\tilde{\mathcal{E}}_k$ and the final state $|\tilde{\psi}_n\rangle$, and it is observed with probability $\tilde{p}_{n,k,m} = \tilde{p}_{n,k|m}\tilde{p}_m^\mathrm{f}$. The ratio of the probability to observe a trajectory $\gamma = n, k, m$ and the probability to observe the reverse trajectory $\tilde{\gamma} = \{m, k, n\}$ in the dual-reverse process is then, from (7.16),

$$\Sigma_\gamma \equiv \ln\frac{p_{n,k,m}}{\tilde{p}_{m,k,n}} = \sigma_{n,m} - \Delta\phi_k, \tag{7.17}$$

where $\sigma_{n,m} \equiv -\ln\tilde{p}_m^\mathrm{f} + \ln p_n^\mathrm{i}$ is a boundary term, only depending on the initial state of the process ρ_i and the initial state of the dual-reverse $\tilde{\rho}_\mathrm{f}$. The quantity Σ_γ is a measure of how different the original and the dual-reverse trajectories are. In particular, when the dual-reverse is the time reversed process (see below), Σ_γ is a measure of the irreversibility of the process for a given trajectory. In the following we will show that it can be identified with an entropy production in many situations of interest. The remarkable feature of Eq. (7.17) is that it splits $\Sigma_{n,k,m}$ into two terms, one depending on the initial states of the process, as in generalized versions of the fluctuation theorem for unitary dynamics [11], and the second one depending *only* on the operation $\mathcal{E}_k$ by virtue of Eq. (7.16).

A (Jarzynski-type) integral fluctuation theorem immediately follows from Eq. (7.17):

$$\left\langle e^{-\Sigma_\gamma}\right\rangle = \sum_{n,k,m} p_{n,k,m}e^{-\Sigma_{n,k,m}} = \sum_{n,k,m}\tilde{p}_{m,k,n} = 1, \tag{7.18}$$

where $\langle\cdot\rangle$ denotes the average over forward trajectories, $p_{n,k,m}$. Finally by Jensen's inequality $\langle e^x\rangle \geq e^{\langle x\rangle}$, we have the second-law-like inequality

$$\langle \Sigma_\gamma \rangle = \langle \sigma_{n,m} \rangle - \langle \Delta\phi_k \rangle \geq 0. \tag{7.19}$$

Here the boundary term $\sigma_{n,m}$ averaged over forward trajectories leads to the following entropic quantity

$$\begin{aligned} \langle \sigma_{n,m} \rangle &= -\sum_{n,k,m} p_{n,k,m} \ln \tilde{p}_m^{\mathrm{f}} + \sum_{n,k,m} \ln p_n^{\mathrm{i}} \\ &= S(\mathcal{E}(\rho_{\mathrm{i}})) - S(\rho_{\mathrm{i}}) + D(\mathcal{E}(\rho_{\mathrm{i}})||\hat{\mathcal{A}}^\dagger \tilde{\rho}_{\mathrm{f}} \hat{\mathcal{A}}), \end{aligned} \tag{7.20}$$

where $D(\rho||\sigma) = -S(\rho) - \mathrm{Tr}[\rho \ln \sigma]$ is the quantum relative entropy as introduced in Sect. 1.1.6. The quantity (7.20) can then be interpreted as the increase in the entropy of the system state due to the map action, $S(\mathcal{E}(\rho_{\mathrm{i}})) - S(\rho_{\mathrm{i}})$, plus the relative entropy between $\mathcal{E}(\rho_{\mathrm{i}})$ and the (inverted) initial state of the backward process $\hat{\mathcal{A}}^\dagger \tilde{\rho}_{\mathrm{f}} \hat{\mathcal{A}}$. On the other hand, the average nonequilibrium potential change during the forward process results in

$$\begin{aligned} \langle \Delta\phi_k \rangle &= \sum_{n,k,m} p_{n,k,m} \Delta\phi_k = \sum_k \mathrm{Tr}[\mathcal{E}_k(\rho_{\mathrm{i}})] \Delta\phi_k \\ &= \mathrm{Tr}[\hat{\Phi}(\mathcal{E}(\rho_{\mathrm{i}}) - \rho_{\mathrm{i}})] \end{aligned} \tag{7.21}$$

where the non-equilibrium potential operator is defined as $\hat{\Phi} \equiv -\ln \pi$, and we used the property $[\hat{\Phi}, \hat{M}_k] = \hat{M}_k \Delta\phi_k$ derived from condition (7.12). This implies that the average potential change $\langle \Delta\phi \rangle$ can be expressed as the change in the expectation value of the operator $\hat{\Phi}$ due to the map.

Equations (7.19) and (7.20) provide a general bound on the changes in the observable $\hat{\Phi}$ induced by the action of the CPTP map $\mathcal{E}$:

$$\langle \Delta\phi_k \rangle \leq S(\mathcal{E}(\rho_{\mathrm{i}})) - S(\rho_{\mathrm{i}}) + D(\mathcal{E}(\rho_{\mathrm{i}})||\hat{\mathcal{A}}^\dagger \tilde{\rho}_{\mathrm{f}} \hat{\mathcal{A}}). \tag{7.22}$$

As the relative entropy is non-negative, the right-hand side of this inequality minimizes for $D(\mathcal{E}(\rho_{\mathrm{i}})||\hat{\mathcal{A}}^\dagger \tilde{\rho}_{\mathrm{f}} \hat{\mathcal{A}}) = 0$, which corresponds to choosing the initial state of the backward process as $\tilde{\rho}_{\mathrm{f}} = \hat{\mathcal{A}} \mathcal{E}(\rho_{\mathrm{i}}) \hat{\mathcal{A}}^\dagger$. The tighter bound $\langle \Delta\phi_k \rangle \leq S(\mathcal{E}(\rho_{\mathrm{i}})) - S(\rho_{\mathrm{i}})$ is hence obtained when choosing the final measurements of the forward process in the eigenbasis of $\mathcal{E}(\rho_{\mathrm{i}})$ and the backward process is initialized by just inverting this state using $\hat{\mathcal{A}}$. This suggests to interpret $\langle \Delta\phi_k \rangle$ as the entropy transferred from the environment to the system during the map action. In Sect. 7.4 we will see how inequality (7.22) [or equivalently (7.19)] extends the second law of thermodynamics to many physical situations of interest, where the nonequilibrium potential changes are associated to thermodynamic entropy flows.

7.3.3 Fluctuation Theorem for Concatenated Maps

Our fluctuation theorems (7.17) – (7.18) can be easily extended to a concatenation of CPTP maps, $\Omega = \mathcal{E}_R\mathcal{E}_{R-1}\ldots\mathcal{E}_r\ldots\mathcal{E}_1$, which is the case of general Markov quantum evolution, unitary evolution punctuated by projective measurements, driven systems in contact with thermal baths, etc. A trajectory now is given by the initial $|\psi_n\rangle$ and final states $|\varphi_m\rangle$ and the outcomes k_r of all the measurements associated to the maps $r = 1, 2, \ldots, R$: $\gamma = \{n, k_1, k_2, \ldots, k_R, m\}$. Each map $\mathcal{E}_r$ has a Kraus representation, given by the operators $\hat{M}_k^{(r)}$, and an invariant state $\pi^{(r)}$ for which the dual-reverse map $\tilde{\mathcal{E}}_r$ and the nonequilibrium potential $\phi_k^{(r)}$ are defined as in Eqs. (7.8) and (7.11).

To derive the fluctuation theorem, we reverse the concatenation of maps. We define the dual-reverse process as $\hat{\Omega} = \tilde{\mathcal{E}}_1\ldots\tilde{\mathcal{E}}_r\ldots\tilde{\mathcal{E}}_{R-1}\tilde{\mathcal{E}}_R$ (notice that, for $R > 1$, in general, $\hat{\Omega} \neq \tilde{\Omega}$, i.e., the dual-reverse process does not coincide with the dual-reverse map of Ω). If each map obeys condition (7.12) [or, equivalently, (7.15)], then we get the following symmetry relation

$$\frac{p(m, k_R, \ldots, k_1|n)}{\tilde{p}(n, k_1, \ldots, k_R|m)} = \frac{|\langle\varphi_m|\hat{M}_{k_R}^{(R)}\ldots\hat{M}_{k_1}^{(1)}|\psi_n\rangle|^2}{|\langle\tilde{\psi}_n|\tilde{\hat{M}}_{k_1}^{(1)}\ldots\tilde{\hat{M}}_{k_R}^{(R)}|\tilde{\varphi}_m\rangle|^2} = \exp\left[-\sum_{r=1}^{R}\Delta\phi_{k_r}^{(r)}\right]. \tag{7.23}$$

A detailed fluctuation theorem can be now obtained by comparing the probability of a trajectory $\gamma = \{n, k_1, \ldots, k_R, m\}$ in the forward process and the probability of the inverse trajectory $\tilde{\gamma} = \{m, k_R, \ldots, k_1, n\}$ in the dual-reverse process:

$$\Sigma_\gamma \equiv \ln\frac{p_\gamma}{\tilde{p}_{\tilde{\gamma}}} = \sigma_{n,m} - \sum_{r=1}^{R}\Delta\phi_{k_r}^{(r)}, \tag{7.24}$$

with a corresponding integral fluctuation theorem that follows readily, like in Eq. (7.18). Thus, for a concatenation of maps implemented in sequence, we merely have to sum up the changes in the nonequilibrium potential along the trajectory. Notice also that we effectively used a Kraus representation for the map Ω where each Kraus operator was labeled with the sequence $\{k_1, \ldots, k_R\}$, requiring possibly many more than the necessary N^2 operators.

A clear interpretation of Σ_γ arises if we consider the concatenation of the same map $\mathcal{E}$, acting on the stationary density matrix π, and the corresponding dual-reverse process acting on $\tilde{\pi}$. In this case $p_n^{\mathrm{i}} = \pi_n$ and $\tilde{p}_m^{\mathrm{f}} = \pi_m$, yielding

$$\Sigma_\gamma = \ln\pi_n - \ln\pi_m - \sum_{r=1}^{R}\Delta\phi_{k_r}^{(r)} = 0 \tag{7.25}$$

for any trajectory γ. This is expected from the (modified) Crooks definition (7.7): the original and the dual-reverse maps acting on π and $\tilde{\pi}$, respectively, produce a trajectory γ and its reverse $\tilde{\gamma}$ with identical probability. Therefore, Σ can be considered as a measure of the distinguishability of the original and the dual-reverse process, but also as a measure of how far the system is from the stationary state. These two equivalent interpretations are familiar in thermodynamics when π is an equilibrium state: the dual-reverse is the operational reverse process and Σ_γ is the entropy production which measures both irreversibility and departure from equilibrium [48]. In more general situations, Σ_γ is the part of the entropy production due to the fact that the state of the system does not coincide with the stationary state. This can occur in the transient from a nonsteady initial condition to the stationary state, or due to a finite-speed driving. In such case, Σ_γ is known as the nonadiabatic [49–51] or excess [33, 52] entropy production, in contrast to the entropy production needed to maintain the stationary state, which is often referred to as adiabatic or house-keeping entropy production [53]. This connection will be clarified in the next chapter, in which we investigate the decomposition of the total entropy production in a quantum process into adiabatic and non-adiabatic contributions.

The fluctuation theorem stated in Eq. (7.24) exploits the dynamical symmetries of the process through the dual-reverse map and the nonequilibrium potential, in the same spirit as the detailed fluctuation theorem for processes connecting nonequilibrium states developed by Esposito and Van den Broeck [49–51]. Finally, the integral theorem (7.18) is the quantum version of the Hatano-Sasa theorem [33], extending the Jarzynski equality to nonequilibrium states. The corresponding second-law-like inequality (7.19) extends to arbitrary boundary conditions the quantum Hatano-Sasa inequality for concatenated CPTP maps proposed by Sagawa [11].

7.4 Applications

Despite their simplicity, the above fluctuation theorems include as special cases many of the known quantum fluctuation relations, extends previous results to more arbitrary situations, and can be used to obtain novel relationships. In the section, we explain how these relations come about in our formalism. We first discuss the boundary term $\sigma_{n,m}$ and then apply the general theorem to different dynamics. Here we specify $\hat{\mathcal{A}} = \hat{\Theta}$, the anti-unitary time-reversal operator.

7.4.1 Boundary Terms

There are two common choices for boundary terms:

1. *Reversible boundaries* setting the initial state of the dual-reverse equal to the final state of the forward process $\tilde{\rho}_\mathrm{f} = \hat{\mathcal{A}}\rho_\mathrm{f}\hat{\mathcal{A}}^\dagger$, with $\rho_\mathrm{f} = \sum_m |\varphi_m\rangle\langle\varphi_m|\mathcal{E}(\rho_\mathrm{i})|\varphi_m\rangle\langle\varphi_m|$ and ρ_i being an arbitrary state.

2. *Equilibrium boundaries* setting the initial state ρ_i of the forward process and the initial state $\tilde{\rho}_\mathrm{f}$ of the dual-reverse process as thermal equilibrium states.

Notice that, by selecting the initial states of the forward and dual-reverse processes, we are also fixing the basis in which the quantum measurements are performed, that is, the basis of the first (second) measurement of the forward (dual-reverse) process corresponds to the eigenbasis of $\rho_\mathrm{i}\,(\hat{\mathcal{A}}\rho_\mathrm{i}\hat{\mathcal{A}}^\dagger)$, and the basis of the first (second) measurement of the dual-reverse (forward) to the one of $\tilde{\rho}_\mathrm{f}\,(\hat{\mathcal{A}}^\dagger\tilde{\rho}_\mathrm{f}\hat{\mathcal{A}})$.

In the first case, the boundary term

$$\sigma_{n,m} = -\ln p^\mathrm{f}_m + \ln p^\mathrm{i}_n = s^\mathrm{f}_m - s^\mathrm{i}_n \tag{7.26}$$

is the increase of the stochastic or trajectory entropy [11, 19, 28, 42, 54]. Its average over forward trajectories yields the increase of von Neumann entropy during the forward process, $\langle\sigma_{n,m}\rangle = S(\rho_\mathrm{f}) - S(\rho_\mathrm{i})$, including the contribution due to the application of the map $\mathcal{E}$ *plus* the average entropy increase due to the projective measurement on the open system at the end of the process. This result follows from Eq. (7.20) by noticing that in this case

$$D(\mathcal{E}(\rho_\mathrm{i})||\hat{\mathcal{A}}^\dagger\tilde{\rho}_\mathrm{f}\hat{\mathcal{A}}) = D(\mathcal{E}(\rho_\mathrm{i})||\rho_\mathrm{f}) = S(\rho_\mathrm{f}) - S(\mathcal{E}(\rho_\mathrm{i})). \tag{7.27}$$

Notice that here the observable being measured at the end of the process is still arbitrary, a particular choice being an observable commuting with the density operator after map action, which implies $\rho_\mathrm{f} = \mathcal{E}(\rho_\mathrm{i})$ (see Sect. 1.3). This choice is relevant from a theoretical point of view, but the resulting dual-reverse process is hard to implement in general, except when the system is small enough to be prepared in an arbitrary state (say, a few qubits or a harmonic oscillator).

The second choice, equilibrium initial states for the forward and dual-reverse dynamics, is more interesting from an operational point of view, since the dual-reverse dynamics can be easily implemented in the laboratory by equilibrating the system with a thermal reservoir and reversing the protocol that drives the Hamiltonian [23, 37, 55, 56]. Let us suppose that, before applying any quantum map, the system Hamiltonian is initially fixed $\hat{H}_\mathrm{i}$, whereas after the Hamiltonian is $\hat{H}_\mathrm{f}$. We further take the initial state of the forward process to be equilibrium at inverse temperature β, that is, $\rho_\mathrm{i} = e^{\beta(F_\mathrm{i} - \hat{H}_\mathrm{i})}$, where F_i is the corresponding free energy. Similarly, we initialize the dual-reverse process in the final equilibrium at the same temperature, $\tilde{\rho}_\mathrm{f} = e^{\beta(F_\mathrm{f} - \hat{H}_\mathrm{f})}$. Then

$$\sigma_{n,m} = \beta(E^\mathrm{f}_m - E^\mathrm{i}_n - F_\mathrm{f} + F_\mathrm{i}) \equiv \beta(\Delta E_{n,m} - \Delta F), \tag{7.28}$$

where the $\{E^\mathrm{i,f}_l\}$ are the eigenvalues of the initial and final Hamiltonians, respectively.

7.4.2 Unital Work Relations

As a first example, we take our quantum map to be *unital* (or bistochastic [57]), that is, the identity is an invariant state of the map, $\mathcal{E}(\mathbb{1}) = \mathbb{1}$ (although the identity may not be the only one). Any unitary evolution $\hat{U}$ is unital, $\hat{U}\mathbb{1}\hat{U}^\dagger = \mathbb{1}$, and its dual-reverse map is the time-reversal $\hat{\tilde{U}} = \hat{\Theta}\hat{U}^\dagger\hat{\Theta}^\dagger$. The *depolarizing channel* acting on finite dimensional systems introduced in Sect. 2.1.4 is also a unital map. Another example of a unital map is the projective measurement of an observable but, more generally, any minimally disturbing measurement is unital [9]. For these maps, the Kraus operators are self-adjoint $\hat{M}_k^\dagger = \hat{M}_k$, leading to dual-reverse operators $\hat{\tilde{M}}_k = \hat{\Theta}\hat{M}_k\hat{\Theta}^\dagger$. Finally, pure decoherence is also implemented with unital maps that remove all the off-diagonal elements in a specified basis. For all unital maps or concatenation of such maps, $\Delta\phi_k = 0$ for all k, and the fluctuation theorem only consists of the boundary term.

Let us now consider a concatenation of unital maps as describing a physical process. An important example is a process consisting of several unitary transformations induced by driven time-dependent Hamiltonians, punctuated by a number of measurements and/or pure decoherence processes. In each map, energy can be transferred to the system. We call the energy input into the system due to the driving w_γ^{drive}, driving work, and w_γ^{meas} the energy input due to the measurements and/or decoherence processes. Whereas the driving work w_γ^{drive} has a clear interpretation as the energy supplied by driving, the origin of the energy input due to measurement is still obscure. This energy transfer occurs, for instance, in a projective measurement of an observable that does not commute with the Hamiltonian. In any case, $\Delta E_{n,m} = w_\gamma^{\text{drive}} + w_\gamma^{\text{meas}}$ and, if we choose equilibrium initial states the boundary term $\sigma_{n,m}$ is given by (7.28) and

$$\Sigma_\gamma = \beta(w_\gamma^{\text{drive}} + w_\gamma^{\text{meas}} - \Delta F) = \beta w_\gamma^{\text{diss}}. \tag{7.29}$$

The fluctuation theorem (7.24), therefore, reproduces the work fluctuation theorems for unital processes derived in Refs. [1, 17, 23, 24] (see also [58, 59]). Notice that, if we allow the system to relax to equilibrium after the maps have been applied, then Σ_γ equals the entropy production along the whole process and Eq. (7.19) reproduces the second law inequality $\langle\Sigma_\gamma\rangle = \beta\langle w_\gamma^{\text{diss}}\rangle \geq 0$ [60]. We stress that this result is valid for any concatenation of unital maps. On the other hand, if we choose the initial state of the dual-reverse process as the final state of the original process, $\Sigma_\gamma = -\ln(p_m^{\text{f}}/p_n^{\text{i}}) = s_m^{\text{f}} - s_n^{\text{i}}$ is just the change in stochastic entropy. When averaged, the entropy production Σ becomes the change in the von Neumann entropy of the system

$$\Delta S_{\text{sys}} = \langle\Sigma_\gamma\rangle \geq 0, \tag{7.30}$$

whose positivity follows from Eq. (7.19). This provides an alternative thermodynamic proof of the well-known property that unital maps can only increase the von Neumann entropy [8].

7.4.3 Thermalization and Heat

Another interesting example is a generic thermalization map [18] at inverse temperature $\beta = 1/(k_B T)$ (or Gibbs-preserving map [43]), that is, a map whose invariant state is the equilibrium density matrix $\pi = e^{\beta(F-\hat{H})}$, where $\hat{H} = \sum_j E_j |e_j\rangle\langle e_j|$ is the Hamiltonian of the system and F its free energy at temperature T. Thus, the nonequilibrium potential is related to the energy as $\phi_j = -\ln \pi_j = \beta(F - E_j)$. To verify our fluctuation theorem, each Kraus operator $\hat{M}_k$ must promote transitions between energy eigenstates involving a given change of energy ΔE_k, that is, $\hat{M}_k = \sum_{ji} m^k_{ji} |e_j\rangle\langle e_i|$, where the sum runs over pairs of energy eigenstates with the same energy difference $\Delta E_k = E_j - E_i$. Now, since the energy is supplied by a thermal reservoir, we can identify these energy exchanges as heat flowing from the reservoir, $q_k = \Delta E_k$, and the change in nonequilibrium potential is hence $\Delta\phi_k = \beta q_k$. The dual-reverse Kraus operators $\tilde{\hat{M}}_k \propto \hat{M}_k^\dagger = \sum_{ji} m^k_{ji} |e_i\rangle\langle e_j|$ (for a time-reversal invariant $\hat{H}$) induce the reverse transitions accompanied by the reverse flow of heat $\tilde{q}_k = -q_k$, and thus can be identified with a Kraus operator in the original map.

One simple example of a Gibbs-preserving map acting on a single qubit system with basis $\{|g\rangle, |e\rangle\}$ and Hamilton operator $\hat{H} = E|e\rangle\langle e|$ is the *generalized amplitude damping channel*, for which a Kraus representation was given in Eq. (2.22) of Sect. 2.1.4

$$\hat{M}_0 = \sqrt{p}\begin{pmatrix} 1 & 0 \\ 0 & \sqrt{1-\lambda} \end{pmatrix}, \qquad \hat{M}_1 = \sqrt{p}\begin{pmatrix} 0 & \sqrt{\lambda} \\ 0 & 0 \end{pmatrix}, \tag{7.31}$$
$$\hat{M}_2 = \sqrt{1-p}\begin{pmatrix} \sqrt{1-\lambda} & 0 \\ 0 & 1 \end{pmatrix}, \quad \hat{M}_3 = \sqrt{1-p}\begin{pmatrix} 0 & 0 \\ \sqrt{\lambda} & 0 \end{pmatrix},$$

with $p \equiv \frac{n_{\text{th}}+1}{2n_{\text{th}}+1}$ and $\lambda \equiv 1 - e^{-t/\tau_R}$, $n_{\text{th}} = 1/(e^{\beta E} - 1)$ being the mean number of excitations in a thermal bosonic reservoir, and $\tau_R = 1/\gamma_0(2n_{\text{th}} + 1)$ a characteristic relaxation time scale (see Sect. 2.1.4). Here the operators $\hat{M}_0$ and $\hat{M}_2$ corresponding to continuous monitoring processes and inducing decoherence in the energy basis produce a null change in the nonequilibrium potential, $\Delta\phi_0 = \Delta\phi_2 = 0$. On the other hand, operators $\hat{M}_1$ and $\hat{M}_3$ correspond to jumps induced by the exchange of an energy quantum E with the environment: from the excited to the ground state, $\Delta\phi_1 = -\beta E$, and from the ground to the excited state, $\Delta\phi_3 = \beta E$. From Eq. (7.8) and the fact that $\hat{\Theta}\hat{H}\hat{\Theta}^\dagger = \hat{H}$, we have that the Kraus operators for the dual-reverse map are

$$\tilde{\hat{M}}_0 = \sqrt{p}\begin{pmatrix} 1 & 0 \\ 0 & \sqrt{1-\lambda} \end{pmatrix}, \qquad \tilde{\hat{M}}_1 = \sqrt{1-p}\begin{pmatrix} 0 & 0 \\ \sqrt{\lambda} & 0 \end{pmatrix},$$
$$\tilde{\hat{M}}_2 = \sqrt{1-p}\begin{pmatrix} \sqrt{1-\lambda} & 0 \\ 0 & 1 \end{pmatrix}, \quad \tilde{\hat{M}}_3 = \sqrt{p}\begin{pmatrix} 0 & \sqrt{\lambda} \\ 0 & 0 \end{pmatrix}, \tag{7.32}$$

which correspond to the time-reversed process in which jumps up are transformed into jumps down and vice-versa. This simple example shows how our theorem is able to link the physical picture provided by the Kraus representation (7.31) with thermodynamics. Taking $t \ll \tau_R$ ($\lambda \ll 1$) the generalized amplitude damping channel describes the action of the environment during a a small time interval, which can be used to model more general coarse-grained evolutions through the concatenation of different maps.

Consider now a thermodynamic process formed by a concatenation of thermalization steps induced by N distinct thermal reservoirs with inverse temperatures $\{\beta_i\}_{i=1}^{N}$ interspersed by unital transformations (unitary drivings, measurements or decoherence). For this setup, the nonequilibrium potential changes during each step of the evolution is either $\Delta\phi_k^{(i)} = \beta_i q_k^{(i)}$ or $\Delta\phi_k = 0$. If we choose the initial state of the dual-reverse process as the final state of the original process [c.f. Eq. (7.26)], we arrive at

$$\Sigma_\gamma = s_m^{\mathrm{f}} - s_n^{\mathrm{i}} - \sum_{i=1}^{N} \beta_i q_\gamma^{(i)}, \tag{7.33}$$

with $q_\gamma^{(i)}$ the total heat flow from the ith reservoir during the whole trajectory γ. In this case, Eq. (7.24) gives a fluctuation theorem for the total irreversible entropy production in the process [28, 61], and Eq. (7.19) results in

$$\langle \Sigma_\gamma \rangle = \Delta S_{\mathrm{sys}} - \sum_{i=1}^{N} \beta_i \langle q_\gamma^{(i)} \rangle \geq 0, \tag{7.34}$$

corresponding to the generalization of the Clausius inequality to many thermal reservoirs, and nonequilibrium initial and final states.

On the other hand, the equilibrium boundary terms are interesting when restricted to one thermal reservoir, leading to

$$\Sigma_\gamma = \beta(\Delta E_{n,m} - \Delta F - q_\gamma) = \beta(w_\gamma - \Delta F), \tag{7.35}$$

where we have used the energy balance $\Delta E_{n,m} = w_\gamma - q_\gamma$. Again, Σ equals the entropy production along the whole process consisting of the map concatenation followed by a thermal relaxation. The detailed and integral fluctuation theorems following from the identification (7.35) are, respectively, the quantum Tasaki-Crooks and Jarzynski fluctuation theorems for thermal maps punctuated by unital maps.

As a final corollary, we point that the identification of nonequilibrium potential changes with the heat transferred from the medium to the open system provides a general formulation of Landauer's principle, valid for any Gibbs-preserving map $\mathcal{E}$. Recalling that $\langle \Delta\phi \rangle = \langle q \rangle / kT$ in Eq. (7.22) we immediately obtain

$$- \langle q_\gamma \rangle \geq kT \left[S(\rho_\mathrm{i}) - S(\mathcal{E}(\rho_\mathrm{i})) \right]. \tag{7.36}$$

This equation states that the heat dissipated *into* a thermal environment as a consequence of the map, $-\langle q_\gamma \rangle$, must be, at least, the reduction in the system's entropy (which can be identified with changes in information processes such as erasure) times kT [62–64].

7.4.4 Generalized Gibbs-Preserving Maps

All the results in the previous subsection can be extended to the more general class of maps preserving *generalized Gibbs ensembles* [65, 66]. This kind of ensembles are relevant in the study of the emerging thermalization properties of isolated quantum many-body systems after a quantum quench [67–74]. Furthermore they are actually attracting increasing attention in more general contexts including the study of work fluctuations in arbitrary out-of-equilibrium integrable systems [75] or resource theories and trade-off relations in processes concerning individual quantum systems [76, 77]. A generalized Gibbs ensemble (GGE) can be defined by means of entropy maximization for an extensive set of conserved quantities (or charges) $\{\hat{I}^{(\alpha)}\}$ and associated Lagrange multipliers $\{\mu_\alpha\}$ [65, 77, 78]

$$\rho_{\mathrm{GGE}} \equiv \frac{e^{-\sum_\alpha \mu_\alpha \hat{I}^{(\alpha)}}}{Z} = e^{-\sum_\alpha \mu_\alpha \hat{I}^{(\alpha)} - F} \tag{7.37}$$

where $Z = \mathrm{Tr}[\exp(-\sum_\alpha \mu_\alpha \hat{I}^{(\alpha)})]$ is the partition function which can be alternatively expressed by means of $F \equiv -\ln Z$, an entropic quantity playing an analogous role to the equilibrium free energy for the canonical ensemble [77]. Notice that the canonical Gibbs ensemble is indeed recovered by setting the only conserved quantity to be the energy $\hat{I}^{(1)} = \hat{H}$, the corresponding Lagrange multiplier being then the inverse temperature $\mu_1 = \beta$.

We are interested in maps preserving the GGE, that is $\pi = \rho_{\mathrm{GGE}}$, and admitting a Kraus representation fulfilling our condition (7.12). A straightforward example can be constructed by considering a reset-like operation of the form

$$\mathcal{E}(\rho) = p\rho_{\mathrm{GGE}} + (1 - p)\rho \tag{7.38}$$

which with probability p substitutes the state of the system by ρ_{GGE}, and with probability $1 - p$ leaves it untouched. For any such map one can construct a Kraus representation $\mathcal{E}(\rho) = \hat{M}_0 \rho \hat{M}_0^\dagger + \sum_{ij} \hat{M}_{ij} \rho \hat{M}_{ij}^\dagger$ with

$$\hat{M}_0 = \sqrt{1-p}\,\hat{\mathbb{1}}, \qquad \hat{M}_{ij} = \sqrt{p\,\pi_i}\,|\pi_i\rangle\langle\pi_j|, \tag{7.39}$$

which are indeed of the form (7.10).

The nonequilibrium potential operator for the GGE (7.37) now reads

$$\hat{\Phi} = \sum_i \phi_i |\pi_i\rangle\langle\pi_i| = \sum_\alpha \mu_\alpha \left(\hat{I}^{(\alpha)}\right) - F. \tag{7.40}$$

Notice that for the simpler case in which $[\hat{I}^{(\alpha)}, \hat{I}^{(\alpha')}] = 0\,\forall\alpha,\alpha'$ we obtain nonequilibrium potential changes associated to each Kraus representation $\Delta\phi_k = \sum_\alpha \mu_\alpha \Delta I_k^{(\alpha)}$, where $\Delta I_k^{(\alpha)}$ is the exchange in the charge α with the environment in a jump k between the $\rho_{\rm GGE}$ eigenstates. This is for instance the case of the grand canonical ensemble in which the two charges are the system Hamiltonian $\hat{I}^{(1)} = \hat{H}$ and the number of particles $\hat{I}^{(2)} = \hat{N}$ and their respective Lagrange multipliers are the inverse temperature $\mu_1 = 1/k_B T$ and the chemical potential divided by temperature $\mu_2 = \mu/k_B T$ [76]. In more general situations with non-commuting charges, $\Delta\phi_k$ may not be decomposed as the previous sum, since the changes $\Delta I_k^{(\alpha)}$ may not be well defined between jumps. This second kind of generalized Gibbs ensembles corresponds e.g. to the case of harmonic oscillators in displaced or squeezed thermal states (see Sects. 1.2.4 and 1.2.5), naturally emerging as steady states of the dynamics when considering general reservoir dynamics as in Sect. 2.3.2.

We can now generalize our previous results by considering a thermodynamic process in which the open system follows a sequence of GGE-preserving steps interspersed by unital transformations (e.g. sudden quenches, measurements, ...). Assuming the same invariant state for each step of the dynamical evolution and initial reversible boundaries, we obtain

$$\Sigma_\gamma = s_m^{\rm f} - s_n^{\rm i} - \Delta\phi_\gamma = s_m^{\rm f} - s_n^{\rm i} - \sum_\alpha \mu_\alpha \Delta I_\gamma^\alpha, \tag{7.41}$$

where we denoted by $\Delta\phi_\gamma$ and ΔI_γ^α the total changes during the whole sequence γ, and in the last equality we considered commuting charges. In the later case, as long as the quantities $\Delta I_\gamma^{(\alpha)}$ correspond to the total transfer of quantity $I^{(\alpha)}$ from the environment, we may immediately identify $\Delta\phi_\alpha$ as the entropy produced in the medium. Equation (7.24) with Eq. (7.41) then provides detailed and integral fluctuation theorems for the total entropy production in driven open systems with commuting charges. When the charges do not commute or when the environment consists of multiple reservoirs, the interpretation of the FT (7.24) may be more involved and depends on the specific situation (see Chaps. 8 and 9). In any case, the second-law-like inequality (7.19) introduces a trade-off relation between currents valid for the general case

$$\Delta S_{\rm sys} \geq \sum_\alpha \mu_\alpha \langle \Delta I_\gamma^{(\alpha)} \rangle, \tag{7.42}$$

which coincides with the inequalities recently derived in Ref. [77].

On the other hand, we may also consider equilibrium boundaries by replacing the initial and final Gibbs distributions previously employed by generalized Gibbs ensembles [75]. Consider now a process as above but for which the arbitrary set of operators $\{\hat{I}^{(\alpha)}\}$ appearing in Eq. (7.37) vary in time throughout the change of some external control parameter. Let us denote $\{\hat{I}_{\mathrm{i}}^{(\alpha)}\}$ the initial charges when the protocol is started and $\{\hat{I}_{\mathrm{f}}^{(\alpha)}\}$ their final value after the sequence of generalized Gibbs-preserving and unital maps has been applied. In this case we assume for simplicity that the set of charges commutes between each others at all times. We obtain for the boundary term

$$\sigma_{n,m} = \sum_{\alpha} \mu_{\alpha} \left(I_m^{\mathrm{f}(\alpha)} - I_n^{\mathrm{i}(\alpha)} \right) - \Delta F, \tag{7.43}$$

where $I_n^{\mathrm{i}(\alpha)}$ and $I_m^{\mathrm{f}(\alpha)}$ are, respectively, the eigenvalues of $\hat{I}_{\mathrm{i}}^{(\alpha)}$ and $\hat{I}_{\mathrm{f}}^{(\alpha)}$. Adding the nonequilibrium potential changes associated to jumps detected during the forward process, the entropy production Σ can be written in the form

$$\Sigma_{\gamma} = \sum_{\alpha} \mu_{\alpha} w_{\gamma}^{I^{(\alpha)}} - \Delta F \tag{7.44}$$

where we employed the first-law-like balance for the αth conserved quantity:

$$w_{\gamma}^{I^{(\alpha)}} \equiv I_m^{\mathrm{f}(\alpha)} - I_n^{\mathrm{i}(\alpha)} - \Delta I_{\gamma}^{\alpha}. \tag{7.45}$$

Here $w_{\gamma}^{I^{(\alpha)}}$ corresponds to a generalized work notion for the quantity $\hat{I}^{(\alpha)}$, taking into account the changes in the charge induced by external driving, in contrast to the changes induced by interaction with the environment akin to heat, $\Delta I_{\gamma}^{\alpha}$ (see also Refs. [75, 77]). In this case Eqs. (7.24) and (7.18) represent generalized versions of the Crooks theorem and Jarzynski equality for open quantum systems in a GGE preserving environment and driven arbitrary far from equilibrium. They will be of crucial relevance to describe irreversibility and work in quenched many-body systems with a set of conserved quantities, extending previous results for isolated driven systems [75] to the open configuration. This extension is of great importance because in typical configurations only (open) subsystems reach a GGE as steady state, while the state of the global system remains pure at all times. Our formalism allows the thermodynamic description of interacting subsystems while subjected to unital measurements or further driving protocols. Finally Eq. (7.22) gives us a bound governing the trade-off between the different conserved quantities. In particular for the case in which energy is one of the conserved charges

$$\beta \langle w_{\gamma} \rangle \geq \Delta F - \sum_{\alpha} \mu_{\alpha} \langle w_{\gamma}^{I^{(\alpha)}} \rangle, \tag{7.46}$$

which predicts work extraction by means of externally fueling the open system with other charges [79] (see also [80]).

7.4.5 Lindblad Master Equations

Another illustration of our results are the Lindblad master equations that model the Markovian dynamic evolution of open quantum systems (see Sect. 2.2). As introduced in Sect. 2.2.1, a master equation in Lindblad form for the evolution of a quantum system in some suitable interaction picture can be specified by an Hermitian (Hamiltonian-like) operator $\hat{H}$, and a collection of Lindblad operators $\{\hat{L}_k\}_{k=1}^K$:

$$\frac{d\rho_t}{dt} = -i[\hat{H}, \rho_t] + \sum_k \mathcal{D}[\hat{L}_k]\rho_t \equiv \mathcal{L}\rho_t, \tag{7.47}$$

where the super-operator $\mathcal{D}$ is defined as

$$\mathcal{D}[\hat{L}]\rho \equiv \hat{L}\rho\hat{L}^\dagger - \frac{1}{2}\left(\hat{L}^\dagger\hat{L}\rho + \rho\hat{L}^\dagger\hat{L}\right). \tag{7.48}$$

To make contact with our fluctuation theorem, we must introduce a *master equation unraveling* introducing a trajectory description of the dynamics as in Sect. 2.5. The solution to Eq. (7.47) can be obtained by concatenating a sequence of maps together that evolve the system forward in small time steps dt:

$$\mathcal{E}(\rho_t) = (\hat{\mathbb{1}} + \mathcal{L}dt)\rho_t = \hat{M}_0\rho_t\hat{M}_0^\dagger + \sum_{k=1}^K \hat{M}_k\rho_t\hat{M}_k^\dagger, \tag{7.49}$$

with Kraus operators

$$\hat{M}_0 = \hat{\mathbb{1}} - \left(i\hat{H} + \frac{1}{2}\sum_k \hat{L}_k^\dagger\hat{L}_k\right)dt \tag{7.50}$$

$$\hat{M}_k = \hat{L}_k\sqrt{dt}, \qquad 1 \le k \le K. \tag{7.51}$$

This map has at least one invariant state π [47], obeying $\mathcal{L}\pi = 0$.

To satisfy our fluctuation theorem, the Kraus operators $\{\hat{M}_k\}$ must be of the form (7.12) and verify the generalized detailed balance relations (7.15). Enforcing these conditions on $\{\hat{M}_k\}_{k\geq 1}$ immediately leads to a restriction on the Lindblad operators similar to (7.12). That is, each Lindblad operator must induce jumps between invariant-state eigenstates, $\hat{L}_k = \sum_{ji} m_{ji}^k |\pi_j\rangle\langle\pi_i|$, where $m_{ji}^k = 0$ for all i, j such that $\Phi_j - \Phi_i \neq \Delta\Phi_k$. In this case, the generalized detailed balance relation (7.15) holds:

$$\tilde{\hat{L}}_k = e^{\Delta\phi_k/2}\, \hat{\Theta}\, \hat{L}_k^\dagger\, \hat{\Theta}^\dagger, \qquad k \geq 1. \tag{7.52}$$

As for the Kraus operators, if the Lindblad operator $\hat{L}_k$ induces jumps where the nonequilibrium potential change equals $\Delta\phi_k$, then they obey commutation relations similar to (7.14):

$$[\hat{L}_k, \hat{\Phi}] = -\Delta\phi_k \hat{L}_k, \quad [\hat{L}_k^\dagger, \hat{\Phi}] = \Delta\phi_k \hat{L}_k^\dagger, \tag{7.53}$$

with $\hat{\Phi} = -\ln \pi$, and $[\hat{L}_k^\dagger \hat{L}_k, \hat{\Phi}] = [\hat{L}_k^\dagger \hat{L}_k, \pi] = 0$.

Let us verify now whether $\hat{M}_0$ also satisfies our conditions. The dual-reverse operator (7.8) reads:

$$\tilde{\hat{M}}_0 = \hat{\Theta}\pi^{\frac{1}{2}} \left[\mathbb{1} - \left(-i\hat{H} + \frac{1}{2}\sum_k \hat{L}_k^\dagger \hat{L}_k \right) dt \right] \pi^{-\frac{1}{2}} \hat{\Theta}^\dagger \tag{7.54}$$

Since $[\hat{L}_k^\dagger \hat{L}_k, \pi] = 0$, for our generalized detailed balance condition to hold, that is $\tilde{\hat{M}}_0 \propto \hat{\Theta}\hat{M}_0^\dagger \hat{\Theta}^\dagger$, we must assume that $[\hat{H}, \pi] = 0$, forcing the invariant state to be diagonal in eigenbasis of the Hermitian operator $\hat{H}$. If $\hat{H}$ is the Hamilton operator of the open system, an immediate consequence of this observation is that $\Delta\phi_k$ must correspond to jumps in the energy. Notice however that in many situations $\hat{H} = \mathbb{1}$, as Eq. (7.47) may be written in interaction picture with respect to the system and environment Hamiltonians. In those situations we will obtain no restrictions in the basis of the jumps. In any case we have

$$\tilde{\hat{M}}_0 = \Theta \left[\hat{\mathbb{1}} - \left(-i\hat{H} + \frac{1}{2}\sum_k \hat{L}_k^\dagger \hat{L}_k \right) dt \right] \hat{\Theta}^\dagger = \hat{\Theta}\hat{M}_0^\dagger \hat{\Theta}^\dagger \tag{7.55}$$

Thus, $\hat{M}_0$ satisfies our generalized detailed balance relations with $\Delta\phi_0 = 0$, as one would expect for a Kraus operator that does not induce transitions.

Consider now the following process. We run the Lindbladian evolution for an interval of time $[0, \tau]$, and measure some observables at time $t = 0$ and $t = \tau$. In this scenario, a trajectory $\gamma = \{n, k_1, k_2, \ldots, k_N, m\}$ is given by the initial and final measurement outcomes, n and m respectively, and a set of jumps k_l occurring at times t_l. Notice that the stochastic trajectory, as defined in the previous sections, should contain a big number of instances $k_r = 0$, i.e., corresponding to operation $\hat{M}_0$, between jumps. However, these operations do not contribute to Σ_γ and we can omit them from the discussion. In this case,

$$\Sigma_\gamma = \sigma_{n,m} - \sum_l \Delta\phi_{k_l}. \tag{7.56}$$

With the entropic boundary conditions (7.26), we arrive at the quantum generalization of the Hatano-Sasa theorem [33] for the nonadiabatic entropy production of Lindblad master equations, as developed in Ref. [19]. Furthermore, if the final projective measurement on the open system after the Lindbladian evolution is performed in its eigenbasis, the average over trajectories yields

$$\langle \Sigma_\gamma \rangle = \Delta S_{\text{sys}} - \langle \Delta\phi_\gamma \rangle = D(\rho_{\text{i}}||\pi) - D(\rho_{\text{f}}||\pi), \tag{7.57}$$

where we denoted $\Delta\phi_\gamma = \sum_l \Delta\phi_{k_l}$. This result coincides with the expression first introduced by Spohn for arbitrary quantum dynamical semigroups [34], then extended by Yukawa to driven quantum Markov processes [10]. The equivalence between our trajectory picture and the average thermodynamic behavior has been discussed in Ref. [13].

So far we have been treating the dissipation in the Lindblad master equation as a whole. When the dissipation can be interpreted as coming from M distinct thermodynamic reservoirs (or Markovian noise processes), we can employ our formula for the entropy production of concatenated maps (7.24) to arrive at a complementary formulation of the thermodynamics of the process. The effect of each of the M reservoirs is captured in the dynamics by a separate collection of Lindblad operators $\{\hat{L}_{k,\alpha}\}_{k=1}^{K_\alpha}$, where $\alpha = 1, \ldots, M$ labels the reservoir:

$$\frac{d\rho_t}{dt} = -\frac{i}{\hbar}[\hat{H}, \rho_t] + \sum_\alpha \sum_k \mathcal{D}[\hat{L}_{k,\alpha}]\rho_t. \tag{7.58}$$

Similar to (7.49), we can implement the evolution of this equation over a small time interval dt by a map, except now it is formed by a *concatenation* of intermediary maps, $\mathcal{E}(\rho_t) = \mathcal{E}_{\alpha_M} \cdots \mathcal{E}_{\alpha_1} \mathcal{E}_0(\rho_t)$, each arising from the different terms in Eq. (7.58). The first map $\mathcal{E}_0(\rho_t) = \rho_t - (i/\hbar)[\hat{H}, \rho_t]dt$ captures the unitary part of the dynamics with a single Kraus operator $\hat{M}_{0,0} = \hat{\mathbb{1}} - (i/\hbar)\hat{H}dt$; the subsequent maps describe the dissipative reservoirs, whose Kraus operators are

$$\hat{M}_{0,\alpha} = \hat{\mathbb{1}} - \left(\frac{1}{2}\sum_k \hat{L}_{k,\alpha}^\dagger \hat{L}_{k,\alpha}\right) dt \tag{7.59}$$

$$\hat{M}_{k,\alpha} = \hat{L}_{k,\alpha}\sqrt{dt}, \qquad 1 \le k \le K_\alpha. \tag{7.60}$$

Notice that the exact sequence of maps $\mathcal{E}_\alpha$ is immaterial as they all commute to first order in dt. Crucially, each reservoir is assumed to have its own invariant state, $\mathcal{E}_\alpha(\pi^{(\alpha)}) = \pi^{(\alpha)}$ (or equivalently $\sum_k \mathcal{D}[\hat{L}_{k,\alpha}]\pi^{(\alpha)} = 0$), while the invariant state of the map $\mathcal{E}_0$ is just the identity, $\mathbb{1}$. For example, a thermal reservoir at inverse temperature $\beta^{(\alpha)}$ would have the equilibrium Boltzmann density matrix $\pi^{(\alpha)} = e^{\beta^{(\alpha)}(F^{(\alpha)} - \hat{H}_\mathcal{S})}$ as invariant state, with $\hat{H}_\mathcal{S}$ the Hamilton operator of the open system. The corresponding Lindblad operators must then induce jumps in that state, $\hat{L}_{k,\alpha} = \sum_{i,j} m_{ji}^{k,\alpha} |\pi_j^{(\alpha)}\rangle\langle\pi_i^{(\alpha)}|$, to satisfy our generalized detailed balance relation (7.16). As a result, the $\{\hat{M}_{0,\alpha}\}_{\alpha=0}^M$ immediately satisfy the generalized detailed balance relations with $\Delta\phi^{(\alpha)} = 0$, which remarkably does not require the invariant state of the whole Lindblad equation to commute with the generic Hermitian operator $\hat{H}$.

Now, a trajectory for this setup corresponds to a list $\gamma = \{n, k_1, k_2, \ldots, k_N, m\}$ given by the initial and final measurement outcomes, n and m, and a set of jumps k_l occurring at times t_l in the α_l reservoir. Notice that only one jump in one of the M reservoirs can happen in any given dt, since the probability to observe two jumps is negligible. The result from (7.24) is then

$$\Sigma_\gamma = \sigma_{n,m} - \sum_l \Delta\phi_{k_l}^{(\alpha_l)}. \tag{7.61}$$

This point of view allows us to treat multiple reservoirs at once, such as an engine operating between hot and cold thermal reservoirs, each represented by a different set of Lindblad operators [81]. Using the entropic boundary conditions (7.26), the resulting average entropy production has long been known from the works of Spohn and Lebowitz [82] and Alicki et al. [81], where again our version generically includes the entropic cost of the final projective measurements on the open system.

It is remarkable that our fluctuation theorem can yield different results for Σ_γ, depending on the resolution of the stochastic trajectory. For instance, in the case of the system in contact with several thermal reservoirs, Σ_γ is given by (7.61) if the trajectory keeps track of the jumps induced by each reservoir separately. On the other hand, if the trajectory only gives information about the jumps of the system in the basis where the stationary density matrix of the entire Lindblad equation is diagonal, we have (7.56). Consequently, for the same map one can have both (7.56) and (7.61). The distinction is the same as the difference between the fluctuation theorem for the entropy production (7.61) and the non-adiabatic entropy production (7.56) [49–51] as we will clarify in more detail in the next chapter.

We stress that condition (7.53) is fulfilled by almost all known examples of driven Lindblad equations for systems weakly coupled to reservoirs. If the Hamiltonian of the open system $\hat{H}_S$ (not necessarily equivalent to the Hermitian operator $\hat{H}$) is constant, the weak coupling limit and rotating wave approximations results in a Lindblad equation where the operators $\hat{L}_\omega$, $\hat{L}_\omega^\dagger$ are labelled by the Bohr frequencies ω which are transition frequencies between the levels of the Hamiltonian, i.e, they are of the form $\omega = \omega_i - \omega_j$, for some pair of levels i, j with energies $\epsilon_i = \hbar\omega_i$ and $\epsilon_j = \hbar\omega_j$, respectively (see Sect. 2.2.2 and the examples in Sect. 2.3). These are ladder operators that lower and raise the energy levels, obeying the commutation relations:

$$[\hat{L}_\omega, \hat{H}_S] = \hbar\omega\hat{L}_\omega, \qquad [\hat{L}_\omega^\dagger, \hat{H}_S] = -\hbar\omega\hat{L}_\omega^\dagger. \tag{7.62}$$

Their commutator with the logarithm of the stationary density operator π can be written as

$$\langle\pi_i|[\hat{L}_\omega, \ln\pi]|\pi_j\rangle = \langle\pi_i|\hat{L}_\omega|\pi_j\rangle \ln\frac{\pi_i}{\pi_j}. \tag{7.63}$$

For (7.53) to be satisfied it is sufficient that the ratio $\pi_i/\pi_j = e^{f(\Delta\epsilon_{ij})}$ is a function of the energy difference $\Delta\epsilon_{ij} = \epsilon_j - \epsilon_i$. In that case

$$[\hat{L}_\omega, \ln\pi] = f(\hbar\omega)\hat{L}_\omega, \tag{7.64}$$

and $\Delta\phi_\omega = f(\hbar\omega)$. In the case of a single thermal reservoir $f(\epsilon) = \beta\epsilon$, and $\Delta\phi_\omega$ is the entropy flow to the reservoir (heat divided by temperature) associated to a transition of frequency ω. Furthermore, the Lindblad operators will come in pairs

$\{\hat{L}_\omega, \hat{L}_{-\omega}\}$ such that $\hat{\tilde{L}}_\omega = \hat{L}_{-\omega} \propto \hat{L}_\omega^\dagger$, and every jump can be undone. As a result, the dual-reverse process is equivalent to the original process. This approach was developed for work fluctuations theorems in Ref. [83] and heat fluctuations in Ref. [84].

The preceding arguments can be naturally extended to the case of a time-dependent Hermitian operator $\hat{H}(t)$ and time-dependent Lindblad operators $\hat{L}_k(t)$, yielding an instantaneous stationary state $\pi(t)$ (or states $\pi^{(\alpha)}(t)$) [18, 85]. This is the case when the open system Hamiltonian $\hat{H}_S(t) = \hat{H}_S(\lambda_t)$ is driven through the slow (not necessarily quasi-static) change of a collection of external parameters λ_t, the Lindblad operators and the Hermitian operator become parameterized by the external parameters $\hat{L}_k(\lambda_t)$ and $\hat{H}(\lambda_t)$, and our generalized detailed balance relation will hold at every time [8, 86, 87]. For fast periodic driving, Floquet theory can be used to derive a Lindblad master equation [46]. This theory picks out as a preferred eigenbasis a collection of time-periodic states, or Floquet states, each with a corresponding quasi-energy or Floquet energy. The collection of Lindblad jump operators $\{\hat{L}_k\}$ then induce transitions between Floquet eigenstates of the periodic Hamiltonian leading again to the generalized detailed balance relation (7.52) with $\Delta\phi_k$ the change in Floquet eigenvalues in the kth jump, which often corresponds to the heat exhausted into the environment [46, 88]. Finally, we recall that our predictions can be used to recover the fluctuation theorems derived for driven Markov dynamics presented in Ref. [19].

7.5 Conclusion

In this chapter we have presented a general fluctuation theorem, Eqs. (7.17) and (7.18), for a large class of CPTP quantum maps and concatenations [Eq. (7.24)] that verify the generalized detailed balance condition in Eq. (7.15). From these relations many of the known quantum fluctuation theorems follow naturally. Included in this family are classical fluctuation theorems for arbitrary stochastic maps, as such maps are special cases of CPTP quantum maps where the dynamics remain diagonal in a particular basis. The theorem exploits the dynamical symmetries of a process and its dual-reverse and can be interpreted as a (generalized) quantum version of the Hatano-Sasa theorem [33]. The most important characteristic of our theorem is that it is fulfilled for general dynamics under simple conditions only depending on the map, and consequently it can be applied to very arbitrary situations without caring about the specific characteristics of the environment. When specialized to maps induced by thermodynamic reservoirs, our results reproduce known quantum fluctuation theorems for work and entropy production, extending preceding results to more general situations such as generalized Gibbs-preserving maps.

We have extended the notion of the dual-reverse process, first introduced by Crooks [18] as a time-reversal, and clarified its relationship with the time-reversal process used by Campisi et al. and Watanabe et al. to derive fluctuation theorems for unitary evolution punctuated by projective measurements [17, 37]. The relation between those processes and the classical dual process used by Esposito and Van den

Broeck to split the entropy production into an adiabatic and nonadiabatic contribution [49–51] will be further clarified in the next chapter.

Our results also help to understand the peculiarity of unital maps regarding entropy exchange, a fact already pointed out in Refs. [17, 23, 24, 26, 37]. The nonequilibrium potential associated to those maps is constant and therefore it does not appear in the fluctuation theorem. The entropy production Σ_γ in this case is only given by the boundary terms, suggesting that unital maps can be induced without any entropy exchange between the system and its surroundings. Thermalization at infinite temperature is an obvious example, but decoherence or, equivalently, projective and minimally disturbing measurements, are relevant examples of unital maps. In all these cases, energy exchange between the system and its surroundings can occur, but this energy exchange does not imply any entropy change in the environment.

For nonunital maps our work shows how quantum fluctuation theorems follow both in detailed and integral versions in very arbitrary situations. We have seen that, by including the nonequilibrium potential as a fluctuating quantity, no correction term is necessary, as we advanced in Sect. 7.1. The resulting quantity fulfilling the fluctuation theorems, Σ_γ, can then be given a clear interpretation as an entropy production in most setups of physical interest, in contrast to σ in Eq. (7.1) which simply represents a boundary term only depending on the initial states of the thermodynamic processes [28]. Furthermore, from our general theorem we obtained a second-law-like inequality, $\langle \Sigma_\gamma \rangle \geq 0$, establishing a general bound on the average nonequilibrium potential changes during the application of a map in terms of information-theoretic entropies, c.f. Eq. (7.22). The precise meaning of the entropy production Σ in our theorem depends on the choice of the Kraus representation, and hence must be clarified specifying the properties of the environment, which in turn determines the nature of the nonequilibrium potential changes. In the next chapter we will further explore the properties of the entropy production in general quantum processes and establish connections between the framework developed here for CPTP maps, and a general picture where the environment appears explicitly in the dynamical description.

References

1. M. Campisi, P. Talkner, P. Hänggi, Fluctuation theorems for continuously monitored quantum fluxes. Phys. Rev. Lett. **105**, 140601 (2010)
2. M. Esposito, U. Harbola, S. Mukamel, Nonequilibrium fluctuations, fluctuation theorems, and counting statistics in quantum systems. Rev. Mod. Phys. **81**, 1665–1702 (2009)
3. B.P. Venkatesh, G. Watanabe, P. Talkner, Transient quantum fluctuation theorems and generalized measurements. New J. Phys. **16**, 015032 (2014)
4. G. Watanabe, B.P. Venkatesh, P. Talkner, Generalized energy measurements and modified transient quantum fluctuation theorems. Phys. Rev. E **89**, 052116 (2014)
5. P. Hänggi, P. Talkner, The other QFT. Nat. Phys. **11**, 108–110 (2015)
6. M.A. Nielsen, I.L. Chuang, *Quantum Computation and Quantum Information* (Cambridge University Press, Cambridge, 2000)
7. K. Kraus, A. Böhm, J.D. Dollard, W.H. Wootters, *States, Effects, and Operations: Fundamental Notions of Quantum Theory*. Lecture Notes in Physics (Springer, Berlin, 1983)

8. H.-P. Breuer, F. Petruccione, *The Theory of Open Quantum Systems* (Oxford University Press, New York, 2002)
9. H.M. Wiseman, G.J. Milburn, *Quantum Measurement and Control* (Cambridge University Press, Cambridge, 2010)
10. S. Yukawa, The Second Law of Steady State Thermodynamics for Nonequilibrium Quantum Dynamics (2001), arXiv:0108421v2
11. T. Sagawa, Second law-like inequalitites with quantum relative entropy: an introduction, in *Lectures on Quantum Computing, Thermodynamics and Statistical Physics*, vol. 8, ed. by M. Nakahara. Kinki University Series on Quantum Computing (World Scientific, New Jersey, USA, 2013)
12. J. Anders, V. Giovannetti, Thermodynamics of discrete quantum processes. New J. Phys. **15**, 033022 (2013)
13. J.M. Horowitz, T. Sagawa, Equivalent definitions of the quantum nonadiabatic entropy production. J. Stat. Phys. **156**, 55–65 (2014)
14. F. Binder, S. Vinjanampathy, K. Modi, J. Goold, Quantum thermodynamics of general quantum processes. Phys. Rev. E **91**, 032119 (2015)
15. G. Manzano, J.M. Horowitz, J.M.R. Parrondo, Nonequilibrium potential and fluctuation theorems for quantum maps. Phys. Rev. E **92**, 032129 (2015)
16. M. Campisi, P. Talkner, P. Hünggi, Influence of measurements on the statistics of work performed on a quantum system. Phys. Rev. E **83**, 041114 (2011)
17. G. Watanabe, B.P. Venkatesh, P. Talkner, M. Campisi, P. Hünggi, Quantum fluctuation theorems and generalized measurements during the force protocol. Phys. Rev. E **89**, 032114 (2014)
18. G.E. Crooks, Quantum operation time reversal. Phys. Rev. A **77**, 034101 (2008)
19. J.M. Horowitz, J.M.R. Parrondo, Entropy production along nonequilibrium quantum jump trajectories. New. J. Phys **15**, 085028 (2013)
20. R. Chetritie, K. Mallick, Quantum fluctuation relations for the Lindblad master equation. J. Stat. Phys. **148**, 480–501 (2012)
21. F. Liu, Equivalence of two Bochkov-Kuzovlev equalities in quantum two-level systems. Phys. Rev. E **89**, 042122 (2014)
22. F. Liu, Calculating work in adiabatic two-level quantum Markovian master equations: a characteristic function method. Phys. Rev. E **90**, 032121 (2014)
23. T. Albash, D.A. Lidar, M. Marvian, P. Zanardi, Fluctuation theorems for quantum processes. Phys. Rev. E **88**, 032146 (2013)
24. A.E. Rastegin, Non-equilibirum equalities with unital quantum channels. J. Stat. Mech.: Theor. Exp. **59**, P06016 (2013)
25. D. Kafri, S. Deffner, Holevo's bound from a gernal quantum fluctuation theorem. Phys. Rev. A **86**, 044302 (2012)
26. A.E. Rastegin, K. Życzkowski, Jarzynski equality for quantum stochastic maps. Phys. Rev. E **89**, 012127 (2014)
27. J. Goold, M. Paternostro, K. Modi, Nonequilibrium quantum Landauer principle. Phys. Rev. Lett. **114**, 060602 (2015)
28. U. Seifert, Entropy production along a stochastic trajectory and an integral fluctuation theorem. Phys. Rev. Lett. **95**, 040602 (2005)
29. M.O. Scully, M.S. Zubairy, G.S. Agarwal, H. Walther, Extracting work from a single heat bath via vanishing quantum coherence. Science **299**, 862–864 (2003)
30. R. Dillenschneider, E. Lutz, Energetics of quantum correlations. Europhys. Lett. **88**, 50003 (2009)
31. J. Roßnagel, O. Abah, F. Schmidt-Kaler, K. Singer, E. Lutz, Nanoscale heat engine beyond the Carnot limit. Phys. Rev. Lett. **112**, 030602 (2014)
32. G. Manzano, F. Galve, R. Zambrini, J.M.R. Parrondo, Entropy production and thermodynamic power of the squeezed thermal reservoir. Phys. Rev. E **93**, 052120 (2016)
33. T. Hatano, S.-I. Sasa, Steady-state thermodynamics of Langevin systems. Phys. Rev. Lett. **86**, 3463 (2001)

34. H. Spohn, Entropy production for quantum dynamical semigroups. J. Math. Phys. **19**, 1227–1230 (1978)
35. J. Prost, J.F. Joanny, J.M.R. Parrondo, Generalized fluctuation- dissipation theorem for steady-state systems. Phys. Rev. Lett. **103**, 090601 (2009)
36. F. Haake, *Quantum Signatures of Chaos*, 3rd edn. Springer Series in Synergetics (Springer, Berlin, 2010)
37. M. Campisi, P. Hünggi, P. Talkner, Colloquium: quantum fluctuation relations: foundations and applications. Rev. Mod. Phys. **83**, 771–791 (2011)
38. J. Anders, Thermal state entanglement in harmonic lattices. Phys. Rev. A **77**, 062102 (2008)
39. C. Maes, The fluctuation theorem as a Gibbs property. J. Stat. Phys. **95**, 367–392 (1999)
40. P.I. Hurtado, C. Perez-Espigares, J.J. del Pozo, P.L. Garrido, Symmetries in fluctuations far from equilibrium. Proc. Natl. Acad. Sci. **108**, 7704–7709 (2011)
41. D. Lacoste, P. Gaspard, Isometric fluctuation relations for equilibrium states with broken symmetry. Phys. Rev. Lett. **113**, 240602 (2014)
42. J.M. Horowitz, Quantum-trajectory approach to the stochastic thermodynamics of a forced harmonic oscillator. Phys. Rev. E **85**, 031110 (2012)
43. P. Faist, J. Oppenheim, R. Renner, Gibbs-preserving maps outperform thermal operations in the quantum regime. New J. Phys. **17**, 043003 (2015)
44. E.B. Davies, *Quantum Theory of Open Systems* (Academic Press, London, 1976)
45. R. Alicki, D.A. Lidar, P. Zanardi, Internal consistency of fault-tolerant quantum error correction in light of rigorous derivations of the quantum Markovian limit. Phys. Rev. A **73**, 052311 (2006)
46. K. Szczygielski, D. Gelbwaser-Klimovsky, R. Alicki, Markovian master equation and thermodynamics of a two-level system in a strong laser field. Phys. Rev. E **87**, 012120 (2013)
47. A. Rivas, S.F. Huelga, *Open Quantum Systems: An Introduction* (Springer, Berlin, 2012)
48. R. Kawai, J.M.R. Parrondo, C. Van den Broeck, Dissipation: the phase-space perspective. Phys. Rev. Lett. **98**, 080602 (2007)
49. M. Esposito, C. Van den Broeck, Three detailed fluctuation theorems. Phys. Rev. Lett. **104**, 090601 (2010)
50. M. Esposito, C. Van den Broeck, Three faces of the second law. I. Master equation formulation. Phys. Rev. E **82**, 011143 (2010)
51. C. Van den Broeck, M. Esposito, Three faces of the second law. II. Fokker-Planck formulation. Phys. Rev. E **82**, 011144 (2010)
52. V.Y. Chernyak, M. Chertkov, C. Jarzynsk, Path-integral analysis of fluctuation theorems for general Langevin processes. J. Stat. Mech.: Theor. Exp. P08001 (2006)
53. T. Speck, U. Seifert, Integral fluctuation theorem for the housekeeping heat. J. Phys. A: Math. Gen. **38**, L581–L588 (2005)
54. T. Monnai, Unified treatment of the quantum fluctuation theorem and the Jarzynski equality in terms of microscopic reversibility. Phys. Rev. E **72**, 027102 (2005)
55. T.B. Batalhão, A.M. Souza, L. Mazzola, R. Auccaise, R.S. Sarthour, I.S. Oliveira, J. Goold, G. De Chiara, M. Paternostro, R.M. Serra, Experimental reconstruction of work distribution and study of fluctuation relations in a closed quantum system. Phys. Rev. Lett. **113**, 140601 (2014)
56. S. An, J.-N. Zhang, M. Um, D. Lv, Y. Lu, J. Zhang, Z.-Q. Yin, H.T. Quan, K. Kim, Experimental test of the quantum Jarzynski equality with a trapped-ion system. Nat. Phys. **11**, 193–199 (2015)
57. I. Bentsoon, K. Zyczkowski, *Geometry of Quantum States: An Introduction to Quantum Entanglement* (University Press, Cambridge, 2006)
58. I. Callens, W. De Roeck, T. Jacobs, C. Maes, K. Netočný, Quantum entropy production as a measure of irreversibility. Phys. D **187**, 383–391 (2004)
59. W. De Roeck, C. Maes, Quantum version of free-energy-irreversiblework relations. Phys. Rev. E **69**, 026115 (2004)
60. C. Jarzynski, Equalities and inequalities: irreversibility and the second law of thermodynamics at the nanoscale. Ann. Rev. Condens. Matter Phys. **2**, 329–351 (2011)

61. S. Deffner, E. Lutz, Nonequilibrium entropy production for open quantum systems. Phys. Rev. Lett. **107**, 140404 (2011)
62. R. Landauer, Irreversibility and heat generation in the computing process. IBM J. Res. Dev. **5**, 183–191 (1961)
63. C.H. Bennett, The thermodynamics of computation-a review. Int. J. Theor. Phys. **21**, 905–940 (1982)
64. D. Reeb, M.M. Wolf, An improved Landauer principle with finite-size corrections. New J. Phys. **16**, 103011 (2014)
65. M. Rigol, V. Dunjko, V. Yurovsky, M. Olshanii, Relaxation in a completely integrable many-body quantum system: an ab initio study of the dynamics of the highly excited states of 1D lattice hard-core bosons. Phys. Rev. Lett. **98**, 050405 (2007)
66. M.A. Cazalilla, Effect of suddenly turning on interactions in the Luttinger model. Phys. Rev. Lett. **97**, 156403 (2006)
67. M. Rigol, V. Dunjko, M. Olshanii, Thermalization and its mechanism for generic isolated quantum systems. Nature **452**, 854–858 (2008)
68. M. Cramer, C.M. Dawson, J. Eisert, T.J. Osborne, Exact relaxation in a class of nonequilibrium quantum lattice systems. Phys. Rev. Lett. **100**, 030602 (2008)
69. P. Calabrese, F.H.L. Essler, M. Fagotti, Quantum quench in the transverse-field ising chain. Phys. Rev. Lett. **106**, 227203 (2011)
70. A.C. Cassidy, C.W. Clark, M. Rigol, Generalized thermalization in an integrable lattice system. Phys. Rev. Lett. **106**, 140405 (2011)
71. J.-S. Caux, R.M. Konik, Constructing the generalized Gibbs ensemble after a quantum quench. Phys. Rev. Lett. **109**, 175301 (2012)
72. M. Fagotti, F.H.L. Essler, Reduced density matrix after a quantum quench. Phys. Rev. B **87**, 245107 (2013)
73. T. Langen, S. Erne, R. Geiger, B. Rauer, T. Schweigler, M. Kuhnert, W. Rohringer, I.E. Mazets, T. Gasenzer, J. Schmiedmayer, Experimental observation of a generalized Gibbs ensemble. Science **348**, 207–211 (2015)
74. R. Hamazaki, T.N. Ikeda, M. Ueda, Generalized Gibbs ensemble in a nonintegrable system with an extensive number of local symmetries. Phys. Rev. E **93**, 032116 (2016)
75. J.M. Hickey, S. Genway, Fluctuation theorems and the generalized Gibbs ensemble in integrable systems. Phys. Rev. E **90**, 022107 (2014)
76. N.Y. Halpern, J.M. Renes, Beyond heat baths: generalized resource theories for small-scale thermodynamics. Phys. Rev. E **93**, 022126 (2016)
77. Y. Guryanova, S. Popescu, A.J. Short, R. Silva, P. Skrzypczyk, Thermodynamics of quantum systems with multiple conserved quantities. Nat. Commun. **7**, 12049 (2016)
78. E.T. Jaynes, Information theory and statistical mechanics. II. Phys. Rev. **108**, 171–190 (1957)
79. J.A. Vaccaro, S.M. Barnett, Information erasure without an energy cost. Proc. Roy. Soc. Lond. A **467**, 1770–1778 (2011)
80. S.M. Barnett, J.A. Vaccaro, Beyond Landauer erasure. Entropy **15**, 4956–4968 (2013)
81. R. Alicki, The quantum open system as a model of a heat engine. J. Phys. A **12**, L103 (1979)
82. H. Spohn, J.L. Lebowitz, Irreversible thermodynamics for quantum systems weakly coupled to thermal reservoirs, in *Advances in Chemical Physics: For Ilya Prigogine*, vol. 38, ed. by S.A. Rice (Wiley, Hoboken, USA, 1978)
83. F.W.J. Hekking, J.P. Pekola, Quantum jump approach for work and dissipation in a two-level system. Phys. Rev. Lett. **111**, 093602 (2013)
84. J. Dereziński, W. De Roeck, C. Maes, Fluctuations of quantum currents and unravelings of master equations. J. Stat. Phys. **131**, 341–356 (2008)
85. G.E. Crooks, On the Jarzynski relation for dissipative quantum dynamics, J. Stat. Mech.: Theor. Exp. **10**, P10023 (2008)
86. T. Albash, S. Boixo, D.A. Lidar, P. Zanardi, Quantum adiabatic Markovian master equations. New J. Phys. **14**, 123016 (2012)
87. S. Suomela, J. Salmilehto, I.G. Savenko, T. Ala-Nissila, M. Möttönen, Fluctuations of work in nearly adiabatically driven open quantum systems. Phys. Rev. E **91**, 022126 (2015)
88. G.B. Cuetara, A. Engel, M. Esposito, Stochastic thermodynamics of rapidly driven systems. New J. Phys. **17**, 055002 (2015)

Chapter 8
Entropy Production Fluctuations in Quantum Processes

In the preceding Chap. 7, we derived a general fluctuation theorem (FT) in detailed and integral form valid for a broad class of CPTP quantum maps, which model a variety of quantum evolutions as we explained in more detail in Chap. 2. In this chapter we clarify and extend these previous results by considering together the system and its surroundings. By tracing over the environment degrees of freedom, we can then recover the quantum map description for the reduced open system dynamics. Making use of general concepts in quantum measurement theory as introduced in Sect. 1.3, quantum correlations (Sect. 1.4), and CPTP maps theory (Sect. 2.1), we will be able to ascribe a precise meaning to the entropy production using von Neumann quantum entropy.

Most of the research on quantum FT's is only valid for equilibrium reservoirs with a focus on the energy exchange between the system and the environment in the form of heat and work [1, 2]. By contrast, classical FT's have been formulated more generally for generic Markov processes [3–5] (see also the review [6]) using the entropy production instead of the heat and work, which are only meaningful in physical situations where a system exchanges energy with equilibrium reservoirs. In light of the success of classical FT's, it is desirable to obtain complementary FT's for generic quantum dynamics. They could be of particular relevance, since quantum mechanics allows for novel and interesting nonequilibrium environments of finite size [7–9], as well as coherent [10, 11], correlated [12], or squeezed [13–15] thermal reservoirs. Such environments can induce striking thermodynamic behavior, such as tighter bounds on Landauer's principle [16, 17], or be used to construct thermal machines able to outperform Carnot efficiency and traditional regimes of operation [18–20].

The task of deriving FT's for generic quantum dynamics also implies a more detailed characterization of entropy production in generic nonequilibrium quantum contexts, a problem that has experimented a growing interest in recent years [19, 21–30]. Nevertheless, the recent theoretical progress has not yet resulted in a satisfactory

G. Manzano Paule, *Thermodynamics and Synchronization in Open Quantum Systems*, Springer Theses, https://doi.org/10.1007/978-3-319-93964-3_8

quantum microscopic framework from which entropy production can be adequately characterized and interpreted in general nonequilibrium situations. Our results contribute to the development of such a general framework by deriving average and stochastic expressions for the total entropy production in system and environment in a two measurement protocol (TMP) scheme. This setup then allows us to split the entropy production into an adiabatic and a nonadiabatic contribution both fulfilling independent FT's, exactly as in classical stochastic thermodynamics [5, 31, 32]. However, contrary to what happens in classical systems, the split is not always possible, and specific conditions on the processes are necessary.

We organize the chapter as follows. In Sect. 8.1 we introduce a thermodynamic process for a generic bipartite system, that models a system and its environment. We will define in this section the entropy production along the process and the concomitant reduced system dynamics. We then develop a FT for this entropy production in Sect. 8.2 using a time-reversed or backward thermodynamic process. In Sect. 8.3, FT's for the adiabatic and nonadiabatic entropy production are derived. Our results are extended to arbitrary quantum trajectories as given by concatenations of CPTP maps in Sect. 8.4. This is applied to the specific case of Lindblad Master Equations in Sect. 8.5. Finally, we conclude in Sect. 8.6 with some final remarks.

8.1 Quantum Operations and Entropy Production

We start by introducing a generic process for a system interacting with some ancilla or environment and possibly subjected to an external driving control. The process is based on some initial and final measurements and a unitary evolution in between, as depicted in Fig. 8.1. We focus on the dynamics of the global bipartite system (system plus environment), but also in the reduced dynamics that affects the system and that can be described as a CPTP map (see Sect. 2.1). For those dynamics we introduce an entropy production based on the change of von Neumann entropy in the global system.

8.1.1 The Process

Along the chapter we consider an isolated quantum system composed of two parts, system and environment, with Hilbert space $\mathcal{H}_S \otimes \mathcal{H}_E$. We will focus our attention on the entropy production along the generic process depicted in Fig. 8.1, consisting of initial and final local projective measurements that bracket a unitary evolution. Notice that this corresponds to an extension of the generalized measurement process introduced in Sect. 1.3.2. The outcomes of the measurements constitute a quantum trajectory, which plays a crucial role in the formulation of FT's, as we have seen in Chap. 7, and will further emphasize in Sects. 8.2 and 8.3.

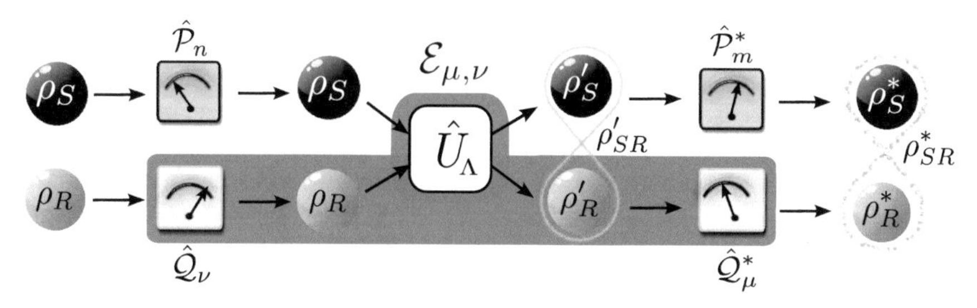

Fig. 8.1 Schematic picture of the forward process presented in the main text. System and environment start from an uncorrelated state $\rho_\mathcal{S} \otimes \rho_E$. A local measurement of observables with projectors $\{\hat{\mathcal{P}}_n, \hat{\mathcal{Q}}_\nu\}$ is carried out, which does not alter the density matrix in the average evolution but selects a pure state $|\psi_n\rangle \otimes |\chi_\nu\rangle$ at the trajectory level. System and environment then interact with each other and some external agent according to the unitary evolution $\hat{U}_\Lambda$, ending in an entangled state denoted as $\rho'_{\mathcal{S}E}$. Finally, we repeat the measurements with arbitrary projectors $\{\hat{\mathcal{P}}^*_m, \hat{\mathcal{Q}}^*_\mu.\}$. In the last measurement quantum correlations in state $\rho'_{\mathcal{S}E}$ are erased, while the final state $\rho^*_{\mathcal{S}E}$ may still have in general non-zero classical correlations. The reduced evolution of the system conditioned to the measurement in the environment are described through the quantum operation $\mathcal{E}_{\mu\nu}$ (shaded green area)

The process begins with the global system in an uncorrelated (product) state $\rho_{\mathcal{S}E} = \rho_\mathcal{S} \otimes \rho_E$, with local states defined by the spectral decompositions

$$\rho_\mathcal{S} = \sum_n p_n \hat{\mathcal{P}}_n, \qquad \rho_E = \sum_\nu q_\nu \hat{\mathcal{Q}}_\nu, \tag{8.1}$$

with eigenvalues p_n and q_ν, and $\hat{\mathcal{P}}_n \equiv |\psi_n\rangle\langle\psi_n|_\mathcal{S}$ and $\hat{\mathcal{Q}}_\nu \equiv |\chi_\nu\rangle\langle\chi_\nu|_E$ (rank-1) orthogonal projectors onto their respective eigenstates.

Our thermodynamic process begins at $t = t_0$ by performing an initial projective measurement on the system and environment, using the eigenprojectors in Eq. (8.1), and obtaining outcomes n and ν. This measurement projects the system and environment onto pure states $|\psi_n\rangle\langle\psi_n|_\mathcal{S}$ and $|\chi_\nu\rangle\langle\chi_\nu|_E$. Notice however that, when averaging over measurement results, the state of the global system does not change ($[\hat{\mathcal{P}}_n \otimes \hat{\mathcal{Q}}_\nu, \rho_{\mathcal{S}E}] = 0$).

Subsequently, we drive the compound system during the time interval $t = t_0$ to $t_0 + \tau$, building up correlations between the system and environment. The corresponding unitary operator $\hat{U}_\Lambda$ is generated by the total Hamiltonian $\hat{H}(t) = \hat{H}(\lambda_t)$, which depends on time through an external parameter λ_t that we vary according to a prescribed protocol $\Lambda = \{\lambda_t : t_0 \leqslant t \leqslant t_0 + \tau\}$:

$$\hat{U}_\Lambda \equiv \hat{T}_+ \exp\left(-\frac{i}{\hbar}\int_{t_0}^{t_0+\tau} dt\, \hat{H}(t)\right), \tag{8.2}$$

where $\hat{T}_+$ denotes the time ordering operator (see Sect. 1.1). As a result, the compound system at time $t = t_0 + \tau$ is described by the new density matrix

$$\rho'_{SE} = \hat{U}_\Lambda(\rho_S \otimes \rho_E)\hat{U}_\Lambda^\dagger. \tag{8.3}$$

The reduced (or local) states of the system and the environment can be obtained by partial tracing: $\rho'_S = \mathrm{Tr}_E[\rho'_{SE}]$ and $\rho'_E = \mathrm{Tr}_S[\rho'_{SE}]$ (see Sect. 1.3.2).

To complete the process, a second local projective measurement is performed at time $t = t_0 + \tau$ on both the system and environment. The measurements are characterized by arbitrary orthogonal projectors (for simplicity we assume rank-1 projectors), denoted as $\{\hat{\mathcal{P}}_m^*\}$ and $\{\hat{\mathcal{Q}}_\mu^*\}$ with outcomes m and μ, corresponding to arbitrary local observables in system and environment. In this case, the average global state is disturbed, transforming into

$$\begin{aligned}\rho^*_{SE} &= \sum_{m,\mu}(\hat{\mathcal{P}}_m^* \otimes \hat{\mathcal{Q}}_\mu^*)\rho'_{SE}(\hat{\mathcal{P}}_m^* \otimes \hat{\mathcal{Q}}_\mu^*)\\ &= \sum_{m,\mu}\rho^*_{m\mu}(\hat{\mathcal{P}}_m^* \otimes \hat{\mathcal{Q}}_\mu^*).\end{aligned} \tag{8.4}$$

Notice that the average state after measurement in Eq. (8.4) is not a product state: the final local measurements do not eliminate the classical correlations contained in ρ^*_{SE} [33]. However, the measurement collapses the local states of the system and environment into pure states $|\psi_m^*\rangle\langle\psi_m^*|_S \equiv \hat{\mathcal{P}}_m^*$ and $|\chi_\mu^*\rangle\langle\chi_\mu^*|_E \equiv \hat{\mathcal{Q}}_\mu^*$. Thus, the spectral decompositions of the reduced states after the final measurement are

$$\begin{aligned}\rho^*_S &\equiv \mathrm{Tr}_E(\rho^*_{SE}) = \sum_m p_m^*\hat{\mathcal{P}}_m^*,\\ \rho^*_E &\equiv \mathrm{Tr}_S(\rho^*_{SE}) = \sum_\mu q_\mu^*\hat{\mathcal{Q}}_\mu^*.\end{aligned} \tag{8.5}$$

where $p_m^* = \sum_\mu \rho^*_{m\mu}$ and $q_\mu^* = \sum_m \rho^*_{m\mu}$ are the corresponding (classical) marginal distributions.

8.1.2 Reduced Dynamics

The global manipulation described in the preceding section corresponds to a particular reduced dynamics of the system alone, whose thermodynamic analysis offers a complementary perspective. This reduced dynamics is illustrated by the shaded area in Fig. 8.1, which can be considered as an effective transformation of the state of the system, $\rho_S \to \rho'_S$, described by the action of a quantum CPTP map $\mathcal{E}$ that admits a Kraus representation [34]

$$\rho'_S = \mathcal{E}(\rho_S) = \sum_{\mu,\nu}\hat{M}_{\mu\nu}\rho_S\hat{M}_{\mu\nu}^\dagger, \tag{8.6}$$

with a set of Kraus operators $\hat{M}_{\mu\nu}$ satisfying

$$\sum_{\mu,\nu} \hat{M}^{\dagger}_{\mu\nu} \hat{M}_{\mu\nu} = \mathbb{1}. \tag{8.7}$$

As discussed in Sects. 2.1.2 and 2.1.3, there exist many Kraus representations, $\{\hat{M}_{\mu\nu}\}$, that reproduce the reduced dynamics on the system. For our purposes a convenient choice is given by

$$\hat{M}_{\mu\nu} = \sqrt{q_\nu} \langle \chi^*_\mu |_E \hat{U}_\Lambda | \chi_\nu \rangle_E. \tag{8.8}$$

This specific representation retains the relevant details of the evolution of the environment, relating unequivocally each Kraus operator $\hat{M}_{\mu\nu}$ with the transition $|\chi_\nu\rangle_E \to |\chi^*_\mu\rangle_E$ in the environment as a result of initial and final local projective measurements. This is a key point in order to describe the thermodynamics of the process at the trajectory level, as we will see shortly. Notice that other Kraus representations can be linked to the introduction of a local unitary operation $\hat{U}_E$ acting only on the environment just after the interaction with the system or, equivalently, to performing the final projective measurement on the environment in a different basis $\{\hat{U}_E \hat{\mathbb{Q}}^*_\mu \hat{U}_E\}$.

Let us finally define the *quantum operation*:

$$\mathcal{E}_{\mu\nu}(\rho_{\mathbb{S}}) = \hat{M}_{\mu\nu}\, \rho_{\mathbb{S}}\, \hat{M}^{\dagger}_{\mu\nu}, \tag{8.9}$$

which describes the conditioned evolution of the system when the environment starts in the pure state $|\chi_\nu\rangle_E$ and ends in the state $|\chi^*_\mu\rangle_E$ after measurement. The above operation can be associated to a conditional state

$$\rho_{S|\mu\nu} = \frac{\mathcal{E}_{\mu\nu}(\rho_{\mathbb{S}})}{P_{\mu\nu}}, \tag{8.10}$$

occurring with probability $P_{\mu\nu} = \mathrm{Tr}_{\mathbb{S}}[\mathcal{E}_{\mu\nu}(\rho_{\mathbb{S}})]$. Therefore we have $\sum_{\mu\nu} P_{\mu\nu}\rho_{S|\mu\nu} = \sum_{\mu\nu} \mathcal{E}_{\mu\nu}(\rho_{\mathbb{S}}) = \mathcal{E}(\rho_{\mathbb{S}}) = \rho'_{\mathbb{S}}$, corresponding to a generalized measurement setup, where the pair (ν, μ) can be considered as the outcome.

8.1.3 Average Entropy Production

We have so far introduced two levels of description for an open quantum system: the global dynamics and the reduced dynamics. Each suggests a different thermodynamics, which we analyze here by studying the variation in the von Neumann entropy $S(\rho)$ (see Sect. 1.1.6) along our process. It should be noticed that von Neumann entropy coincides with the thermodynamic entropy for equilibrium states (setting the Boltzmann constant $k_{\mathrm{B}} = 1$). Furthermore for the nonequilibrium states analyzed here, there are some situations where it can be also interpreted as a thermodynamic entropy [35]. However, we will refrain from identifying $S(\rho)$ with a thermodynamic entropy in general, and refer to it simply as the entropy or the quantum entropy of state ρ.

Along our thermodynamic process, the quantum entropy of the global system changes as

$$\Delta_{\rm i} S_{\rm inc} \equiv S(\rho^*_{\mathcal{S}E}) - S(\rho_{\mathcal{S}E}) \geqslant 0. \tag{8.11}$$

We will refer to $\Delta_{\rm i} S_{\rm inc}$ as the *inclusive* entropy production to distinguish it from the entropy production when the system and the environment are separated at the end of the process and the final classical correlations are lost (see below). The inclusive entropy production is always non-negative, since the von Neumann entropy cannot decrease in every projective measurement (see Sect. 1.3.1) and stays constant along any unitary evolution, i.e., $S(\rho_{\mathcal{S}E}) = S(\rho'_{\mathcal{S}E}) \leqslant S(\rho^*_{\mathcal{S}E})$.

Our analysis relies on taking account of the entropy production associated to the classical and quantum correlations built up between our system and its surroundings. The quantum mutual information, as defined in Sect. 1.4.2, assesses the total amount of such correlations. For any arbitrary bipartite state $\sigma_{\mathcal{S}E}$ with reduced states $\sigma_{\mathcal{S}}$ and σ_E, we will denote the mutual information

$$I(\sigma_{\mathcal{S}E}) \equiv S(\sigma_{\mathcal{S}}) + S(\sigma_E) - S(\sigma_{\mathcal{S}E}) = S(\sigma_{\mathcal{S}E}||\sigma_{\mathcal{S}} \otimes \sigma_E), \tag{8.12}$$

which becomes zero for product states $\sigma_{\mathcal{S}E} = \sigma_{\mathcal{S}} \otimes \sigma_E$.

Let us denote $\Delta S_{\mathcal{S}} = S(\rho^*_{\mathcal{S}}) - S(\rho_{\mathcal{S}})$ and $\Delta S_E = S(\rho^*_E) - S(\rho_E)$ the local entropy changes in system and environment during the whole process. Using relative entropies and mutual information, the inclusive entropy production $\Delta_{\rm i} S_{\rm inc}$ can be rewritten as

$$\begin{aligned} \Delta_{\rm i} S_{\rm inc} &= \Delta S_{\mathcal{S}} + \Delta S_E - I(\rho^*_{\mathcal{S}E}) \\ &= \Delta S_{\mathcal{S}}^{\rm meas} + \Delta S_E^{\rm meas} + I(\rho'_{\mathcal{S}E}) - I(\rho^*_{\mathcal{S}E}) \geqslant 0 \end{aligned} \tag{8.13}$$

where we have used the definition of the quantum mutual information in Eq. (8.12), and the fact that the initial state is uncorrelated. In the second equality, we introduced the local entropy changes $\Delta S_{\mathcal{S}}^{\rm meas} \equiv S(\rho^*_{\mathcal{S}}) - S(\rho'_{\mathcal{S}})$ and $\Delta S_E^{\rm meas} \equiv S(\rho^*_E) - S(\rho'_E)$ which allow us to explicitly show the two main consequences introduced by the local projective measurements at the end of the protocol in entropic terms. The first one, captured by the term $\Delta S_{\mathcal{S}}^{\rm meas} + \Delta S_E^{\rm meas} \geqslant 0$, corresponds to entropy production due to the disturbance induced by the final measurement in the local states $\rho'_{\mathcal{S}} \to \rho^*_{\mathcal{S}}$ and $\rho'_E \to \rho^*_E$. The second one, $I(\rho'_{\mathcal{S}E}) - I(\rho^*_{\mathcal{S}E}) \geqslant 0$, is the erasure of the quantum correlations in the state $\rho'_{\mathcal{S}E}$, due to the local character of the measurements [36]. If the final measurement is performed in the eigenbasis of the local states after interaction, that is $[\hat{\mathcal{P}}^*_m, \rho'_{\mathcal{S}}] = 0$ and $[\hat{\mathcal{Q}}^*_\mu, \rho'_E] = 0$, the first term vanishes and second one reduces to the so-called *measurement induced disturbance* introduced by Luo in Ref. [37] as a measure of the quantumness of correlations akin to discord [36]. Therefore the process is completely reversible, i.e. it generates exactly zero entropy production, if and only if the global state after interaction $\rho'_{\mathcal{S}E}$ is *classical* in the sense [37]

$$\rho'_{\mathcal{S}E} = \sum_{m,\mu} \rho'_{m,\mu} \hat{\mathcal{P}}'_m \otimes \hat{\mathcal{Q}}'_\mu, \tag{8.14}$$

where we introduced the eigenprojectors $\{\hat{\mathcal{P}}'_m\}$ and $\{\hat{\mathcal{Q}}'_\mu\}$ of the reduced states $\rho'_{\mathcal{S}} = \sum_{m,\mu} \rho'_{m,\mu} \hat{\mathcal{P}}'_m$, and $\sum_{m,\mu} \rho'_{m,\mu} \hat{\mathcal{Q}}'_\mu$ respectively. However, we notice that the tradefoff between local and global disturbances should be examined in more detail in order to obtain minimum (non-zero) entropy production, as given by Eq. (8.13).

Moreover, in most situations the classical correlations remaining after the final measurement are irreversibly lost, with an entropic cost equal to the mutual information $I(\rho^*_{\mathcal{S}E})$. This is case if we separate system and environment after the process and all subsequent manipulations are local.[1] The entropy production in those situations is

$$\Delta_{\text{i}} S \equiv \Delta S_{\mathcal{S}} + \Delta S_E = \Delta S_{\mathcal{S}}^{\text{meas}} + \Delta S_E^{\text{meas}} + I(\rho'_{\mathcal{S}E}) \geqslant 0. \tag{8.15}$$

We will refer to $\Delta_{\text{i}} S$ as the non-inclusive entropy production or simply the entropy production. Notice finally that $\Delta_{\text{i}} S \geqslant \Delta_{\text{i}} S_{\text{inc}} \geqslant 0$, since the mutual information $I(\rho^*_{\mathcal{S}E})$ is always non-negative.

8.2 Backward Process and Fluctuation Theorem

So far we have introduced our thermodynamic process and discussed possible definitions of the average entropy production in the whole setup. Our main goal here is to find the corresponding stochastic entropic changes at the level of quantum trajectories and the quantum FT's that they satisfy. Most FT's are based on the ratio of the probability to find a trajectory in a process with the probability to find the time-reversal trajectory when the process is run backwards in time [1, 2]. We, therefore, need to define trajectories and the backward process associated to the one introduced in the previous section (Fig. 8.1), which we will call the forward process.

A trajectory γ of the forward process is simply given by the outcome of the four measurements, i.e., $\gamma = \{n, \nu, \mu, m\}$. This trajectory corresponds to the following transition between pure states,

$$|\psi_n\rangle_{\mathcal{S}} \otimes |\chi_\nu\rangle_E \rightarrow |\psi^*_m\rangle_{\mathcal{S}} \otimes |\chi^*_\mu\rangle_E. \tag{8.16}$$

Notice that, in virtue of our choice of the Kraus representation for the reduced dynamics Eq. (8.8) a trajectory γ is also a trajectory of the reduced dynamics, where

[1] We also exclude here the possibility of further implementation of feedback protocols using local measurements and classical communication of the results.

the pair (ν, μ) now indicates the quantum operation affecting the system instead of the initial and final states of the environment (which is otherwise hidden in the reduced dynamics). The probability to observe that trajectory γ is hence

$$P_\gamma = p_n q_\nu \mathrm{Tr}[(\hat{\mathcal{P}}^*_m \otimes \hat{\mathcal{Q}}^*_\mu)\hat{U}_\Lambda(\hat{\mathcal{P}}_n \otimes \hat{\mathcal{Q}}_\nu)\hat{U}^\dagger_\Lambda]. \tag{8.17}$$

To introduce the backward process, we make use of the anti-unitary time-reversal operator in quantum mechanics, $\hat{\Theta}$ introduced previously (see Sects. 1.1 and 7.2). This operator changes the sign of odd variables under time reversal, like linear and angular momenta or magnetic field [38]. We will consider the separate time reversal operators for system, $\hat{\Theta}_\mathcal{S}$, and environment, $\hat{\Theta}_E$, as well as the one for the total system $\hat{\Theta} = \hat{\Theta}_\mathcal{S} \otimes \hat{\Theta}_E$.

The backward process is defined by implementing the control actions of the forward process in reverse under the action of the time-reversal operator. Thus, we start with an initial state of the form

$$\tilde{\rho}_{\mathcal{S}E} = \sum_{m,\mu} \tilde{p}_{m\mu}\, \hat{\Theta}_\mathcal{S} \hat{\mathcal{P}}^*_m \hat{\Theta}^\dagger_\mathcal{S} \otimes \hat{\Theta}_E \hat{\mathcal{Q}}^*_\mu \hat{\Theta}^\dagger_E \tag{8.18}$$

As in the forward process, the first step at time $t = t_0$ is a local measurement of the family of projectors $\{\hat{\Theta}_\mathcal{S} \hat{\mathcal{P}}^*_m \hat{\Theta}^\dagger_\mathcal{S}, \hat{\Theta}_E \hat{\mathcal{Q}}^*_\mu \hat{\Theta}^\dagger_E\}$. According to Eq. (8.18), the outcomes m and μ are obtained with probability $\tilde{p}_{m\mu}$.

We then let the global system evolve under the Hamiltonian $\hat{H}(\lambda_t)$ used in the forward process, but inverting the time-dependent protocol to $\tilde{\Lambda} \equiv \{\tilde{\lambda}_t | \, t_0 \leqslant t \leqslant t_0 + \tau\}$, which follows exactly the inverse sequence of values for the control parameter with respect to Λ. The evolution is thus given by the unitary transformation

$$\hat{U}_{\tilde{\Lambda}} \equiv T_+ \exp\left(-\frac{i}{\hbar}\int_{t_0}^{t_0+\tau} dt\, \hat{\Theta}\hat{H}(2t_0 + \tau - t)\hat{\Theta}^\dagger\right). \tag{8.19}$$

Following Sect. 1.1.4 (see also [39, 40]), the microreversibility principle for non-autonomous systems relates forward and backward unitary evolutions by:

$$\hat{\Theta}^\dagger \hat{U}_{\tilde{\Lambda}} \hat{\Theta} = \hat{U}^\dagger_\Lambda \tag{8.20}$$

which is the key property we need for relating probabilities of trajectories γ and $\tilde{\gamma}$. Finally, at time $t = t_0 + \tau$ we perform new local measurements on the system and environment using projectors $\{\hat{\Theta}_\mathcal{S} \hat{\mathcal{P}}_n \hat{\Theta}^\dagger_\mathcal{S}, \hat{\Theta}_E \hat{\mathcal{Q}}_\nu \hat{\Theta}^\dagger_E\}$. The outcome induces a quantum jump

$$\hat{\Theta}|\psi^*_m\rangle_\mathcal{S} \otimes |\chi^*_\mu\rangle_E \rightarrow \hat{\Theta}|\psi_n\rangle \otimes |\chi_\nu\rangle_E, \tag{8.21}$$

and the corresponding backward trajectory $\tilde{\gamma} = \{m, \mu, \nu, n\}$ occurs with probability

$$\tilde{P}_{\tilde{\gamma}} = \tilde{p}_{m\mu} \mathrm{Tr}[\hat{\Theta}\left(\hat{\mathcal{P}}_n \otimes \hat{\mathcal{Q}}_\nu\right)\hat{\Theta}^\dagger \hat{U}_{\tilde{\Lambda}} \hat{\Theta}\left(\hat{\mathcal{P}}^*_m \otimes \hat{\mathcal{Q}}^*_\mu\right)\hat{\Theta}^\dagger \hat{U}^\dagger_{\tilde{\Lambda}}]. \tag{8.22}$$

The FT now follows readily by comparing the probabilities (8.17) and (8.22), using the microreversibility property (8.20) and the cyclic property of the trace. The result is

$$\Delta_{\mathrm{i}} s_\gamma \equiv \ln \frac{P_\gamma}{\tilde{P}_{\tilde{\gamma}}} = \ln \frac{p_n q_\nu}{\tilde{\rho}_{m,\mu}} = \sigma^S_{nm} + \sigma^E_{\nu\mu} - \tilde{I}_{m\mu}, \tag{8.23}$$

with the quantities

$$\sigma^{\mathrm{S}}_{nm} = \ln p_n - \ln \tilde{p}_m, \tag{8.24}$$

$$\sigma^E_{\mu\nu} = \ln q_\nu - \ln \tilde{q}_\mu, \tag{8.25}$$

$$\tilde{I}_{m,\mu} = \ln \tilde{\rho}_{m,\mu} - \ln \tilde{p}_m \tilde{q}_\mu. \tag{8.26}$$

The two first terms, σ^{S}_{nm} and $\sigma^E_{\mu\nu}$, can be interpreted as local boundary terms depending on the choice of the local states for system and environment at the beginning of forward and backward processes, in analogy to the previous chapter (see Sect. 7.4) and to entropy fluctuation theorems for the classical case [4]. In particular, if $\tilde{p}_m = p^*_m$ and $\tilde{q}_\mu = q^*_\mu$, then $\sigma^{\mathrm{S}}_{nm} = \Delta s^{\mathrm{S}}_{nm}$ and $\sigma^E_{\nu\mu} = \Delta s^E_{\nu\mu}$ correspond to stochastic entropic changes per trajectory in system and environment, as introduced in Refs. [4, 23, 24, 41, 42]. The third term, $\tilde{I}_{m,\mu}$ corresponds to the stochastic version of the mutual information [43] in the initial state of the backward process, c.f. Eq. (8.18). From the detailed FT in Eq. (8.23), we immediately have the integral version

$$\langle e^{-\Delta_{\mathrm{i}} s_\gamma} \rangle = \sum_\gamma P_\gamma e^{-\Delta_{\mathrm{i}} s_\gamma} = \sum_\gamma \tilde{P}_{\tilde{\gamma}} = 1. \tag{8.27}$$

Furthermore, concavity of the exponential function (Jensen's inequality) implies $\langle e^x \rangle \geqslant e^{\langle x \rangle}$, yielding the second-law-like inequality

$$\langle \Delta_{\mathrm{i}} s_\gamma \rangle = \langle \sigma^S \rangle + \langle \sigma^E \rangle - \langle \tilde{I} \rangle \geqslant 0. \tag{8.28}$$

The interpretation of $\Delta_{\mathrm{i}} s_\gamma$ depends on the choice of $\tilde{\rho}_{\mathrm{S}E}$, the initial global state of the backward process. If we set reversible boundaries of the form $\tilde{\rho}_{\mathrm{S}E} = \hat{\Theta} \rho^*_{\mathrm{S}E} \hat{\Theta}^\dagger$, then $\tilde{\rho}_{m\mu} = \rho^*_{m\mu}$, which implies $\tilde{I}_{m,\mu} = \ln \rho^*_{m,\mu} - \ln p^*_m q^*_\mu \equiv I^*_{m,\mu}$, and $\Delta_{\mathrm{i}} s_\gamma$ is the inclusive entropy production per trajectory. Its average

$$\begin{aligned}
\langle \Delta_{\mathrm{i}} s_\gamma \rangle &= -\sum_{m,\mu} \rho^*_{m\mu} \ln \rho^*_{m\mu} + \sum_n p_n \ln p_n + \sum_\nu q_\nu \ln q_\nu \\
&= S(\rho^*_{\mathrm{S}E}) - S(\rho_{\mathrm{S}}) - S(\rho_E) = \Delta_{\mathrm{i}} S_{\mathrm{inc}}
\end{aligned} \tag{8.29}$$

equals the inclusive entropy production defined in (8.11). If the initial condition for the backward process is instead the uncorrelated state $\tilde{\rho}_{\mathrm{S}E} = \hat{\Theta}(\rho^*_{\mathrm{S}} \otimes \rho^*_E)\hat{\Theta}^\dagger$, then $\tilde{\rho}_{m\mu} = p^*_m q^*_\mu$ implying $\tilde{I}_{m,\mu} = 0$, and $\Delta_{\mathrm{i}} s_\gamma = \Delta s^{\mathrm{S}}_{nm} + \Delta s^E_{\nu\mu}$ is the non inclusive entropy production per trajectory, whose average yields the entropy production in

Eq. (8.15)

$$\langle \Delta_i s_\gamma \rangle = S(\rho_S^*) - S(\rho_S) + S(\rho_E^*) - S(\rho_E) = \Delta_i S. \tag{8.30}$$

For equilibrium canonical initial conditions both in the forward and in the backward processes, the entropy per trajectory equals the dissipative work and one recovers the celebrated Crooks work theorem and the original Jarzynski equality [40].

8.3 Adiabatic and Non-adiabatic Entropy Production

We now focus on the reduced dynamics. Our aim is to obtain FT's involving only the quantum trajectory defined in Sect. 8.2 and the initial and final states of the system. To do that, we follow our previous work in Chap. 7, where we derived a FT for CPTP maps, basing on the dual dynamics introduced by Crooks in Ref. [44] and the introduction of a nonequilibrium potential depending on the invariant state of the map. Interestingly, the resulting FT goes beyond the one that we have obtained considering the global dynamics, Eq. (8.23), and will reveal an interesting split of the total entropy production per trajectory $\gamma = \{n, \nu, m, \mu\}$ into two terms:

$$\Delta_i s_\gamma = \Delta_i s_{\mu\nu}^{\mathrm{a}} + \Delta_i s_\gamma^{\mathrm{na}}, \tag{8.31}$$

the adiabatic entropy production $\Delta_i s_{\mu\nu}^{\mathrm{a}}$ which accounts for the irreversibility of evolution in the stationary regime, and the non-adiabatic entropy production $\Delta_i s_\gamma^{\mathrm{na}}$ which measures how far the system is from the stationary state of the dynamics.

We will apply the formalism discussed in Chap. 7 to $\mathcal{E}$, the map governing the reduced dynamics of our process, as well as to the map corresponding to the backward dynamics. We first need to introduce the reduced dynamics in the backward process, which will be described by a new CPTP map denoted by $\tilde{\mathcal{E}}$. To do that, it is necessary that the system and the environment start the backward process in an uncorrelated state $\tilde{\rho}_{SE} = \tilde{\rho}_S \otimes \tilde{\rho}_E$, i.e., we have to impose $\tilde{I}_{m\mu} = 0$ (see Eq. (8.26)). In that case, similarly to our choice (8.8) for the forward process, a useful representation of $\tilde{\mathcal{E}}$ is

$$\tilde{\mathcal{E}}_{\nu\mu}(\tilde{\rho}_S) = \hat{\tilde{M}}_{\nu\mu} \tilde{\rho}_S \hat{\tilde{M}}_{\nu\mu}^\dagger \tag{8.32}$$

where the backward Kraus operators are given by

$$\hat{\tilde{M}}_{\nu\mu} = \sqrt{\tilde{q}_\mu} \langle \chi_\nu |_E \hat{\Theta}_E^\dagger \, \hat{U}_{\tilde{\Lambda}} \, \hat{\Theta}_E | \chi_\mu^* \rangle_E. \tag{8.33}$$

Notice that here we have swapped the subscripts with respect to the definition of the forward operators given by Eq. (8.8). This can be done since the pair (μ, ν) is just a label of the Kraus operator. The choice in Eq. (8.33) means that the operation $\tilde{\mathcal{E}}_{\nu\mu}$ is equivalent to obtaining μ in the initial measurement of the backward process followed ν at the end. Now, microreversibility (8.20) implies an intimate relationship between the forward and backward Kraus operators:

$$\hat{\Theta}_{\mathcal{S}}^{\dagger} \hat{\tilde{M}}_{\nu\mu} \hat{\Theta}_{\mathcal{S}} = \sqrt{\tilde{q}_{\mu}} \langle \chi_{\nu} |_E \hat{U}_{\Lambda}^{\dagger} | \chi_{\mu}^{*} \rangle_E = e^{-\sigma_{\mu\nu}^{E}/2} \hat{M}_{\mu\nu}^{\dagger}. \tag{8.34}$$

Forward and backward processes can then be completely associated to the maps $\mathcal{E}$, and $\tilde{\mathcal{E}}$ respectively, together with its initial condition for the system. Each one induces an evolution onto the system characterized by trajectories. We can compute the probability of observing a trajectory $\gamma = \{n, \nu, m, \mu\}$ in the forward process or its reverse $\tilde{\gamma} = \{m, \mu, \nu, n\}$ in the backward process as

$$P_{\gamma} = p_n \mathrm{Tr}[\hat{\mathcal{P}}_m^{*} \ \hat{M}_{\mu\nu}(\hat{\mathcal{P}}_n) \hat{M}_{\mu\nu}^{\dagger}], \tag{8.35}$$

$$\tilde{P}_{\tilde{\gamma}} = \tilde{p}_m \mathrm{Tr}[\hat{\Theta}_{\mathcal{S}} \hat{\mathcal{P}}_n \hat{\Theta}_{\mathcal{S}}^{\dagger} \ \hat{\tilde{M}}_{\nu\mu} (\hat{\Theta}_{\mathcal{S}} \hat{\mathcal{P}}_m^{*} \hat{\Theta}_{\mathcal{S}}^{\dagger}) \hat{\tilde{M}}_{\nu\mu}^{\dagger}]. \tag{8.36}$$

Notice that constructing the log-ratio between P_{γ} and $\tilde{P}_{\tilde{\gamma}}$ immediately gives the FT for the total entropy production in Eq. (8.23) by virtue of the relationship (8.34). In other words, Eq. (8.34) expresses the fundamental symmetry under time reversal yielding the FT for the total entropy production.

8.3.1 *The Dual-Reverse Process*

In order to go beyond the FT for the total entropy production, we proceed as in Chap. 7, introducing the dual-reverse dynamics that reveals the irreversibility associated to a map when starting from a positive-definite invariant state $\pi = \mathcal{E}(\pi)$. The dual-reverse dynamics is defined as a map $\tilde{\mathcal{D}}(\rho)$ such that $\tilde{\pi} \equiv \hat{\Theta}_{\mathcal{S}} \pi \ \hat{\Theta}_{\mathcal{S}}^{\dagger}$ is an invariant state, i.e., $\tilde{\mathcal{D}}(\tilde{\pi}) = \tilde{\pi}$. Furthermore, we require that when the map is applied several times starting in the stationary state $\tilde{\pi}$, it generates trajectories $\tilde{\gamma}$ distributed as $\tilde{P}_D(\tilde{\gamma}|\tilde{\pi}) = P(\gamma|\pi)$. Here the trajectories are given by $\gamma = \{n, (\nu_1, \mu_1), \ldots, (\nu_N, \mu_N), m\}$ and $\tilde{\gamma} = \{m, (\mu_N, \nu_N), \ldots, (\mu_1, \nu_1), n\}$, corresponding to N applications of the maps.

Summarizing, in the stationary regime the dual-reverse generates the same ensemble of trajectories as the forward process, but reversed in time. For instance, if the map describes the dynamics of a system in contact with a single thermal bath (thermalization), then the forward process generates reversible trajectories (indistinguishable from their reversal) and the dual-reverse coincides with the forward map. In nonequilibrium situations, the dual generically inverts flows. For instance, for a system in contact with two thermal baths at different temperatures, the dual-reverse is usually obtained by swapping the temperatures of the baths, hence inverting the flow of heat.

In any case, one can prove that a Kraus representation of the dual-reverse map is given by the operators:

$$\hat{\tilde{D}}_{\nu\mu} = \hat{\Theta}_{\mathcal{S}} \ \pi^{\frac{1}{2}} \hat{M}_{\mu\nu}^{\dagger} \pi^{-\frac{1}{2}} \ \hat{\Theta}_{\mathcal{S}}^{\dagger}. \tag{8.37}$$

Here again we have swapped the subscripts ν and μ with respect to the Kraus operators of the forward map in Eq. (8.8). Finally, the dual-reverse process is the dual-reverse map complemented by a specific choice of the initial condition for the system (the environment does not appear explicitly in the map, which acts only on the system). The appropriate initial condition for the dual-reverse process is $\tilde{\rho}_{\mathcal{S}}$, i.e., the same as in the backward one. Therefore we can now compute the probability in the dual-reverse process to obtain the (reverse) trajectory $\tilde{\gamma} = \{m, \mu, \nu, n\}$ as

$$\tilde{P}_{\tilde{\gamma}}^{D} = \tilde{p}_m \text{Tr}[\hat{\Theta}_{\mathcal{S}} \hat{\mathcal{P}}_n \hat{\Theta}_{\mathcal{S}}^{\dagger} \,\hat{\tilde{D}}_{\nu\mu} (\hat{\Theta}_{\mathcal{S}} \hat{\mathcal{P}}_m^{*} \hat{\Theta}_{\mathcal{S}}^{\dagger}) \hat{\tilde{D}}_{\nu\mu}^{\dagger}]. \tag{8.38}$$

To obtain a FT from P_γ and $\tilde{P}_{\tilde{\gamma}}^{D}$ is necessary a condition of proportionality between operators $\hat{M}_{\mu\nu}^{\dagger}$, and $\hat{\tilde{D}}_{\nu\mu}$, similar to the relationship (8.34) between $\hat{M}_{\mu\nu}^{\dagger}$, and $\hat{\tilde{M}}_{\nu\mu}$.

In the previous chapter, we found that a necessary and sufficient condition for that proportionality is the following. We first define the nonequilibrium potential $\hat{\Phi} = -\ln \pi$, from the invariant state π. Its spectral decomposition reads:

$$\hat{\Phi} = \sum_i \phi_i |\pi_i\rangle\langle\pi_i| \tag{8.39}$$

where $\phi_i = -\ln \pi_i$, and π_i and $\{|\pi_i\rangle\}$ are, respectively, the eigenvalues and eigenstates of the invariant density matrix π. Now we require that each Kraus operator $\hat{M}_{\mu\nu}$ is unambiguously related to a nonequilibrium potential change $\Delta\phi_{\mu\nu}$.[2] In the invariant state eigenbasis:

$$\hat{M}_{\mu\nu} = \sum_{i,j} m_{ij}^{\mu\nu} |\pi_j\rangle\langle\pi_i| \tag{8.40}$$

that condition is equivalent to:

$$m_{ij}^{\mu\nu} = 0 \qquad \text{whenever } \phi_j - \phi_i \neq \Delta\phi_{\mu\nu}. \tag{8.41}$$

As pointed in Chap. 7 this condition does not imply single jumps between pairs of π eigenstates, but it could account for any set of correlated transitions between different pairs with same associated $\Delta\phi_{\mu\nu}$. An extreme example are unital maps, where π is proportional to the identity matrix. In that case, $\Delta\phi_{\mu\nu} = 0$ and any complex coefficients $m_{ij}^{\mu\nu}$ satisfy Eq. (8.41). It can also be show that condition (8.41) is equivalent to $[\hat{\Phi}, \hat{M}_{\mu,\nu}] = \ \Delta\phi_{\mu,\nu}\hat{M}_{\mu,\nu}$. This alternative formulation of (8.41) indicates that, when $\Delta\phi_k \neq 0$, $\hat{M}_k(\lambda_t)$ can be interpreted as ladder operators in the eigenbasis of the invariant state π.

Introducing condition (8.41) in Eq. (8.37), one easily derives the relationship (7.15) in Chap. 7 between the forward and the dual-reverse Kraus operators, which here reads

[2] Note however that the converse statement is not necessarily true, i.e. we may have for different values of μ and ν the same value of $\Delta\phi_{\mu\nu}$

$$\hat{\Theta}_{\mathrm{S}}^{\dagger} \hat{\tilde{D}}_{\nu\mu} \hat{\Theta}_{\mathrm{S}} = e^{\Delta\phi_{\mu\nu}/2} \hat{M}_{\mu\nu}^{\dagger}. \tag{8.42}$$

Finally, inserting (8.42) in the expressions for the probability of trajectories (8.35)–(8.38) we reproduce the FT derived in Chap. 7 for the present setup:

$$\Delta_{\mathrm{i}} s_{\gamma}^{\mathrm{na}} \equiv \ln \frac{P_{\gamma}}{\tilde{P}_{\tilde{\gamma}}^{D}} = \sigma_{nm}^{\mathrm{S}} - \Delta\phi_{\mu\nu}. \tag{8.43}$$

We call $\Delta_{\mathrm{i}} s_{\gamma}^{\mathrm{na}}$ the *non-adiabatic* entropy production, following the terminology used in classical stochastic thermodynamics [5, 31, 32]. Its precise meaning can be seen by assuming that both the forward and dual-reverse process start the evolution in its invariant states: π, and $\tilde{\pi} = \hat{\Theta}_{\mathrm{S}} \pi \, \hat{\Theta}_{\mathrm{S}}^{\dagger}$. In such case, by construction, $\Delta\phi_{\mu\nu} = \Delta\phi_{mn} = -\ln \pi_m + \ln \pi_n = \sigma_{nm}^{\mathrm{S}}$, and hence $\Delta_{\mathrm{i}} s_{\gamma}^{\mathrm{na}} = 0 \; \forall \gamma$, that is, the forward and the dual-reverse processes are time-symmetric, being any trajectory γ in the forward process equally probable than its reverse $\tilde{\gamma}$ in the dual-reverse process. We conclude that the non-adiabatic entropy production $\Delta_{\mathrm{i}} s_{\gamma}^{\mathrm{na}}$ then captures the irreversibility in terms of time-symmetry breaking in any single trajectory γ due to the distance between the actual state of the system and the invariant state π. Below we will discuss the average version of the non-adiabatic entropy production in some cases, clarifying its origin.

8.3.2 The Dual Process

Let us now apply the same procedure to the backward process. In this way, we will obtain the dual-reverse of the backward map, which we simply call the dual map $\mathcal{D}$. If condition (8.41) is satisfied, then, by virtue of (8.34), the backward Kraus operators can be written as:

$$\hat{\tilde{M}}_{\nu\mu} = e^{-\sigma_{\mu\nu}^{E}/2} \sum_{i,j} (m_{ij}^{\mu\nu})^{*} \hat{\Theta}_{\mathrm{S}} |\pi_i\rangle\langle\pi_j| \hat{\Theta}_{\mathrm{S}}^{\dagger} = \sum_{i,j} \tilde{m}_{ij}^{\nu\mu} |\tilde{\pi}_j\rangle\langle\tilde{\pi}_i|$$

with $\tilde{m}_{ij}^{\nu\mu} \equiv e^{-\sigma_{\mu\nu}^{E}/2} (m_{ji}^{\mu\nu})^{*}$. We observe that, setting $\Delta\tilde{\phi}_{\nu\mu} = -\Delta\phi_{\mu\nu}$, condition (8.41) is recovered for the backward process. However, an additional requirement to apply this theoretical framework is that $\tilde{\pi} = \hat{\Theta}_{\mathrm{S}} \pi \hat{\Theta}_{\mathrm{S}}^{\dagger}$ is an invariant state of the backward map

$$\tilde{\mathcal{E}}(\tilde{\pi}) = \tilde{\pi}. \tag{8.44}$$

This is not guaranteed by the definition of $\tilde{\mathcal{E}}$, not even when the Kraus operators are of the form (8.41). In particular, it is satisfied when the driving protocol is time-symmetric, the Hamiltonian of the environment is invariant under time reversal, and we perform the same measurements at the beginning and the end of the process on the environment.

Therefore, adding this extra assumption, we now obtain the dual operators $\hat{D}_{\mu\nu}$, applying transformation (8.37) to the backward Kraus operators $\hat{\tilde{M}}_{\nu\mu}$ (with the role of $\hat{\Theta}_{\mathrm{S}}$ and $\hat{\Theta}_{\mathrm{S}}^{\dagger}$ swapped). Similarly to (8.42), condition (8.41) on the backward operators implies

$$\hat{\Theta}_{\mathrm{S}} \hat{D}_{\mu\nu} \hat{\Theta}_{\mathrm{S}}^{\dagger} = e^{\Delta\tilde{\phi}_{\nu\mu}/2} \hat{\tilde{M}}_{\nu\mu}^{\dagger} = e^{-\Delta\phi_{\mu\nu}/2} \hat{\tilde{M}}_{\nu\mu}^{\dagger} \tag{8.45}$$

and, using Eq. (8.34),

$$\hat{D}_{\mu\nu} = e^{-(\sigma_{\mu\nu}^{E} + \Delta\phi_{\mu\nu})/2} \hat{M}_{\mu\nu}. \tag{8.46}$$

The dual process is given by the dual map $\mathcal{D}$ with initial condition ρ_{S}. The trajectories generated by this process are distributed as

$$P_{\gamma}^{D} = p_n \mathrm{Tr}_{\mathrm{S}} \left[\hat{\mathcal{P}}_m^{*} \hat{D}_{\mu\nu} \hat{\mathcal{P}}_n \hat{D}_{\mu\nu}^{\dagger} \right], \tag{8.47}$$

to be compared with the probability of obtaining the same trajectory, $\gamma = \{n, \nu, m, \mu\}$, in the forward process.

Combining Eqs. (8.35) and (8.47), and using condition (8.34), we get a third FT:

$$\Delta_{\mathrm{i}} s_{\mu\nu}^{\mathrm{a}} = \ln \frac{P_{\gamma}}{P_{\gamma}^{D}} = \sigma_{\mu\nu}^{E} + \Delta\phi_{\mu\nu}. \tag{8.48}$$

where we call $\Delta_{\mathrm{i}} s_{\mu\nu}^{\mathrm{a}}$ the *adiabatic* entropy production [5, 31, 32]. Notice that, unlike the non-adiabatic entropy production, the adiabatic contribution is independent of the measurement results performed on the open system, stressing the fact that it is independent of its state. Indeed the adiabatic entropy production $\Delta_{\mathrm{i}} s_{\mu\nu}^{\mathrm{a}}$ captures the complementary time-symmetry breaking in a trajectory γ not accounted from by $\Delta_{\mathrm{i}} s_{\gamma}^{\mathrm{na}}$. When the initial states of forward and dual processes are the invariant state π, which implies $\Delta\phi_{\mu\nu} = \Delta\phi_{mn} = \sigma_{nm}^{\mathrm{S}}$ as before, we have that $\Delta_{\mathrm{i}} s_{\mu\nu}^{\mathrm{a}} = \sigma_{\mu\nu}^{E} + \sigma_{nm}^{\mathrm{S}} = \Delta_{\mathrm{i}} s_{\gamma}$ $\forall \gamma$, i.e. the adiabatic entropy production becomes the total entropy production $\Delta_{\mathrm{i}} s_{\gamma}$ for any trajectory γ. On the other hand, the adiabatic entropy production vanishes when the changes in the nonequilibrium potential become (minus) the changes in stochastic entropy of the environment, i.e. $\Delta_{\mathrm{i}} s_{\mu\nu}^{\mathrm{a}} = -\Delta\phi_{\mu,\nu}$. In that case the situation simplifies, being the forward process equal to the dual process and the backward process equal to the dual-reverse.

8.3.3 *Second-Law-Like Equalities and Inequalities*

We have hence obtained two detailed fluctuation theorems for the adiabatic and non-adiabatic entropy production, Eqs. (8.48) and (8.43) respectively, which contribute the total entropy production per trajectory, $\Delta_{\mathrm{i}} s_{\gamma} = \Delta_{\mathrm{i}} s_{\mu\nu}^{\mathrm{a}} + \Delta_{\mathrm{i}} s_{\gamma}^{\mathrm{na}}$. We can now derive integral FT's for both contributions:

$$\langle e^{-\Delta_i s^{na}} \rangle = 1, \qquad \langle e^{-\Delta_i s^{a}} \rangle = 1, \tag{8.49}$$

which follow from the detailed versions by averaging over trajectories γ. Finally, convexity of the exponential function provides the following two second-law-like inequalities as a corollary $\langle \Delta_i s_\gamma^{na} \rangle \geq 0$ and $\langle \Delta_i s_\gamma^{a} \rangle \geq 0$.

As for the FT for the total entropy production (8.23), the meaning of these average entropies becomes clearer if the initial condition of the backward process is specified. Setting reversible boundaries without correlations $\tilde{\rho}_{\mathcal{S}E} = \hat{\Theta}(\rho_{\mathcal{S}}^* \otimes \rho_E^*)\hat{\Theta}^\dagger$, the averages of the adiabatic and non-adiabatic entropy productions defined by (8.42) and (8.46) read

$$\Delta_i S_{na} \equiv \langle \Delta_i s_\gamma^{na} \rangle = \Delta S_{\mathcal{S}} - \langle \Delta \phi \rangle \geq 0, \tag{8.50}$$

$$\Delta_i S_a \equiv \langle \Delta_i s_\gamma^{a} \rangle = \Delta S_E + \langle \Delta \phi \rangle \geq 0, \tag{8.51}$$

and the sum equals the total non-inclusive average entropy production $\Delta_i S$ in Eq. (8.15).

It is interesting to notice that the average change of the nonequilibrium potential

$$\langle \Delta \phi \rangle = \sum_{\mu,\nu} P_\gamma \Delta \phi_{\mu\nu} = \sum_{\mu,\nu} \text{Tr}[\hat{M}_{\mu\nu} \rho_{\mathcal{S}} \hat{M}_{\mu\nu}^\dagger] \Delta \phi_{\mu\nu}, \tag{8.52}$$

can be alternatively written in terms of averages over the states of the system, $\rho'_{\mathcal{S}}$ and $\rho_{\mathcal{S}}$ if condition (8.41) is fulfilled. That condition implies $[\hat{\Phi}, \hat{M}_{\mu\nu}] = \hat{M}_{\mu\nu} \Delta \phi_{\mu\nu}$ (see Sect. 7.3), and introducing the commutator in (8.52), we obtain

$$\langle \Delta \phi \rangle = \sum_{\mu,\nu} \text{Tr}\big[[\hat{\Phi}, \hat{M}_{\mu\nu}] \rho_{\mathcal{S}} \hat{M}_{\mu\nu}^\dagger\big] = \text{Tr}[\hat{\Phi}\,(\rho'_{\mathcal{S}} - \rho_{\mathcal{S}})], \tag{8.53}$$

where we have used the cyclic property of the trace and Eq. (8.6). Therefore, the average potential change $\langle \Delta \phi \rangle$ can be expressed as the change in the expected value of the operator $\hat{\Phi}$ due to the map. Recall that, as commented in Sect. 7.3, the operator $\hat{\Phi}$ acts on the Hilbert space of the system $\mathcal{H}_{\mathcal{S}}$, i.e., is a local observable on the system which captures an effective thermodynamic action of the environment. Eqs. (8.50) and (8.51) now provide upper and lower entropic bounds on the change in this key quantity during the evolution

$$\Delta S_{\mathcal{S}} \geqslant \langle \Delta \phi \rangle \geqslant -\Delta S_E, \tag{8.54}$$

which may be interpreted as the effective transfer of entropy from the environment to the open system.

Furthermore, if the final measurement does not alter the state of the system, i.e., if $\rho_{\mathcal{S}}^* = \rho'_{\mathcal{S}}$, or if the final measurement is just skipped, as it is the case when we concatenate maps and the system is measured only after the whole concatenation (see Sect. 8.4 below), we can write the average non-adiabatic entropy production in an appealing form:

$$\Delta_{\rm i} S_{\rm na} = \Delta S_{\rm S} - \langle \Delta \phi \rangle = {\rm Tr}[\rho'_{\rm S}(\ln \rho'_{\rm S} + \hat{\Phi})] - {\rm Tr}[\rho'_{\rm S}(\ln \rho'_{\rm S} - \hat{\Phi})]$$
$$= S(\rho_{\rm S}||\pi) - S(\rho'_{\rm S}||\pi) \geqslant 0, \quad (8.55)$$

where we have used the definition $\hat{\Phi} = -\ln \pi$ of the potential operator in terms of the invariant state π. Here we see that the non-adiabatic entropy production is related to the distance between the state of the system and the invariant state π. During the evolution, the state of the system can only approximate the invariant state and the non-adiabatic entropy production is a measure of the irreversibility in the system state associated to such convergence. In fact, inequality in Eq. (8.55) follows from direct application of Ulhman's inequality (monotonicity of quantum relative entropy) holding for general CPTP evolutions [23, 45] (see Sect. 1.1). We stress that this expression coincides with entropy production introduced by Spohn for quantum dynamical semi-groups [46], and with the non-adiabatic entropy production appearing in Refs. [5, 24, 31, 32, 43, 47].

8.3.4 *Multipartite Environments*

The results obtained in this section and the formalism developed in the previous one are applicable as well for the case in which the environment consists of a multipartite system. The environment Hilbert space is here decomposed as $\mathcal{H}_E = \bigotimes_{r=1}^{R} \mathcal{H}_r$, corresponding to R ancillas or reservoirs not interacting between them but only with the open system. We assume the initial state of the environment is uncorrelated, $\rho_E = \rho_1 \otimes \cdots \otimes \rho_R$, and that the measurements are performed locally in each environmental ancilla. The local density operators of the environmental ancilla r at the beginning and at the end of the process are

$$\rho_r = \sum_\nu q_\nu^{(r)} \hat{\mathcal{Q}}_\nu^{(r)}, \qquad \rho_r^* = \sum_\mu q_\mu^{(r)*} \hat{\mathcal{Q}}_\mu^{(r)*}, \quad (8.56)$$

with eigenvalues $q_\nu^{(r)}$ and $q_\mu^{(r)*}$, and orthogonal projectors onto its eigenstates $\hat{\mathcal{Q}}_\nu^{(r)} = |\chi_\nu^{(r)}\rangle\langle\chi_\nu^{(r)}|_E$ and $\hat{\mathcal{Q}}_\mu^{(r)*} = |\chi_\mu^{(r)*}\rangle\langle\chi_\mu^{(r)*}|_E$.

The generalization of the results is then straightforward by considering the same steps and assumptions as before. The reduced system dynamics is again given by Eq. (8.6), but the operators $\hat{M}_{\mu\nu}$ now using collective indices

$$(\mu, \nu) = \{(\nu^{(1)}, \mu^{(1)}), \ldots, (\nu^{(R)}, \mu^{(R)})\}, \quad (8.57)$$

representing the set of transitions obtained in the projective measurements of all environmental ancillas:

$$|\chi^{(r)}_{\nu^{(r)}}\rangle_E \to |\chi^{(r)*}_{\mu^{(r)}}\rangle_E \quad \text{for } r = 1, \ldots, R. \tag{8.58}$$

That is, the Kraus operators of the forward process are given by

$$\hat{M}_{\mu\nu} = \left(\prod_{r=1}^{R} \sqrt{q^{(r)}_{\nu^{(r)}}} \right) \langle \chi^{(1)*}_{\mu^{(1)}} \ldots \chi^{(R)*}_{\mu^{(R)}} |_E \hat{U}_\Lambda | \chi^{(1)}_{\nu^{(1)}} \ldots \chi^{(R)}_{\nu^{(R)}} \rangle_E, \tag{8.59}$$

and analogously for the Kraus operators of the backward process (8.33) we have

$$\hat{\tilde{M}}_{\nu\mu} = \left(\prod_{r=1}^{R} \sqrt{\tilde{q}^{(r)}_{\mu^{(r)}}} \right) \langle \chi^{(1)}_{\nu^{(1)}} \ldots \chi^{(R)}_{\nu^{(R)}} |_E \hat{\Theta}^\dagger_E \hat{U}_{\tilde{\Lambda}} \hat{\Theta}_E | \chi^{(1)*}_{\mu^{(1)}} \ldots \chi^{(R)*}_{\mu^{(R)}} \rangle_E.$$

The key relation (8.34) necessary to obtain the fluctuation theorem for the total entropy production (8.23) hence follows as well in this case, with a decomposition of the environment boundary term

$$\sigma^E_{\mu\nu} = \sum_{r=1}^{R} \sigma^{(r)}_{\mu^{(r)}\nu^{(r)}}, \quad \text{being } \sigma^{(r)}_{\mu^{(r)}\nu^{(r)}} \equiv -\ln \tilde{q}^{(r)}_{\mu^{(r)}} + \ln q^{(r)}_{\nu^{(r)}}. \tag{8.60}$$

The application of the above formalism introducing the dual and dual-reverse processes follows immediately in the same manner, leading to the fluctuation theorems for the adiabatic and non-adiabatic entropy production in detailed and integral versions, Eqs. (8.43), (8.48) and (8.49). The adiabatic entropy production per trajectory and its average then read in this case:

$$\Delta_{\rm i} s^{\rm a}_{\mu\nu} = \sum_{r=1}^{R} \sigma^{r}_{\mu^{(r)}\nu^{(r)}} + \Delta\phi_{\mu\nu}, \tag{8.61}$$

$$\Delta_{\rm i} S_{\rm a} = \sum_{r=1}^{R} S(\rho^*_r) - S(\rho_r) + \langle \Delta\phi \rangle \geqslant 0, \tag{8.62}$$

where in the averaged version we set again (uncorrelated) reversible boundaries, $\tilde{\rho}_{SE} = \hat{\Theta}(\rho^*_{\rm S} \otimes \rho^*_1 \otimes \cdots \otimes \rho^*_R)\hat{\Theta}^\dagger$.

8.4 Concatenation of CPTP Maps

Up to now, we have considered a single interaction between system and environment of duration τ (see Eq. (8.2)). The CPTP map $\mathcal{E}$ describes the evolution of the open system when the environment is measured before and after interaction. This framework is well suited to be extended to the more general case of quantum trajectories [48] as we did for the fluctuation theorem introduced in Chap. 7. In this section we consider a collisional model, in which the system interacts sequentially with the environment. The environment consists of many quantum ancillas which interact once at a time with the system, while being monitored. Each single collision between time t and $t + \tau$, is described by a single CPTP map like $\mathcal{E}$, but which now can differ from one time to another (See Fig. 8.2). To be more specific, consider such dynamics for an interval consisting in $N \gg 1$ interactions. The map describing the reduced dynamical evolution from $t = 0$ to $t = N\tau$, is then the concatenation

$$\hat{\Omega} = \mathcal{E}^{(N)} \circ \cdots \circ \mathcal{E}^{(l)} \circ \cdots \circ \mathcal{E}^{(1)}, \tag{8.63}$$

where, in particular, each map $\mathcal{E}^{(l)}$ may have a different (positive-definite) invariant state $\pi^{(l)}$.

As it is customary in the theory of open quantum systems to achieve a Markovian evolution (see e.g. Sect. 2.3.2), we assume that the system interacts from time $t_{l-1} \equiv (l-1)\tau$ to time $t_l \equiv l\tau$ with a 'fresh' (uncorrelated) environmental ancilla in a generic state

$$\rho_E^{(l)} \equiv \sum_\alpha q_\alpha^{(l)} \hat{\mathcal{Q}}_\alpha^{(l)}. \tag{8.64}$$

As in the single map case, the environment is measured before and after interaction with the system by projective measurements. The outcomes of the measurements are labeled ν_l and μ_l, respectively. They are specified by the rank-one projective operators $\{\hat{\mathcal{Q}}_{\nu_k}^{(l)} \equiv |\chi_{\nu_l}^{(l)}\rangle\langle\chi_{\nu_l}^{(l)}|\}$ for the initial measurement and $\{\hat{\mathcal{Q}}_{\mu_l}^{(l)*} \equiv |\chi_{\mu_l}^{(l)*}\rangle\langle\chi_{\mu_l}^{(l)*}|\}$ for the final one. Under this conditions, each map in the concatenation can be written as:

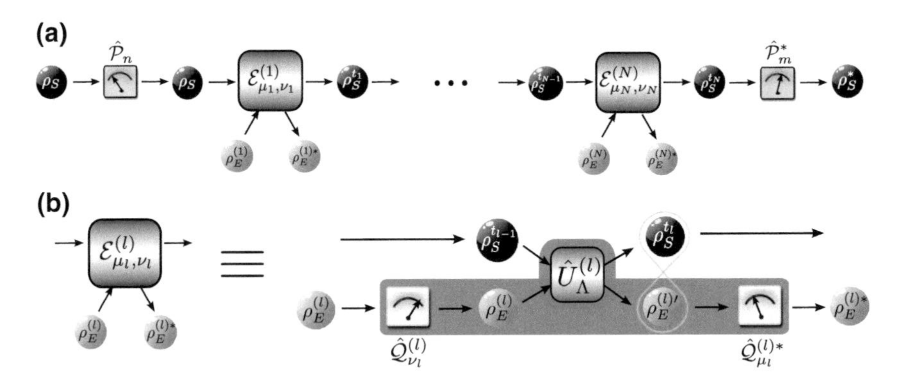

Fig. 8.2 **a** Schematic diagram of the maps concatenation introduced in the text, where projective measurements on the system are only performed at the begging and at the end of the concatenation. **b** Any operation $\mathcal{E}_{\mu_l,\nu_l}^{(l)}$ in the concatenation consists in the interaction of the system with an environmental ancilla in the state $\rho_E^{(l)}$ via the unitary $\hat{U}_\Lambda^{(l)}$ depending on the protocol Λ_l. The ancilla is measured before and after interaction generating outcomes ν_l and μ_l respectively

$$\mathcal{E}^{(l)}(\cdot) = \sum_{\mu_l, \nu_l} \hat{M}^{(l)}_{\mu_l, \nu_l}(\cdot)\, \hat{M}^{(l)\dagger}_{\mu_l \nu_l} \tag{8.65}$$

$$\hat{M}^{(l)}_{\mu_l \nu_l} \equiv \sqrt{q^{(l)}_{\nu_l}} \langle \chi^{(l)*}_{\mu_l} | \hat{U}^{(l)}_{\Lambda} | \chi^{(l)}_{\nu_l} \rangle \tag{8.66}$$

where the unitary evolution $\hat{U}^{(l)}_{\Lambda}$ is given in Eq. (8.2) for $t_0 = t_{l-1}$. Here we consider always the same total time-dependent Hamiltonian $\hat{H}(t)$, following an arbitrary driving protocol $\Lambda = \{\lambda_t |\, 0 \leqslant t \leqslant N\tau\}$. For convenience the latter can also be split into N intervals; hence the partial protocol $\Lambda_l = \{\lambda_t |\, t_{l-1} \leqslant t \leqslant t_l\}$ generates the unitary operator $\hat{U}^{(l)}_{\Lambda}$ (Fig. 8.2).

A quantum trajectory in this context is defined as follows. At time $t = 0$ we start with our system in $\rho_{\mathcal{S}}$, which is measured with eigenprojectors $\{\hat{\mathcal{P}}_n\}$, obtaining outcome n. Then the sequence of maps $\hat{\Omega}$ defined in Eq. (8.63) is applied, obtaining outcomes $\{\mu_l, \nu_l\}$ from each of the $l = 1, \ldots, N$ pairs of measurements in the environment. Finally at time $t = N\tau$ the system is measured again with arbitrary (rank-one) projectors $\{\hat{\mathcal{P}}^*_m\}$ giving outcome m. A quantum trajectory is now completely specified by the set of outcomes, $\gamma = \{n, (\nu_1, \mu_1), \ldots, (\nu_N, \mu_N), m\}$, and occurs with probability

$$P_\gamma = p_n \operatorname{Tr}[\hat{\mathcal{P}}^*_m\, \mathcal{E}^{(N)}_{\mu_N \nu_N} \circ \cdots \circ \mathcal{E}^{(1)}_{\mu_1 \nu_1}(\hat{\mathcal{P}}_n)]. \tag{8.67}$$

Now we can apply the same arguments in previous sections to construct the three different processes used to state the FT's. For the initial state of the backward process, we consider again an arbitrary initial state of the system $\tilde{\rho}_{\mathcal{S}} = \sum_m \tilde{p}_m \hat{\Theta}_{\mathcal{S}} \hat{\mathcal{P}}^*_m \hat{\Theta}^\dagger_{\mathcal{S}}$, uncorrelated from the environment initial states $\tilde{\rho}^{(l)}_E = \sum_\alpha \tilde{q}^{(l)}_\alpha \hat{\Theta}_E \hat{\mathcal{Q}}^*_\alpha \hat{\Theta}^\dagger_E$, and apply the sequence of maps

$$\tilde{\Omega} = \tilde{\mathcal{E}}^{(1)} \circ \cdots \circ \tilde{\mathcal{E}}^{(l)} \circ \cdots \circ \tilde{\mathcal{E}}^{(N)}, \tag{8.68}$$

generating a trajectory $\tilde{\gamma} = \{m, (\mu_1, \nu_1), \ldots, (\mu_N, \nu_N), n\}$ which occurs with probability:

$$\tilde{P}_{\tilde{\gamma}} = \tilde{p}_m \operatorname{Tr}[\hat{\Theta}_{\mathcal{S}} \hat{\mathcal{P}}_n \hat{\Theta}^\dagger_{\mathcal{S}}\, \tilde{\mathcal{E}}^{(1)}_{\nu_1 \mu_1} \circ \cdots \circ \tilde{\mathcal{E}}^{(N)}_{\nu_N \mu_N}(\hat{\Theta}_{\mathcal{S}} \hat{\mathcal{P}}^*_m \hat{\Theta}^\dagger_{\mathcal{S}})]. \tag{8.69}$$

Here the backward maps, $\tilde{\mathcal{E}}^{(l)}$, and their corresponding operations, are defined from each map $\mathcal{E}^{(l)}$ in the concatenation $\hat{\Omega}$ by applying Eqs. (8.32) and (8.33).

Dual and dual-reverse maps and operations also follow from its definitions in Sect. 8.3 when conditions (8.41) and $\tilde{\mathcal{E}}^{(l)}(\tilde{\pi}^{(l)}) = \tilde{\pi}^{(l)}$ are met for each map in the sequence. The probabilities of sampling trajectory γ in the dual process, and trajectory $\tilde{\gamma}$ in the dual-reverse read

$$P^D_\gamma = p_n \operatorname{Tr}[\hat{\mathcal{P}}^*_m\, \mathcal{D}^{(N)}_{\mu_N \nu_N} \circ \cdots \circ \mathcal{D}^{(1)}_{\mu_1 \nu_1}(\hat{\mathcal{P}}_n)], \tag{8.70}$$

$$\tilde{P}^D_{\tilde{\gamma}} = \tilde{p}_m \operatorname{Tr}[\hat{\Theta}_{\mathcal{S}} \hat{\mathcal{P}}_n \hat{\Theta}^\dagger_{\mathcal{S}}\, \tilde{\mathcal{D}}^{(1)}_{\nu_1, \mu_1} \circ \cdots \circ \tilde{\mathcal{D}}^{(N)}_{\nu_N \mu_N}(\hat{\Theta}_{\mathcal{S}} \hat{\mathcal{P}}^*_m \hat{\Theta}^\dagger_{\mathcal{S}})], \tag{8.71}$$

where in the dual-reverse trajectories we took again the sequence of maps in inverted order, this is, we applied $\tilde{\mathcal{D}}^{(1)} \circ \cdots \circ \tilde{\mathcal{D}}^{(N)}$ over the system initial state $\tilde{\rho}_{\mathbb{S}}$.

Again, the Kraus operators for the backward, dual, and dual-reverse processes, fulfill the set of operator detailed-balance relations:

$$\hat{\tilde{M}}^{(l)}_{\nu\mu} = e^{-\sigma^E_{\mu_l,\nu_l}/2}\, \hat{\Theta}_{\mathbb{S}} \hat{M}^{(l)\dagger}_{\mu\nu} \hat{\Theta}^\dagger_{\mathbb{S}}, \tag{8.72}$$

$$\hat{\tilde{D}}^{(l)}_{\nu\mu} = e^{\,\Delta\phi^{(l)}_{\mu\nu}/2}\, \hat{\Theta}_{\mathbb{S}} \hat{M}^{(l)\dagger}_{\mu\nu} \hat{\Theta}^\dagger_{\mathbb{S}}, \tag{8.73}$$

$$\hat{D}^{(l)}_{\mu\nu} = e^{-(\sigma^E_{\mu_l,\nu_l}+\Delta\phi^{(l)}_{\mu\nu})/2}\, \hat{M}^{(l)}_{\mu\nu}, \tag{8.74}$$

where the nonequilibrium potential changes are defined with respect to the invariant state $\pi^{(l)}$ of each map $\mathcal{E}^{(l)}$ as in the single map case:

$$\Delta\phi^{(l)}_{\mu\nu} = -\ln \pi^{(l)}_\mu + \ln \pi^{(l)}_\nu. \tag{8.75}$$

The set of Eqs. (8.72)–(8.74) immediately implies the detailed FT's for quantum trajectories

$$\Delta_{\mathrm{i}} s^{\mathrm{na}}_\gamma = \ln \frac{P_\gamma}{\tilde{P}^D_{\tilde{\gamma}}} = \sigma^{\mathbb{S}}_{nm} - \sum_{l=1}^N \Delta\phi^{(l)}_{\mu_l\nu_l}, \tag{8.76}$$

$$\Delta_{\mathrm{i}} s^{\mathrm{a}}_\gamma = \ln \frac{P_\gamma}{P^D_\gamma} = \sum_{l=1}^N \left(\sigma^E_{\mu_l\nu_l} + \Delta\phi^{(l)}_{\mu_l\nu_l}\right), \tag{8.77}$$

$$\Delta_{\mathrm{i}} s_\gamma = \ln \frac{P_\gamma}{\tilde{P}_{\tilde{\gamma}}} = \Delta_{\mathrm{i}} s^{\mathrm{na}}_\gamma + \Delta_{\mathrm{i}} s^{\mathrm{a}}_\gamma, \tag{8.78}$$

where $\sigma^{\mathbb{S}}_{nm} = \ln p_n - \ln \tilde{p}_m$ is the boundary term in the system, and $\sigma^E_{\mu_l\nu_l} = \ln q^{(l)}_{\nu_l} - \ln \tilde{q}^{(l)}_{\mu_l}$ the boundary term in the l-th environmental ancilla.

From the detailed FT's in Eqs. (8.76) and (8.77), their corresponding integral versions and second-law-like inequalities follow immediately as a corollary. Considering reversible boundaries $\tilde{\rho}_{\mathbb{S}} = \hat{\Theta}_{\mathbb{S}} \rho^*_{\mathbb{S}} \hat{\Theta}^\dagger_{\mathbb{S}}$ and $\tilde{\rho}^{(l)}_E = \hat{\Theta}_E \rho^{*(l)}_E \hat{\Theta}^\dagger_E\ \forall l$, we obtain $\sigma^{\mathbb{S}}_{nm} = \Delta s^{\mathbb{S}}_{mn}$, and $\sigma^E_{\mu_l\nu_l} = \Delta s^E_{\mu_l\nu_l}$ is the trajectory entropy change in the environment during the l-th map. Therefore we have

$$\Delta_{\mathrm{i}} S_{\mathrm{na}} = \Delta S_{\mathbb{S}} - \sum_l \langle \Delta\phi^{(l)} \rangle \geqslant 0, \tag{8.79}$$

$$\Delta_{\mathrm{i}} S_{\mathrm{a}} = \sum_l S(\rho^{(l)*}_E) - S(\rho^{(l)}_E) + \langle \Delta\phi^{(l)} \rangle \geqslant 0. \tag{8.80}$$

Finally, it is interesting to consider the expression of the average nonequilibrium potential change during the whole sequence $\langle \Delta\phi \rangle$. By denoting $\rho_{\mathbb{S}}(t_l)$ the reduced state of the system at time t_l

$$\langle \Delta\phi \rangle = \sum_{l=1}^{N} \langle \Delta\phi^{(l)} \rangle = \sum_{l=1}^{N} \mathrm{Tr}[\mathcal{E}^{(l)}_{\mu_l \nu_l}(\rho_{\mathrm{S}}(t_{l-1}))] \Delta\phi^{(l)}_{\mu_l \nu_l} =$$

$$= \sum_{l=1}^{N} \mathrm{Tr}[\hat{\Phi}_l \big(\rho_{\mathrm{S}}(t_l) - \rho_{\mathrm{S}}(t_{l-1})\big)], \tag{8.81}$$

where $\hat{\Phi}_l = -\ln \pi^{(l)}$. The above expression can be decomposed into the following *boundary* and *path* contributions:

$$\langle \Delta\phi \rangle_{\mathrm{b}} = \mathrm{Tr}[\rho'_{\mathrm{S}} \hat{\Phi}_N] - \mathrm{Tr}[\rho_{\mathrm{S}} \hat{\Phi}_1], \tag{8.82}$$

$$\langle \Delta\phi \rangle_{\mathrm{p}} = -\sum_{l=1}^{N-1} \mathrm{Tr}[\rho_{\mathrm{S}}(t_l)(\hat{\Phi}_{l+1} - \hat{\Phi}_l)]. \tag{8.83}$$

When all the maps in the concatenation have the same invariant state, $\hat{\Phi}_{l+1} = \hat{\Phi}_l \equiv \hat{\Phi}\ \forall l$, we obtain $\langle \Delta\phi \rangle_{\mathrm{p}} = 0$, while $\langle \Delta\phi \rangle_{\mathrm{b}} = \mathrm{Tr}[(\rho'_{\mathrm{S}} - \rho_{\mathrm{S}})\hat{\Phi}]$ and we recover the expression for the single map case, c.f. Eq. (8.53). In the other hand the boundary term only vanishes for cyclic processes, such that $\rho'_{\mathrm{S}} = \rho_{\mathrm{S}}$, implemented by cyclic concatenations with $\hat{\Phi}_N = \hat{\Phi}_1$. In this case $\langle \Delta\phi \rangle_{\mathrm{b}} = 0$ while $\langle \Delta\phi \rangle_{\mathrm{p}}$ gives in general a non-zero contribution.

8.5 Lindblad Master Equations

The generalization introduced in the last section can be applied to situations in which a dynamical description is available, e.g. given by a Markovian master equation, for which unravellings in terms of quantum trajectories were introduced in Sect. 2.5. This case represents the limit in which an infinite number of maps, $N \to \infty$, are applied in infinitesimal time steps $\Delta t = t_l - t_{l-1} \to dt$. The system density operator change then becomes $\rho_{\mathrm{S}}(t_l) - \rho_{\mathrm{S}}(t_{l-1}) \to d\rho_{\mathrm{S}}$ in the continuous limit, and the map $\hat{\Omega}$ in Eq. (8.63) can be described by a quantum dynamical semi-group (see Sect. 2.2). Furthermore, in the general case, the steady state of the dynamics may depend on the external control parameter λ_t, such that its modification prevents the system from relaxation towards a steady state $\pi(\lambda_t)$. Here we will assume that at any infinitesimal time interval the nonequilibrium potential remains constant at $\hat{\Phi}(\lambda_t)$ while changing from one step to the next.

As in the previous chapter, we consider the system evolution to be given by the following master equation in Lindblad form:

$$\dot{\rho}_t = -\frac{i}{\hbar}[\hat{H}, \rho] + \sum_{k=1}^{K} \left(\hat{L}_k \rho_t \hat{L}_k^\dagger - \frac{1}{2}\{\hat{L}_k^\dagger \hat{L}_k, \rho_t\} \right) \equiv \mathcal{L}\rho_t \tag{8.84}$$

where $\hat{H}(\lambda_t)$ is an Hermitian Hamiltonian like term, and the set $\{\hat{L}_k \equiv \hat{L}_k(\lambda_t)\}$ are positive Lindblad operators, which generally may depend on the control parameter, λ_t, describing jumps in some (possibly time-dependent) basis. When the driving is frozen, $\lambda_t \equiv \lambda_*$, the above Lindblad master equation has at least one invariant state, π_*, given by $\mathcal{L}(\pi_*) = 0$ [49].

The stochastic description we provided in the previous section is here naturally recovered within the formalism introduced in Sect. 2.5. The underlying idea is monitoring the interactions with the environment to unveil the jumps induced by the Lindblad operators $\{\hat{L}_k\}$, which now play the role of the Kraus operators in the CPTP maps. An infinitesimal time evolution step of the dynamics provided by Eq. (8.84) is here identified with a generic map $\mathcal{E}$ in the sequence $\hat{\Omega}$ defined in Eq. (8.63) [omitting the superscript (l) for the order in the sequence]:

$$\rho_{t+dt} = \left(\mathbb{1}_{\mathcal{S}} + dt\mathcal{L}\right)\rho_t \equiv \mathcal{E}(\rho_t) = \sum_{k=0}^{K} \hat{M}_k \rho_t \hat{M}_k^\dagger \tag{8.85}$$

for which a generic set of Kraus operators can be written as

$$\hat{M}_0(\lambda_t) \equiv \mathbb{1}_{\mathcal{S}} - dt\left(\frac{i}{\hbar}\hat{H}(\lambda_t) + \frac{1}{2}\sum_{k=1}^{K} \hat{L}_k^\dagger(\lambda_t)\hat{L}_k(\lambda_t)\right) \tag{8.86}$$

$$\hat{M}_k(\lambda_t) \equiv \sqrt{dt}\;\hat{L}_k(\lambda_t) \quad k = 1, \ldots, K. \tag{8.87}$$

Notice that the map $\mathcal{E}$ have as invariant state the instantaneous steady-state of the dynamics, π_λ. Furthermore, as pointed in Sect. 2.5, this Kraus representation is not unique for the open system dynamics because of the symmetry

$$\hat{H}' = \hat{H} - \frac{i\hbar}{2}\sum_{k=1}^{K}\left(l_k^* \hat{L}_k - l_k \hat{L}_k^\dagger\right) + \hbar r, \quad \hat{L}_k' = \hat{L}_k + l_k, \tag{8.88}$$

leaving invariant Eq. (8.84), but it is related to a specific detection scheme for the jumps. The specific form of the set $\{\hat{L}_k\}_{k=1}^K$ appearing in Eq. (8.84) hence fixes the measurement scheme proposed in the previous section, that is, it implies a specific choice on the local observables being monitored in the environmental ancillas at the beginning and at the end of the system-environment interaction (though the set of orthogonal projectors $\{\hat{\mathcal{Q}}_\nu\}$ and $\{\hat{\mathcal{Q}}_\mu^*\}$).

The Kraus representation (8.86) is based on a family of operations $\hat{M}_k$ with $k = 1, \ldots, K$ that induce jumps in the state of the system and occur with probabilities of order dt, and a single operation $\hat{M}_0$ that induces a smooth evolution in the state of the system and occurs with probability of order 1 (see Sect. 2.5). This implies that the trajectories γ consists of a large number of zeros punctuated by a few jumps $\hat{M}_k$ with $k = 1, \ldots, K$. An alternative way of describing the trajectory is to specify the jumps k_j and the times τ_j where they occur, i.e.,

$\gamma = \{n, (k_1, \tau_1), \ldots, (k_j, \tau_j), \ldots, (k_N, \tau_N), m\}$, where, as before, n and m denote the outcomes of the initial and final measurements in the system. Jump k is given by the operation $\mathcal{E}_k(\rho) \equiv \hat{M}_k \rho \hat{M}_k^\dagger$, whereas between two consecutive jumps at t_j and t_{j+1} is given by the repeated application of the operation corresponding to the Kraus operator $\hat{M}_0(\lambda_t)$ in (8.86). This results in a smooth evolution given by the operator:

$$\hat{U}_{\text{eff}}(t_{j+1}, t_j) = \hat{T}_+ \exp\left(-\frac{i}{\hbar} \int_{t_j}^{t_{j+1}} ds\ \hat{H}_{\text{eff}}(\lambda_s) \right), \tag{8.89}$$

with an effective non-hermitian Hamiltonian that reads

$$\hat{H}_{\text{eff}}(\lambda_t) = \hat{H}(\lambda_t) - \frac{i\hbar}{2} \sum_{k=1}^{K} \hat{L}_k^\dagger(\lambda_t) \hat{L}_k(\lambda_t). \tag{8.90}$$

In this representation, the probability to measure a trajectory $\gamma = \{n, (k_1, \tau_1), \ldots, (k_j, \tau_j), \ldots, (k_N, \tau_N), m\}$ is given by

$$\begin{aligned} P_\gamma = \text{Tr}[\hat{\mathcal{P}}_m^* \mathcal{U}_{t_f, t_N} \mathcal{E}_{k_N} \mathcal{U}_{t_N, t_{N-1}} \ldots\ \mathcal{E}_{k_l} \\ \ldots \mathcal{U}_{t_2, t_1} \mathcal{E}_{k_1} \mathcal{U}_{t_1, t_0} (\hat{\mathcal{P}}_n \rho_0 \hat{\mathcal{P}}_n)], \end{aligned} \tag{8.91}$$

with $\mathcal{U}_{t_{j+1}, t_j}(\rho) = \hat{U}_{\text{eff}}(t_{j+1}, t_j) \rho\, \hat{U}_{\text{eff}}^\dagger(t_{j+1}, t_j)$.

Consider now the backward dynamics. Here time-inversion of the global system evolution correspond to a time-reversed version of the Lindblad master equation in Eq. (8.84). As in the previous section, the backward process is generated by inverting the sequence of operations together with time-inversion of each operation in the sequence. The mapping producing an infinitesimal time-step in the time-reversed dynamics, $\tilde{\rho}_{t+dt} = \tilde{\mathcal{E}}(\tilde{\rho}_t)$, admits a Kraus representation $\{\hat{\tilde{M}}_k(\lambda_t)\}$ analogous to (8.86) and (8.87). Our previous analysis allows us to write these operators without knowing any detail about the interaction between the system and the environment, since they have to obey condition Eq. (8.34), that is:

$$\hat{\tilde{M}}_0 = e^{-\sigma_0^E/2} \hat{\Theta} \hat{M}_0^\dagger \hat{\Theta}^\dagger, \qquad \hat{\tilde{M}}_k = e^{-\sigma_k^E/2} \hat{\Theta} \hat{M}_k^\dagger \hat{\Theta}^\dagger. \tag{8.92}$$

Imposing the backward maps to be trace-preserving, $\sum_k \hat{\tilde{M}}_k^\dagger \hat{\tilde{M}}_k = \mathbb{1}$, we obtain $\sigma_0^E = 0$, and the consistency condition

$$\sum_{k=1}^{K} \left(\hat{L}_k^\dagger \hat{L}_k - \hat{L}_k \hat{L}_k^\dagger e^{-\sigma_k^E} \right) = 0. \tag{8.93}$$

This is a completely general result: for any Lindblad master equation one can find a set of numbers $\{\sigma_k^E\}_{k=1}^K$ such that (8.93) is fulfilled. The specific meaning of the quantities σ_k^E was given in Eq. (8.25), which depends on the initial state of the environment in the backward process. In practical applications this can be deduced from the environmental modeling and the measurement scheme used to detect the quantum jumps, leading to a specific set of Lindblad operators $\{\hat{L}_k\}_{k=1}^K$ in the master equation (8.84).

In many applications, the Lindblad operators come in pairs and the corresponding pair of terms in the sum (8.93) cancel. This occurs if, for a specific pair of operators $\{\hat{L}_i, \hat{L}_j\}$, we have $\hat{L}_i = \sqrt{\Gamma_i}\hat{L}$ and $\hat{L}_j = \sqrt{\Gamma_j}\hat{L}^\dagger$, being $\Gamma_i(\lambda_t)$ and $\Gamma_j(\lambda_t)$ some positive rates, and $\hat{L}(\lambda_t)$ some arbitrary (possibly time-dependent) system operator. Then, the condition (8.93) implies $\sigma_i^E(\lambda_t) = \ln(\Gamma_i/\Gamma_j)$ and $\sigma_j^E(\lambda_t) = \ln(\Gamma_j/\Gamma_i) = -\sigma_i^E(\lambda_t)$.

As in the previous section, for any trajectory $\gamma = \{n, k_1, \ldots, k_N, m\}$ generated in the forward process with probability P_γ, there exist a backward trajectory $\tilde{\gamma} = \{m, k_N, \ldots, k_1, n\}$ occurring in the backward process with probability $\tilde{P}_{\tilde{\gamma}}$. The backward trajectory can here be identified by the times of successive jumps as well. In this representation, the probability of trajectory $\tilde{\gamma}$ can hence be written as:

$$\begin{aligned}\tilde{P}_{\tilde{\gamma}} = \mathrm{Tr}[&\hat{\Theta}\hat{\mathcal{P}}_n\hat{\Theta}^\dagger\, \tilde{\mathcal{U}}_{t_1,t_0}\tilde{\mathcal{E}}_{k_1}\tilde{\mathcal{U}}_{t_2,t_1}\ \ldots\ \tilde{\mathcal{E}}_{k_l}\ \ldots\\ &\ldots\ \tilde{\mathcal{U}}_{t_N,t_{N-1}}\tilde{\mathcal{E}}_{k_N}\tilde{\mathcal{U}}_{t_f,t_N}(\hat{\Theta}\hat{\mathcal{P}}_m^*\rho_{t_f}\hat{\mathcal{P}}_m^*\hat{\Theta}^\dagger)],\end{aligned} \tag{8.94}$$

where $\tilde{\mathcal{E}}_k(\tilde{\rho}) = \hat{\tilde{M}}_k\tilde{\rho}\hat{\tilde{M}}_k^\dagger$. The smooth evolution between jumps, here $\tilde{\mathcal{U}}_{t',t}(\tilde{\rho}_t) = \tilde{U}_{\text{eff}}(t',t)\tilde{\rho}_t\tilde{U}_{\text{eff}}^\dagger(t',t)$, is given by the operator

$$\hat{\tilde{U}}_{\text{eff}}(t',t) = \hat{T}_+ \exp\left(\frac{i}{\hbar}\int_t^{t'} ds\ \hat{\Theta}\hat{H}_{\text{eff}}^\dagger(\tilde{\lambda}_s)\hat{\Theta}^\dagger\right), \tag{8.95}$$

where $\{\tilde{\lambda}_t\}$ now corresponds to the inverse sequence of values for the control parameter. It can be easily shown that the forward and backward smooth evolutions obey the microreversibility relationship $\hat{\Theta}^\dagger\hat{\tilde{U}}_{\text{eff}}(t',t)\hat{\Theta} = \hat{U}_{\text{eff}}(t',t)^\dagger$.

At this point we particularize condition (8.41), necessary to decompose the entropy production into adiabatic and non-adiabatic contributions. It can be written as:

$$[\hat{\Phi}, \hat{L}_k] = \Delta\phi_k\hat{L}_k \quad ; \quad [\hat{\Phi}, \hat{L}_k^\dagger] = -\Delta\phi_k\hat{L}_k^\dagger. \tag{8.96}$$

These commutation relationships indicate that the Lindblad operators $\hat{L}_k(\lambda_t)$ promote jumps between the eigenstates of $\pi(\lambda_t)$ at any time of the dynamics. Furthermore, as the condition must be fulfilled for the operator $\hat{M}_0$ in Eq. (8.86) as well, we need $[\hat{H}, \sum_k \hat{L}_k^\dagger\hat{L}_k] = [\hat{H}, \hat{\Phi}] = 0$, which in turn implies $\Delta\phi_0 = 0$. As explained in Sect. 7.4 of the previous chapter, this means that the steady state of the dynamics must be diagonal in the basis of the Hamiltonian term appearing in Eq. (8.84). This is not very restrictive since in most situations the operator $\hat{H}$ is just the identity operator, $\mathbb{1}$, when we move to an appropriate interaction picture. In second place, we recall that the fluctuation theorem for the adiabatic entropy production can be stated when the backward maps $\tilde{\mathcal{E}}$ admit $\tilde{\pi}_\lambda \equiv \hat{\Theta}\pi_\lambda\hat{\Theta}^\dagger$ as an invariant state. We stress that this condition is immediately fulfilled when the Lindblad operators come in pairs $\{\hat{L}_i, \hat{L}_j\}$ with $\hat{L}_i = \sqrt{\Gamma_i/\Gamma_j}\hat{L}_j^\dagger$ as before. In such case the (inverted) Kraus operators of the backward map also pertain to the forward map:

$$\hat{\Theta}^\dagger \hat{\tilde{M}}_i \hat{\Theta} = e^{-\sigma_i^E/2}\hat{M}_i^\dagger = \sqrt{dt}e^{-\sigma_i^E/2}\hat{L}_i^\dagger = \sqrt{dt}\hat{L}_j = \hat{M}_j,$$

where we used Eq. (8.92), and hence:

$$\sum_k \hat{\tilde{M}}_k \hat{\Theta}\pi_\lambda\hat{\Theta}^\dagger \hat{\tilde{M}}_k^\dagger = \sum_k \hat{\Theta}\hat{M}_k\pi_\lambda\hat{M}_k^\dagger\hat{\Theta}^\dagger = \tilde{\pi}_\lambda. \tag{8.97}$$

In such circumstances, with the help of the instantaneous stationary state of the dynamics $\pi(\lambda_t)$, the dual and dual-reverse processes can be constructed as well. For the dual process, the probability of trajectory γ, P_γ^D, can be calculated from Eq. (8.91) by using the same map $\mathcal{U}_{t',t}$ for the no-jump time evolution intervals, and replacing the operations $\mathcal{E}_k$ by the dual operations $\mathcal{D}_k(\cdot) = \hat{D}_k(\cdot)\hat{D}_k^\dagger$ with corresponding Kraus operators $\{\hat{D}_k\}$ as defined in Eq. (8.46):

$$\hat{D}_k = e^{-(\sigma_k^E + \Delta\phi_k)/2}\ \hat{M}_k. \tag{8.98}$$

Analogously, for the dual-reverse process the probability of trajectory $\tilde{\gamma}$, $\tilde{P}_{\tilde{\gamma}}^D$, can be constructed from Eq. (8.94) with $\tilde{\mathcal{U}}_{t',t}$ for the no-jump evolution, and dual-reverse operations $\tilde{\mathcal{D}}_k = \hat{\tilde{D}}_k(\cdot)\hat{\tilde{D}}_k$ with Kraus operators $\{\hat{\tilde{D}}_k\}$ as in Eq. (8.42):

$$\hat{\tilde{D}}_k = e^{\ \Delta\phi_k/2}\hat{\Theta}\ \hat{M}_k^\dagger\hat{\Theta}^\dagger. \tag{8.99}$$

We further notice that in general $\hat{D}_k \neq \hat{M}_k$, and $\hat{\tilde{D}}_k \neq \hat{\tilde{M}}_k$, this is, $\sigma_k^E \neq -\Delta\phi_k$.

The above considerations let us reproduce the three detailed FT's in Eqs. (8.76)–(8.77), for quantum trajectories generated by Lindblad master equations:

$$\Delta_{\mathrm{i}} s_\gamma^{\mathrm{na}} = \ln\frac{P_\gamma}{\tilde{P}_{\tilde{\gamma}}^D} = \sigma_{nm}^S - \sum_{l=1}^N \Delta\phi_{k_l}(\lambda_{t_l}), \tag{8.100}$$

$$\Delta_{\mathrm{i}} s_\gamma^{\mathrm{a}} = \ln\frac{P_\gamma}{P_\gamma^D} = \sum_{l=1}^N \left(\sigma_{k_l}^E(\lambda_{t_l}) + \Delta\phi_{k_l}(\lambda_{t_l})\right), \tag{8.101}$$

$$\Delta_{\mathrm{i}} s_\gamma = \ln\frac{P(\gamma)}{\tilde{P}_{\tilde{\gamma}}} = \sigma_{nm}^S - \sum_{l=1}^N \sigma_{k_l}^E(\lambda_{t_l}), \tag{8.102}$$

with $\Delta_i s_\gamma = \Delta_i s_\gamma^{na} + \Delta_i s_\gamma^{a}$. The integral versions of the three FTs follow readily from the fact that $\tilde{P}_\gamma^D$, P_γ^D, and $\tilde{P}_\gamma$ are well defined probability distributions.

In addition we may derive for this case the dynamical version of the second-law-like inequalities, taking the continuous limit from Eqs. (8.79) and (8.80). Considering reversible boundaries in the system such that $\tilde{\rho}_S = \hat{\Theta}(\rho'_S)\hat{\Theta}^\dagger$ we obtain:

$$\dot{S}_{na} = \dot{S} - \dot{\Phi} \geqslant 0, \tag{8.103}$$

$$\dot{S}_{a} = \dot{\sigma}_E + \dot{\Phi} \geqslant 0, \tag{8.104}$$

$$\dot{S}_{i} = \dot{S} + \dot{\sigma}_E \geqslant 0, \tag{8.105}$$

where again $\dot{S}_i = \dot{S}_{na} + \dot{S}_a$. We will refer to this quantities as the *entropy production rates*, where $\dot{S} = -\mathrm{Tr}[\dot{\rho}_t \ln \rho_t]$ is the derivative of the von Neumann entropy of the system, $\dot{\Phi} = \mathrm{Tr}[\dot{\rho}_t \hat{\Phi}(\lambda_t)]$ the nonequilibrium potential flow, and we obtain a dynamical version of the boundary term in the environment

$$\begin{aligned} \dot{\sigma}_E(\lambda_t)\, dt &= \sum_{k_l} \mathrm{Tr}[\mathcal{E}_{k_l}(\rho_t)]\sigma_{k_l}^E(\lambda_t) \\ &= \frac{d}{dt}\left[\Delta S_E + D(\rho_E^* || \hat{\Theta}^\dagger \tilde{\rho}_E \hat{\Theta})\right] dt. \end{aligned} \tag{8.106}$$

Notice that when reversible boundaries are chosen in the environment, $\hat{\Theta}^\dagger \tilde{\rho}_E \hat{\Theta} = \rho_E^*$, we simply obtain $\dot{\sigma}_E(\lambda_t) = \dot{S}_E$, the rate at which the entropy of the environment varies. On the other hand, if equilibrium conditions are considered, $\hat{\Theta}^\dagger \tilde{\rho}_E \hat{\Theta} = \rho_E = e^{-\beta \hat{H}_E}/Z_E$, being $\hat{H}_E$ the environment Hamiltonian and $\beta = 1/k_B T$ its inverse temperature, we have $\dot{\sigma}_E(\lambda_t) = \beta \frac{d}{dt}\left(\mathrm{Tr}[\hat{H}_E(\rho_E^* - \rho_E)]\right)$, that is, we obtain the heat flow dissipated into the environment divided by temperature. Both expressions coincides when the environment is considered to be a thermal reservoir (or thermal bath), i.e. a large system in thermal equilibrium for which $\rho_E^* \simeq \rho_E = e^{-\beta \hat{H}_E}/Z_E$, and hence $\dot{S}_E \simeq \beta \frac{d}{dt}\left(\mathrm{Tr}[\hat{H}_E(\rho_E^* - \rho_E)]\right)$. Finally, we stress that the above inequalities (8.103)–(8.105) follow from the fact the FT's apply for any single map applied at infinitesimal time-step dt (Sect. 8.3). They guarantee the monotonicity of the average entropy production, $\Delta_i S$, and its adiabatic and non-adiabatic contributions, $\Delta_i S_{na}$ and $\Delta_i S_a$ during the whole Markovian evolution.

The physical interpretation of the adiabatic and non-adiabatic entropy production now becomes clear. In a process where the initial state is $\pi(\lambda_0)$ and the control parameter is quasi-statically modified, $\rho_t \simeq \pi(\lambda_t)$. Therefore, the non-adiabatic entropy production $\dot{S}_{na} \simeq -\dot{S}(\rho_t || \pi(\lambda_t)) \simeq 0$ vanishes. This is in agreement with the classical non-adiabatic contribution introduced in Refs. [5, 31, 32]. On the other hand, the contribution $\dot{S}_a$ is in general different from zero even if the driving is extremely slow, which is the reason why it is called adiabatic. We finally provide the dynamical versions of the nonequilibrium potential boundary and path terms:

$$\dot{\Phi}_{\rm b} = \frac{d}{dt}\left(\mathrm{Tr}[\rho_t \hat{\Phi}(\lambda_t)]\right), \qquad \dot{\Phi}_{\rm p} = -\mathrm{Tr}[\rho_t \dot{\hat{\Phi}}(\lambda_t)], \tag{8.107}$$

in analogy to the classical case [5, 31, 32].

8.6 Conclusions

In this chapter we have analyzed the (von Neumann) entropy production in general processes embedded in a two measurement protocol, with local measurements performed in both system and environment. We obtained three different fluctuation theorems for the adiabatic, the non-adiabatic, and the total entropy production in detailed and integral forms, which apply for quantum trajectories in different situations, generalizing to the quantum regime previous results reported for classical Markov processes [5].

We first discussed how the total entropy production depends on both the classical and the quantum correlations generated between system and environment, as they are irreversibly lost when introducing quantum measurements, c.f. Eqs. (8.13) and (8.15). The first expression Eq. (8.13) has been called *inclusive* entropy production and measures the quantum correlations lost in the measurement process and the local measurement-induced disturbance. If the remaining classical correlations are also lost hence the entropy production is called *non-inclusive*, and given in Eq. (8.15), which corresponds to the prototypical identification as the sum of the entropy changes in system and environment during the process.

In this context, we derived a fluctuation theorem in detailed Eq. (8.23) and integral forms Eq. (8.27) by comparing the statistics of the local measurement results in the process with its time-reversed version. This theorem is universal and includes a double boundary term depending on system and environment initial states for the process and its time-reverse, Eqs. (8.24) and (8.25), and a third one depending on the initial correlations of the time-reverse process, Eq. (8.26). We also notice that our theorem may be alternatively derived from previous versions of FT's for quantum systems following unitary dynamics [2, 23] by introducing bipartite systems and local measurements. Our main contribution here is hence the identification of the quantity fulfilling the FT as the total entropy production per trajectory in both inclusive and non-inclusive versions when taking adequate boundary conditions. Our results then generalize the classical FT derived by Seifert in Ref. [4] for the total entropy production to arbitrary (open) quantum dynamics.

Once derived the total entropy production per trajectory, we have identified *adiabatic* and *non-adiabatic* contributions, accounting for different sources of irreversibility in processes with a steady state [5, 31, 32]. Extending the general formalism introduced in Chap. 7, we were able to link the global thermodynamics in system and environment with the reduced thermodynamics experimented by the open system. This required the introduction of three different thermodynamic processes: the backward (or time-reverse) process, the dual process, and the dual-reverse process, as described by different CPTP maps and Kraus operators exploiting the symmetries of the setup, c.f. Eqs. (8.34), (8.42) and (8.46). This allows the derivation of two more FT's, Eqs. (8.43), (8.48), and (8.49) and its identification with the adiabatic and non-adiabatic entropy productions. Remarkably, those FT's are not always fulfilled but the map describing the open system dynamics must verify the condition (8.41) and, additionally, the backward (or time-reverse) dynamics must preserve the (inverted) invariant state of the original process (8.44). Those requirements are generally fulfilled by classical Markov dynamics but may lead to the breaking of the entropy production decomposition in more general quantum processes. An example of a situation in which the split is broken is given in the next chapter.

The results obtained for a process described by a single CPTP map have been also extended to the case of concatenations Eqs. (8.76) and (8.77), in which the different maps appearing in the sequence may have different invariant states, and to quantum trajectories generated by unraveling the driven Lindblad master equation (8.84). In the latter case we developed a general method to identify the environmental entropy change during the trajectories induced by the quantum jumps associated to the Lindblad operators (see Eq. (8.93) and below), which allowed us to recover the FTs in Eqs. (8.100)–(8.102). The meaning of the terms adiabatic and non-adiabatic become clear in this situation as the non-adiabatic contribution becomes zero for quasi-static drivings following the instantaneous steady state of the dynamics.

In the next chapter we illustrate the results obtained here for three different and relevant cases of quantum dynamical evolutions. We will specifically see the differences between the backward, the dual, and the dual-reverse process, and the resulting expressions for the adiabatic, non-adiabatic and total entropy productions. Furthermore, we will show how the formalism developed in this chapter and the previous one, provides a natural thermodynamical description, both at the trajectory and at the averaged levels, of specific quantum processes when coherence comes into play.

References

1. M. Campisi, P. Talkner, P. Hänggi, Fluctuation theorems for continuously monitored quantum fluxes. Phys. Rev. Lett. **105**, 140601 (2010)
2. M. Esposito, U. Harbola, S. Mukamel, Nonequilibrium fluctuations, fluctuation theorems, and counting statistics in quantum systems. Rev. Mod. Phys. **81**, 1665–1702 (2009)
3. T. Hatano, S.-I. Sasa, Steady-state thermodynamics of Langevin systems. Phys. Rev. Lett. **86**, 3463 (2001)
4. U. Seifert, Entropy production along a Stochastic trajectory and an integral fluctuation theorem. Phys. Rev. Lett. **95**, 040602 (2005)
5. M. Esposito, C. Van den Broeck, Three detailed fluctuation theorems. Phys. Rev. Lett. **104**, 090601 (2010)
6. U. Seifert, Stochastic thermodynamics, fluctuation theorems and molecular machines. Rep. Prog. Phys. **75**, 126001 (2012)
7. J.P. Pekola, P. Solinas, A. Shnirman, D.V. Averin, Calorimetric measurement of work in a quantum system. New J. Phys. **15**, 115006 (2013)

8. S. Gasparinetti, K.L. Viisanen, O.-P. Saira, T. Faivre, M. Arzeo, M. Meschke, J.P. Pekola, Fast electron thermometry for ultrasensitive calorimetric detection. Phys. Rev. Appl. **3**, 014007 (2015)
9. S. Suomela, A. Kutvonen, T. Ala-Nissila, Quantum jump model for a system with a finite-size environment. Phys. Rev. E **93**, 062106 (2016)
10. M.O. Scully, M.S. Zubairy, G.S. Agarwal, H. Walther, Extracting work from a single heat bath via vanishing quantum coherence. Science **299**, 862–864 (2003)
11. A.Ü.C. Hardal, Ö.E. Müstecaplıoğlu, Superradiant quantum heat engine. Sci. Rep. **5**, 12953 (2015)
12. R. Dillenschneider, E. Lutz, Energetics of quantum correlations. Europhys. Lett. **88**, 50003 (2009)
13. X.L. Huan, T. Wang, X.X. Yi, Effects of reservoir squeezing on quantum systems and work extraction. Phys. Rev. E **86**, 051105 (2012)
14. J. Roßnagel, O. Abah, F. Schmidt-Kaler, K. Singer, E. Lutz, Nanoscale heat engine beyond the carnot limit. Phys. Rev. Lett. **112**, 030602 (2014)
15. L.A. Correa, J.P. Palao, D. Alonso, G. Adesso, Quantumenhanced absorption refrigerators. Sci. Rep., 3949 (2014)
16. D. Reeb, M.M. Wolf, An improved Landauer principle with finite-size corrections. New J. Phys. **16**, 103011 (2014)
17. J. Goold, M. Paternostro, K. Modi, Nonequilibrium quantum landauer principle. Phys. Rev. Lett. **114**, 060602 (2015)
18. O. Abah, E. Lutz, Efficiency of heat engines coupled to nonequilibrium reservoirs. Europhys. Lett. **106**, 20001 (2014)
19. G. Manzano, F. Galve, R. Zambrini, J.M.R. Parrondo, Entropy production and thermodynamic power of the squeezed thermal reservoir. Phys. Rev. E **93**, 052120 (2016)
20. W. Niedenzu, D. Gelbwaser-Klimovsky, A.G. Kofman, G. Kurizki, On the operation of machines powered by quantum nonthermal baths. New J. Phys. **18**, 083012 (2016)
21. M. Esposito, K. Lindenberg, C. Van den Broeck, Entropy production as correlation between system and reservoir. New J. Phys. **12**, 013013 (2010)
22. S. Deffner, E. Lutz, Nonequilibrium entropy production for open quantum systems. Phys. Rev. Lett. **107**, 140404 (2011)
23. T. Sagawa, Second law-like inequalitites with quantum relative entropy: an introduction, in lectures on quantum computing, thermodynamics and statistical physics, in *Kinki University Series on Quantum Computing*, ed. by M. Nakahara (World Scientific, USA, 2013)
24. J.M. Horowitz, J.M.R. Parrondo, Entropy production along nonequilibrium quantum jump trajectories. New. J. Phys **15**, 085028 (2013)
25. B. Leggio, A. Napoli, A. Messina, H.P. Breuer, Entropy production and information fluctuations along quantum trajectories. Phys. Rev. A **88**, 042111 (2013)
26. S. Deffner, Quantum entropy production in phase space. Europhys. Lett. **103**, 30001 (2013)
27. K. Funo, Y. Watanabe, M. Ueda, Integral quantum fluctuation theorems under measurement and feedback control. Phys. Rev. E **88**, 052121 (2013)
28. F. Fagnola, R. Rebolledo, Entropy production for quantum Markov semigroups. Commun. Math. Phys. **335**, 547–570 (2015)
29. M. Esposito, M.A. Ochoa, M. Galperin, Quantum thermodynamics: a nonequilibrium Green's function approach. Phys. Rev. Lett. **114**, 080602 (2015)
30. G. Manzano, J.M. Horowitz, J.M.R. Parrondo, Nonequilibrium potential and fluctuation theorems for quantum maps. Phys. Rev. E **92**, 032129 (2015)
31. M. Esposito, C. Van den Broeck, Three faces of the second law. I. Master equation formulation. Phys. Rev. E **82**, 011143 (2010)
32. C. Van den Broeck, M. Esposito, Three faces of the second law. II. Fokker-Planck formulation. Phys. Rev. E **82**, 011144 (2010)
33. B. Groisman, S. Popescu, A. Winter, Quantum, classical, and total amount of correlations in a quantum state. Phys. Rev. A **72**, 032317 (2005)

34. K. Kraus, A. Böhm, J.D. Dollard, W.H. Wootters, *States, Effects, and Operations : Fundamental Notions of Quantum Theory*, Lecture notes in physics (Springer, Berlin, 1983)
35. J.M.R. Parrondo, J.M. Horowitz, T. Sagawa, Thermodynamics of information. Nat. Phys. **11**, 131–139 (2015)
36. K. Modi, A. Brodutch, H. Cable, T. Paterek, V. Vedral, The classical-quantum boundary for correlations: discord and related measures. Rev. Mod. Phys. **84**, 1655–1707 (2012)
37. S. Luo, Using measurement-induced disturbance to characterize correlations as classical or quantum. Phys. Rev. A **77**, 022301 (2008)
38. F. Haake, *Quantum Signatures of Chaos*, 3rd edn. Springer series in synergetics (Springer, Berlin, 2010)
39. J. Anders, Thermal state entanglement in harmonic lattices. Phys. Rev. A **77**, 062102 (2008)
40. M. Campisi, P. Hänggi, P. Talkner, Colloquium: Quantum fluctuation relations: Foundations and applications. Rev. Mod. Phys. **83**, 771–791 (2011)
41. T. Monnai, Unified treatment of the quantum fluctuation theorem and the Jarzynski equality in terms of microscopic reversibility. Phys. Rev. E **72**, 027102 (2005)
42. J.M. Horowitz, Quantum-trajectory approach to the stochastic thermodynamics of a forced harmonic oscillator. Phys. Rev. E **85**, 031110 (2012)
43. T. Sagawa, M. Ueda, Role of mutual information in entropy production under information exchanges. New J. Phys. **15**, 125012 (2013)
44. G.E. Crooks, Quantum operation time reversal. Phys. Rev. A **77**, 034101 (2008)
45. M.A. Nielsen, I.L. Chuang, *Quantum Computation and Quantum Information* (Cambridge University Press, Cambridge, 2000)
46. H. Spohn, Entropy production for quantum dynamical semigroups. J. Math. Phys. **19**, 1227–1230 (1978)
47. J.M. Horowitz, T. Sagawa, Equivalent definitions of the quantum nonadiabatic entropy production. J. Stat. Phys. **156**, 55–65 (2014)
48. H.M. Wiseman, G.J. Milburn, *Quantum Measurement and Control* (Cambridge University Press, Cambridge, 2010)
49. A. Rivas, S.F. Huelga, *Open Quantum Systems : An Introduction* (Springer, Berlin, 2012)

Chapter 9
Simple Applications of the Entropy Production FT's

In Chap. 8 we analyzed the production of entropy in generic quantum processes by explicitly identifying how the relevant environmental properties can be introduced in a thermodynamic description. This allowed us to introduce the split of the total entropy production into adiabatic and non-adiabatic contributions, together with the derivation of FT's for the three quantities when some general conditions are satisfied.

In this chapter we give three examples of relevant situations where those fluctuation theorems for the entropy production can be applied. Our purpose is to clarify the meaning of the concepts introduced above, as the backward, the dual, and the dual-reverse dynamics in Sect. 8.3, the nonequilibrium potential, the specific meaning of the entropy production split, and the conditions needed to obtain it. We first consider the case of autonomous quantum thermal machines in Sect. 9.1, as composed by a three-level system selectively coupled to three thermal reservoirs at different temperatures. This example contains the necessary elements to interpret our results in a simple thermal situation, in which quantum effects do not play a fundamental role. Next we consider in Sect. 9.2 the case of resonant periodic modulation of an open quantum system by means of a classical field. Here we will see how the entropy production split is broken, discussing its consequences in energetic and entropic terms. As a third example, we consider a Maxwell's demon toy model in Sect. 9.3, designed to study the interplay between purely informational quantities with thermal properties. In addition, this configuration allows us to introduce nonequilibrium thermal reservoirs and discuss the meaning of our formalism in this situation. Finally in Sect. 9.4 we present some general conclusions.

G. Manzano Paule, *Thermodynamics and Synchronization in Open Quantum Systems*, Springer Theses, https://doi.org/10.1007/978-3-319-93964-3_9

9.1 Autonomous Quantum Thermal Machines

We first consider an autonomous three-level thermal machine powered by three thermal reservoirs at different temperatures [1–5] (see also Sect. 3.3). Each bath mediates a different transition between the energy levels, $\{|g\rangle, |e_A\rangle, |e_B\rangle\}$. The Hamiltonian of the system is

$$\hat{H}_S = \hbar\omega_1 |e_A\rangle\langle e_A| + \hbar(\omega_1 + \omega_2)|e_B\rangle\langle e_B|, \tag{9.1}$$

and the three possible transitions $g \leftrightarrow e_A$, $e_A \leftrightarrow e_B$ and $g \leftrightarrow e_B$ have frequency gaps ω_1, ω_2, and $\omega_3 \equiv \omega_1 + \omega_2$, respectively. Each transition is weakly coupled to a bosonic thermal reservoir in equilibrium at different inverse temperatures, $\beta_r = 1/k_B T_r$ with $r = 1, 2, 3$, where we assume $\beta_1 \geqslant \beta_3 \geqslant \beta_2$ for concreteness. A schematic representation of the model is depicted in Fig. 9.1.

The dynamics of the three-level thermal machine can be described by a Lindblad master equation, obtained in the weak coupling limit by applying standard techniques from open quantum systems theory (see Sects. 2.2 and 2.3 in Chap. 2). It reads

$$\dot{\rho}_t = -\frac{i}{\hbar}[\hat{H}_S, \rho_t] + \mathcal{L}_1(\rho_t) + \mathcal{L}_2(\rho_t) + \mathcal{L}_3(\rho_t), \tag{9.2}$$

where ρ_t is the density operator of the three level system and Lamb-Stark shifts have been neglected. The three dissipative terms in the above equation describe the irreversible dynamical contributions induced by each of the three thermal reservoirs:

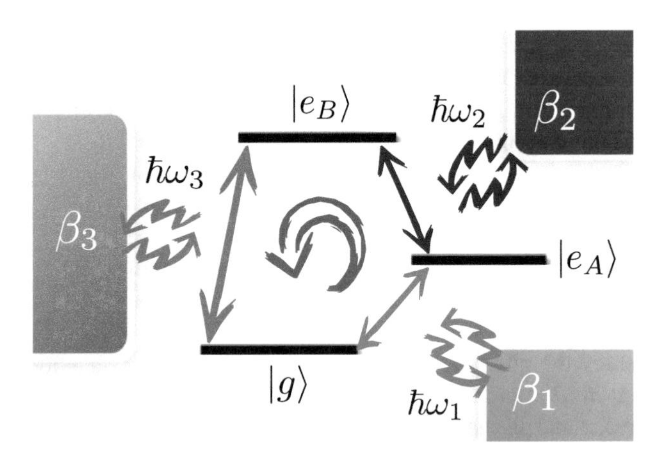

Fig. 9.1 Schematic diagram of a three-level thermal machine acting as a refrigerator. The three transitions of the machine are weakly coupled to thermal reservoirs at temperatures β_1, β_2 and β_3, inducing jumps between the machine energy levels (double arrows). In a refrigeration cycle the machine performs a sequence of three jumps $|g\rangle \to |e_A\rangle \to |e_B\rangle \to |g\rangle$, where it absorbs a quantum of energy $\hbar\omega_1$ from the cold reservoir, together with a quantum $\hbar\omega_2$ from the hot one, while emitting a quantum $\hbar\omega_3$ into the reservoir at intermediate temperature

$$\mathfrak{L}_r(\rho_t) = \Gamma_{\downarrow}^{(r)} \left(\hat{\sigma}_r \rho_t \hat{\sigma}_r^{\dagger} - \frac{1}{2}\{\hat{\sigma}_r^{\dagger}\hat{\sigma}_r, \rho_t\} \right) + \\ + \Gamma_{\uparrow}^{(r)} \left(\hat{\sigma}_r^{\dagger} \rho_t \hat{\sigma}_r - \frac{1}{2}\{\hat{\sigma}_r\hat{\sigma}_r^{\dagger}, \rho_t\} \right), \tag{9.3}$$

where $\hat{\sigma}_1 = |g\rangle\langle e_A|$, $\hat{\sigma}_2 = |e_A\rangle\langle e_B|$ and $\hat{\sigma}_3 = |g\rangle\langle e_B|$ are the ladder operators of the three-level system, inducing jumps in the corresponding transitions $r = 1, 2, 3$. The above Eq. (9.3) describes the emission and absorption of excitations of energy $\hbar\omega_r$ to (or from) the reservoir r, at rates fulfilling detailed balance

$$\frac{\Gamma_{\downarrow}^{(r)}}{\Gamma_{\uparrow}^{(r)}} = \frac{\gamma_r(n_r^{\text{th}} + 1)}{\gamma_r n_r^{\text{th}}} = e^{\beta_r \hbar\omega_r}, \tag{9.4}$$

where $n_r^{\text{th}} = (e^{\beta_r \hbar\omega_r} - 1)^{-1}$ is the mean number of excitations of energy $\hbar\omega_r$ in the reservoir r, and $\gamma_r \ll \omega_{r'}$ $\forall r, r' = 1, 2, 3$, the spontaneous emission decay rate associated to the transition.

The three average heat fluxes entering from the reservoirs associated to the imbalance in emission and absorption processes, $\dot{Q}_r = \text{Tr}[\hat{H}_S \mathfrak{L}_r(\rho_t)]$, read:

$$\dot{Q}_1 = \hbar\omega_1 \left(\Gamma_{\uparrow}^{(1)} p_g(t) - \Gamma_{\downarrow}^{(1)} p_{e_A}(t) \right), \\ \dot{Q}_2 = \hbar\omega_2 \left(\Gamma_{\uparrow}^{(2)} p_{e_A}(t) - \Gamma_{\downarrow}^{(2)} p_{e_B}(t) \right), \\ \dot{Q}_3 = \hbar\omega_3 \left(\Gamma_{\uparrow}^{(3)} p_g(t) - \Gamma_{\downarrow}^{(3)} p_{e_B}(t) \right), \tag{9.5}$$

where $p_i(t) \equiv \text{Tr}[|i\rangle\langle i|\rho_t]$ for $|i\rangle = \{|g\rangle, |e_A\rangle, |e_B\rangle\}$ are the instantaneous populations of the machine energy levels, $\sum_i p_i(t) = 1$. The first law of thermodynamics in the model consequently reads

$$\dot{U}_S \equiv \text{Tr}[\hat{H}_S \dot{\rho}_t] = \dot{Q}_1 + \dot{Q}_2 + \dot{Q}_3. \tag{9.6}$$

The master equation (9.2) describes the relaxation dynamics from any initial state of the machine to the stationary state, $\dot{\pi} = 0$, reading:

$$\pi = \pi_g |g\rangle\langle g| + \pi_{e_A} |e_A\rangle\langle e_A| + \pi_{e_B} |e_B\rangle\langle e_B|, \tag{9.7}$$

which is diagonal in the energy basis. For the simpler case in which $\gamma_1 = \gamma_2 = \gamma_3 \equiv \gamma$, we obtain:

$$\pi_g = \left[e^{\beta_3 \hbar\omega_3} \left(2e^{\beta_1 \hbar\omega_1 + \beta_2 \hbar\omega_2} - 1 \right) - e^{\beta_1 \hbar\omega_1 + \beta_2 \hbar\omega_2} \right] / Z_\pi, \\ \pi_{e_A} = \left[e^{\beta_2 \hbar\omega_2} \left(e^{\beta_1 \hbar\omega_1} - 2 \right) + e^{\beta_3 \hbar\omega_3} \left(2e^{\beta_2 \hbar\omega_2} - 1 \right) \right] / Z_\pi, \\ \pi_{e_B} = \left[e^{\beta_3 \hbar\omega_3} + e^{\beta_1 \hbar\omega_1 + \beta_2 \hbar\omega_2} - 2 \right] / Z_\pi, \tag{9.8}$$

where we defined

$$Z_\pi \equiv e^{\beta_2 \hbar\omega_2}\left(-2 + e^{\beta_1 \hbar\omega_1}\right) - 2 \\ + e^{\beta_3 \hbar\omega_3}\left(2e^{\beta_2 \hbar\omega_2}\left(1 + e^{\beta_1 \hbar\omega_1}\right) - 1\right).$$

The present setup constitutes the simplest model of an ideal quantum absorption heat pump and refrigerator, usually considered to operate at steady-state conditions [3–5]. We now focus on the fridge configuration, but similar conclusions follow as well in the heat pump mode of operation. The cooling mechanism exploits the average heat flow entering from the reservoir at the hottest temperature T_2, $\dot{Q}_2 > 0$, to continuously extract heat from the reservoir at the lowest temperature T_1, $\dot{Q}_1 > 0$, while draining $\dot{Q}_3 < 0$ to the reservoir at the intermediate temperature, T_3 (see Fig. 9.1). At steady state conditions, it can be easily checked that this is indeed the case by properly tuning the energy level spacings [Eq. (9.10)]:

$$\dot{Q}_1^{\mathrm{ss}} = \gamma \hbar\omega_1 \frac{\Delta}{Z_\pi} \geqslant 0, \qquad \dot{Q}_2^{\mathrm{ss}} = \gamma \hbar\omega_2 \frac{\Delta}{Z_\pi} \geqslant 0, \tag{9.9}$$

and $\dot{Q}_3^{\mathrm{ss}} = -(\dot{Q}_1^{\mathrm{ss}} + \dot{Q}_2^{\mathrm{ss}}) \leqslant 0$, where $Z_\pi \geqslant 0$ and the quantity $\Delta \equiv (e^{\beta_3 \hbar\omega_3} - e^{\beta_1 \hbar\omega_1 + \beta_2 \hbar\omega_2})$, remains positive when the following design condition is met:

$$\omega_2 \geqslant \left(\frac{\beta_1 - \beta_3}{\beta_3 - \beta_2}\right)\omega_1. \tag{9.10}$$

Notice also that, when inequality (9.10) is inverted, we obtain $\Delta \leqslant 0$, and the three heat flows invert signs, hence generating a heat pump mode of operation (see Fig. 9.2 below).

9.1.1 Quantum Trajectories and Entropy Production

We now apply our trajectory formalism to unravel the thermodynamics of the above three-level thermal machine at the stochastic level. We analyze the complete transient dynamics of the model when the machine starts the evolution in some arbitrary initial state, with respect to which the steady state regime is a particular case.

Let us start by identifying the operations introduced by the dynamics (9.2) for an infinitesimal time-step dt. Following Sect. 8.5, the coarse-grained dynamical evolution is given by the action of a CPTP map $\mathcal{E}$ [Eq. (8.85)], with Kraus operators:

$$\hat{M}_0 = \mathbb{1} - dt\left(\frac{i}{\hbar}\hat{H}_{\mathrm{S}} + \frac{1}{2}\sum_{r=1}^{3}\sum_{k=\downarrow,\uparrow}\hat{L}_k^{(r)\dagger}\hat{L}_k^{(r)}\right), \tag{9.11}$$

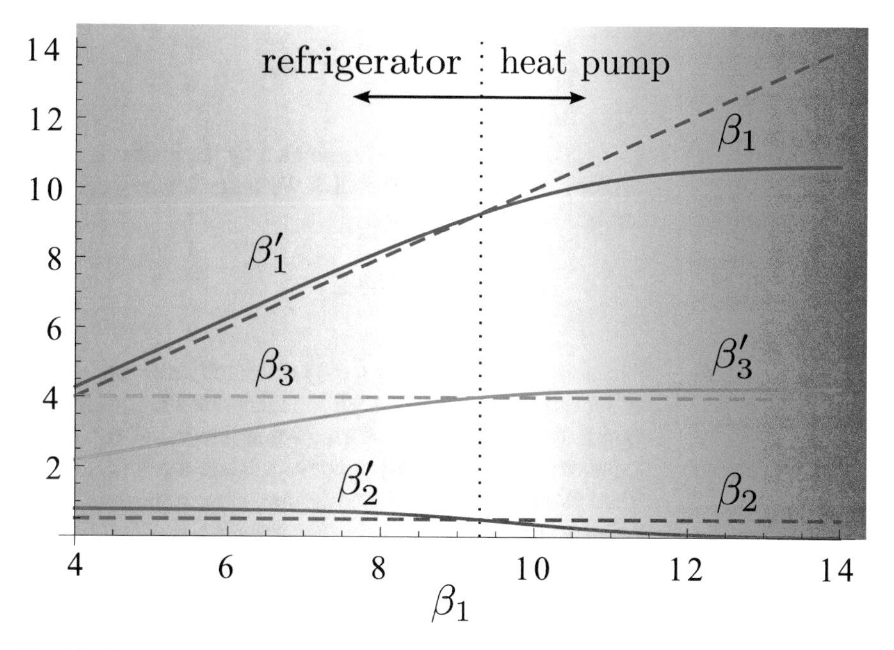

Fig. 9.2 Comparison between the inverse effective (or virtual) temperatures β'_r (solid lines) and the real inverse temperatures of the reservoirs β_r (dashed lines) for $r = 1, 2, 3$ (blue, red, orange), as a function of β_1 when $\omega_1 = \hbar^{-1}$, and $\omega_2 = 1.5\hbar^{-1}$. In the plot we stressed the two modes of operation of the autonomous three-level machine separated by a dotted line corresponding to the equality case in Eq. (9.10), implying reversible environmental conditions, $\Delta_{\rm i} s_\gamma^{\rm a} = 0 \;\; \forall \gamma$ in Eq. (9.28). In the refrigerator regime, the transition $g \leftrightarrow e_A$ is at an effective temperature colder than the coldest reservoir, $\beta'_1 \geqslant \beta_1$, inducing heat extraction from it, while the other transitions induce dissipation of heat to the reservoir at intermediate temperature, $\beta_2 \geqslant \beta'_2$, and absorption of heat in the hotter one $\beta'_2 \geqslant \beta_2$. In the heat pump regime the three heat flows change its directions as the previous inequalities become inverted

corresponding to the absence of jumps during dt, and:

$$\hat{M}_{\downarrow}^{(r)} = \sqrt{dt}\, \hat{L}_{\downarrow}^{(r)} = \sqrt{dt\, \Gamma_{\downarrow}^{(r)}}\, \hat{\sigma}_r, \tag{9.12}$$

$$\hat{M}_{\uparrow}^{(r)} = \sqrt{dt}\, \hat{L}_{\uparrow}^{(r)} = \sqrt{dt\, \Gamma_{\uparrow}^{(r)}}\, \hat{\sigma}_r^{\dagger}, \tag{9.13}$$

where $\hat{L}_{\downarrow}^{(r)}$ and $\hat{L}_{\uparrow}^{(r)}$ are the Lindblad operators promoting jumps down and up in the transition r of the three-level system. Furthermore, the concatenation of the no-jump operator, $\hat{M}_0$ between arbitrary times t and t', gives us [see Eq. (8.89)]:

$$\hat{U}_{\rm eff}(t', t) = \exp\left(-\frac{i}{\hbar} \hat{H}_{\rm eff}(t' - t)\right), \tag{9.14}$$

as the effective evolution operator between jumps along the dynamics, with

$$\hat{H}_{\text{eff}} = \hat{H}_{\text{S}} - (i\hbar/2)\sum_{r=1}^{3}\sum_{k=\downarrow,\uparrow} \hat{L}_k^{(r)\dagger}\hat{L}_k^{(r)} . \tag{9.15}$$

A trajectory $\gamma = \{n, (k_1, t_1), \ldots, (k_N, t_N), m\}$ generated by the master equation (9.2) can be hence constructed as explained in Sect. 8.5. We start with the three-level thermal-machine in an arbitrary initial state

$$\rho_0 = \sum_n p_n |\psi_n\rangle\langle\psi_n|, \tag{9.16}$$

which is projectively measured in its eigenbasis $\{|\psi_n\rangle\}$ obtaining outcome n at the initial time, t_0. Then we let the machine evolve while registering the sequence of jumps induced by the operators $\hat{M}_\downarrow^{(r)}$ and $\hat{M}_\uparrow^{(r)}$ during some interval of time τ. In this way, we obtain a sequence of N jumps $\{k_l\}$ at corresponding times $\{t_l\}$. At the final time $t_{\text{f}} = t_0 + \tau$, the system is measured again, now using projectors in the eigenbasis of $\rho_{t_{\text{f}}} = \exp(\mathcal{L}\tau)\rho_0$, that is

$$\rho_{t_{\text{f}}} = \sum_m p'_m |\psi'_m\rangle\langle\psi'_m|, \tag{9.17}$$

obtaining outcome m. Notice that here the stochastic jumps during the evolution correspond to simple transitions between the energy levels $\{|g\rangle, |e_A\rangle, |e_B\rangle\}$, which enforce the state of the system to lose all coherences in the energy basis after the first jump. Therefore, the stochastic dynamics is mostly classical during the transient dynamics.

The backward trajectory $\tilde{\gamma} = \{m, (k_N, t_N), \ldots, (k_1, t_1), n\}$, represents the inverse sequence of events with respect to γ, occurring in the backward process. Following Sect. 8.5, we can construct the backward process by specifying the initial state [here the inverted final state of the forward process, $\hat{\Theta}\rho_{t_{\text{f}}}\hat{\Theta}^\dagger$], and obtaining the operations governing the time-reversed dynamics from its forward counterparts. The backward map, $\tilde{\mathcal{E}}$, can be obtained from the generalized detailed-balance relation in Eq. (8.92):

$$\hat{\tilde{M}}_0 = e^{-\sigma_0^E/2}\hat{\Theta}\hat{M}_0^\dagger\hat{\Theta}^\dagger, \qquad \hat{\tilde{M}}_k^{(r)} = e^{-\sigma_k^{E_r}/2}\hat{\Theta}\hat{M}_k^\dagger\hat{\Theta}^\dagger, \tag{9.18}$$

where $\{\sigma_k^{E_r}\}$ are the boundary terms for the reservoir $r = 1, 2, 3$. Furthermore, by noticing that the Lindblad operators in this case come in pairs, that is

$$\hat{L}_\downarrow^{(r)} = \sqrt{\frac{\Gamma_\downarrow^{(r)}}{\Gamma_\uparrow^{(r)}}}\,\hat{L}_\uparrow^{(r)\dagger} = e^{\beta_r\hbar\omega_r/2}\hat{L}_\uparrow^{(r)\dagger}, \tag{9.19}$$

we can easily determine the quantities $\{\sigma_k^{E_r}\}$ from Eq. (8.93):

$$\sigma_0^E = 0, \quad \sigma_\downarrow^{E_r} = \beta_r\hbar\omega_r, \quad \sigma_\uparrow^{E_r} = -\beta_r\hbar\omega_r. \tag{9.20}$$

Notice that the jumps induced by the operators $\hat{L}_{\downarrow}^{(r)}$ ($\hat{L}_{\uparrow}^{(r)}$) in the forward trajectories γ, are associated to the emission (absorption) of a quantum of energy $\hbar\omega_r$ in the transition $r = 1, 2, 3$. Hence the boundary terms $\sigma_{\downarrow}^{E_r}$ ($\sigma_{\uparrow}^{E_r}$) can be interpreted as the entropy produced (annihilated) in reservoir r, when a quantum $\hbar\omega_r$ of heat is transferred to (from) the reservoir at inverse temperature β_r. As long as the reservoirs have been here assumed to be large systems at equilibrium state ρ_E during the whole evolution, this corresponds to choosing an initial state for the environment in the backward process of the form $\hat{\Theta}^\dagger \tilde{\rho}_E \hat{\Theta} = \rho_E \simeq \rho_E^*$ (see Sect. 8.5).

Introducing the expressions for $\sigma_k^{E_r}$ [Eq. (9.20)] into Eq. (9.18), we obtain the Kraus operators of the backward map:

$$\begin{aligned} \hat{\tilde{M}}_{\downarrow}^{(r)} &= \sqrt{dt}\; \hat{\tilde{L}}_{\downarrow}^{(r)} = \sqrt{dt}\; \hat{\Theta}\hat{L}_{\uparrow}^{(r)}\hat{\Theta}^\dagger = \hat{M}_{\uparrow}^{(r)}, \\ \hat{\tilde{M}}_{\uparrow}^{(r)} &= \sqrt{dt}\; \hat{\tilde{L}}_{\uparrow}^{(r)} = \sqrt{dt}\; \hat{\Theta}\hat{L}_{\downarrow}^{(r)}\hat{\Theta}^\dagger = \hat{M}_{\downarrow}^{(r)}, \end{aligned} \tag{9.21}$$

together with $\hat{\tilde{M}}_0 = \hat{\Theta}\hat{M}_0\hat{\Theta}^\dagger = \hat{M}_0$ for the no-jump evolution. Indeed, by exploiting the symmetries of the smooth no-jump evolution [see Eq. (8.95) and below], we obtain $\hat{\tilde{U}}_{\text{eff}} = \hat{\Theta}\hat{U}_{\text{eff}}^\dagger\hat{\Theta}^\dagger = \hat{U}_{\text{eff}}$ for the effective evolution operator describing the dynamics between jumps in the backward process. From the above equations we explicitly see that the forward and backward maps, $\mathcal{E}$ and $\tilde{\mathcal{E}}$, are equivalent, while the jumps up in the forward process are related with jumps down in the backward process (and vice-versa). We also notice that, consequently, the backward map $\tilde{\mathcal{E}}$ admits the time-reversed steady state $\tilde{\pi} = \hat{\Theta}\pi\hat{\Theta}^\dagger = \pi$ as an invariant state.

We next construct the dual and dual-reverse processes for the model. We note that condition (8.96) is here fulfilled, together with (8.44). Indeed, the nonequilibrium potential, $\hat{\Phi} = -\ln\pi$, obeys $[\hat{\Phi}, \hat{H}_S] = 0$ and

$$[\hat{\Phi}, \hat{L}_k^{(r)}] = \Delta\phi_k^{(r)}\hat{L}_k^{(r)}, \qquad [\hat{\Phi}, \hat{L}_k^{(r)\dagger}] = -\Delta\phi_k^{(r)}\hat{L}_k^{(r)\dagger}, \tag{9.22}$$

where the nonequilibrium potential changes associated to each jump in the trajectory read

$$\Delta\phi_0 = 0, \quad \Delta\phi_{r\downarrow} = -\beta_r'\hbar\omega_r, \quad \Delta\phi_{r\uparrow} = \beta_r'\hbar\omega_r. \tag{9.23}$$

Here the quantities $\beta_1' = \ln\left(\frac{\pi_g}{\pi_{e_A}}\right)/\hbar\omega_1$, $\beta_2' = \ln\left(\frac{\pi_{e_A}}{\pi_{e_B}}\right)/\hbar\omega_2$ and $\beta_3' = \ln\left(\frac{\pi_g}{\pi_{e_B}}\right)/\hbar\omega_3$ are effective inverted temperatures (or virtual temperatures [6, 7]) associated to each of the transitions in the steady state π in Eq. (9.7). Each time a jump down is detected in transition r, corresponding to the emission of a quantum of energy $\hbar\omega_r$, an amount $\beta_r'\hbar\omega_r$ of entropy (heat divided by temperature) is transferred from the system to the reservoir r (and vice-versa for the jumps up).

The Kraus operators for dual and dual-reverse maps, $\mathcal{D}$ and $\tilde{\mathcal{D}}$, can be obtained as well from Eqs. (8.98) and (8.99), by using Eqs. (9.20) and (9.23). They read:

$$\hat{D}_{\downarrow}^{(r)} = \sqrt{dt}\, e^{(\beta_r' - \beta_r)\hbar\omega_r/2}\, \hat{L}_{\downarrow}^{(r)} \propto \hat{M}_{\downarrow}^{(r)}, \tag{9.24}$$

$$\hat{D}_{\uparrow}^{(r)} = \sqrt{dt}\, e^{-(\beta_r' - \beta_r)\hbar\omega_r/2}\, \hat{L}_{\uparrow}^{(r)} \propto \hat{M}_{\uparrow}^{(r)}, \tag{9.25}$$

$$\hat{\tilde{D}}_{\downarrow}^{(r)} = \sqrt{dt}\, e^{-(\beta_r' - \beta_r)\hbar\omega_r/2}\, \hat{L}_{\uparrow}^{(r)} \propto \hat{\tilde{M}}_{\downarrow}^{(r)}, \tag{9.26}$$

$$\hat{\tilde{D}}_{\uparrow}^{(r)} = \sqrt{dt}\, e^{(\beta_r' - \beta_r)\hbar\omega_r/2}\, \hat{L}_{\downarrow}^{(r)} \propto \hat{\tilde{M}}_{\uparrow}^{(r)}. \tag{9.27}$$

Notice that the dual and dual-reverse maps are similar to maps $\mathcal{E}$ and $\tilde{\mathcal{E}}$ respectively. In particular, all the Kraus operators corresponding to the dual and dual-reversed dynamics are proportional to the original ones, inducing the same jumps in the three-level system, but at modified rates depending on the difference $\beta_r' - \beta_r$. Only when $\beta_r' = \beta_r$ for each r the dual process becomes equal to the forward process, and hence the dual-reverse process equals the backward process (see Fig. 9.2).

As stressed in the previous chapter, Eq. (9.22) together with the backward map having $\tilde{\pi}$ as an invariant state, are sufficient conditions to ensure the existence of the three fluctuation theorems for the adiabatic, non-adiabatic and total entropy productions during trajectory γ. They explicitly read:

$$\Delta_{\mathrm{i}} s_{\gamma}^{\mathrm{a}} = \sum_{r=1}^{3} (\beta_r' - \beta_r) q_{\gamma}^{(r)}, \tag{9.28}$$

$$\Delta_{\mathrm{i}} s_{\gamma}^{\mathrm{na}} = \sigma_{nm}^{\mathcal{S}} - \sum_{r=1}^{3} \beta_r' q_{\gamma}^{(r)}, \tag{9.29}$$

$$\Delta_{\mathrm{i}} s_{\gamma} = \sigma_{nm}^{\mathcal{S}} - \sum_{r=1}^{3} \beta_r q_{\gamma}^{(r)}, \tag{9.30}$$

where we stress that $\sigma_{nm}^{\mathcal{S}} = -\ln p_m' + \ln p_n = \Delta s_{nm}^{\mathcal{S}}$ is the change in the entropy of the system during the trajectory [8–12], and

$$q_{\gamma}^{(r)} = \hbar\omega_r (n_{\uparrow}^{(r)} - n_{\downarrow}^{(r)}), \tag{9.31}$$

is the stochastic heat entering the system from reservoir r during the jumps, $n_{\uparrow}^{(r)}$ ($n_{\downarrow}^{(r)}$) being the total number of jumps up (down) in transition r. It is easy to check from the above equations that $\Delta_{\mathrm{i}} s_{\gamma} = \Delta_{\mathrm{i}} s_{\gamma}^{\mathrm{a}} + \Delta_{\mathrm{i}} s_{\gamma}^{\mathrm{na}}$. Moreover, they provide us the following interpretation of the entropy production decomposition in the model. The adiabatic term, $\Delta_{\mathrm{i}} s_{\gamma}^{\mathrm{a}}$, captures the entropy produced in the process which does not modify the local state of the machine. On the other hand, the non-adiabatic term accounting for the entropy production generated in the relaxation of the system to the steady state π, can be viewed as the entropy produced in the thermalization process of the three-level machine if each transition were coupled to a thermal reservoir at temperatures β_r', instead of β_r, respectively.

Turning to the averaged behavior as given by the master equation (9.2), we can calculate the average flow of nonequilibrium potential and the entropy changes in the reservoirs:

$$\dot{S}_r = \sum_{k=\uparrow,\downarrow} \mathrm{Tr}[\hat{L}_k^{(r)\dagger} \hat{L}_k^{(r)} \rho_t] \sigma_k^{(r)} = -\beta_r \dot{Q}_r, \tag{9.32}$$

$$\dot{\Phi}_r = \sum_{k=\uparrow,\downarrow} \mathrm{Tr}[\hat{L}_k^{(r)\dagger} \hat{L}_k^{(r)} \rho_t] \Delta\phi_k^{(r)} = \beta_r' \dot{Q}_r, \tag{9.33}$$

where we split in three parts the nonequilibrium potential flow $\dot{\Phi} = \dot{\Phi}_1 + \dot{\Phi}_2 + \dot{\Phi}_3 = -\mathrm{Tr}[\dot{\rho}_t \ln \pi]$.

The entropy production rates hence read:

$$\dot{S}_{\mathrm{a}} = \sum_r (\beta_r' - \beta_r) \dot{Q}_r \geqslant 0, \tag{9.34}$$

$$\dot{S}_{\mathrm{na}} = \dot{S} - \sum_r \beta_r' \dot{Q}_r \geqslant 0, \tag{9.35}$$

$$\dot{S}_{\mathrm{i}} = \dot{S}_{\mathrm{a}} + \dot{S}_{\mathrm{na}} = \dot{S} - \sum_r \beta_r \dot{Q}_r \geqslant 0, \tag{9.36}$$

with $\dot{S} = -\mathrm{Tr}[\dot{\rho}_t \ln \rho_t]$ the derivative of the von Neumann entropy of the three-level machine. They show the same structure as the trajectory versions in Eqs. (9.28)–(9.30).

Finally, notice that in the steady state we have $\dot{S}_{\mathrm{na}} = 0$, and the first law (9.6) becomes $\sum_r \dot{Q}_r^{\mathrm{ss}} = 0$. This implies that the adiabatic entropy production rate in Eq. (9.34) equals the total entropy production rate [Eq. (9.36)]:

$$\dot{S}_{\mathrm{a}} = \dot{S}_{\mathrm{i}} = (\beta_3 - \beta_2) \dot{Q}_2^{\mathrm{ss}} - (\beta_1 - \beta_3) \dot{Q}_1^{\mathrm{ss}} \geqslant 0, \tag{9.37}$$

which rules the stationary heat fluxes passing through the system, and sets a bound for the thermal machine efficiency in any regime of operation. Indeed, combining Eqs. (9.37) and (9.6) we obtain that the efficiency of the refrigeration process, as captured by the coefficient of performance (COP) [5], is bounded by

$$\epsilon = \frac{\dot{Q}_1^{\mathrm{ss}}}{\dot{Q}_2^{\mathrm{ss}}} \leqslant \frac{\beta_3 - \beta_2}{\beta_1 - \beta_3} \equiv \epsilon_C \tag{9.38}$$

where ϵ_C represents the Carnot COP, that is, the equivalent of the Carnot efficiency for fridges [6]. The above bound can be alternatively obtained by noticing from Eq. (9.9) that $\epsilon = \omega_1/\omega_2$, and hence the bound $\epsilon \leqslant \epsilon_C$ directly follows from condition Eq. (9.10), ensuring that the machine is acting as a refrigerator. This means that the Carnot COP can be reached by properly tuning the spacing of the transitions ω_1 and ω_2, i.e. approaching $\omega_2 \to \omega_2^C = \eta_C \omega_1$ from above. In such case the stationary currents tend to vanish, $Q_1^{\mathrm{ss}} \to 0$ and $Q_2^{\mathrm{ss}} \to 0$, and the entropy production rate $\dot{S}_{\mathrm{a}} \to 0$, a well-known feature of reversible 'Carnot conditions' characterized by extreme slowness in the energy exchange processes.

9.2 Periodically Driven Cavity Mode at Resonance

As a second simple example we consider a single electromagnetic field mode with frequency ω in a microwave cavity with slight losses in one of their mirrors. The losses of the cavity are produced by the weak coupling of the cavity mode to a bosonic thermal reservoir in equilibrium at some inverse temperature $\beta = 1/k_B T$. In addition, an external laser of same frequency ω and weak intensity drives the cavity mode producing excitations. The Hamilton operator for the system consists of two terms, $\hat{H}_S(t) = \hat{H}_0 + \hat{V}_S(t)$, the first one representing the Hamiltonian of the undriven mode $\hat{H}_0 = \hbar\omega\hat{a}^\dagger\hat{a}$ and

$$\hat{V}_S(t) = i\hbar(\epsilon\hat{a}^\dagger e^{-i\omega t} - \epsilon^*\hat{a}e^{i\omega t}) \tag{9.39}$$

describing the effect of the classical resonant laser field $\epsilon = |\epsilon|e^{i\varphi_\epsilon}$ with amplitude $|\epsilon|$ and phase φ_ϵ (see Sect. 1.2). In Fig. 9.3 we show a schematic picture of the setup.

The reduced evolution of the cavity mode can be described by a Lindblad master equation in the interaction picture with respect to $\hat{H}_0$, of the form [13]:

$$\dot{\rho}_t = -\frac{i}{\hbar}[\hat{V}, \rho_t] + \mathcal{L}(\rho_t) \tag{9.40}$$

where $\hat{V} = i\hbar(\epsilon\hat{a}^\dagger - \epsilon^*\hat{a})$ is the driving Hamiltonian in the interaction picture, and the dissipative part of the dynamics is assumed to take the form of the undriven case (see Sect. 2.3)

$$\mathcal{L}(\rho) = \Gamma_\downarrow\left(\hat{a}\rho\hat{a}^\dagger - \frac{1}{2}\{\hat{a}^\dagger\hat{a}, \rho\}\right) + \Gamma_\uparrow\left(\hat{a}^\dagger\rho\hat{a} - \frac{1}{2}\{\hat{a}\hat{a}^\dagger, \rho\}\right)$$

accounting for emission and absorption of photons by the cavity mode to (from) the equilibrium reservoir at respective rates $\Gamma_\downarrow = \gamma_0(n^{\text{th}} + 1)$ and $\Gamma_\uparrow = \gamma_0\, n^{\text{th}}$. As usual

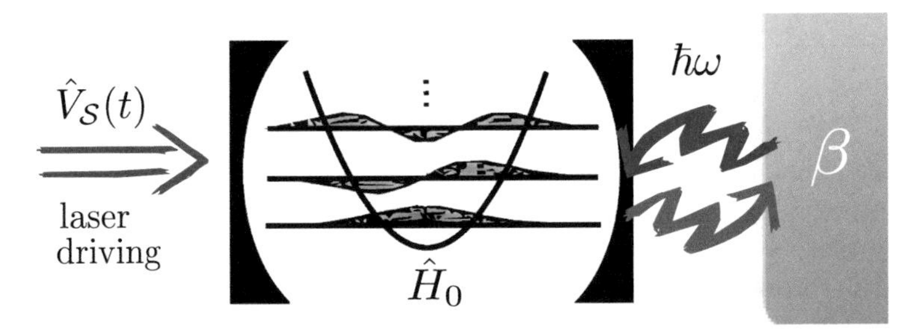

Fig. 9.3 Schematic picture of the setup. The intracavity mode $\hat{H}_0$ is externally driven by a resonant laser field $\hat{V}_S(t)$, while in weak contact with the environment at inverse temperature β, producing the emission and absorption of photons

$n^{\text{th}} = (e^{-\beta\hbar\omega} - 1)^{-1}$ is the mean number of excitations of frequency ω in the reservoir at inverse temperature β, and γ_0 is the spontaneous emission decay rate in absence of driving. We require for consistency that $\gamma_0, |\epsilon| \ll \omega$ but $|\epsilon|/\gamma_0 =$ constant, which implies that terms $|\epsilon|\gamma \to 0$ can be safely neglected in the master equation. This guarantees the decoupling of the dynamics induced by the driving laser and that generated by the presence of the thermal reservoir.

The steady state of the dynamics (9.40) can be easily obtained analytically, resulting in the displaced thermal state

$$\pi = \hat{\mathcal{D}}(\alpha)\,\frac{e^{-\beta\hat{H}_0}}{Z_0}\,\hat{\mathcal{D}}^\dagger(\alpha) = \frac{e^{-\beta(\hat{H}_0 - \mu\hat{X}_\alpha)}}{Z_\alpha} \tag{9.41}$$

where $\alpha \equiv 2\epsilon/\gamma_0$, $\hat{\mathcal{D}}(\alpha) = \exp(\alpha\hat{a}^\dagger - \alpha^*\hat{a})$ is the unitary displacement operator in optical phase space, $\hat{\mathcal{D}}(\alpha)\hat{a}\hat{\mathcal{D}}^\dagger(\alpha) = \hat{a} - \alpha$ (see Sect. 1.2.4), and $Z_0 = \text{Tr}[\exp(-\beta\hat{H}_0)]$. In the second equality, we have written the state π as a generalized Gibbs ensemble (see our analysis of generalized Gibbs-preserving maps in Sect. 7.4) by defining a Lagrange multiplier for the field φ_ϵ induced by the external laser:

$$\mu \equiv \sqrt{2}\hbar\omega|\alpha|, \qquad \hat{X}_\alpha = \frac{1}{\sqrt{2}}\left(\hat{a}e^{-i\varphi_\epsilon} + \hat{a}^\dagger e^{i\varphi_\epsilon}\right), \tag{9.42}$$

with $[\hat{H}_0, \hat{X}_\alpha] \neq 0$, and $Z_\alpha = e^{\beta\hbar\omega|\alpha|^2} Z_0$ in Eq. (9.41). In contrast to the undriven case, here the cavity does not reach thermal equilibrium with the reservoir, as coherences in the energy basis do not decay to zero, being connected with the work performed by the external laser. The fact that the steady state π can be written as a generalized Gibbs ensemble is a consequence of the presence of an extra conserved quantity in global dynamics, the field quadrature $\hat{X}_\alpha$. Notice also that the state π defines a limit cycle (unitary orbit) in the Schrödinger picture by $\pi_S(t) = e^{-\frac{i}{\hbar}\hat{H}_0 t}\,\pi\,e^{\frac{i}{\hbar}\hat{H}_0 t}$, rotating in optical phase space, according to the free evolution $i\hbar\dot{\pi}_S = [\hat{H}_0, \pi_S]$.

The energy balance during the evolution can be stated by noticing that the total power input from the laser drive results $\dot{W} = \text{Tr}[\frac{d\hat{H}_{\mathcal{S}}}{dt}\rho_S]$, where $\rho_S(t) = e^{-\frac{i}{\hbar}\hat{H}_0 t}\,\rho_t\,e^{\frac{i}{\hbar}\hat{H}_0 t}$ is the density operator of the cavity mode in the Schrödinger picture. If we further identify the internal energy of the cavity mode as $U_{\mathcal{S}} = \text{Tr}[\hat{H}_{\mathcal{S}}\rho_S]$, the first law of thermodynamics in this configuration just reads

$$\dot{U}_{\mathcal{S}} = \dot{W} + \dot{Q} = \text{Tr}\left[\frac{d\hat{H}_{\mathcal{S}}}{dt}\rho_S\right] + \text{Tr}\left[\hat{H}_{\mathcal{S}}\frac{d\rho_S}{dt}\right], \tag{9.43}$$

and therefore the heat flow is $\dot{Q} = \text{Tr}[\hat{H}_{\mathcal{S}}\frac{d\rho_S}{dt}] = \text{Tr}[\hat{H}_{\mathcal{S}}\mathcal{L}(\rho_S)]$ in analogy to the previous section. In the heat expression we used that the dissipative dynamics is invariant when moving between the Sch-rödinger picture and the interaction picture with respect to $\hat{H}_0$.

9.2.1 Failure of the FT for Adiabatic Entropy Production

As in the previous example, we may explicitly construct the thermodynamic description for this model at the trajectory level. The master equation (9.40) provides us the Kraus operators for the map $\mathcal{E}$ in Eq. (8.85):

$$\hat{M}_0 = \mathbb{1} - dt\left(\frac{i}{\hbar}\hat{V} + \frac{1}{2}\sum_{k=\downarrow,\uparrow}\hat{L}_k^\dagger\hat{L}_k\right), \tag{9.44}$$

for the no jump evolution, and

$$\hat{M}_\downarrow = \sqrt{dt}\,\hat{L}_\downarrow = \sqrt{dt\,\Gamma_\downarrow}\,\hat{a}, \tag{9.45}$$

$$\hat{M}_\uparrow = \sqrt{dt}\,\hat{L}_\uparrow = \sqrt{dt\,\Gamma_\uparrow}\,\hat{a}^\dagger, \tag{9.46}$$

for the jumps. The effective evolution between jumps, $\hat{U}_{\mathrm{eff}}(t', t)$ in Eq. (8.89), is again given by Eq. (9.14) where the effective non-hermitian operator is in this case

$$\hat{H}_{\mathrm{eff}}(t) = \hat{V} - (i\hbar/2)\sum_{k=\downarrow,\uparrow}\hat{L}_k^\dagger\hat{L}_k, \tag{9.47}$$

and the trajectory $\gamma = \{n, (k_1, t_1), \ldots, (k_N, t_N), m\}$ is constructed as in the previous example by counting the jumps $\{\hat{L}_\downarrow, \hat{L}_\uparrow\}$ induced by the reservoir and registering the times at which jumps occurred.

The backward evolution is specified as well by Kraus operators for the backward map, $\tilde{\mathcal{E}}$, which can be obtained from the detailed balance relation (8.92) in Sect. 8.5 as in the previous example. As the forward dynamics, it is governed by a single pair of Lindblad operators:

$$\hat{L}_\downarrow = \sqrt{\Gamma_\downarrow}\,\hat{a}, \qquad \hat{L}_\uparrow = \sqrt{\Gamma_\uparrow}\,\hat{a}^\dagger. \tag{9.48}$$

The stochastic entropy change in the environment (again a thermal reservoir in equilibrium) associated to each Kraus operator in Eqs. (9.44) and (9.45) is given by

$$\sigma_0^E = 0, \qquad \sigma_\downarrow^E = \beta\hbar\omega, \qquad \sigma_\uparrow^E = -\beta\hbar\omega. \tag{9.49}$$

That is, when a jump down (up) occurs, the entropy in the environment increases (decreases) by $\beta\hbar\omega$, associated to the emission (absorption) of a quantum of energy $\hbar\omega$ from the reservoir. Therefore, the Kraus operators for the backward map $\tilde{\mathcal{E}}$ read

$$\begin{aligned}
\hat{\tilde{M}}_0 &= \hat{\Theta}\hat{M}_0\hat{\Theta}^\dagger = \hat{M}_0, \\
\hat{\tilde{M}}_\downarrow &= \sqrt{dt}\ \hat{\tilde{L}}_\downarrow = \sqrt{dt}\ \hat{\Theta}\hat{L}_\uparrow\hat{\Theta}^\dagger = \hat{M}_\uparrow, \\
\hat{\tilde{M}}_\uparrow &= \sqrt{dt}\ \hat{\tilde{L}}_\uparrow = \sqrt{dt}\ \hat{\Theta}\hat{L}_\downarrow\hat{\Theta}^\dagger = \hat{M}_\downarrow,
\end{aligned} \tag{9.50}$$

implying again that forward and backward maps, $\mathcal{E}$ and $\tilde{\mathcal{E}}$, are equivalent.

However it is worth noticing that in this case the condition (8.41) needed for the existence of a dual and dual-reverse dynamics is not fulfilled. Indeed the nonequilibrium thermodynamic potential, from Eq. (9.41), reads in this case

$$\begin{aligned}
\hat{\Phi} &= -\ln\pi = \beta\ \hat{\mathcal{D}}(\alpha)\ \hat{H}_0\ \hat{\mathcal{D}}^\dagger(\alpha) + \ln Z_0 \\
&= \beta\left(\hat{H}_0 - \mu\hat{X}_\alpha\right) + \ln Z_\alpha,
\end{aligned} \tag{9.51}$$

which does not obey the condition (8.96), because the Lindblad operators appearing in the dynamics (9.40) do not promote jumps in the steady state basis, but in the unperturbed Hamiltonian ($\hat{H}_0$) basis. This implies that we cannot associate a single change in the nonequilibrium potential to each Lindblad jump operator. As a consequence, the entropy production per trajectory cannot be decomposed in adiabatic and non-adiabatic contributions, and the fluctuation theorems for the adiabatic and non-adiabatic entropy production are not valid.

We indeed anticipate the breaking of the entropy production decomposition whenever the (possibly many) dissipative contributions to a dynamical evolution

$$\dot{\rho}_t = -\frac{i}{\hbar}[\hat{H}, \rho_t] + \sum_n \mathcal{L}_n(\rho_t) \equiv \mathcal{L}(\rho_t) \tag{9.52}$$

possesses a steady state $\mathcal{L}_n(\pi^{(n)}) = 0$ which does not commute with the actual steady state generated by the whole dynamics, $[\pi, \pi^{(n)}] \neq 0$ with $\mathcal{L}(\pi) = 0$. This is a purely quantum feature arising when coherences are introduced in the thermodynamical description.

9.2.2 *Implications to the Second-Law-Like Inequalities*

Let us finally calculate the average total entropy production by obtaining the average entropy change in the environment [Eq. (8.106)]

$$\begin{aligned}
\dot{S}_E &= \sum_k \mathrm{Tr}[\mathcal{E}_k(\rho_t)]\sigma_k^E/dt = \sum_{k=\downarrow,\uparrow} \mathrm{Tr}[\hat{L}_k^\dagger\hat{L}_k\rho_t]\sigma_k^E \\
&= -\ \beta\ \mathrm{Tr}[\mathcal{L}(\rho_t)\hat{H}_0] = -\beta\ \mathrm{Tr}[\mathcal{L}(\rho_S)\hat{H}_S] = -\beta\dot{Q}
\end{aligned} \tag{9.53}$$

where in the second line we have safely neglected the contribution coming from the driving term in the heat, that is $\mathrm{Tr}[\mathcal{L}(\rho_S)\hat{H}_S] \simeq \mathrm{Tr}[\mathcal{L}(\rho_S)\hat{H}_0]$, as $|\epsilon|\gamma \to 0$. The average total entropy production rate [Eq. (8.105)] hence reads

$$\dot{S}_\mathrm{i} = \dot{S} - \beta\dot{Q} = \beta(\dot{W} - \dot{\mathfrak{F}}) \geqslant 0, \tag{9.54}$$

where $\dot{\mathfrak{F}} = \dot{U}_S - T\dot{S}$ is the time derivative of the nonequilibrium free energy (see Chap. 3). In Eq. (9.54) the term $\dot{W} - \dot{\mathfrak{F}}$ can be fully interpreted as a dissipative power, i.e. the rate at which work is irreversibly dissipated in the process. We indeed note that at steady state conditions we have $\dot{U}_S = \dot{S} = 0$, and then $\dot{S}_\mathrm{i} = \beta\dot{W}_\mathrm{ss} \geqslant 0$, that is, work from the external drive is needed to maintain the nonequilibrium steady state $\pi_S(t)$, producing entropy at a constant rate

$$\beta\dot{W}_{ss} \equiv \beta\hbar\omega\mathrm{Tr}[(\epsilon\hat{a}^\dagger + \epsilon^*\hat{a})\pi] = \beta\mu\sqrt{2}|\epsilon| \geqslant 0. \tag{9.55}$$

On the other hand, even if the non-adiabatic entropy production cannot be defined at the trajectory level, we can always calculate its averaged expression [see Eq. (8.55) in Sect. 8.3], and then the non-adiabatic entropy production rate [14]:

$$\dot{S}_\mathrm{na} = -\frac{d}{dt}D(\rho_t||\pi) = \dot{S} - \beta\left(\dot{U}_S - \mu\dot{X}_S\right) \geqslant 0 \tag{9.56}$$

where $\dot{X}_S \equiv \mathrm{Tr}[\hat{X}_\alpha\,\dot{\rho}_t]$ is the rate at which the cavity field is displaced in the external field direction (as given by φ_ϵ), until the steady state is reached $\langle\hat{X}_{\alpha_\infty}\rangle = \mathrm{Tr}[\hat{X}_\alpha\pi] = \sqrt{2}|\alpha|$. The transient evolution of $X_S \equiv \langle\hat{X}_{\alpha_t}\rangle = \mathrm{Tr}[\hat{X}_{\alpha\rho_t}]$ is simply given by

$$\dot{X}_S = -\frac{\gamma_0}{2}(X_S - \langle\hat{X}_{\alpha_\infty}\rangle), \tag{9.57}$$

that is, it exponentially converges to its steady state value. Therefore $\dot{X}_S$ will be either positive or negative during the evolution depending on the displacement of the initial state. If $\langle\hat{X}_{\alpha_0}\rangle \leqslant \langle\hat{X}_{\alpha_\infty}\rangle$ then $\dot{X}_S \geqslant 0\ \forall t$, and the system state increases its coherence in the energy basis, while if $\langle\hat{X}_{\alpha_0}\rangle \geqslant \langle\hat{X}_{\alpha_\infty}\rangle$, we have $\dot{X}_S \leqslant 0\ \forall t$ and the coherence decreases. It is worth noticing that the rate $\dot{X}_S$ modifies the velocity at which the system converges to the steady state, c.f. Eq. (9.56) but, like the work performed by the external drive, does not produce any entropy change in the reservoir, which in this model only exchanges heat with the cavity mode [see Eq. (9.53)]. This situation can be understood in the framework of generalized Gibbs ensembles by looking at the external drive as a coherent thermal reservoir at infinite temperature, which exchanges both energy (work) and coherence (displacement in the field quadrature $\hat{X}_\alpha$) without modifying its entropy.

By using the expressions for the total and non-adiabatic entropy production rates from Eqs. (9.54) and (9.56), the adiabatic entropy production rate may be defined as

$$\dot{S}_{\rm a} \equiv \dot{S}_{\rm i} - \dot{S}_{\rm na} = \beta(\dot{W} - \mu\dot{X}_{\mathbb{S}}), \quad (9.58)$$

proportional to the input power not being used to generate a displacement in the cavity field. This quantity provides the correct expression for the entropy production in the steady state, $\dot{S}_{\rm a} \rightarrow \beta\dot{W}_{\rm ss} = -\beta\dot{Q}_{\rm ss} \geqslant 0$, corresponding to the input power dissipated as heat to maintain the cavity field out of equilibrium. However, its positivity is not guaranteed at arbitrary times. By noticing that $\dot{W} + \mu\dot{X}_{\mathbb{S}} = \dot{W}_{\rm ss}$ holds, we can explicitly evaluate the adiabatic entropy production rate as

$$\dot{S}_{\rm a} = \beta\dot{W}_{\rm ss} + \beta\mu\gamma_0\left(X_{\mathbb{S}} - \langle\hat{X}_{\alpha_\infty}\rangle\right), \quad (9.59)$$

with $X_{\mathbb{S}} = \langle\hat{X}_{\alpha_{t_0}}\rangle e^{-\gamma_0 t/2} + \langle\hat{X}_{\alpha_\infty}\rangle(1 - e^{-\gamma_0 t/2})$ from Eq. (9.57). We notice that Eq. (9.59) is negative for any initial transient for which $\mu X_{\mathbb{S}} < \mu\langle\hat{X}_{\alpha_\infty}\rangle + \dot{W}_{\rm ss}/\gamma_0$. In particular, if the dynamics starts in any state diagonal in the $\hat{H}_0$ basis, this happens for $t < 2\ln(2)/\gamma_0$ (see Fig. 9.4). During this transient, the relaxation of the cavity mode to its periodic steady state is boosted by the gain in $\dot{X}_{\mathbb{S}}$ on the top of the entropy produced in the process, i. e. $\dot{S}_{\rm na} \geqslant \dot{S}_{\rm i}$.

In Fig. 9.4a we show the dynamical evolution of the three entropy production rates, Eqs. (9.54), (9.56) and (9.58) when the cavity mode starts the evolution in a Gibbs thermal state in equilibrium with the reservoir temperature, $\rho_S(t_0) = \exp(-\beta\hat{H}_0)/Z_0$. In this case, we find that the entropy of the mode is kept constant during the evolution, $\dot{S} = 0\ \forall t$, which implies $\dot{S}_{\rm i} = -\beta\dot{Q} \geqslant 0$, and $\dot{S}_{\rm na} = \beta(\mu\dot{X}_{\mathbb{S}} - \dot{U}_{\mathbb{S}}) \geqslant 0$, while the adiabatic entropy production rate $\dot{S}_{\rm a} = \beta(\dot{W} - \mu\dot{X}_{\mathbb{S}})$ becomes negative in the initial transient dynamics.

The energetics of the relaxation process is shown in Fig. 9.4b. The cavity mode absorbs energy at constant entropy from the external laser until the periodic steady state is reached, $\dot{U}_{\mathbb{S}} = \dot{W}e^{-\gamma_0 t/2}$, where $\dot{W} = \dot{W}_{\rm ss}(1 - e^{-\gamma_0 t/2}) \geqslant 0$, while dissipating an increasing part of the input power as heat, $\dot{Q}_{\rm diss} \equiv -\dot{Q} = \dot{W}(1 - e^{-\gamma_0 t/2}) \geqslant 0$. At this point the input laser power starts to be fully dissipated into the reservoir, i. e. $\dot{Q}_{\rm ss} = -\dot{W}_{\rm ss}$. We notice that the energy absorbed by the cavity mode during the evolution is fully employed to generate the (unitary) displacement α, that is, $\Delta U_{\mathbb{S}} = \mu\Delta X_{\mathbb{S}} = \hbar\omega|\alpha|^2$. However, the transient dynamics ruling this process is far from being trivial. The cavity mode is always displaced (gaining coherence) at a higher rate than energy, $U_{\mathbb{S}} = \mu\dot{X}_{\mathbb{S}}(1 - e^{-\gamma_0 t/2})$, in accordance with the positive non-adiabatic entropy production rate. In addition, by comparing Figs. 9.4a and b the energetic meaning of the adiabatic entropy production rate can be clarified. In the initial transient where $\dot{S}_{\rm na} < 0$ the coherence gain surpass the input power, i.e. $\mu\dot{X}_{\mathbb{S}} > \dot{W}$, which in turn implies that the rate at which the cavity mode gains energy speeds-up in this period $\ddot{U}_{\mathbb{S}} > 0$. At time $\gamma_0 t = 2\ln 2$, when $\dot{S}_{\rm a} = 0$, we have $\dot{W} = \mu\dot{X}_{\mathbb{S}} = \dot{W}_{\rm ss}/2$, and $\dot{U}$ peaks at its maximum. After this time, the adiabatic entropy production rate is positive $\dot{S}_{\rm a} > 0$, implying $\mu\dot{X}_{\mathbb{S}} < \dot{W}$, and $\dot{U}_{\mathbb{S}}$ decreases until it becomes zero in the long time run, when the periodic steady state is reached. In conclusion, we obtained that the sign of the adiabatic entropy production rate spotlights the acceleration in the internal energy changes of the cavity mode.

9.3 Squeezing in a Maxwell Fridge Toy Model

As a third example we envisage a model acting as a Maxwell demon in which coherences can be naturally introduced in by nonequilibrium quantum reservoirs. The model consist of a small thermal device operating between two resonant bosonic reservoirs at different (inverse) temperatures $\beta_r = 1/k_B T_r$ $(r = 1, 2)$, and an external

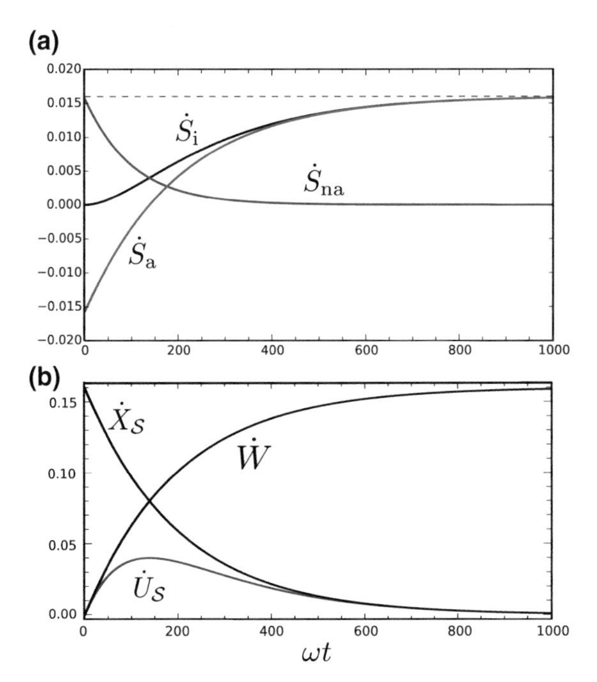

Fig. 9.4 Time evolution of (**a**) adiabatic ($\dot{S}_{\mathrm{a}}$), non-adiabatic ($\dot{S}_{\mathrm{na}}$), and total ($\dot{S}_{\mathrm{i}}$) entropy production rates represented by color solid lines, and (b) input power ($\dot{W}$), rate at which the cavity mode absorbs energy ($\dot{U}_{\mathrm{S}}$), and rate at which it gets displaced ($\dot{X}_{\mathrm{S}}$). The cavity mode starts in equilibrium with the thermal reservoir, $\rho_0 = e^{-\beta \hat{H}_0} Z$, and the laser driving is suddenly switched on at $t = 0$ without any energy cost, and during the whole evolution the entropy of the cavity mode does not change. In the initial dynamical transient the adiabatic entropy production rate becomes negative, implying $\dot{S}_{\mathrm{na}} \geqslant \dot{S}_{\mathrm{i}}$, while it tends to $\beta \dot{W}_{\mathrm{ss}}$ (dashed line) in the long time run [see Eq. (9.55)]. The sign of the adiabatic entropy production rate is related to the interplay between the input power $\dot{W}$ and the displacement rate $\dot{X}_{\mathrm{S}}$. For the initial transient where $\dot{X}_{\mathrm{S}} \geqslant \dot{W}$, an acceleration of the rate at which energy is absorbed by the cavity mode occurs (see the main text for more details). In the figure we used the parameters $\epsilon = 0.02\omega$, $\gamma_0 = 0.01\omega$, and reservoir's temperature $T = 10 k_B^{-1} \hbar\omega$

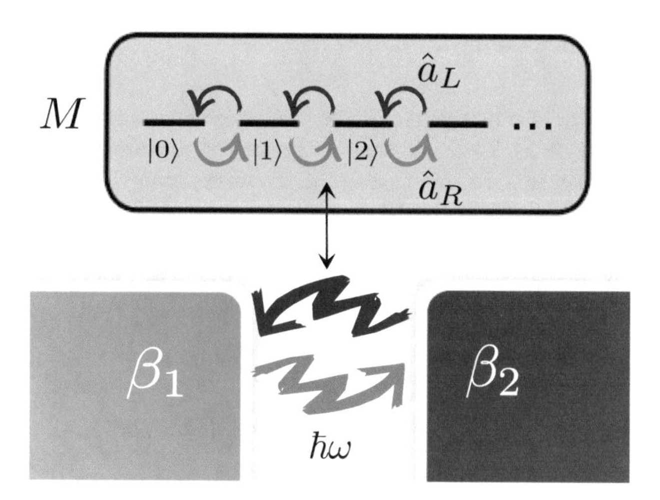

Fig. 9.5 Schematic diagram of the Maxwell refrigerator. Two reservoirs of resonant bosonic modes at different (inverse) temperatures $\beta_1 \geqslant \beta_2$ exchange energy by inducing jumps between the degenerated energy levels of the external memory (M). Each time a quantum $\hbar\omega$ of heat is transferred from the hot (cold) to the cold (hot) reservoirs, the memory performs a collective jump to the left (right) as given by the operator $\hat{a}_L$ ($\hat{a}_R$)

memory system, M, in which information can be erased or stored (see Fig. 9.5). The memory is a semi-infinite set of quantum levels $\{|0\rangle, |1\rangle, \ldots, |n\rangle, \ldots\}$ with degenerated energies $\hat{H}_M = 0$ (conveniently set to zero), and ladder operators $[\hat{a}_L, \hat{a}_R] = \mathbb{1}$, producing jumps between the degenerated levels to the left ($\hat{a}_L$) or to the right ($\hat{a}_R = \hat{a}_L^\dagger$). The device is characterized by an interaction Hamiltonian term weakly coupling the memory and the reservoir modes throughout a three-body interaction:

$$\hat{H}_{\text{int}} = \hbar g \left(\hat{a}_L \hat{b}^\dagger \hat{c} \; + \hat{a}_R \hat{b} \; \hat{c}^\dagger \right) \tag{9.60}$$

where $g \ll \omega$, being ω the natural frequency of the reservoir modes, with Hamiltonians $\hat{H}_1 = \hbar\omega b^\dagger b$ and $\hat{H}_2 = \hbar\omega c^\dagger c$, and $[\hat{b}, \hat{b}^\dagger] = \mathbb{1}$ ($[\hat{c}, \hat{c}^\dagger] = \mathbb{1}$) ladder operators of the two reservoir bosonic modes. The above interaction Hamiltonian preserves energy and induces jumps on the memory to the left (right) when an energy quantum is transferred from the reservoir 2 (1) to the reservoir 1 (2). The underlying idea of the model is to profit from the heat flows induced by the environment throughout the device in order to push the state of the memory as much as possible to its leftmost level $|0\rangle$ (Landauer's erasure), or alternatively, use the memory as a battery in order to induce a heat flow against the constraints imposed by the environment (Maxwell fridge).

9.3.1 Thermal Reservoirs Case

We first consider the case in which both reservoirs are ideal and at thermal equilibrium at temperatures $\beta_1 \geqslant \beta_2$. The environmental bosonic modes are hence assumed to be always in a Gibbs state. We are interested in the relaxation dynamics of this model when starting from an arbitrary initial state in the memory. Using standard techniques form open quantum system theory (see Sects. 2.2 and 2.3), one arrives to the next master equation (ME) for the dissipative dynamics of the memory density operator:

$$\dot{\rho}_t = \mathcal{L}(\rho_t) = \Gamma_{\leftarrow}\left(\hat{a}_L \rho_t \hat{a}_R - \frac{1}{2}\{\hat{a}_R \hat{a}_L, \rho_t\}\right) + \Gamma_{\rightarrow}\left(\hat{a}_R \rho_t \hat{a}_L - \frac{1}{2}\{\hat{a}_L \hat{a}_R, \rho_t\}\right), \tag{9.61}$$

where we neglected Lamb-Stark frequency shifts. The above equation describes incoherent jump processes in the memory to the left at rate $\Gamma_{\leftarrow}$, and to the right at rate $\Gamma_{\rightarrow}$, related to heat fluxes from reservoir 2 to reservoir 1, and from reservoir 1 to reservoir 2 respectively. We have

$$\Gamma_{\leftarrow} = \gamma_0 (n_1^{\text{th}} + 1) n_2^{\text{th}}, \qquad \Gamma_{\rightarrow} = \gamma_0 (n_2^{\text{th}} + 1) n_1^{\text{th}}, \tag{9.62}$$

being γ_0 a constant depending on the interaction strength and $n_r^{\text{th}} \equiv (e^{\beta_r \hbar\omega} - 1)^{-1}$ denotes the mean number of excitations of energy $\hbar\omega$ in reservoir r. Notice that the contact with the thermal reservoirs implies a detailed balance relation between jumps to the left and jumps to the right in the memory

$$\frac{\Gamma_{\leftarrow}}{\Gamma_{\rightarrow}} = e^{(\beta_1 - \beta_2)\hbar\omega} = e^{\mu}, \tag{9.63}$$

where we introduced the parameter

$$\mu \equiv (\beta_1 - \beta_2)\hbar\omega \geqslant 0. \tag{9.64}$$

The memory steady state density operator $[\mathcal{L}(\pi) = 0]$ is, from the ME (9.61)

$$\pi = \frac{e^{-\mu \hat{N}_M}}{Z}, \tag{9.65}$$

for which jumps to the left and jumps to the right are equally probable. Here $\hat{N}_M = \hat{a}_R \hat{a}_L$ is the number operator in the memory, and $Z = \text{Tr}[\exp(-\mu \hat{N}_M)] = (1 - e^{-\mu})^{-1}$. Therefore, the quantity μ fully determines the steady-state occupation in the degenerated energy levels of the memory, together with its entropy $S(\pi) = \mu \langle \hat{N}_{M_\pi} \rangle + \ln Z$, being $\langle \hat{N}_M \rangle_\pi = (e^{\mu} - 1)^{-1}$. Consequently, the greater the temperature gradient between the reservoirs, the greater μ, the more peaked the distribution in the level leftmost, and the lower the entropy of the steady state. On the contrary, if the temperatures of the reservoirs are very similar $\beta_1 \to \beta_2$, we have $\mu \to 0$, and the steady state of the external system approaches the fully mixed state.

This simple toy model has all the necessary elements to act as a Maxwell demon. On the one hand, if the initial state of the memory is a low entropic state (in particular if it has lower entropy than π), the memory acts as an information battery powering a flux of heat against the environment temperature gradient. This flow is maintained until the memory reaches the steady state π, moment at which it should be replaced with a fresh initial state if one wants to maintain the flow. On the other hand, if the initial state is very mixed (it has greater entropy than π), the device acts as a Landauer's eraser, which profits from the spontaneous heat flow from the hot to the cold reservoirs to purify the memory. We notice that a different model for a Maxwell refrigerator operating at steady state conditions, and showing the above mentioned regimes, has been recently proposed by Mandal and Jarzynski in Ref. [15], and extended to the quantum regime in Ref. [16].

Let us now analyze the full stochastic thermodynamics of the model and its entropy production. As in the previous examples, we start by identifying the following Kraus operators for the map $\mathcal{E}$ in Eq. (8.85)

$$\begin{aligned} \hat{M}_0 &= \mathbb{1} - \frac{1}{2} \sum_{k=\leftarrow,\rightarrow} \hat{L}_k^\dagger \hat{L}_k, \\ \hat{M}_\leftarrow &= \sqrt{dt}\, \hat{L}_\leftarrow = \sqrt{dt\, \Gamma_\leftarrow}\, \hat{a}_L, \\ \hat{M}_\rightarrow &= \sqrt{dt}\, \hat{L}_\rightarrow = \sqrt{dt\, \Gamma_\rightarrow}\, \hat{a}_R. \end{aligned} \tag{9.66}$$

In the above equations we identified a single pair of Lindblad operators, $\{\hat{L}_\leftarrow, \hat{L}_\rightarrow\}$, fulfilling

$$\hat{L}_\leftarrow = \sqrt{\Gamma_\leftarrow}\, \hat{a}_L, \quad \hat{L}_\rightarrow = \sqrt{\Gamma_\rightarrow}\, \hat{a}_L^\dagger. \tag{9.67}$$

Exploiting this fact in the trace preserving condition for the backward maps [Eq. (8.93)], we obtain the following boundary terms in the environment associated to each Kraus operator in Eq. (9.66)

$$\sigma_0^E = 0, \quad \sigma_\leftarrow^E = \mu, \quad \sigma_\rightarrow^E = -\mu. \tag{9.68}$$

As the environment is here again given by two uncorrelated thermal reservoirs in equilibrium, the above quantities can be interpreted as the stochastic entropy changes in the environment during the jumps. That is, when a jump to the left (right) occurs, the entropy in the environment increases (decreases) by $\mu = (\beta_1 - \beta_2)\hbar\omega$, associated to the exchange of a quantum of energy $\hbar\omega$ from the hot (cold) to the cold (hot) reservoir.

The backward evolution is analogously specified by the backward map, $\tilde{\mathcal{E}}$, with Kraus operators reading

$$
\begin{aligned}
\hat{\tilde{M}}_0 &= \hat{\Theta}\hat{M}_0^\dagger\hat{\Theta}^\dagger = \hat{\Theta}\hat{M}_0\hat{\Theta}^\dagger = \hat{M}_0 \\
\hat{\tilde{M}}_\leftarrow &= \sqrt{dt}\ \hat{\tilde{L}}_\leftarrow = \hat{\Theta}\sqrt{dt}\ \hat{L}_\rightarrow\hat{\Theta}^\dagger = \hat{M}_\rightarrow \\
\hat{\tilde{M}}_\rightarrow &= \sqrt{dt}\ \hat{\tilde{L}}_\rightarrow = \hat{\Theta}\sqrt{dt}\ \hat{L}_\leftarrow\hat{\Theta}^\dagger = \hat{M}_\leftarrow .
\end{aligned} \tag{9.69}
$$

As in the previous example, the forward and the backward maps are essentially equal, and operations corresponding to a jump to the left in the forward process transforms in a jump to the right in the backward process, and vice-versa.

Notice that the conditions for the adiabatic and non-adiabatic decomposition of the entropy production hold. From the steady state (9.65) it is easy to see that the Kraus operators in Eq. (9.66) are related with an unique change in the nonequilibrium potential

$$
\hat{\Phi} = -\ln\pi = \mu\hat{N}_M + \ln Z, \tag{9.70}
$$

that is, $[\hat{\Phi}, \hat{L}_k] = \Delta\phi_k\hat{L}_k$ for $k = \{\leftarrow, \rightarrow\}$, with associated potential changes

$$
\Delta\phi_0 = 0, \quad \Delta\phi_\leftarrow = -\mu, \quad \Delta\phi_\rightarrow = \mu. \tag{9.71}
$$

On the other hand, Eq. (9.69) ensure that the map $\tilde{\mathcal{E}}$ has as invariant state $\tilde{\pi} = \hat{\Theta}\pi\hat{\Theta}^\dagger$ as required to define the dual map (see Sect. 8.5). Comparing Eqs. (9.68) and (9.71) we see that in this case the changes in the nonequilibrium potential produced by the jumps exactly coincide with the decrease in stochastic entropy in the reservoirs, that is $\Delta\phi_{\leftarrow,\rightarrow} = -\sigma_{\leftarrow,\rightarrow}$. Therefore we can conclude that the dual-reverse and backward processes are exactly equal, and hence the dual process is just the original forward one, which implies zero adiabatic entropy production per trajectory. A trajectory $\gamma = \{n, k_1, \ldots, k_N, m\}$ is again defined by the initial and final measurements on the system performed in the instantaneous eigenbasis of ρ_t (with outcomes n and m respectively) and the N jumps $\{k_l\}$, registered at times $\{t_{k_l}\}$ during the evolution. The total entropy production per trajectory hence corresponds here to a single non-adiabatic contribution

$$
\begin{aligned}
\Delta_{\mathrm{i}} s_\gamma^{\mathrm{na}} = \Delta_{\mathrm{i}} s_\gamma &= \sigma_{nm}^{\mathcal{S}} - \Delta\phi_\gamma \\
&\equiv \sigma_{nm}^{\mathcal{S}} - (\beta_1 - \beta_2) q_\gamma ,
\end{aligned} \tag{9.72}
$$

which fulfills the detailed and integral fluctuation theorems. As in the previous example, $\sigma_{nm}^S = \ln p_n^0 - \ln p_m^t$ is the stochastic entropy change in the system, where p_n^0 and p_m^t are, respectively, the eigenvalues of ρ_0 and ρ_t corresponding to outcomes n and m. Notice that in the last equality of Eq. (9.72) we identified the net heat during the trajectory, q_γ, flowing from the cold to the hot reservoir

$$
q_\gamma = \hbar\omega(n_\rightarrow - n_\leftarrow), \tag{9.73}
$$

where $n_\rightarrow$ ($n_\leftarrow$) is the total number of right (left) jumps detected during the trajectory γ. The absence of adiabatic entropy production can be understood from the fact that in this model, any transfer of heat between reservoirs is achieved by means of jumps in the memory, cf. Eq. (9.60). This implies that no heat can flow without modifying the system density operator, and hence no entropy can be produced irrespective of the local changes in the state of the memory ρ. As a consequence, any flow ceases in the long-time run, when the steady state π is reached, blocking at this point the heat transfer between the reservoirs.

The entropy production rate can be finally obtained by averaging over trajectories [see Eqs. (8.105) and (8.103)]

$$\dot{S}_{\rm i} = \dot{S} - \mu \langle \dot{N}_M \rangle = \dot{S} - (\beta_1 - \beta_2)\dot{Q} \geqslant 0 \tag{9.74}$$

where $\dot{S} = -\mathrm{Tr}[\dot{\rho}_t \ln \rho_t]$, and $\dot{Q} = \mathrm{Tr}[\dot{\rho}_t \hbar\omega \hat{N}_M]$ is the heat flow from the cold to the hot reservoir. The second law inequality (9.74), can be now used to discuss the performance of the two different regimes of operations of the device: Landauer's eraser and Maxwell refrigerator. In the first case we see that when $\dot{Q} < 0$, that is, the heat flows spontaneously from the hot to the cold reservoir, the entropy in the memory system is allowed to decrease, $\dot{S} < 0$, until the entropy produced by the spontaneous heat flow is compensated. The heat dissipated in the erasure process is then lower bounded by

$$|\dot{Q}| \geqslant (\beta_1 - \beta_2)^{-1} \, |\dot{S}|, \tag{9.75}$$

corresponding to a manifestation of Landauer's principle in our setting. On the other hand, if $\dot{S} > 0$, now the flux of heat can be inverted against the thermal gradient, $\dot{Q} > 0$, the cold reservoir is refrigerated at the price of entropy production in the memory system. The performance of this Maxwell fridge is then analogously bounded by

$$\dot{Q} \leqslant (\beta_1 - \beta_2)^{-1} \, \dot{S}. \tag{9.76}$$

9.3.2 *Squeezed Thermal Reservoir Enhancements*

Once the thermodynamic behavior of the model has been analyzed for the case of ideal thermal reservoirs, we now move to the case of nonequilibrium reservoirs. We replace the ideal thermal reservoirs at inverse temperatures β_1 and β_2, by squeezed thermal reservoirs at the same temperatures, with additional parameters $\{r_1, r_2\}$ and $\{\theta_1, \theta_2\}$, characterizing the squeezing (see Sect. 1.2.5). Squeezing constitutes a useful resource not only from the perspective of quantum information, with applications in quantum metrology, imaging, computation, and cryptography [17], but also from the perspective of quantum thermodynamics, for the design of enhanced thermal machines [4, 18, 19].

In Sect. 2.3.2 of Chap. 2 we derived the open system dynamics induced by a squeezed thermal reservoir on a bosonic mode, which can be easily adapted to the present situation. The master equation in Eq. (9.61) changes to

$$\dot{\rho}_t = \mathcal{L}^*(\rho_t) = \Gamma_- \left(\hat{R}\, \rho_t \hat{R}^\dagger - \frac{1}{2}\{\hat{R}^\dagger \hat{R}, \rho_t\} \right) + \Gamma_+ \left(\hat{R}^\dagger \rho_t \hat{R} - \frac{1}{2}\{\hat{R}\hat{R}^\dagger, \rho_t\} \right) \tag{9.77}$$

where $\hat{R}$ is the ladder operator of a Bogoliubov mode defined by the canonical transformation

$$\hat{R} \equiv \hat{a}_L \cosh(r) + \hat{a}_R \sinh(r)\, e^{i\theta} = \hat{\mathbb{S}}(\xi)\hat{a}_L\hat{\mathbb{S}}^\dagger(\xi), \tag{9.78}$$

being $\xi = re^{i\theta}$, and $\hat{\mathbb{S}}(\xi) = \exp\left(\frac{r}{2}[\hat{a}_L^2 e^{-i\theta} - \hat{a}_R^2 e^{i\theta}]\right)$ the squeezing operator on the memory system, with $\theta \equiv \theta_1 - \theta_2$ and r depending on the reservoir temperatures and squeezing parameters (r_1 and r_2) thought the relation

$$\tanh(2r) \equiv \frac{2M_1M_2}{(N_1+1)N_2 + (N_2+1)N_1}, \tag{9.79}$$

where $M_i \equiv -\sinh(r_i)\cosh(r_i)(2n_i^{\text{th}} + 1)$ and $N_i \equiv n_i^{\text{th}} \cosh(2r_i) + \sinh(r_i)$ for $i = 1, 2$. The above equation is only well defined for the right hand side taking values in between -1 and 1, to which we will restrict from now on. We also stress that $r \to 0$ when either $r_1 \to 0$ or $r_2 \to 0$.

The operators $\hat{R}$ and $\hat{R}^\dagger$ in the master equation (9.77), promote jumps to the left and to the right, respectively, between the states of the squeezed basis of the memory system, $\{\hat{\mathbb{S}}(\xi)|0\rangle, \ldots, \hat{\mathbb{S}}(\xi)|n\rangle, \ldots\}$, at rates

$$\Gamma_\mp = \frac{\gamma_0}{2}\left(\delta N \pm (N_2 - N_1)\right), \tag{9.80}$$

where $\delta N \equiv \sqrt{((N_1+1)N_2 + (N_2+1)N_1)^2 - 4|M_1M_2|^2}$. It is worth noticing that the rates $\Gamma_\mp$ no longer fulfill the detailed balance relation in Eq. (9.63) but now $\Gamma_- = \Gamma_+\, e^{\mu_*}$ with a new parameter

$$\mu_* \equiv \ln\left(\frac{N_1 \sinh^2(r) + N_2 \cosh^2(r) + N_1N_2}{N_1 \cosh^2(r) + N_2 \sinh^2(r) + N_1N_2}\right). \tag{9.81}$$

This can be both greater or lower than μ depending on the squeezing parameters $\{r_1, r_2\}$ (inside the allowed range) of the two squeezed thermal reservoirs. In particular, it is worth noticing that $\mu_* \to \mu$ only when both $r_1 \to 0$ *and* $r_2 \to 0$.

Crucially, the master equation (9.77) now induces the following steady state in the long time run

$$\pi_* = \hat{\mathbb{S}}(\xi)\, \frac{e^{-\mu_* \hat{N}_M}}{Z_*}\, \hat{\mathbb{S}}^\dagger(\xi), \tag{9.82}$$

with μ_* defined in Eq. (9.81), and $Z_* = \text{Tr}[e^{-\mu_* \hat{N}_M}] = (1 - e^{-\mu_*})^{-1}$. This is equivalent to a squeezed thermal state for the energyless memory system, with same entropy than $e^{-\mu_* \hat{N}_M}/Z_*$, but modified occupations in the degenerated levels, showing second-order coherences in the $\hat{N}_M$ basis (see Sect. 1.2.5). As in the example of the driven cavity mode, the steady state π_* in Eq. (9.82) can be rewritten as a generalized Gibbs ensemble (see Sect. 7.4) $\pi_* \propto \exp[-\mu_*(\cosh(2r)\hat{N}_M - \sinh(2r)\hat{A}_\theta)]$, where we introduced the operator

$$\hat{A}_M \equiv -\frac{1}{2}\left(\hat{a}_R^2 e^{i\theta} + \hat{a}_L^2 e^{-i\theta}\right) = \frac{1}{2}\left(\hat{P}_{\theta/2}^2 - \hat{X}_{\theta/2}^2\right), \tag{9.83}$$

measuring the asymmetry between the conjugated memory quadratures in the direction given by $\theta/2$, $[\hat{X}_{\theta/2}, \hat{P}_{\theta/2}] = i$ (see Eq. (1.75) in Sect. 1.2.4). Notice that $[\hat{N}_M, \hat{A}_\theta] \neq 0$, so that π_* corresponds to a generalized Gibbs ensemble with non-commuting charges. We will return to the properties of the operator $\hat{A}_\theta$ in Chap. 10.

Following the trajectory formalism, the Kraus operators for the map $\mathcal{E}$ in the forward process now read

$$\begin{aligned}
\hat{M}_0 &= \mathbb{1} - \frac{1}{2}\sum_{k=-,+} \hat{L}_k^\dagger \hat{L}_k, \\
\hat{M}_- &= \sqrt{dt}\ \hat{L}_- = \sqrt{dt\ \Gamma_-}\ \hat{R}, \\
\hat{M}_+ &= \sqrt{dt}\ \hat{L}_+ = \sqrt{dt\ \Gamma_+}\ \hat{R}^\dagger.
\end{aligned} \tag{9.84}$$

with the new Lindblad operators, $\{\hat{L}_-, \hat{L}_+\}$. Notice that the map $\mathcal{E}$ is a generalized Gibbs-Preserving map (see Sect. 7.4). The boundary terms in the environment associated to the operators $\hat{M}_k$ become

$$\sigma_0^E = 0, \quad \sigma_-^E = \mu_*, \quad \sigma_+^E = -\mu_*. \tag{9.85}$$

They can be interpreted again as stochastic entropy changes in the environment, as they correspond to choosing $\tilde{\rho}_E = \hat{\Theta}\rho_E\hat{\Theta}^\dagger$, ρ_E being a product of squeezed thermal states in the two reservoirs. Here $\rho_E^* \simeq \rho_E$ follows from the fact that they are large systems which do not substantially modify their state during the evolution. Notice however that they have no longer a clear interpretation in terms of exchange of energy quanta between the reservoirs, as in this case both energy and coherence (or asymmetry) are exchanged between the reservoirs in each jump, producing (or annihilating) an entropy quantum $\pm\mu_*$ in the whole environment. Yet, the Kraus operators for the backward map fulfill the same structure than in the previous case

$$\begin{aligned}
\hat{\tilde{M}}_0 &= \hat{\Theta}\hat{M}_0^\dagger\hat{\Theta}^\dagger = \hat{\Theta}\hat{M}_0\hat{\Theta}^\dagger = \hat{M}_0, \\
\hat{\tilde{M}}_- &= \sqrt{dt}\hat{\tilde{L}}_- = \hat{\Theta}\sqrt{dt}\hat{L}_+\hat{\Theta}^\dagger = \hat{M}_+, \\
\hat{\tilde{M}}_+ &= \sqrt{dt}\hat{\tilde{L}}_+ = \hat{\Theta}\sqrt{dt}\hat{L}_-\hat{\Theta}^\dagger = \hat{M}_-,
\end{aligned} \tag{9.86}$$

and the conditions for the adiabatic and non-adiabatic decomposition of the entropy production hold again. Indeed, from the steady state (9.82) the nonequilibrium potential now reads

$$\hat{\Phi} = -\ln \pi_* = \mu_* \hat{\mathcal{S}} \hat{N}_M \hat{\mathcal{S}}^\dagger + \ln Z_*, \tag{9.87}$$

which verifies $[\hat{\Phi}, \hat{L}_k] = \Delta\phi_k \hat{L}_k$ for $k = \{-, +\}$, being

$$\Delta\phi_0 = 0, \quad \Delta\phi_- = -\mu_*, \quad \Delta\phi_+ = \mu_*, \tag{9.88}$$

as required to develop the dual-reverse process. Moreover, the map $\tilde{\mathcal{E}}$ has as invariant state $\tilde{\pi}_* = \hat{\Theta}\pi_*\hat{\Theta}^\dagger$ as required to define the dual process. Therefore, from Eqs. (9.85) and (9.88) we see that also in this case the changes in the nonequilibrium potential produced by the jumps exactly coincides with the decrease in stochastic entropy in the reservoirs, $\Delta\phi_\mp = -\sigma_\mp$, and hence the adiabatic entropy production is again zero for any trajectory. This means that when the steady state π_* is reached, no further entropy production is needed in order to maintain the nonequilibrium steady state π_*.

The average entropy production rate reads now

$$\begin{aligned} \dot{S}_\mathrm{i} &= \dot{S} - \mu_* \mathrm{Tr}[\hat{\mathcal{S}}(\xi) \hat{N}_M \hat{\mathcal{S}}^\dagger(\xi) \dot{\rho}_t] = \\ &= \dot{S} - \frac{\mu_*}{\hbar\omega}(\cosh(2r)\dot{Q} - \sinh(2r)\dot{\mathcal{A}}_M) \geqslant 0 \end{aligned} \tag{9.89}$$

where $\dot{Q} = \mathrm{Tr}[\hbar\omega\hat{N}_M\dot{\rho}_t]$ is again the energy flux from the cold to the hot reservoir, and $\dot{\mathcal{A}}_M \equiv -\hbar\omega\mathrm{Tr}[\hat{A}_\theta\dot{\rho}_t]$, with $\hat{A}_\theta$ in Eq. (9.83). The later is a flow of second-order coherences (asymmetry in the fluctuations) from the reservoirs (see Chap. 10), obeying an exponential law, $\dot{\mathcal{A}}_M = -\gamma_0(\mathcal{A}_M - \hbar\omega\langle\hat{A}_{\theta_\infty}\rangle)$, where $\mathcal{A}_M \equiv \hbar\omega\langle\hat{A}_{\theta_t}\rangle$, and $\langle\mathcal{A}_{\theta_\infty}\rangle = (1/2)\sinh(2r)\coth(\mu_*/2) \geqslant 0$ in the steady state π_*.

Comparing Eqs. (9.74) and (9.89) we see two main effects of reservoir squeezing. The first one is the appearance of the parameter μ_* instead of μ, which implies that the interplay between the energy flux from the cold to the hot reservoirs and the entropy in the memory can be modified without varying the temperature gradient. The second one is the appearance of an extra entropy flow related to the exchange between system and environment of the quantity $\dot{\mathcal{A}}$, proportional to the second-order coherences in the memory system. This may induce an extra reduction (or increase) of the memory entropy independently of the heat exchanged between the reservoirs. The two effects have indeed a deep impact in the performance of the Maxwell demon device (see Fig. 9.6). To make the discussion more precise, we will look at the machine operation when the memory system starts in the state π and is then connected to the device with the squeezed thermal reservoirs until it reaches the steady state π_*. The extra increase in entropy and energy pumped from the cold to the hot reservoirs due to reservoir squeezing read in this case

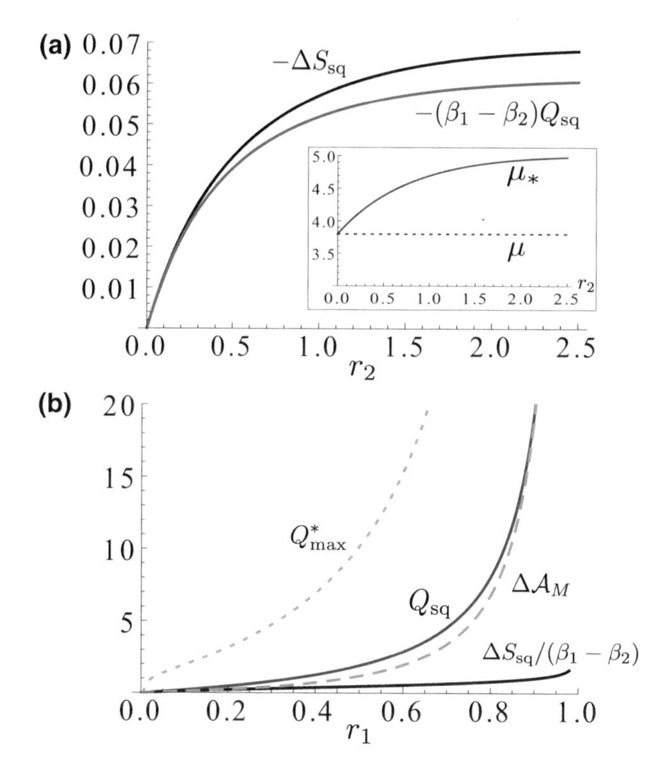

Fig. 9.6 **a** Enhancements in the entropy erased $-\Delta S_{\text{sq}}$ in the memory, together with the (scaled) heat flow from the hot to the cold reservoirs, $-\dot{Q}_{\text{sq}}$, as a function of the squeezing parameter r_2, in the case of no squeezing in the cold reservoir, $r_1 = 0$. In the inset figure we show the change in the parameter μ_* characterizing the steady state of the memory. **b** Enhancements in the energy extracted from the cold reservoir (blue line), together with the (scaled) entropy produced in the memory ΔS_{sq} (black line), the coherences flow to the memory $\Delta \mathcal{A}_M$ (orange-dashed line), and the maximum extractable heat from the second law-like inequality in Eq. (9.89), $\dot{Q}_{\text{max}}$ (pink-dotted line) as a function of the squeezing parameter r_1, when both reservoirs are squeezed ($r_2 = 0.5$). In both plots we used $\beta_1 = 5/\hbar\omega$, and $\beta_2 = 1.2/\hbar\omega$

$$\Delta S_{\text{sq}} \equiv S(\pi_*) - S(\pi) = \frac{\mu_*}{e^{\mu_*} - 1} - \frac{\mu}{e^{\mu} - 1} + \ln \frac{1 - e^{-\mu}}{1 - e^{-\mu_*}},$$
$$Q_{\text{sq}} \equiv \hbar\omega \left(\langle N_M \rangle_{\pi_*} - \langle N_M \rangle_{\pi} \right) = \frac{\hbar\omega}{e^{\mu_*} - 1} - \frac{\hbar\omega}{e^{\mu} - 1}. \tag{9.90}$$

In Fig. 9.6a we show the break of the Landauer's bound in Eq. (9.75) when the device acts as a Landauer's eraser, by just considering squeezing in the hot reservoir, $r_2 > 0$, while the cold one remains in a thermal state ($r_1 = 0$). In this case we have $r = 0$, meaning that the steady state in Eq. (9.82) reduces to

$$\pi_* = \exp(\mu_* \hat{N}_M)/Z_*, \tag{9.91}$$

with $\mu_* \geqslant \mu$, which corresponds to a lower entropic state than π in Eq. (9.65). Introducing squeezing only in the cold thermal reservoir we can therefore erase a greater amount of entropy in the memory which overcomes the bound (9.75), at the cost of inducing some more energy flowing from the hot to the cold reservoirs.

On the other hand, in Fig. 9.6b, we show the Maxwell refrigerator regime when both reservoirs contain squeezing, $r_1, r_2 > 2$. In this case extra energy can be extracted from the cold reservoir. The asymmetry induced in the memory state, together with the modification of the parameter μ_* are the responsible of allowing refrigeration on the top of the entropy produced in the memory, overcoming again the bound in Eq. (9.76) for the thermal reservoirs case.

9.4 Conclusions

In this chapter we have studied the decomposition of the total entropy production into adiabatic and non-adiabatic contributions in three specific situations of interest: an autonomous three-level thermal machine, a dissipative cavity mode resonantly driven by a classical field, and a Maxwell's demon toy model.

The first example illustrates the differences between adiabatic, non-adiabatic, and total entropy production in a purely thermal situation [Eqs. (9.28)–(9.30) and (9.34)–(9.36)]. We have seen that the forward and backward processes in this case are equivalent inverting the direction of the jumps. The dual and the dual-reverse processes are very similar to the forward and backward ones, but the rates at which the different quantum jumps occurs are modified. The entropy changes in the environment are the sum of the heat flow dissipated in each reservoir divided by its temperature [see Eqs. (9.20) and (9.32)], and the changes in the nonequilibrium potential are, analogously, the sum of the heat flowing into the system in each transition divided by it effective (or virtual) temperature [Eqs. (9.23) and (9.33)]. The non-adiabatic entropy production hence describes the entropy changes in the three-level machine not accounted for by the entropy flow due to the heat absorbed in each transition at its effective temperature. On the other hand, the adiabatic entropy production is the sum of the irreversible heat exchanged between each reservoir and its corresponding transition in the machine. Furthermore, we stress that the differences between the actual temperatures of the reservoirs and the effective (virtual) temperatures of the three-level machine transitions determine the direction of the heat flows in the steady state (see Fig. 9.2), together with its performance properties, Eq. (9.38).

In the second example we discussed a configuration in which the adiabatic and non-adiabatic decomposition of the entropy production is broken, and only the FT for the total entropy production holds. The backward process is also equivalent to the forward one, with the quantum jumps inverted in the time-reversed trajectories with respect to the forward ones. The dual and dual-reverse processes cannot be defined in the setup, as the condition for the Kraus operators in Eq. (8.96) is not fulfilled. This break of the split will occur whenever the different dynamical contributions may promote jumps between eigenstates of different system observables [see Eq. (9.52) and the discussion above]. This can be seen from the non-commutativity between the unitary and dissipative contributions in the master equation (9.40), i.e. $[\hat{V}, \hat{a}^\dagger \hat{a}] \neq 0$. As a consequence, the non-adiabatic entropy production rate, Eq. (9.56), which measures the convergence of the system to its steady state, can be greater than the total entropy production rate, Eq. (9.54). This implies a negative adiabatic entropy production rate, Eq. (9.58), in some transient dynamics during which the cavity mode experiences an accelerated energy gain (see Fig. 9.4).

Finally, the third example allowed us to apply our formalism in a simple setup with a pure informational component, and in which nonequilibrium squeezed thermal reservoirs can be easily introduced. We first explored the simpler case in which a memory system controls the heat flow between thermal reservoirs at different temperatures. We have seen that in this case the dual and the forward process are exactly the same, and no adiabatic entropy production is generated. Equivalently the backward and the dual-reverse processes coincide, with nonequilibrium potential changes equivalent to (minus) the stochastic entropy changes in the environment (see Eqs. (9.68) and (9.71)). This entropy changes are produced by the flow of heat between the reservoirs at different temperatures. The total entropy production and the non-adiabatic ones are hence equal in this case, Eq. (9.72), and from the corresponding entropy production rate (9.74) the Landauer's bound is recovered in the model. As a second step we replaced the regular thermal reservoirs by squeezed thermal reservoirs. Our formalism applies as well for this case. Now the entropy changes in the environment are produced by the exchange of both energy and coherences between the nonequilibrium reservoirs. This induces squeezing in the memory system at the steady state (9.82), and modifies the total entropy production rate, Eq. (9.89), which now includes a term proportional to the asymmetry induced by the memory quadratures. The enhancements in the performance of the device due to the squeezing in the reservoirs, Eq. (9.90), implies the energetic overtaking of the Landauer's bound as exemplified in Fig. 9.6.

References

1. H.E.D. Scovil, E.O. Schulz-DuBois, Three-level masers as heat engines. Phys. Rev. Lett. **2**, 262 (1959)
2. J.E. Geusic, E.O. Schulz-DuBois, H.E.D. Scovil, Quantum equivalent of the Carnot cycle. Phys. Rev. **156**, 343 (1967)
3. J.P. Palao, R. Kosloff, Quantum thermodynamic cooling cycle. Phys. Rev. E **64**, 056130 (2001)
4. L.A. Correa, J.P. Palao, D. Alonso, G. Adesso, Quantum-enhanced absorption refrigerators. Sci. Rep. **4**, 3949 (2014)
5. R. Kosloff, A. Levy, Quantum heat engines and refrigerators: continuous devices. Annu. Rev. Phys. Chem. **65**, 365–393 (2014)
6. N. Brunner, N. Linden, S. Popescu, P. Skrzypczyk, Virtual qubits, virtual temperatures, and the foundations of thermodynamics. Phys. Rev. E **85**, 051117 (2012)
7. P. Skrzypczyk, R. Silva, N. Brunner, Passivity, complete passivity, and virtual temperatures. Phys. Rev. E **91**, 052133 (2015)
8. U. Seifert, Entropy production along a stochastic trajectory and an integral fluctuation theorem. Phys. Rev. Lett. **95**, 040602 (2005)

9. T. Monnai, Unified treatment of the quantum fluctuation theorem and the Jarzynski equality in terms of microscopic reversibility. Phys. Rev. E **72**, 027102 (2005)
10. J.M. Horowitz, Quantum-trajectory approach to the stochastic thermodynamics of a forced harmonic oscillator. Phys. Rev. E **85**, 031110 (2012)
11. T. Sagawa, Second law-like inequalitites with quantum relative entropy: an introduction, in *Lectures on Quantum Computing, Thermodynamics and Statistical Physics*, vol. 8, Kinki University Series on Quantum Computing, ed. by M. Nakahara (World Scientific, New Jersey, 2013)
12. J.M. Horowitz, J.M.R. Parrondo, Entropy production along nonequilibrium quantum jump trajectories. New. J. Phys. **15**, 085028 (2013)
13. H.M. Wiseman, G.J. Milburn, *Quantum Measurement and Control* (Cambridge University Press, Cambridge, 2010)
14. H. Spohn, Entropy production for quantum dynamical semigroups. J. Math. Phys. **19**, 1227–1230 (1978)
15. D. Mandal, H.T. Quan, C. Jarzynski, Maxwell's refrigerator: an exactly solvable model. Phys. Rev. Lett. **111**, 030602 (2013)
16. A. Chapman, A. Miyake, How an autonomous quantum Maxwell demon can harness correlated information. Phys. Rev. E **92**, 062125 (2015)
17. E.S. Polzik, The squeeze goes on. Nature **453**, 45–46 (2008)
18. J. Roßnagel, O. Abah, F. Schmidt-Kaler, K. Singer, E. Lutz, Nanoscale heat engine beyond the carnot limit. Phys. Rev. Lett. **112**, 030602 (2014)
19. O. Abah, E. Lutz, Efficiency of heat engines coupled to nonequilibrium reservoirs. Europhys. Lett. **106**, 20001 (2014)

Part IV
Quantum Thermal Machines

Chapter 10
Thermodynamic Power of the Squeezed Thermal Reservoir

The last part of this thesis is related to the performance of quantum thermal machines. Quantum thermal machines have been introduced in Chap. 3 as a topic which has attracted increasing attention within the new field of quantum thermodynamics. They generically consist of small quantum devices performing some useful thermodynamic task, such as refrigeration, heat pumping, or work extraction, while powered by out-of-equilibrium thermodynamic or mechanical forces. Their importance comes from the fact that they can be used to investigate fundamental questions related to the laws of thermodynamics as well as being useful in practical applications.

One of the most interesting open questions concerning quantum thermal machines (and more generally quantum thermodynamics) is understanding the implications of quantum features, such as quantum measurement [1–4], coherence [5–8], or quantum correlations [9–14]. In this context, inspired by the breakthrough on the photo-Carnot engine driven by quantum fuel by Scully et al. [5], different theoretical studies recently focused on the implications for work extraction associated to nonequilibrium quantum reservoirs. In particular it has been shown that using coherent [15–17], correlated [18], or squeezed thermal reservoirs [19–22], power and efficiency of heat engines can be improved, even surpassing the Carnot bound. However, a general framework providing a deeper understanding of such quantum nonequilibrium phenomena is still an open challenge [23, 24].

In this chapter we will apply our analysis on the entropy production in quantum processes developed in Chap. 8 to clarify the role of nonequilibrium quantum reservoirs in work extraction.[1] We stress that entropy production is one of the most fundamental concepts in nonequilibrium thermodynamics, which quantifies the degree of irreversibility of a dynamical evolution [26]. For a quantum system relaxing to thermal equilibrium at inverse temperature $\beta = 1/k_B T$, it simply reads [27–29]:

$$\Delta_{\mathrm{i}} S \equiv \Delta S - \beta Q \geqslant 0 \tag{10.1}$$

[1] Most of the results in the chapter have been published in Ref. [25].

G. Manzano Paule, *Thermodynamics and Synchronization in Open Quantum Systems*, Springer Theses, https://doi.org/10.1007/978-3-319-93964-3_10

where ΔS is the change in the von Neumann entropy of the system, and Q is the heat released from the reservoir. The positivity of the entropy production (10.1) is a particular case of the second law. However, in more general situations, different processes others than heat flows may produce an exchange of entropy between the system and its surroundings, modifying (10.1). In Chap. 9 we have already seen some examples in which Eq. (10.1) is modified by the presence of coherences, yielding new bounds on the performance of thermodynamic tasks.

Squeezing has been introduced in Sect. 1.2.5 as a property intimately related with Heisenberg's uncertainty principle: It is the result of reducing the variance of an observable with respect to its conjugate (see also Ref. [30]). Nowadays it constitutes a central tool in quantum information with several applications in quantum metrology, computation, cryptography and imaging [31]. Most commonly considered squeezed states are coherent but also thermal ones have been largely studied [32, 33]. Experimental realizations of squeezed thermal states range from microwaves [34] to present squeezing of motional degrees of freedom in optomechanical oscillators [35, 36].

In Sect. 10.1 we explicitly address the characterization of the entropy production for the case of a bosonic mode interacting with a single squeezed thermal reservoir. This analysis is then applied to discuss work extraction in two models of non-autonomous quantum thermal machines (see Sect. 3.3). In Sect. 10.2 the maximum irreversible work cyclically extractable from a single squeezed reservoir is obtained. Further, in Sect. 10.3 we discuss an Otto cycle which can operate as a heat engine converting the heat entering from both reservoirs into work at one hundred per cent efficiency, or as a refrigerator pumping energy from the cold to the hot reservoir while producing a positive amount of output work at the same time. It is important to stress that our results do not contradict the second law of thermodynamics, which is modified by the inclusion of squeezing as an available resource in the reservoir. Indeed in Sect. 10.4 this point is developed by providing an interpretation of the squeezed thermal reservoir as a source of free energy. An experimental proposal for implementing our results is given in Sect. 10.5 by constructing on previous works on a single-ion heat engine [37, 38]. The main conclusions of the chapter are presented in Sect. 10.6, while some further technical details are provided in Appendix.

10.1 Thermodynamics of the Squeezed Thermal Reservoir

Consider a quantum system consisting of a single bosonic mode with Hamiltonian $\hat{H}_S = \hbar\omega\hat{a}^\dagger\hat{a}$, weakly dissipating into a bosonic reservoir $\hat{H}_R = \sum_k \hbar\Omega_k \hat{b}_k^\dagger \hat{b}_k$, prepared in a squeezed thermal state at inverse temperature β with squeezing parameter $\xi = re^{i\theta}$ ($r \geqslant 0$ and $\theta \in [0, 2\pi]$), see Sect. 1.2.5). In Sect. 2.3.2 we analyzed the dynamical evolution of such a system throughout the development of a Markovian collisional model, where the interaction between mode and reservoir in the rotating wave approximation (RWA) reads

$$\hat{H}_{int} = \sum_k i g_k (\hat{a}\,\hat{b}_k^\dagger - \hat{a}^\dagger \hat{b}_k). \tag{10.2}$$

This yields an open system dynamics described by the following Lindblad master equation in interaction picture

$$\dot{\rho}_t = \mathcal{L}(\rho_t) = \sum_{i=\pm} \hat{R}_i \rho_t \hat{R}_i^\dagger - \frac{1}{2}\{\hat{R}_i^\dagger \hat{R}_i, \rho_t\}, \tag{10.3}$$

where Lamb–Stark shifts have been neglected (an alternative derivation can be found in Ref. [39]). The two Lindblad operators in (10.3) read:

$$\hat{R}_- = \sqrt{\gamma_0(n_{\text{th}} + 1)}\,\hat{R}, \qquad \hat{R}_+ = \sqrt{\gamma_0 n_{\text{th}}}\,\hat{R}^\dagger, \tag{10.4}$$

with $\hat{R} = \hat{a}\cosh(r) + \hat{a}^\dagger \sinh(r) e^{i\theta}$. This corresponds to the ladder operator of a Bogoliubov mode $\hat{R} = \hat{\mathbb{S}}(\xi)\,\hat{a}\,\hat{\mathbb{S}}^\dagger(\xi)$ where

$$\hat{\mathbb{S}}(\xi) \equiv \exp\left(\frac{r}{2}(\hat{a}^2 e^{-i\theta} - \hat{a}^{\dagger 2} e^{i\theta})\right) \tag{10.5}$$

denotes the unitary squeezing operator on the system mode. In addition we introduced γ_0 the spontaneous emission decay rate and $n_{\text{th}} = (e^{\beta\hbar\omega} - 1)^{-1}$ the mean number of bosons of energy $\hbar\omega$ in a thermal reservoir at inverse temperature β.

The Lindblad operators $\hat{R}_\mp$ in Eq. (10.4), promote jumps associated to the correlated emission and absorption of bosons

$$\hat{R}_\mp \hat{\mathbb{S}}(\xi)|n\rangle \to \hat{\mathbb{S}}(\xi)|n \mp 1\rangle, \tag{10.6}$$

which fit into the formalism for quantum fluctuation theorems developed in Chaps. 7 and 8. The steady state solution, $\mathcal{L}(\pi) = 0$, is no longer diagonal in the $\hat{H}_{\mathbb{S}}$ basis

$$\pi = \hat{\mathbb{S}}(\xi)\frac{e^{-\beta\hat{H}_{\mathbb{S}}}}{Z}\hat{\mathbb{S}}^\dagger(\xi), \tag{10.7}$$

with $Z = \text{Tr}[e^{-\beta\hat{H}_{\mathbb{S}}}]$. As we already stressed in Sect. 1.2.5, the squeezed thermal state π has the same entropy as the Gibbs state, but higher mean energy. A crucial property is that its variance in the quadrature $\hat{X}_{\theta/2} \equiv (\hat{a}^\dagger e^{i\theta/2} + \hat{a}e^{-i\theta/2})/\sqrt{2}$ has been squeezed by a factor e^{-r}, while the variance of the conjugate quadrature $\hat{P}_{\theta/2}$ (with $[\hat{X}_{\theta/2}, \hat{P}_{\theta/2}] = i$) is multiplied by e^r. Notice that when turning to the Schrödinger picture, the steady state (10.7) acquires a time-dependent phase inducing a rotation in phase-space, which has to be accounted for in applications.

The Lindblad master equation (10.3) describes the relaxation from any initial state of the mode to π. The irreversibility of the process is well captured by the non-adiabatic entropy production introduced in Chap. 8 (see also [40–43]):

$$\dot{S}_{\text{na}} \equiv -\frac{d}{dt} D(\rho_t || \pi) = \dot{S} - \dot{\Phi} \geqslant 0 \tag{10.8}$$

where $D(\rho||\sigma) = \text{Tr}[\rho(\ln \rho - \ln \sigma)] \geqslant 0$ is the quantum relative entropy (Sect. 1.1.6). The term $\dot{\Phi} = \text{Tr}[\hat{\Phi}\dot{\rho}_t]$ defines the effective rate at which entropy is transferred from the surroundings into the system throughout the nonequilibrium potential, $\hat{\Phi} = -\ln \pi$. The positivity of $\dot{S}_{\text{na}}$ is always guaranteed for quantum dynamical semigroups [40], while the emerging second-law-like inequality in Eq. (10.8), has been derived in Sects. 7.4 and 8.5 as a corollary from the general fluctuation theorem for quantum CPTP maps. Recall that the effective entropy flow $\dot{\Phi}$ becomes zero for unital maps and reproduces the heat flow divided by temperature in the case of thermalization or Gibbs-preserving maps. Remarkably, in this case it can further be shown that it equals the rate at which entropy decreases in the reservoir during the relaxation process, $\dot{\Phi} = -\dot{S}_E$ (see Appendix), so that we can identify Eq. (10.8) with the total entropy production rate in system and environment during the process, i.e. $\dot{S}_{\text{na}} = \dot{S}_{\text{i}} = \dot{S} + \dot{S}_E$ (see Sect. 8.3).

Using the steady state π in Eq. (10.7) we obtain:

$$\dot{\Phi} = \beta\, \text{Tr}[\hat{\mathbb{S}}(\xi)\hat{H}_{\mathbb{S}}\hat{\mathbb{S}}^{\dagger}(\xi)\dot{\rho}_t] = \beta \left(\cosh(2r)\dot{Q} - \sinh(2r)\dot{\mathcal{A}}\right) \tag{10.9}$$

where we identify the heat flux entering the system from the reservoir as the energy absorbed by the bosonic mode (no external driving), $\dot{Q} = \text{Tr}[\hat{H}_{\mathbb{S}}\dot{\rho}_t] = \dot{U}_{\mathbb{S}}$, and obtain the extra non-thermal contribution:

$$\dot{\mathcal{A}} = \hbar\omega \text{Tr}[\hat{A}_\theta \dot{\rho}_t] = -\frac{\hbar\omega}{2}\text{Tr}[(\hat{a}^{\dagger 2} e^{i\theta} + \hat{a}^2 e^{-i\theta})\dot{\rho}_t]. \tag{10.10}$$

Rewriting $\hat{A}_\theta = (1/2)(\hat{p}_{\theta/2}^2 - \hat{x}_{\theta/2}^2)$, we see that it measures the asymmetry in the second order moments of the mode quadratures, which includes both the relative shape of the variances and the relative displacements in optical phase space. From the Lindblad master equation (10.3) we obtain (see details in Appendix)

$$\dot{Q}(t) = -\gamma\left(U_{\mathbb{S}}(t) - \langle \hat{H}_{\mathbb{S}} \rangle_\pi\right), \tag{10.11}$$

$$\dot{\mathcal{A}}(t) = -\gamma(\mathcal{A}(t) - \hbar\omega\langle \hat{A}_\theta \rangle_\pi), \tag{10.12}$$

where $U_{\mathbb{S}}(t) = \text{Tr}[\hat{H}_{\mathbb{S}}\rho_t]$, $\mathcal{A}(t) = \hbar\omega \text{Tr}[\hat{A}_\theta \rho_t]$, and the expected value of $\hat{A}_\theta$ in the stationary state reads $\langle \hat{A}_\theta \rangle_\pi = \sinh(2r)(n_{\text{th}} + 1/2) \geqslant 0$. Therefore, the evolution of $\mathcal{A}(t)$ is rather simple: it increases (decreases) exponentially when the interaction with the reservoir induces (reduces the) asymmetry in the $\theta/2$ phase-selected quadratures. Analogously $Q(t) = U_{\mathbb{S}}(t) - U_{\mathbb{S}}(0)$ increases (decreases) exponentially when the energy in the initial state $U_{\mathbb{S}}(0) = \langle \hat{H}_{\mathbb{S}} \rangle_{\rho_0}$ is lower (greater) than in the steady state $\langle \hat{H}_{\mathbb{S}} \rangle_\pi = \hbar\omega[\cosh(2r)n_{\text{th}} + \sinh^2(r)]$.

As an illustrative example of the entropic implications of the above asymmetry flow, consider an initial state ρ_0 with $\mathcal{A}(0) = 0$, e.g. a Gaussian state without displacement, but with diagonal elements in the $\hat{H}_{\mathrm{S}}$ basis as those in π, so that $\langle \hat{H}_{\mathrm{S}} \rangle_{\rho_0} = \langle \hat{H}_{\mathrm{S}} \rangle_{\pi}$. Clearly, during the relaxation to the steady state π, $\dot{\mathcal{A}} > 0$, while $\dot{Q} = 0$, the uncertainty in $\hat{X}_{\theta/2}$ is reduced with respect to the one in $\hat{P}_{\theta/2}$ at constant energy until the steady state is reached. In this case, according to (10.9), $\Delta\Phi < 0$, meaning that entropy is transferred from the system to the reservoir. This entropy flux indeed overcomes the entropy produced in the process, $\Delta_{\mathrm{i}} S_{\mathrm{na}} > 0$, which corresponds to a net reduction in the system local entropy $\Delta S = \Delta_{\mathrm{i}} S_{\mathrm{na}} + \Delta\Phi < 0$. That is, the bosonic mode is purified by contact with the squeezed thermal reservoir without any (average) exchange of heat.

10.2 Extracting Work from a Single Reservoir

As a first consequence of reservoir squeezing, we point out the possibility of cyclic work extraction from a single reservoir. This operation is forbidden by the second law of thermodynamics in the thermal reservoir case. Nevertheless it becomes possible when including extra sources of coherence [5], neg-entropy [44], or additional information reservoirs [45, 46]. We consider a two-stroke cyclic process operated as sketched in Fig. 10.1a. In the first step we start with the state π in Eq. (10.7), and Hamiltonian $\hat{H}_{\mathrm{S}} = \hbar\omega \hat{a}^\dagger \hat{a}$, implementing a unitary (isentropic) evolution $\hat{U}$, which drives the system detached from the reservoir (e.g. by modulating the frequency $\omega(t)$ as explained in Sect. 10.5). The bosonic mode ends up in some state $\rho = \hat{U}\pi\hat{U}^\dagger$ with the same Hamiltonian $\hat{H}_{\mathrm{S}}$. In this process work can be extracted by the external driving $W_{\mathrm{out}} = \mathrm{Tr}[\hat{H}_{\mathrm{S}}\pi] - \mathrm{Tr}[\hat{H}_{\mathrm{S}}\rho]$, while no heat is produced. In the second step the system is put in contact with the squeezed thermal reservoir until it relaxes back to π. This produces a heat flow entering from the reservoir, which equals the work extracted in the first step, $Q = \mathrm{Tr}[\hat{H}_{\mathrm{S}}\pi] - \mathrm{Tr}[\hat{H}_{\mathrm{S}}\rho] = W_{\mathrm{out}}$, as required from energy conservation. The entropy production in Eq. (10.8), integrated over a whole cycle, yields $-\Delta\Phi \geqslant 0$. Using Eq. (10.9), we find:

$$W_{\mathrm{out}} \leqslant \tanh(2r)\Delta\mathcal{A} \tag{10.13}$$

where $\Delta\mathcal{A} = \langle\mathcal{A}\rangle_{\pi_{\mathrm{S}}} - \langle\mathcal{A}\rangle_{\rho_{\mathrm{S}}}$. Hence positive work may be extracted from the reservoir whenever $\Delta\mathcal{A} > 0$, e.g. by having ρ_{S} less squeezed than π_{S}. Maximum work is extracted by requiring $\rho_{\mathrm{S}} = e^{-\beta\hat{H}_{\mathrm{S}}}/Z$ (which means that $\hat{U} = \hat{\mathbb{S}}^\dagger(\xi)$), as it minimizes the mean energy for a fixed entropy. In that particular case:

$$W_{\max} = \hbar\omega(2n_{\mathrm{th}} + 1)\sinh^2(r) \geqslant 0, \tag{10.14}$$

which vanishes in the thermal case, $r = 0$, as expected. It is worth mentioning that this process does not saturate inequality (10.13), meaning that it is not reversible, but an amount $\Delta_{\mathrm{i}} S_{\mathrm{na}} = \beta W_{\max}$ of entropy is produced in each cycle. Indeed reversibility conditions ($\Delta_{\mathrm{i}} S_{\mathrm{na}} = 0$) can only be achieved, following Eq. (10.8), in the trivial case $\rho_{\mathrm{S}} = \pi_{\mathrm{S}}$, implying $W_{\mathrm{out}} = \Delta\mathcal{A} = 0$.

10.3 Heat Engine with a Squeezed Thermal Reservoir

10.3.1 Optimal Otto Cycle

As a second application of interest we consider a quantum heat engine operating between two reservoirs: a cold equilibrium thermal bath at inverse temperature β_1, and a hot squeezed thermal reservoir at $\beta_2 \leqslant \beta_1$ with squeezing parameter $\xi = re^{i\theta}$. The bosonic mode performs a thermodynamic four-stroke cycle (Fig. 10.1b) as in traditional quantum Otto cycles [47–49], while the isentropic expansion is allowed to unsqueezed the mode.

Quantum Otto heat engines are characterized by the implementation on the working fluid of a four-stroke cycle in which isentropic and isochoric processes are alternated. In the case of a bosonic mode, the isentropic (unitary) strokes are performed by external modulation of the mode frequency. The isochoric steps are obtained by keeping a constant frequency, while relaxing in contact with thermal reservoirs at different temperatures. In such case, adiabatic modulation of the frequency leads to both maximum work extraction and high efficiencies. This fact can be understood from a simple argument: as long as the mode state before the isentropic stroke, say $\rho_{\rm i}$, is fixed by the previous thermalization step, the work extracted in the process, $W_{\rm stroke} = {\rm Tr}[\hat{H}_{\rm i}\rho_{\rm i}] - {\rm Tr}[\hat{H}_{\rm f}\rho_{\rm f}]$, is minimized when $\rho_{\rm f}$ (the state after modulation) has minimum energy for a fixed entropy. This occurs when it has Gibbs form $\rho_{\rm f} = \exp(-\beta\hat{H}_{\rm f})/Z_{\rm f}$ for some β, which is the case if the modulation is implemented adiabatically. Moreover, the quantum friction in such case is zero, as the non-diagonal elements of the mode state in its instantaneous Hamiltonian basis are zero during the whole cycle. However, in the case of squeezed thermal reservoirs, the above situation is slightly modified. Here we will introduce a modification in the traditional Otto cycle which maximizes the work extracted by applying the above argument to this new situation (see also Ref. [24]). In contrast to Refs. [20, 23], we will require

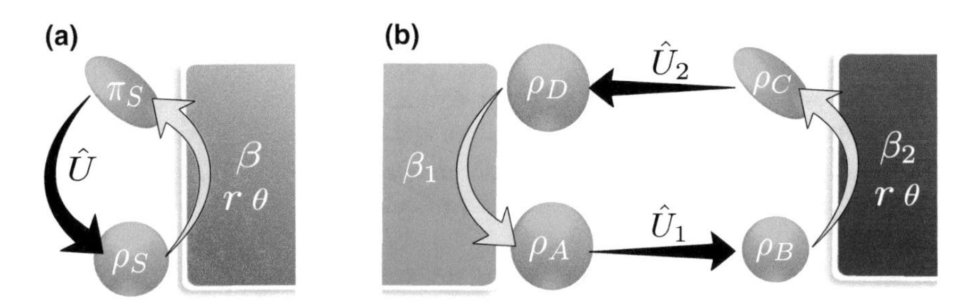

Fig. 10.1 Schematic diagrams of **a** the two-step protocol introduced to extract work from a single squeezed reservoir and **b** the four-step Otto-like cycle operating between reservoirs at different temperatures. The unitary $\hat{U}_1$ represents the adiabatic frequency modulation from ω_1 to ω_2, while $\hat{U}_2$ represents the convolution of the unitary unsqueezing the bosonic mode, $\hat{S}^\dagger(\xi)$, followed by adiabatic modulation from ω_2 to ω_1

a isentropic stroke driving the state after relaxation in the presence of the squeezed thermal reservoir to a perfect Gibbs state with respect to the final Hamiltonian at the end of the stroke. This operation can be achieved by first unsqueezing the mode and then applying regular adiabatic modulation, or by an unique taylored modulation [50]. As a consequence, the power output defined as the work extracted in a single cycle, see Eq. (10.19) below, divided by its duration is maximized.

We start with our system in point A, in equilibrium with the cold thermal reservoir,

$$\rho_A = \exp(-\beta_1 \hat{H}_1)/Z_A, \tag{10.15}$$

being $Z_A = \mathrm{Tr}[e^{-\beta_1 \hat{H}_1}]$, and the initial Hamiltonian is $\hat{H}_1 = \hbar\omega_1 \hat{a}_1^\dagger \hat{a}_1$. During the first step the system is isolated from the reservoirs, and its frequency adiabatically modulated from ω_1 to $\omega_2 \geqslant \omega_1$, without changing the populations of the energy eigenstates. The density matrix at point B is

$$\rho_B = \hat{U}_1 \rho_A \hat{U}_1^\dagger = \exp\left(-\beta_1 \frac{\omega_1}{\omega_2} \hat{H}_2\right)/Z_B \tag{10.16}$$

where $\hat{U}_1$ represents the adiabatic modulation, $Z_B = Z_A$, and the Hamiltonian is changed to $\hat{H}_2 = \hbar\omega_2 \hat{a}_2^\dagger \hat{a}_2$ during the process. The work extracted during this isentropic compression is negative (external work is needed to perform it), and reads $W_{AB} = \mathrm{Tr}[\hat{H}_1 \rho_A] - \mathrm{Tr}[\hat{H}_2 \rho_B] = -\hbar(\omega_2 - \omega_1) n_{\text{th}}^{(1)}$, where $n_{\text{th}}^{(1)} = (e^{\beta_1 \hbar\omega_1} - 1)^{-1}$. The Gibbs form of the state ρ_B minimizes the work lost in the compression and, as long as the system is isolated, no heat is produced in this step. In the second stroke, the bosonic mode is put in contact with the squeezed thermal reservoir while the frequency stays constant, resulting in an isochoric process where the mode relaxes to the steady-state

$$\rho_C = \hat{\mathbb{S}}(\xi)\, \exp(-\beta_2 \hat{H}_2)\, \hat{\mathbb{S}}^\dagger(\xi)/Z_C. \tag{10.17}$$

The heat entering the system from the squeezed thermal bath in the relaxation is $Q_{BC} = \mathrm{Tr}[\hat{H}_2 \rho_C] - \mathrm{Tr}[\hat{H}_2 \rho_B] = \hbar\omega_2 (n_{\text{th}}^{(2)} \cosh(2r) + \sinh^2(r) - n_{\text{th}}^{(1)})$, with $n_{\text{th}}^{(2)} = (e^{\beta_2 \hbar\omega_2} - 1)^{-1}$. In addition, the reservoir induces an asymmetry in the $\theta/2$ system quadratures which, following Eq. (10.10), reads $\Delta\mathcal{A}_{BC} = \hbar\omega_2 \sinh(2r)(n_{\text{th}}^{(2)} + 1/2)$.

In the third stroke, the bosonic mode is again detached from the reservoirs, we apply the unitary unsqueezing the mode, $\hat{\mathbb{S}}^\dagger(\xi)$, and then we change its frequency adiabatically back to ω_1, as has been also considered in Ref. [24]. This process can alternatively be done by a unique taylored modulation $\omega(t)$ [50]. The system state at point D is then

$$\rho_D = \hat{U}_2 \rho_C \hat{U}_2^\dagger = \exp\left(-\beta_2 \frac{\omega_2}{\omega_1} \hat{H}_1\right)/Z_D, \tag{10.18}$$

where $\hat{U}_2$ represents the two operations, and $Z_D = Z_C$. Consequently, the work extracted in this isentropic expansion reads $W_{CD} = \text{Tr}[\hat{H}_2\rho_C] - \text{Tr}[\hat{H}_1\rho_D] = \hbar\omega_2[n_{\text{th}}^{(2)}\cosh(2r) + \sinh^2(r)] - \hbar\omega_1 n_{\text{th}}^{(2)}$. The state ρ_D has been chosen to maximize the work extracted, as indicated by our previous example and Eq. (10.14). The cycle is closed by putting the bosonic mode in contact with the cold thermal reservoir, and hence relaxing back to ρ_A without varying its frequency. During the last isochoric process, the heat transferred from the cold reservoir to the system is $Q_{DA} = \text{Tr}[\hat{H}_1\rho_A] - \text{Tr}[\hat{H}_1\rho_D] = \hbar\omega_1(n_{\text{th}}^{(1)} - n_{\text{th}}^{(2)})$. The total work extracted in the cycle is given by the contributions of the two isentropic strokes:

$$\begin{aligned} W_{\text{out}} \equiv W_{AB} + W_{CD} = \hbar(\omega_2 - \omega_1)(n_{\text{th}}^{(2)} - n_{\text{th}}^{(1)}) + \\ + \hbar\omega_2(2n_{\text{th}}^{(2)} + 1)\sinh^2(r), \end{aligned} \tag{10.19}$$

which is nothing but the sum of the work extractable from an ideal quantum Otto cycle between two regular thermal reservoirs (first term), plus the work extractable from a single squeezed thermal reservoir (last term), as given by Eq. (10.14). Notice that $W_{\text{out}} = Q_{BC} + Q_{DA}$, as required by the first law.

In Fig. 10.2 we plot the work output of the cycle as a function of the frequency modulation ω_2 (in units of ω_1) for different values of the squeezed parameter. As we can see in the plot, the maximum power with respect to ω_2 is no longer confined to the low-frequency modulation region if moderate values of the squeezing parameter are considered. This opens the possibility of increasing the power by frequency modulation. However the local maximum is placed at the same point as for the traditional cycle for the high-temperature regime, given by $\omega_2/\omega_1 = \sqrt{\beta_1[1 + 2\sinh^2(r)]/\beta_2}$ [20].

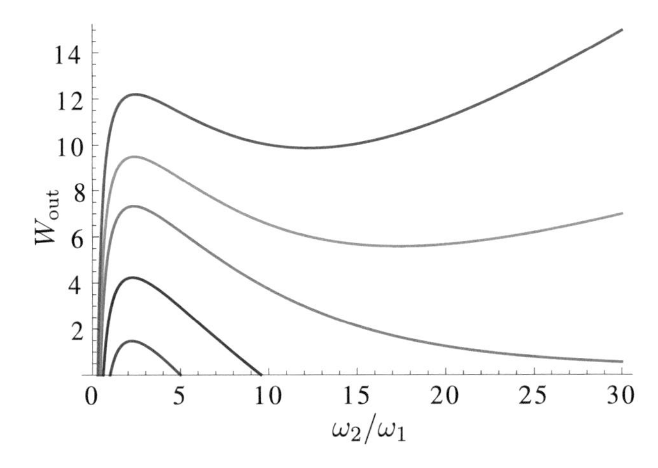

Fig. 10.2 Total work output, W_{out}, (in units of $\hbar\omega_1$) generated in a single cycle as a function of the frequency modulation, ω_2/ω_1, for different values of the squeezed parameter (from bottom to top) $r = (0.0, 0.5, 0.7, 0.8, 0.9)$. We used $\beta_1 = (\hbar\omega_1)^{-1}$ and $\beta_2 = 0.2(\hbar\omega_1)^{-1}$

10.3.2 Regimes of Operation

The above introduced cycle presents different regimes of operation depending on the squeezing parameter r and on the final frequency after modulation ω_2, some of them *forbidden* in the regular Otto cycle. They are summarized in the phase diagram of Fig. 10.3.

- *Region I* corresponds to a regular heat engine, for which work is extracted from the heat released by the hot (squeezed) reservoir, while dissipating some part in the cold thermal one. In this regime, a small frequency modulation, $\omega_2 \leqslant \omega_2^* \equiv \omega_1\beta_1/\beta_2 \Leftrightarrow n_{\text{th}}^{(2)} \geqslant n_{\text{th}}^{(1)}$, guarantees $W_{\text{out}} \geqslant 0$, $Q_{BC} \geqslant 0$ and $Q_{DA} \leqslant 0$. The energetic efficiency, defined as the total work output, W_{out}, divided by the input heat, Q_{BC}, reads:

$$\eta = 1 - \frac{\omega_1}{\omega_2}\left(\frac{n_{\text{th}}^{(2)} - n_{\text{th}}^{(1)}}{(2n_{\text{th}}^{(2)} + 1)\sinh^2(r) + n_{\text{th}}^{(2)} - n_{\text{th}}^{(1)}}\right) \tag{10.20}$$

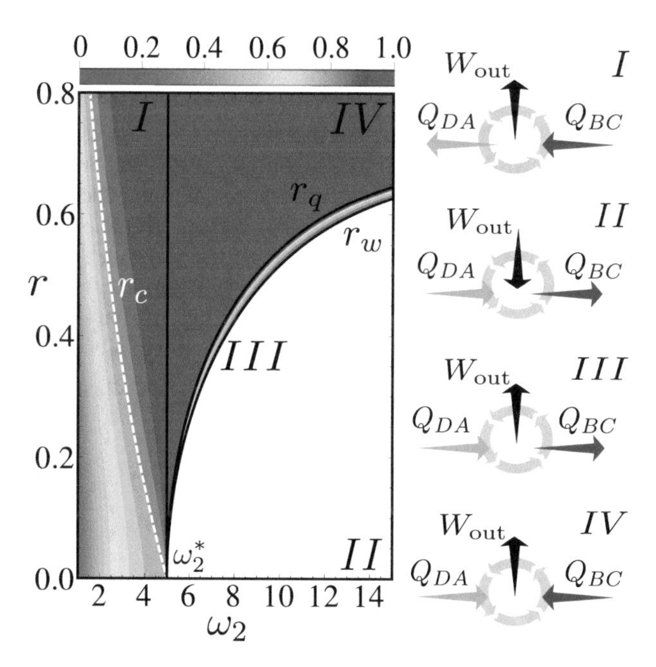

Fig. 10.3 Phase diagram with the four regimes of operation of the cycle (I, II, III, IV) as a function of ω_2 (in units of ω_1) and r. The color scale corresponds to the *energetic* efficiency of the cycle $\eta = W_{\text{out}}/Q_{\text{in}}$ as a heat engine, for $\beta_1 = (\hbar\omega_1)^{-1}$ and $\beta_2 = 0.2(\hbar\omega_1)^{-1}$, yielding $\eta_c = 0.8$. In the right side the direction of the arrows represents the sign of the energy fluxes for each regime

which differs from the traditional Otto cycle efficiency for adiabatic strokes, $\eta_q = 1 - \omega_1/\omega_2$ [47]. Indeed the efficiency (10.20) can surpass Carnot efficiency, $\eta \geqslant \eta_c = 1 - \beta_2/\beta_1$, for sufficient large squeezing, $r \geqslant r_c(\omega_2)$.
The Carnot line, $r_c(\omega_2)$ is defined by

$$\sinh^2(r_c) = \left(\omega_2^*/\omega_2 - 1\right)(n_{\rm th}^{(2)} - n_{\rm th}^{(1)})/(2n_{\rm th}^{(2)} + 1) \tag{10.21}$$

when $\omega_2 \leqslant \omega_2^*$, and depicted in Fig. 10.3 (white dashed line). Furthermore we see from Eq. (10.20) that $\eta \to 1$ when $\omega_2 \to \omega_2^*$ while maintaining a finite work output in the cycle, $W_{\rm out} \to \hbar\omega_2^*(2n_{\rm th}^{(2)} + 1)\sinh^2(r)$, which is the same result as in the single reservoir case.

- *Region II* (white area in Fig. 10.3) corresponds to the well-known case of a driven refrigerator: external input work is needed to pump heat from the cold to the hot reservoir ($W_{\rm out} \leqslant 0$ and $Q_{BC} \leqslant 0$). Notice that for large frequency modulation, $\omega_2 \geqslant \omega_2^* \Leftrightarrow n_{\rm th}^{(1)} \geqslant n_{\rm th}^{(2)}$, we have always a positive amount of heat extracted from the cold reservoir, i.e. $Q_{DA} \geqslant 0$.
- *Regions III and IV* are the most striking regimes, implying refrigeration and work extraction at the same time, as has been also independently suggested in Ref. [24]. From Eq. (10.19) one can obtain the conditions for $W_{\rm out}$ and Q_{BC} to vanish, $r_w(\omega_2)$ and $r_q(\omega_2)$, respectively. Then $r \geqslant r_w(\omega_2)$ implies a positive amount of output work, whereas the heat flux entering the hot reservoir, Q_{BC}, is positive when $r \geqslant r_q(\omega_2)$. We then distinguish two regions (see Fig. 10.3). *Region III* is the narrow strip between the two boundaries, $r_q \geqslant r \geqslant r_w$, where we obtain a refrigerator producing a positive work output while pumping heat from the cold to the hot reservoir ($W_{\rm out} \geqslant 0$ and $Q_{BC} \leqslant 0$). Its energetic efficiency as a heat engine is given by $\eta = W_{\rm out}/Q_{DA} = 1 - (\omega_2/\omega_1)[1 - \sinh^2(r)/\sinh^2(r_q)]$, which varies from 0 to 1 between the two boundaries. Finally in *region IV* ($r \geqslant r_q$), we obtain a heat engine which absorbs heat from both reservoirs, transforming it into useful work ($W_{\rm out} \geqslant 0$ and $Q_{BC} \geqslant 0$) at efficiency $\eta = W_{\rm out}/Q_{\rm in} = 1$, as guaranteed by the first law. The explicit expressions for the curve r_c and the boundaries r_q and r_w are given by:

$$\begin{aligned}\sinh^2(r_q) &= (n_{\rm th}^{(1)} - n_{\rm th}^{(2)})/(2n_{\rm th}^{(2)} + 1),\\ \sinh^2(r_w) &= (1 - \omega_1/\omega_2)\sinh^2(r_q),\end{aligned} \tag{10.22}$$

which are well defined for $\omega_2 \geqslant \omega_2^*$ ensuring $n_{\rm th}^{(1)} \geqslant n_{\rm th}^{(2)}$ and hence refrigeration of the cold reservoir.

It is worth noticing that our results do not contradict the second law of thermodynamics, when correctly generalized to this situation, Eq. (10.8). Indeed, it can be written as the positivity of the entropy production for a single cycle of the engine:

$$\Delta_{\rm i} S_{\rm cyc} = -\beta_1 Q_{DA} - \beta_2\left[\cosh(2r)Q_{BC} - \sinh(2r)\Delta\mathcal{A}_{BC}\right] \geqslant 0, \tag{10.23}$$

which follows from Eq. (10.9). Using the explicit expressions of Q_{BC}, Q_{DA} and $\Delta\mathcal{A}_{BC}$ for the cycle, we obtain that reversibility conditions ($\Delta_{\rm i} S_{\rm cyc} = 0$) can be only reached when $\omega_2 = \omega_2^*$ and $r = 0$, hence implying $W_{\rm out} = 0$. Finally, when the second law (10.23) is combined with the first law, $W_{\rm out} = Q_{BC} + Q_{DA}$, we obtain bounds on the energetic efficiency for the heat engine regimes, $\eta \leqslant \min(\eta_{\rm max}, 1.0)$, where:

$$\eta_{\rm max} = \begin{cases} 1 - \frac{\beta_2}{\beta_1}\left(\cosh(2r) - \sinh(2r)\frac{\Delta\mathcal{A}_{BC}}{Q_{BC}}\right) & \text{(I)} \\ 1 - \frac{\beta_1}{\beta_2\cosh(2r)} + \tanh(2r)\,\frac{\Delta\mathcal{A}_{BC}}{Q_{DA}} & \text{(III)} \\ 1 & \text{(IV)} \end{cases}$$

As can be easily checked, $\eta_{\rm max} \to \eta_{\rm c}$ when $r \to 0$ in *region I*, while *regions III and IV* disappear in such case. The above equation is exact and generalizes previous efficiency bounds [20, 23] (only valid in the high-temperature limit) to any temperatures and frequencies. The explicit formulas for $\eta_{\rm max}$ are:

$$\eta_{\rm max}^{(I)} = 1 - \frac{\beta_2}{\beta_1}\frac{(2n_{\rm th}^{(2)}+1) - \cosh(2r)(2n_{\rm th}^{(1)}+1)}{\cosh(2r)(2n_{\rm th}^{(2)}+1) - (2n_{\rm th}^{(1)}+1)}, \tag{10.24}$$

for our cycle operating in the regime $\omega_2 \leqslant \omega_2^*$ (*region I*). Notice that it collapses to Carnot efficiency, when $r \to 0$. On the other hand, for *region III* we obtain:

$$\eta_{\rm max}^{(III)} = 1 - \frac{\beta_1}{\beta_2\cosh(2r)} + \frac{\omega_2}{\omega_1}\frac{\tanh(2r)\sinh(2r)}{2\sinh^2(r_q)}, \tag{10.25}$$

only valid when $\omega_2 \geqslant \omega_2^*$ and $r_w \leqslant r \leqslant r_q$. Finally we stress that in *region IV* we have

$$\eta_{\rm max}^{(IV)} = \eta = \frac{W_{\rm out}}{Q_{BC} + Q_{DA}} = 1, \tag{10.26}$$

which follows from energy conservation.

We show in Fig. 10.4 how the energetic efficiency η of our cycle, even when working as a normal heat engine [Eq. (10.20)], can overcome the so-called generalized Carnot efficiency obtained in Refs. [20, 23] by using the high-temperature approximation ($\beta_i\hbar\omega_i \ll 1$ for $i = 1, 2$):

$$\eta_{\rm ht} = 1 - \frac{\beta_2}{\beta_1[1 + 2\sinh^2(r)]}, \tag{10.27}$$

which verifies $\eta_{\rm ht} \geqslant \eta_c = 1 - \beta_2/\beta_1$. In contrast, our general bound, $\eta_{\rm max} \geqslant \eta_{\rm ht}$, obtained by applying the second law of thermodynamics in the full quantum regime, cannot be surpassed in any case. A complementary interpretation of the generalized second law in Eq. (10.23) in terms of the free energy released from the hot squeezed thermal reservoir, is further given in the next section.

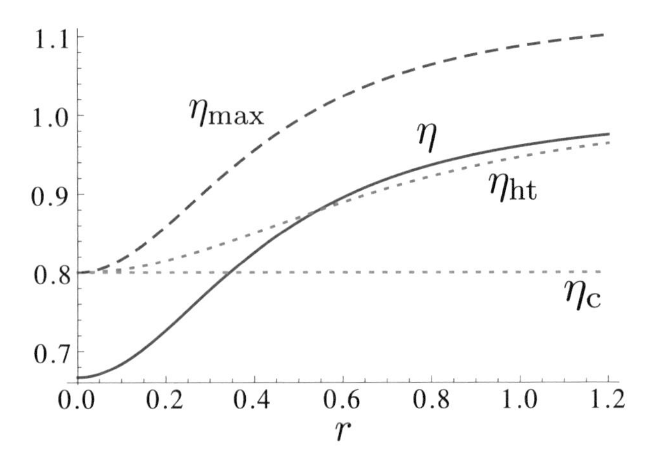

Fig. 10.4 Comparison of the energetic efficiency of the heat engine, η, the maximum efficiency allowed by the second law, $\eta_{\max}$, the Carnot efficiency, η_c, and the high-temperature generalized Carnot efficiency, η_{ht}, as a function of the squeezing parameter r. The high-temperature efficiency fails to bound correctly the efficiency of the cycle for moderate values of the squeezing parameter. Here we used $\omega_2 = 3\omega_1$ (i.e. $\omega_2 < \omega_2^* = 5\omega_1$, corresponding to *region I*) and again $\beta_1 = (\hbar\omega_1)^{-1}$ and $\beta_2 = 0.2(\hbar\omega_1)^{-1}$

10.4 Squeezing as a Source of Free Energy

Here we provide an interpretation of the squeezed thermal reservoir as a free energy source, which enables work extraction in the quantum Otto cycle discussed in Sect. 10.3. The nonequilibrium free energy, already introduced in Sect. 3.1, is a powerful concept in nonequilibrium thermodynamics and specifically in thermodynamics of information [51]. We recall that it is defined as a property of a system in some arbitrary state ρ with Hamiltonian $\hat{H}$, with respect to a thermal reservoir at temperature T, as

$$\mathcal{F}(T) = \langle \hat{H} \rangle_\rho - k_B T S(\rho), \tag{10.28}$$

being $S(\rho)$ the von Neumann entropy of the system state for the quantum case. The most important property of the nonequilibrium free energy is that its variation measures the maximum work which can be extracted when letting the system equilibrate to temperature T in an optimal way [7, 51] (see details in Sect. 3.1).

In order to apply this concept in our situation we proceed by using the fact that the entropy transfer between system and reservoir equals the decrease in entropy in the squeezed reservoir during the corresponding relaxation stroke of the Otto cycle, that is $\Delta\Phi_{BC} = -\Delta S_{R_2}$, as we show in Appendix. When this is combined with the first law in the cycle, $W_{\text{out}} = Q_{DA} + Q_{BC}$, we can rewrite the second law inequality in Eq. (10.23) as

$$W_{\text{out}} \leqslant \Delta\mathcal{F}_2(T_1), \qquad \eta_{\text{th}} \equiv \frac{W_{\text{out}}}{\Delta\mathcal{F}_2(T_1)} \leqslant 1, \tag{10.29}$$

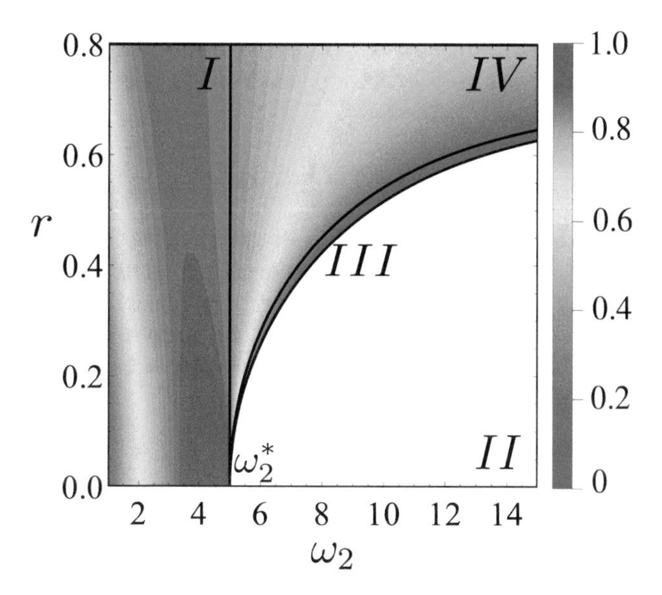

Fig. 10.5 Plot of the thermodynamic efficiency in Eq. (10.29) as a function of the frequency modulation ω_2 and the squeezing parameter r. The black thick lines represent the different regimes of operation introduced in Sect. 10.3.2. We assumed the rest of the parameters as in Fig. 10.3

where $\Delta\mathcal{F}_2(T_1) = Q_{BC} + k_B T_1 \Delta S_{R_2}$ is the loss of (nonequilibrium) free energy experimented by the hot squeezed thermal reservoir in a cycle, with respect to the cold thermal reservoir at temperature T_1. This allows to define a work extraction *thermodynamic* efficiency always bounded by 1, in contrast to the *energetic* efficiency considered in the previous section. In Fig. 10.5 we provide a map for the thermodynamic efficiency (10.29) analogous to the one analyzed for the energetic one in the previous section. There, higher thermodynamic efficiencies are shown to be achieved in regime I of operation, while unit thermodynamic efficiency is only approached from $r \to 0$ and $\omega_2 \to \omega_2^*$, corresponding to zero output work.

Furthermore, the input free energy from the squeezed thermal reservoir can be decomposed into two separate contributions by using the explicit expression of the entropy flow, Eq. (10.9)

$$\Delta\mathcal{F}_2(T_1) = \left(1 - \frac{T_1}{T_2}\right) Q_{BC} + \frac{T_1}{T_2}\left(\sinh(2r)\Delta\mathcal{A}_{BC} - 2\sinh^2(r) Q_{BC}\right). \tag{10.30}$$

The two terms correspond respectively to the free energy available as a consequence of the temperature gradient between two thermal reservoirs (first term), and the one provided by the nonequilibrium squeezing effects (second term). The first term is always positive when $Q_{BC} > 0$, meaning that free energy is available from the spontaneous flux of heat from a hot reservoir to a colder one. The second term, purely due to squeezing in the reservoir, is instead positive when squeezing is present, $r > 0$, and the following inequality is verified:

$$\Delta \mathcal{A}_{BC} \geqslant \tanh(r) Q_{BC}. \tag{10.31}$$

This implies that the entropic flux of second order coherences from the squeezed thermal reservoir [see Eq. (10.10)] acts as an independent source of free energy when the above inequality is fulfilled, increasing the work that can be extracted in the cycle. Furthermore it can be positive even if $Q_{BC} \leqslant 0$, and compensate the thermal term (which in this case would be negative), in order to enable work extraction, as is the case of *region III* of the phase diagram in Fig. 10.3.

10.5 Experimental Realization

The single trapped-ion Otto cycle proposed in Ref. [37] has been realized experimentally in Ref. [38] only recently. There, a trapped ion in a tapered Paul trap is subjected to adiabatic frequency modulations for the isentropic strokes of the cycle. The thermalization strokes are implemented by laser cooling with variable detuning (and thus final temperature). The same authors proposed theoretically to enhance the cycle by having a hot bath which is squeezed [20], finding an increase of the energetic efficiency at maximum power. The squeezed hot reservoir was effectively implemented by rather having the ion thermalize and then squeezing it, resulting thus in a final thermal squeezed state (as if the bath were squeezed). Such squeezing operation consists in quenching the ion frequency from ω to $\omega + \Delta\omega$ "For a quarter of the oscillation period", then to $\omega - \Delta\omega$ "For another quarter, before it is returned to its initial value" ω (notice that the authors of Ref. [20] are talking about periods of different duration, since the frequency of oscillations differ by $2\Delta\omega$, and this has to be carefully accounted for in the experiment). This operation can be easily understood from Fig. 1 in Ref. [52], by noting that suddenly increasing (decreasing) the frequency squeezes (stretches) the x variance, while at constant frequency the Wigner function just rotates at that frequency. Finally, the authors propose to output the work of the cycle (done in the radial coordinate of the ion) into the axial coordinate (the two motions are coupled due to the tapered geometry of the trap). In this sense, the engine does work on the axial motion and the working substance is the radial motion.

In our cycle, we are adding an extra step which tries to use the squeezing absorbed from the hot reservoir to produce work. In terms of operations we could modify the CD-branch (operation $\hat{U}_2$ in Sect. 10.3.1), reversing the modulation, which would remove the squeezing from the system. In this way, though, the work would be wasted into the frequency quencher (the electronics of the experiment). In order to profit from the squeezing absorbed from the reservoir, we should be able to transfer it to some fruitful target. One possibility is to wait for the axial-radial coupling to induce an exchange of squeezing until the axial absorbs all energy from the radial. The detailed dynamics should be studied thoroughly to check for limitations, though. Another possibility, seemingly involved, would be to transfer this squeezing to an optical mode. This process has been considered in Ref. [53], where three electronic levels of an ion trapped inside a cavity would be used to transfer the motional squeezing to light squeezing of the cavity mode. A fiber collecting the output light from the cavity could be used to transfer this squeezing to the target.

10.6 Conclusions

Squeezing is a quantum thermodynamic resource from which useful work can be delivered. When squeezing is present in an otherwise thermal reservoir, it does not only modifies the entropy flow associated to the heat, but induces an extra term proportional to second order coherences [Eq. (10.10)] with a specific meaning.

The nonequilibrium second law-inequality, Eq. (10.8) with (10.9), introduces remarkable modifications which may give rise to novel phenomena and applications as squeezing-fueled batteries, multi-task (refrigerator, heat pump, *and* heat engine) thermal machines, or a perfect heat-to-work transformer working at 100% efficiency. The extra non-thermal contribution to the entropy transfer hints also at possible erasure devices operating below Landauer's limit (see e.g. the toy model discussed in Sect. 9.3).

In this chapter the squeezed thermal reservoir has been considered as a given thermodynamic resource. Consequently, we did not consider any extra energetic or thermodynamic cost associated to its creation, in the same manner as thermal reservoirs at different temperatures are considered as resources for the operation of traditional heat engines. The thermodynamic cost for generating squeezing may in general depend on the specific configuration employed, and has been investigated e.g. in Refs. [50, 54].

Alternatively the squeezed thermal reservoir can be seen as a source of free energy powering work extraction. This interpretation leads to a natural definition of the *thermodynamic* efficiency of the engine, namely, the output work divided by the input free energy, which is bounded by one by the positivity of the entropy production.

Finally, our results may be tested as in the recent experiment of a single-ion Otto heat engine [20, 38], with an added modification to additionally exploit the squeezing absorbed from the hot reservoir, as we detailed in Sect. 10.5.

Appendix

A.1 Reservoir Entropy Changes

In the main text we claim that the effective entropy flow, $\dot{\Phi}$, appearing in the generalized second law inequality, Eq. (10.9) in Sect. 10.1, equals the entropy decrease in the reservoir due to the interaction with the bosonic mode. We demonstrate here this relation from the collisional model introduced in Sect. 2.3.2, where the system bosonic mode interacts sequentially with a 'fresh' reservoir mode k in the same squeezed thermal state at inverse temperature β, and squeezing parameter $\xi = re^{i\theta}$ with $r \geqslant 0$ and $\theta \in [0, 2\pi]$:

$$\begin{aligned} \rho_R^{(k)} &= \hat{\mathbb{S}}_k(\xi)\frac{e^{-\beta \hat{H}_R(\Omega_k)}}{Z_R}\hat{\mathbb{S}}_k^\dagger(\xi) \\ &= \sum_\nu \left(\frac{e^{-\beta\hbar\Omega_k\nu}}{Z_R}\right)\hat{\mathbb{S}}_k(\xi)|\nu_k\rangle\langle\nu_k|\hat{\mathbb{S}}_k^\dagger(\xi) \end{aligned} \tag{A.1}$$

where $\hat{\mathbb{S}}_k(\xi) \equiv \exp\frac{r}{2}(b_k^2 e^{-i\theta} - b_k^{\dagger 2}e^{i\theta})$, stands for the squeezing operator on the reservoir mode k, and in the last equality we decomposed the Gibbs state in its Fock basis $\{|\nu_k\rangle\}$. It's easy to see from the above equation that the eigenvalues and eigenvectors of $\hat{\rho}_R$ are given by:

$$\epsilon_\nu^{(k)} = \frac{e^{-\beta\hbar\Omega_k\nu}}{Z_R}, \quad |\epsilon_\nu^{(k)}\rangle = \hat{\mathbb{S}}_k(\xi)|\nu_k\rangle, \tag{A.2}$$

i.e. the state $\rho_R^{(k)}$ can be viewed as a classical mixture of squeezed Fock states $|\epsilon_\nu^{(k)}\rangle$ with Boltzmann weights $\epsilon_\nu^{(k)}$.

We can estimate the reservoir entropy change during the evolution by constructing, analogously to what have been done for the system bosonic mode, a coarse-grained time derivative by partial tracing Eq. (2.63) over the system degrees of freedom:

$$\dot{\rho}_R^{(k)} \simeq \frac{1}{\delta t}[\rho_R^{(k)}(t+\delta t) - \rho_R^{(k)}] = \mathcal{R}\,[\rho_R^{(k)}(t+\tau) - \rho_R^{(k)}] \tag{A.3}$$

for the interaction of duration $\tau \ll g_k^{-1}$ between system and a particular mode k in the reservoir.

Using Eqs. (2.64) and (2.65) we obtain:

$$\begin{aligned} \dot{\rho}_R^{(k)} &= -i[\Delta\hat{H}_R(\Omega_k), \rho_R^{(k)}] + [\epsilon_k^*\langle\hat{a}\rangle_t\hat{b}_k^\dagger - \epsilon_k\langle\hat{a}^\dagger\rangle_t\hat{b}_k, \rho_R^{(k)}] \\ &+ c_k\langle\hat{a}\hat{a}^\dagger\rangle_t\left(\hat{b}_k\rho_R^{(k)}\hat{b}_k^\dagger - \frac{1}{2}\{\hat{b}_k^\dagger\hat{b}_k, \rho_R^{(k)}\}\right) \end{aligned}$$

$$+ c_k \langle \hat{a}^\dagger \hat{a} \rangle_t \left(\hat{b}_k^\dagger \rho_R^{(k)} \hat{b}_k - \frac{1}{2} \{ \hat{b}_k \hat{b}_k^\dagger, \rho_R^{(k)} \} \right)$$
$$- c_k e^{-i\Delta_k(2t+\tau)} \langle \hat{a}^2 \rangle_t \left(\hat{b}_k^\dagger \rho_R^{(k)} \hat{b}_k^\dagger - \frac{1}{2} \{ \hat{b}_k^{\dagger 2}, \rho_R^{(k)} \} \right)$$
$$- c_k e^{i\Delta_k(2t+\tau)} \langle \hat{a}^{\dagger 2} \rangle_t \left(\hat{b}_k \rho_R^{(k)} \hat{b}_k - \frac{1}{2} \{ \hat{b}_k^2, \rho_R^{(k)} \} \right), \tag{A.4}$$

where $\langle \hat{O} \rangle_t = \mathrm{Tr}_{\mathbb{S}}[\hat{O} \rho_t]$ are the system expectation values at time t, and $\Delta_k = \omega - \Omega_k$. In the above equation we defined

$$\epsilon_k \equiv \mathcal{R}\, \tau\, g_k\, \mathrm{sinc}(\Delta_k \tau/2)\, e^{i\Delta_k(t+\tau/2)},$$
$$c_k \equiv \mathcal{R}\, \tau^2 g_k^2\, \mathrm{sinc}^2(\Delta_k \tau/2), \tag{A.5}$$

together with the mode dependent frequency-shift in the reservoir

$$\Delta \hat{H}_R(\Omega_k) \equiv \mathcal{R}\, \frac{g_k^2 \tau}{\Delta_k} \big[\hat{b}^\dagger \hat{b}\, (\mathrm{sinc}(\Delta_k \tau/2) \cos(\Delta_k \tau/2) - 1) + 1$$
$$- \mathrm{sinc}(\Delta_k \tau/2) \left(2 \langle \hat{a}^\dagger \hat{a} \rangle_t (\cos(\Delta_k \tau/2) - 1) + e^{i\Delta_k \tau/2} \right) \big],$$

which is analogous to the system frequency shift, and will be neglected as well. Notice that Eq. (A.4) give us the average evolution of the reservoir modes k when it interacts once at a time with the system at random times (as specified by the rate $\mathcal{R}$). However, we don't know the frequency of the reservoir mode interacting with the system in each collision, so we must assume that the system interacts with all modes in the reservoir with certain probability, given by the density of states in the reservoir $\vartheta(\Omega_k)$. Therefore the average reservoir entropy change due to the entropy change in all reservoir modes during the evolution should read

$$\dot{S}_R = \sum_k \vartheta(\Omega_k) \dot{S}_R^{(k)} = -\sum_k \vartheta(\Omega_k) \mathrm{Tr}_R[\dot{\rho}_R^{(k)} \ln \rho_R^{(k)}]. \tag{A.6}$$

In the following we introduce the explicit form of $\rho_R^{(k)}$ as given in Eq. (A.1) into the above expression for the average reservoir entropy change, and exploit Eq. (A.4). We obtain:

$$\dot{S}_R = \beta \sum_k \vartheta(\Omega_k)\, \mathrm{Tr}_R[\dot{\rho}_R^{(k)}\, \hat{\mathbb{S}}_k(\xi) \hat{H}_R(\Omega_k) \hat{\mathbb{S}}_k^\dagger(\xi)] =$$
$$= -\beta\, \mathrm{Tr}_{\mathbb{S}}[\dot{\rho}_t \hat{\mathbb{S}}(\xi) \hat{H}_{\mathbb{S}} \hat{\mathbb{S}}^\dagger(\xi)] = -\dot{\Phi} \tag{A.7}$$

where the second line follows after a little of operator algebra, by expanding $\hat{\mathbb{S}}_k(\xi) \hat{H}_R(\Omega_k) \hat{\mathbb{S}}_k^\dagger(\xi)$ and using Eqs. (A.4) and (10.3). As a hint, first notice that the first order term in Eq. (A.4) does not contribute to the entropy. Secondly notice that once the trace over the reservoir degrees of freedom have been performed, one can take the continuum limit over the reservoir spectra by introducing the spectral density, $J(\Omega)$, to recover the system master equation decay factors in Eq. (2.73) after integrating over frequencies.

Henceforth the entropy flow entering the system during the evolution, as given by $\dot{\Phi}(t) = -\text{Tr}[\dot{\rho}_t \ln \pi]$, Eq. (10.9) in Sect. 10.1, is the average entropy lost in the the reservoir in the sequence of collisions. This implies that the non-adiabatic entropy production [41–43, 55], $\Delta_i S_{na}$ in Eq. (10.8), corresponds indeed the total entropy produced in the process. In terms of the rates:

$$\dot{S}_{na} \equiv -\frac{d}{dt} D(\rho_t || \pi) = \dot{S} + \dot{S}_R \geqslant 0 \tag{A.8}$$

where $D(\rho||\sigma) = \text{Tr}[\rho(\ln \rho - \ln \sigma)]$ is the quantum relative entropy. As a consequence the adiabatic (or house-keeping) contribution due to non-equilibrium external constraints [41, 42] is always zero in the present case. An important consequence of the above finding is that no entropy is produced in order to maintain the non-equilibrium steady state π, Eq. (10.7), provided we have access to an arbitrarily big ensemble of reservoir modes in the state ρ_R.

A.2 Equations of Motion

From the Master Equation (10.3) in Sect. 10.1, one can derive the following equations of motion for the expectation values of the Lindblad operators expectation values and its combinations:

$$\frac{d}{dt}\langle \hat{R} \rangle_t = -\frac{\gamma_0}{2} \langle \hat{R} \rangle_t \tag{A.9}$$

$$\frac{d}{dt}\langle \hat{R}^2 \rangle_t = -\gamma_0 \langle \hat{R}^2 \rangle_t, \tag{A.10}$$

$$\frac{d}{dt}\langle \hat{R}^\dagger \hat{R} \rangle_t = -\gamma_0 \left(\langle \hat{R}^\dagger \hat{R} \rangle_t - n_{\text{th}}(\omega) \right). \tag{A.11}$$

They can then be employed to explicitly asses the dynamics of the different contributions appearing in the effective entropy flow, $\dot{\Phi}$ in Eq. (10.9). Indeed by rewriting

$$\hat{a} = \hat{R} \cosh(r) - \hat{R}^\dagger \sinh(r) e^{i\theta}, \tag{A.12}$$

$$\hat{a}^\dagger = \hat{R}^\dagger \cosh(r) - \hat{R} \sinh(r) e^{-i\theta}, \tag{A.13}$$

and substituting into the expressions $\dot{Q}(t) = \dot{U}_S(t) = \text{Tr}[\hat{H}_S \dot{\rho}_t]$ for the heat flux entering from the reservoir, and $\dot{\mathcal{A}}(t) = \hbar\omega \text{Tr}[\hat{A}_\theta \dot{\rho}_t]$ with $\hat{A}_\theta = -\frac{1}{2}(\hat{a}^{\dagger 2} e^{i\theta} + \hat{a}^2 e^{-i\theta})$, for the extra non-thermal contribution, we obtain the following equations

$$\dot{Q}(t) = -\gamma_0 \left(Q(t) + \langle \hat{H}_{\mathrm{S}} \rangle_{\rho_0} - \langle \hat{H}_{\mathrm{S}} \rangle_\pi \right),$$
$$\dot{A}(t) = -\gamma_0 \left(\mathcal{A}(t) - \hbar\omega \langle \hat{A}_\theta \rangle_\pi \right). \tag{A.14}$$

In the above equations we introduced the steady state expectation values $\langle \hat{H}_{\mathrm{S}} \rangle_\pi = \hbar\omega N_\omega$ and $\langle \hat{A}_\theta \rangle_\pi = |M_\omega|$, being π given in Eq. (10.7), with the reservoir expectation values, $N_\omega = \langle \hat{b}_k^\dagger \hat{b}_k \rangle_{\rho_R}$ and $M_\omega = \langle \hat{b}_k^2 \rangle_{\rho_R}$ as defined in (2.74) for a mode with resonant frequency $\Omega_k = \omega$ in the state ρ_R. We notice that both flows behave monotonically, yielding to an exponential decay as discussed in Sect. 10.1.

References

1. S. Hormoz, Quantum collapse and the second law of thermodynamics. Phys. Rev. E **87**, 022129 (2013)
2. J.M. Horowitz, K. Jacobs, Quantum effects improve the energy efficiency of feedback control. Phys. Rev. E **89**, 042134 (2014)
3. K. Brandner, M. Bauer, M.T. Schmid, U. Seifert, Coherenceenhanced efficiency of feedback-driven quantum engines. New J. Phys. **17**, 065006 (2015)
4. P. Kammerlander, J. Anders, Coherence and measurement in quantum thermodynamics. Sci. Rep. **6**, 22174 (2016)
5. M.O. Scully, M.S. Zubairy, G.S. Agarwal, H. Walther, Extracting work from a single heat bath via vanishing quantum coherence. Science **299**, 862–864 (2003)
6. M.O. Scully, Quantum photocell: using quantum coherence to reduce radiative recombination and increase efficiency. Phys. Rev. Lett. **104**, 207701 (2010)
7. P. Skrzypczyk, A.J. Short, S. Popescu, Work extraction and thermodynamics for individual quantum systems. Nat. Commun. **5**, 4185 (2014)
8. J. Äberg, Catalytic coherence. Phys. Rev. Lett. **113**, 150402 (2014)
9. J. Oppenheim, M. Horodecki, P. Horodecki, R. Horodecki, Thermodynamical approach to quantifying quantum correlations. Phys. Rev. Lett. **89**, 180402 (2002)
10. W.H. Zurek, Quantum discord and Maxwell's demons. Phys. Rev. A **67**, 012320 (2003)
11. L. del Rio, J. Äberg, R. Renner, O. Dahlsten, V. Vedral, The thermodynamic meaning of negative entropy. Nature **474**, 61–63 (2011)
12. J.J. Park, K.-H. Kim, T. Sagawa, S.W. Kim, Heat engine driven by purely quantum information. Phys. Rev. Lett. **111**, 230402 (2013)
13. N. Brunner, M. Huber, N. Linden, S. Popescu, R. Silva, P. Skrzypczyk, Entanglement enhances cooling in microscopic quantum refrigerators. Phys. Rev. E **89**, 032115 (2014)
14. M. Perarnau-Llobet, K.V. Hovhannisyan, M. Huber, P. Skrzypczyk, N. Brunner, A. Acín, Extractable work from correlations. Phys. Rev. X **5**, 041011 (2015)
15. H.T. Quan, P. Zhang, C.P. Sun, Quantum-classical transition of photon-Carnot engine induced by quantum decoherence. Phys. Rev. E **73**, 036122 (2006)
16. H. Li, J. Zou, W.-L. Yu, B.-M. Xu, J.-G. Li, B. Shao, Quantum coherence rather than quantum correlations reflect the effects of a reservoir on a system's work capability. Phys. Rev. E **89**, 052132 (2014)
17. A.Ü.C. Hardal, Ö.E. Müstecaplıoĝlu, Superradiant quantum heat engine. Sci. Rep. **5**, 12953 (2015)
18. R. Dillenschneider, E. Lutz, Energetics of quantum correlations. Europhys. Lett. **88**, 50003 (2009)
19. X.L. Huan, T. Wang, X.X. Yi, Effects of reservoir squeezing on quantum systems and work extraction. Phys. Rev. E **86**, 051105 (2012)

20. J. Roßnagel, O. Abah, F. Schmidt-Kaler, K. Singer, E. Lutz, Nanoscale heat engine beyond the carnot limit. Phys. Rev. Lett. **112**, 030602 (2014)
21. L.A. Correa, J.P. Palao, D. Alonso, G. Adesso, Quantumenhanced absorption refrigerators. Sci. Rep. 3949 (2014)
22. R. Long, W. Liu, Performance of quantum Otto refrigerators with squeezing. Phys. Rev. E **91**, 062137 (2015)
23. O. Abah, E. Lutz, Efficiency of heat engines coupled to nonequilibrium reservoirs. Europhys. Lett. **106**, 20001 (2014)
24. W. Niedenzu, D. Gelbwaser-Klimovsky, A.G. Kofman, G. Kurizki, On the operation of machines powered by quantum nonthermal baths. New J. Phys. **18**, 083012 (2016)
25. G. Manzano, F. Galve, R. Zambrini, J.M.R. Parrondo, Entropy production and thermodynamic power of the squeezed thermal reservoir. Phys. Rev. E **93**, 052120 (2016)
26. D. Kondepudi, I. Prigogine, *Modern Thermodynamics From Heat Engines to Dissipative Structures* (Wiley, Chichester, 1998)
27. H. Spohn, J.L. Lebowitz, Irreversible thermodynamics for quantum systems weakly coupled to thermal reservoirs, in *Advances in Chemical Physics: For Ilya Prigogine*, vol. 38, ed. by S.A. Rice (Wiley, Hoboken, 1978)
28. R. Alicki, The quantum open system as a model of a heat engine. J. Phys. A **12**, L103 (1979)
29. S. Deffner, E. Lutz, Nonequilibrium entropy production for open quantum systems. Phys. Rev. Lett. **107**, 140404 (2011)
30. P.D. Drummond, Z. Ficek, *Quantum Squeezing* (Springer, Berlin, 2008)
31. E.S. Polzik, The squeeze goes on. Nature **453**, 45–46 (2008)
32. R. Loudon, P.L. Knight, Squeezed light. J. Mod. Opt. **34**, 709–759 (1987)
33. H. Fearn, M.J. Collett, Representations of squeezed states with thermal noise. J. Mod. Opt. **35**, 553–564 (1988)
34. B. Yurke, P.G. Kaminsky, R.E. Miller, E.A. Whittaker, A.D. Smith, A.H. Silver, R.W. Simon, Observation of 4.2- K equilibrium-noise squeezing via a Josephson-parametric amplifier. Phys. Rev. Lett. **60**, 764 (1988)
35. E.E. Wollman, C.U. Lei, A.J. Weinstein, J. Suh, A. Kronwald, F. Marquardt, A.A. Clerk, K.C. Schwab, Quantum squeezing of motion in a mechanical resonator. Science **349**, 952–955 (2015)
36. J.-M. Pirkkalainen, E. Damskägg, M. Brandt, F. Massel, M.A. Sillanpää, Squeezing of quantum noise of motion in a micromechanical resonator. Phys. Rev. Lett. **115**, 243601 (2015)
37. O. Abah, J. Roßnagel, G. Jacob, S. Deffner, F. Schmidt-Kaler, K. Singer, E. Lutz, Single-Ion heat engine at maximum power. Phys. Rev. Lett. **109**, 203006 (2012)
38. J. Roßnagel, S.T. Dawkins, K.N. Tolazzi, O. Abah, E. Lutz, F. Schmidt-Kaler, K. Singer, A single-atom heat engine. Science **352**, 325–329 (2016)
39. M.O. Scully, M.S. Zubairy, *Quantum Optics* (Cambridge University Press, Cambridge, 1997)
40. H. Spohn, Entropy production for quantum dynamical semigroups. J. Math. Phys. **19**, 1227–1230 (1978)
41. M. Esposito, C. Van den Broeck, Three detailed fluctuation theorems. Phys. Rev. Lett. **104**, 090601 (2010)
42. J.M. Horowitz, J.M.R. Parrondo, Entropy production along nonequilibrium quantum jump trajectories. New. J. Phys **15**, 085028 (2013)
43. J.M. Horowitz, T. Sagawa, Equivalent definitions of the quantum nonadiabatic entropy production. J. Stat. Phys. **156**, 55–65 (2014)
44. M.O. Scully, Extracting work from a single thermal bath via quantum negentropy. Phys. Rev. Lett. **87**, 220601 (2001)
45. D. Mandal, H.T. Quan, C. Jarzynski, Maxwell's refrigerator: an exactly solvable model. Phys. Rev. Lett. **111**, 030602 (2013)
46. S. Deffner, C. Jarzynski, Information processing and the second law of thermodynamics: an inclusive. Hamiltonian Approach. Phys. Rev. X **3**, 041003 (2013)
47. T.D. Kieu, The second law, Maxwell's Demon, and work derivable from quantum heat engines. Phys. Rev. Lett. **93**, 140403 (2004)

48. Y. Rezek, R. Kosloff, Irreversible performance of a quantum harmonic heat engine. New J. Phys. **8**, 83 (2006)
49. H.T. Quan, Y.-X. Liu, C.P. Sun, F. Nori, Quantum thermodynamic cycles and quantum heat engines. Phys. Rev. E **76**, 031105 (2007)
50. F. Galve, E. Lutz, Nonequilibrium thermodynamic analysis of squeezing. Phys. Rev. A **79**, 055804 (2009)
51. J.M.R. Parrondo, J.M. Horowitz, T. Sagawa, Thermodynamics of information. Nat. Phys. **11**, 131–139 (2015)
52. J. Janszky, P. Adam, Strong squeezing by repeated frequency jumps. Phys. Rev. A **46**, 6091–6092 (1992)
53. E. Massoni, M. Orszag, Squeezing transfer from vibrations to a cavity field in an ion-trap laser. Opt. Commun. **190**, 239–243 (2001)
54. A.M. Zagoskin, E. Il'ichev, F. Nori, Heat cost of parametric generation of microwave squeezed states. Phys. Rev. A **85**, 063811 (2012)
55. T. Sagawa, M. Ueda, Role of mutual information in entropy production under information exchanges. New J. Phys. **15**, 125012 (2013)

Chapter 11
Performance of Autonomous Quantum Thermal Machines

In Chap. 10 we investigated the effects of nonequilibrium squeezed quantum reservoirs in the performance of some non-autonomous quantum thermal machines. Those machines, in analogy to classical thermal machines, are operated in cycles consisting of different strokes implemented by an external agent who performs or extracts work. In contrast, the present chapter is devoted to the analysis of autonomous thermal machines, where all the components can be explicitly modeled using time-independent Hamiltonians, that is, without the need of any external driving [1–3].

An autonomous thermal machine consists of a set of quantum levels, some of which are selectively coupled to different thermal baths as well as to an object to be acted upon (see the example provided in Sect. 3.3). Various models of thermal baths and thermal couplings can be considered and formalized via master equations, which usually involve many different parameters, including coupling factors or bath spectral densities, to precisely characterize the machine and its interaction with the environment. In this context, the standard method is the derivation of a Lindblad master equation from the microscopic Hamiltonian that includes the interaction between the machine and various bosonic thermal reservoirs [1, 3–6]. Other approaches consider more phenomenological models, as e.g. the reset model proposed in Refs. [2, 7–9] (see also the autonomous heat engine introduced in Sect. 3.3). Nevertheless, the basic functioning of these machines can be captured in much simpler terms. In particular, the notion of 'virtual qubits' and 'virtual temperatures' [8] (see also Ref. [10]), essentially associating a temperature to a transition via its population ratio, was developed in order to capture the fundamental limitations of the simplest machines. Some of the features of the machine can be deduced from simple considerations about its *static* configuration, i.e. without requiring any specific knowledge of the dynamics of the thermalization process induced by contact with the baths. In the following, we discuss the performance of general autonomous thermal machines involving an arbitrary number of levels.[1]

[1]The results in this chapter have been published in Ref. [11].

G. Manzano Paule, *Thermodynamics and Synchronization in Open Quantum Systems*, Springer Theses, https://doi.org/10.1007/978-3-319-93964-3_11

Exploiting the notions of virtual qubits and virtual temperatures, we characterize fundamental limits of such machines, based on their level structure and the way they are coupled to the reservoirs. This allows us to explore the relation between the size of the machine, as given by its Hilbert space dimension (or equivalently the number of its available levels), and its performance. We find that machines with more levels can outperform simpler machines. In particular, considering fixed thermodynamic resources (two heat baths at different temperatures), we show that lower temperatures, as well as higher cooling power, can always be engineered using higher dimensional refrigerators. By characterizing the range of virtual qubits and virtual temperatures that can be reached with fixed resources, we propose optimal designs for single-cycle, multi-cycle and concatenated machines featuring an arbitrary number of levels. Furthermore, our considerations lead to a formulation of the third law in terms of the Hilbert space dimension of the machine: reaching absolute zero temperature requires infinite dimension.

This chapter is organized as follows. We begin in Sect. 11.1 by discussing the role of the swap operation as the primitive operation for the functioning of autonomous quantum thermal machines, allowing an extremely simple characterization of their performance in terms of virtual qubits and virtual temperatures. Section 11.2 is devoted to reviewing the basic functioning of a three-level quantum thermal machine, helping us to identify various *resources and limitations* when optimizing its design. Higher dimensional thermal machines are presented in Sect. 11.3, where we point out the existence of two different strategies for improving performance. The first strategy consists in adding energy levels to the original thermal cycle, and is analyzed in detail in Sect. 11.4, while the extension to the case of multi-cycle machines is presented in Sect. 11.5. The second strategy, based on concatenating three-level (or qutrit) machines, is analyzed in Sect. 11.6. Furthermore, in Sect. 11.7 we discuss the third law of thermodynamics in terms of Hilbert space dimension, while Sect. 11.8 is devoted to characterizing the trade-off between the power and speed of operation of the thermal machine, given an explicit model of thermalization. Our conclusions are presented in Sect. 11.9. Further details of the calculations are given in Appendix A.

11.1 The Primitive Operation

Generally speaking, the working of an autonomous quantum thermal machine can be divided into two steps which are continuously repeated. For clarity, we discuss the case of a fridge powered by two thermal baths at different temperatures. In the first step, a temperature colder than the cold bath is engineered on a subspace of the machine, i.e. on a subset of the levels comprising the machine. This can be done by selectively coupling levels in the machine to the thermal baths. The second step consists in interacting the engineered subspace with an external physical system to be cooled. We will consider a pair of levels of the machine to constitute our engineering subspace, the population ratio of which can be tuned in order to correspond to a cold temperature. Here we shall refer to this pair of levels as the *virtual qubit*, and

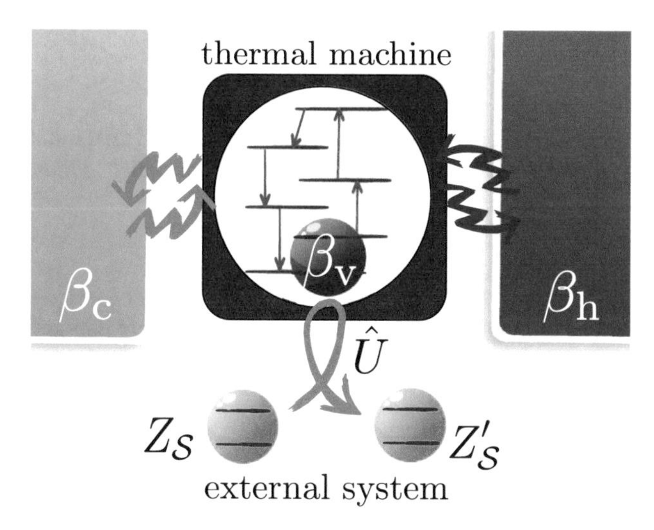

Fig. 11.1 The different transitions of a thermal machine comprising an arbitrary number of levels are selectively coupled to two thermal baths at inverse temperatures $\beta_c \geqslant \beta_h$ (blue and red boxes). This allows engineering an effective inverse temperature (the virtual temperature) β_v in a inner subspace of the machine (the virtual qubit). The virtual qubit (purple circle) then interacts via the unitary swap operation $\hat{U}$ with an external system (orange circle), changing its bias from Z_S to Z'_S in the operation

its associated temperature as its *virtual temperature* [8]. Typically the virtual qubit is chosen to be resonant with the system to be cooled in order to avoid non energy conserving interactions. Notably, the swap operation between the virtual qubit and the external physical system, can thus be considered as the primitive operation of quantum fridges, and more generally of all quantum thermal machines (see Fig. 11.1).

Let us consider a machine comprised of n levels, with associated Hilbert space $\mathcal{H}$ such that $\dim\mathcal{H} = n$, and Hamiltonian $\hat{H}_M$. Within this machine, we will refer to any pair of levels ($|k\rangle_M$ and $|l\rangle_M$) as a *transition*, denoted $\mathcal{T}_{k,l}$. Among the $n(n-1)/2$ possible transitions, we focus our attention on a particular pair of levels $|i\rangle$ and $|j\rangle$ with populations λ_i and λ_j and energies E_i and $E_j > E_i$. Assume the transition $\mathcal{T}_{i,j}$ is coupled to the external system to be cooled, hence representing the virtual qubit. Here it will be useful to introduce two quantities to fully characterize the virtual qubit, namely its normalization N_v and its (normalized) bias Z_v defined by

$$N_v \equiv \lambda_i + \lambda_j, \qquad Z_v \equiv \frac{\lambda_i - \lambda_j}{N_v}. \tag{11.1}$$

As we focus here on the case where the density operator of the machine is diagonal in the energy basis,[2] we may define the virtual temperature via the Gibbs relation $\lambda_j = \lambda_i e^{-E_v/k_B T_v}$. That is

[2]More generally, one could also consider the case of virtual qubits with coherences.

$$T_{\mathrm{v}} \equiv \frac{E_{\mathrm{v}}}{k_{\mathrm{B}} \ln \left(\lambda_i / \lambda_j\right)}, \tag{11.2}$$

where we defined $E_{\mathrm{v}} \equiv E_j - E_i$ as the energy gap of the virtual qubit. The virtual temperature is then monotonically related to the above introduced bias by

$$Z_{\mathrm{v}} = \tanh(\beta_{\mathrm{v}} E_{\mathrm{v}}/2), \tag{11.3}$$

where $\beta_{\mathrm{v}} = 1/k_{\mathrm{B}} T_{\mathrm{v}}$ is the inverse virtual temperature. Notice that $-1 \leqslant Z_{\mathrm{v}} \leqslant 1$, where the lower bound represents a virtual qubit with complete population inversion ($\beta_{\mathrm{v}} \to -\infty$) and the upper bound corresponds to the virtual qubit in its ground state $|i\rangle_M$ ($\beta_{\mathrm{v}} \to 0$).

Next, the virtual qubit interacts with the physical system via the swap operation. For simplicity, the physical system is taken here to be a qubit with energy gap E_{v}, hence resonant with the virtual qubit. We denote the levels of the physical system by $|0\rangle_{\mathbb{S}}$ and $|1\rangle_{\mathbb{S}}$, with corresponding populations p_0 and p_1, and hence bias $Z_{\mathbb{S}} = p_0 - p_1$ (note that $N_{\mathbb{S}} = 1$). The swap (energy-conserving) operation is given by the unitary

$$\begin{aligned} \hat{U} = \hat{\mathbb{1}}_{M\mathbb{S}} &- |i\rangle\langle i|_M \otimes |1\rangle\langle 1|_{\mathbb{S}} - |j\rangle\langle j|_M \otimes |0\rangle\langle 0|_{\mathbb{S}} \\ &+ |i\rangle\langle j|_M \otimes |1\rangle\langle 0|_{\mathbb{S}} + |j\rangle\langle i|_M \otimes |0\rangle\langle 1|_{\mathbb{S}}. \end{aligned} \tag{11.4}$$

The effect of the swap operation is to modify the bias of the physical system, which changes from $Z_{\mathbb{S}}$ to

$$Z'_{\mathbb{S}} = N_{\mathrm{v}} Z_{\mathrm{v}} + (1 - N_{\mathrm{v}}) Z_{\mathbb{S}}. \tag{11.5}$$

The above equation can be intuitively understood as follows. With probability N_{v}, the virtual qubit is available (i.e. the machine is in the subspace of the virtual qubit), and the swap replaces the initial bias of the system with the bias of the virtual qubit. With the complementary probability, $1 - N_{\mathrm{v}}$, the virtual qubit is not available, hence the swap cannot take place and the bias of the system remains unchanged. Consequently, the virtual temperature fundamentally limits the temperature the external system can reach. A complete derivation of Eq. (11.5) can be found in Appendix A.

Finally, it is worth noticing that the virtual qubit must be refreshed in order to ensure the continuous operation of the machine. Indeed, after interaction with the system, the virtual qubit is left with the initial bias of the system, $Z_{\mathbb{S}}$, and must be therefore reset to the desired bias, Z_{v}, in order to continue operating. Moreover, the setup can be straightforwardly generalized to the cooling of a higher dimensional system. For systems featuring a single energy gap, e.g. harmonic oscillators, the virtual qubit is coupled to all resonant transitions. For systems with several different energy gaps, one will use one virtual qubit for each different energy gap.

Within this picture two different directions to improve the performance of a machine emerge. The first consists in optimizing the properties of the virtual qubit (N_{v} and Z_{v}) in order to achieve the desired bias $Z'_{\mathbb{S}}$ in the external system ($Z'_{\mathbb{S}} \to 1$ in

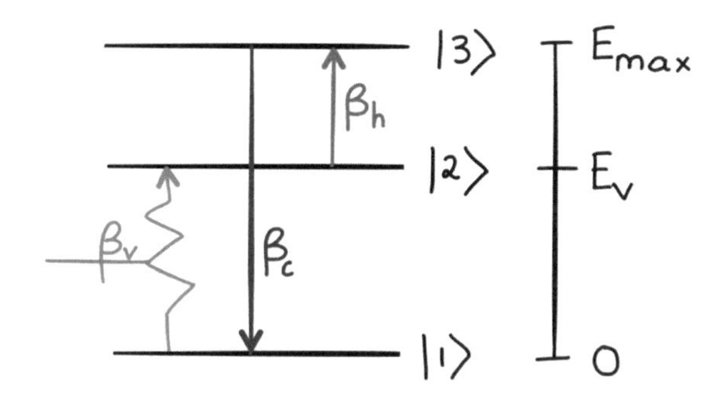

Fig. 11.2 The smallest possible fridge comprising three energy levels. We denote couplings to β_c by (blue) downward arrows, couplings to β_h by (red) upward arrows, and the virtual qubit by an (orange) arrow in the direction consistent with the machine (upward for the fridge, downward for the heat engine)

the case of a fridge), which represents the *statics* of the machine. The second consists in optimizing the dynamics of the machine, in particular the rate of interaction with the external system and the rate at which the virtual qubit is refreshed by contact with the thermal baths. Crucially, whereas the dynamics is model dependent, the statics is model independent, and hence a universal property of the machine.

In the following sections, we shall see how the performance of thermal machines can be optimized in the presence of natural constraints, such as limits on the available energy gaps or on the dimension of its Hilbert space. We focus primarily on the statics: we will see that increasing the number of levels of the machine improves the performance (for instance to be able to cool to lower temperatures). In the last sections, we will move beyond purely static considerations, and discuss the interplay between statics and dynamics. Again we find that machines with more levels can lead to enhanced performance.

11.2 Warm-Up: Three-Level Machine

In order to better illustrate the main concepts, we start our analysis with the smallest possible quantum thermal machine, comprising only three energy levels $|1\rangle_M$, $|2\rangle_M$ and $|3\rangle_M$, working between two thermal baths at different temperatures. This machine can be operated as a fridge or as a heat engine depending on which transitions are coupled to the hot and cold baths. For simplicity, our presentation will focus on the former. In this case, the transition $\mathcal{T}_{1,3}$ is coupled to the cold bath at inverse temperature β_c, while transition $\mathcal{T}_{2,3}$ is coupled to the hot bath at $\beta_h < \beta_c$. Finally, the transition $\mathcal{T}_{1,2}$ is chosen to be the virtual qubit (see Fig. 11.2).

The operation of the three-level fridge can be understood as a simple thermal cycle:

$$|2\rangle_M \xrightarrow{\beta_h} |3\rangle_M \xrightarrow{\beta_c} |1\rangle_M. \tag{11.6}$$

in which a quantum of energy $\Delta E_{23} \equiv E_3 - E_2$ is absorbed from the hot bath making the machine jump from state $|2\rangle_M$ to $|3\rangle_M$, followed by a jump from $|3\rangle_M$ to $|1\rangle_M$ while emitting a quantum of energy ΔE_{13} to the cold bath. The cycle is closed by swap of the virtual qubit, $\mathcal{T}_{1,2}$, with the external qubit to be cooled as described in Sect. 11.1. This cycle involves 3 states, and is thus of length 3. It represents the basic building block of the machine.

The fact that transitions $\mathcal{T}_{1,3}$ and $\mathcal{T}_{2,3}$ are coupled to baths at different temperatures will allow us to control the (inverse) temperature of the virtual qubit, β_v. While there exist many different possible models for representing the coupling to a thermal bath, the only feature that we will consider here is that, after sufficient time, each transition connected to a bath will thermalize. That is, in the steady-state of the machine, the population ratio of a transition $\mathcal{T}_{i,j}$ coupled to a thermal bath, will be equal to $e^{-\Delta E_{ij}\beta_{\mathrm{bath}}}$, where ΔE_{ij} is the energy gap of the transition, and β_{bath} the inverse temperature of the bath. Under such conditions, the inverse temperature of the virtual qubit and its norm are given by

$$\beta_v = \beta_c + (\beta_c - \beta_h)\left(\frac{\Delta E_{13}}{E_v} - 1\right), \tag{11.7}$$

$$N_v = \frac{1 + e^{-\beta_v E_v}}{1 + e^{-\beta_v E_v} + e^{-\beta_c \Delta E_{13}}} \tag{11.8}$$

where $E_v \equiv \Delta E_{12}$ is the virtual qubit energy gap, chosen to match the energy gap of the qubit to be cooled. Note that we have $\beta_v > \beta_c$ (since $\Delta E_{13} > E_v$), implying that the machine works as a refrigerator.

At this point, one can already identify various *resources* for the control of the virtual temperature β_v. The first is the range of available temperatures, captured by β_c and β_h. The second is the largest energy gap, ΔE_{13} coupled to a thermal bath. Clearly if ΔE_{13} is unbounded, then we can cool arbitrarily close to absolute zero, i.e. $\beta_v \to \infty$ as $\Delta E_{13} \to \infty$ while $N_v \to 1$, implying $Z'_{\mathcal{S}} \to 1$, c.f. Eq. (11.5). However, it is reasonable to impose a bound on this quantity, which we label $E_{\max}$. From physical considerations, one expects that thermal effects play a role only up to a certain energy scale. Indeed, as we have seen along Chap. 2 a thermal bath is characterized by a spectral density with a cutoff for high frequencies. This implies the existence of an energy above which there exist a negligible number of systems in the bath interacting with the machine. The coldest achievable temperature given this maximum energy is then given by

$$\beta_v = \beta_c + (\beta_c - \beta_h)\left(\frac{E_{\max}}{E_v} - 1\right). \tag{11.9}$$

As mentioned above, the three-level machine can also work as a heat pump or heat engine, if one switches the hot and cold baths. Imposing again a maximum energy gap, $E_{\max}$, we obtain the following lower bound in the inverse virtual temperature

$$\beta_{\rm v} = \beta_{\rm h} - (\beta_{\rm c} - \beta_{\rm h}) \left(\frac{E_{\rm max}}{E_{\rm v}} - 1 \right). \tag{11.10}$$

Notice that in this case $\beta_{\rm v} < \beta_{\rm h}$, that is, the virtual temperature is greater than the temperature of the hot bath. Moreover, when $\beta_{\rm c}/(\beta_{\rm c} - \beta_{\rm h}) < E_{\rm max}/E_{\rm v}$, then $\beta_{\rm v} < 0$, and the machine transitions from a heat pump to a heat engine (see the example in Sect. 3.3).

11.3 Multi-level Machines

We have seen that imposing a bound on the maximum energy gap which can be coupled to a heat reservoir, the performance of the simplest three-level machine becomes limited through the range of accessible virtual temperatures. The general question investigated below is whether we can engineer colder temperatures (or hotter ones, as well as achieving a greater population inversion) by using more sophisticated machines.

Following Eq. (11.5), in order to optimize the effect the machine has on the physical system, there are two important features the virtual qubit should have. First, it should have a high bias $Z_{\rm v}$. Second, the norm $N_{\rm v}$ should be as close to one as possible. Below we discuss different classes of multilevel machines, and investigate the range of available virtual qubits as a function of the number of levels n of the machine. First we will see that the range of accessible virtual temperatures (or equivalently bias $Z_{\rm v}$) increases as n increases. Hence machines with more levels allow one to reach lower temperatures, given fixed thermal resources. However, this usually comes at the price of having a relatively low norm $N_{\rm v}$ for the virtual qubit, which is clearly a detrimental feature. Nevertheless we will see that it is always possible to bring the norm back to one by adding extra levels.

We will discuss two natural ways to generalize the three-level machines to more levels, sketched in Fig. 11.3. The first one consists in adding levels and thermal couplings in order to extend the length of the thermal cycle. In other words, while the three-level machine represents a machine with one cycle of length three, we now consider machines with a single cycle of length n (see Sect. 11.4). This will allow us to improve both the bias $Z_{\rm v}$ and the normalization $N_{\rm v}$ of the virtual qubit. We first characterize the optimal single-cycle machine, which in the limit of large n, approaches perfect bias (i.e. zero virtual temperature, or perfect population inversion). However, while the norm $N_{\rm v}$ does not vanish, it is bounded away from one in this case. We then show how the norm can be further increased to one by extending the optimal single-cycle machine to a multi-cycle machine (Sect. 11.5). This procedures requires the addition of $n - 2$ levels, while maintaining the same bias $Z_{\rm v}$. In Fig. 11.4 we show the range of available virtual qubits (as characterized by its norm $N_{\rm v}$ and bias $Z_{\rm v}$) as a function of the number of levels n, for single cycle machines (green dots) and multi-cycle machines (blue dots).

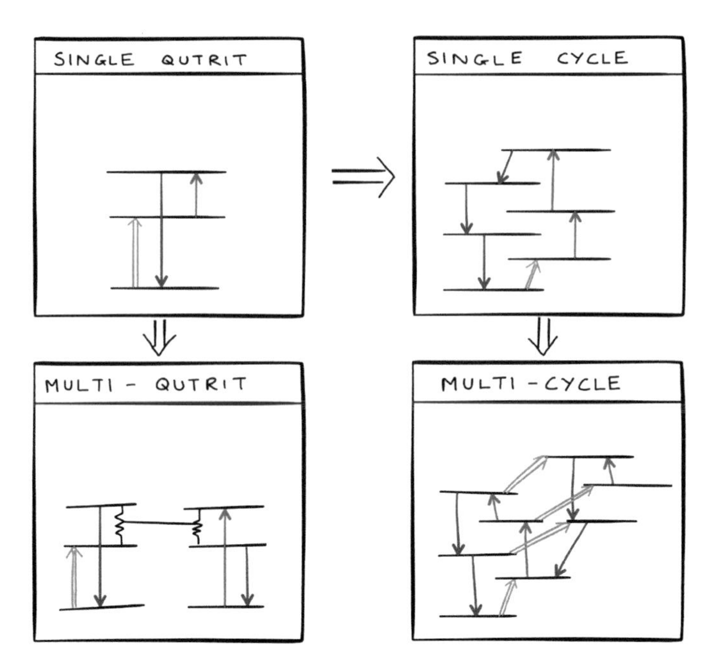

Fig. 11.3 Sketch of multi-level machines as discussed here. We consider several generalizations of the simplest three-level machine (top left). We first discuss single cycle machine (top right), which can then be extended to multi-cycle machines (bottom right). Second, we study concatenated three-level machines (bottom left)

Next, in Sect. 11.6, we follow a second possibility which consists in concatenating k three-level machines. The main idea is that the hot bath is now effectively replaced by an even hotter bath or source of work, engineered via the use of an additional three-level heat pump or heat engine. In the limit of k large, we can also approach a perfect bias Z_{v} and the norm N_{v} tends to one (see red dots on Fig. 11.4), similarly to the multi-cycle machine. It is however worth mentioning that in this case the machine has now $n = 3^k$ levels, while the multi-cycle machine used only a number of levels linear in n.

The above results, which are summarized in Fig. 11.4, demonstrate that machines with a larger Hilbert space can outperform smaller ones, which implies that the Hilbert space dimension should be considered a thermodynamical resource. Note that, for clarity, results are generally discussed for the case of fridges, but hold also for heat engines *mutatis mutandis*.

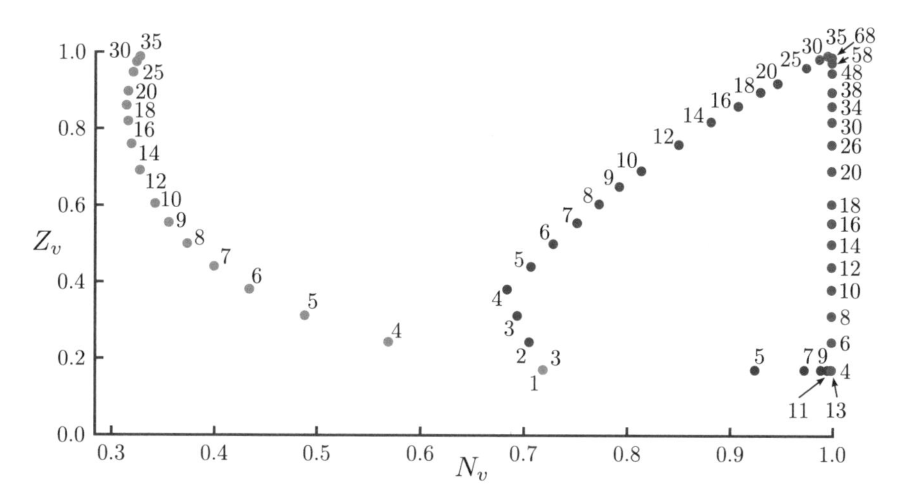

Fig. 11.4 Performance of machines as a function of dimension. The accessible virtual qubit, characterized by the bias Z_v and the norm N_v [see Eq. (11.1)], is shown for single cycle machine (green dots), multi-cycle machine (blue dots), and concatenated three-level machines (red dots). As a comparison we also show the machines discussed in Ref. [12] (purple dots). The dimension of the machine (i.e. the number of levels) is indicated next to each point, for all machines except the three-level; there, the number k of concatenated machines is given (hence the dimension is exponentially larger, 3^k)

11.4 Single-Cycle Machines

We start by discussing thermal machines featuring an arbitrary number of levels, n, but only a single thermal cycle. We then define a *n-level (thermal) cycle machine* as a quantum system with Hilbert space $\mathcal{H}_M$ of dimension n, and Hamiltonian $\hat{H}_M = \sum_{j=1}^{n} E_j |j\rangle\langle j|_M$, where every transition $\mathcal{T}_{j,j+1}$, is coupled to a thermal bath. It is worth mentioning that the levels $\{|j\rangle_M\}$, with $1 \leqslant j \leqslant n$, are not necessarily ordered with respect to its associated energies E_j. We further denote the energy gap of the transition $\mathcal{T}_{j,j+1}$ as $\Delta E_{j,j+1} = E_{j+1} - E_j$, and the temperature of the bath coupled to this transition is labeled as $\beta_{j,j+1}$. We choose the transition $\mathcal{T}_{1,n}$ to correspond to the virtual qubit of the machine, whose energy gap, E_v, obeys the following consistency relation

$$E_v = \sum_{j=1}^{n-1} E_{j+1} - E_j = \sum_{j=1}^{n-1} \Delta E_{j,j+1}. \tag{11.11}$$

In the absence of any additional couplings, the machine approaches a steady state, as each transition tends to equilibrate with the thermal bath to which it is coupled (notice also that each level is involved in at least one thermal coupling). This implies that the density matrix of the steady state must be diagonal in the energy basis, as all off-diagonal elements decay away due to the thermal interactions. Additionally, the populations of the two levels in each transition are given by the Gibbs ratio corresponding to the temperature of the bath. Labeling the population of the $|j\rangle_M$ state as p_j, we have

$$p_{j+1} = p_j \, e^{-\beta_{j,j+1}\Delta E_{j,j+1}} \quad \text{for } 1 \leqslant j \leqslant n-1. \tag{11.12}$$

The above $n-1$ thermal couplings determine the ratios between all of the populations $\{p_j\}$. Together with the normalization condition $\sum_j p_j = 1$, this completely determines the steady state of the machine.[3] The virtual temperature corresponding to transition $\mathcal{T}_{1,n}$ can hence be obtained from

$$e^{-\beta_v E_v} = \frac{p_n}{p_1} = \frac{p_n}{p_{n-1}} \frac{p_{n-1}}{p_{n-2}} \cdots \frac{p_2}{p_1}, \tag{11.13}$$

leading to the following result:

$$\beta_v = \sum_{j=1}^{n-1} \beta_{j,j+1} \frac{\Delta E_{j,j+1}}{E_v}. \tag{11.14}$$

Similarly one may calculate the norm of the virtual qubit,

$$N_v = \left(\frac{1 + e^{-\beta_v E_v}}{1 + \sum_{j=1}^{n-1} \prod_{k=1}^{k=j} e^{-\beta_{k,k+1}\Delta E_{k,k+1}}} \right). \tag{11.15}$$

We are interested in the best single cycle machine, that is, the one which using a limited set of *resources*, achieves the largest change in bias of the system acted upon, $Z'_S - Z_S$, as given in Eq. (11.5). This corresponds to the one that achieves the largest possible bias, Z_v, together with the largest norm, N_v, given this optimized bias. In what follows we determine the optimal single cycle machine with n levels, given bath temperatures and bound on the energy of a coupled transition $E_{\max}$.

11.4.1 Optimal Single-Cycle Machine

The optimal arbitrary single cycle fridge, sketched in Fig. 11.5, has a rather simple structure. All but one of its transitions are at the maximal allowed energy, $E_{\max}$. Roughly, the first half of the transitions (starting from the upper state of the virtual qubit) are all connected to the hot bath, while the second half of the transitions are connected to the cold bath. A complete proof of optimality can be found in Appendix A. Furthermore, explicit expressions for the inverse virtual temperature and norms in this case can be easily obtained from Eqs. (11.14) and (11.15). For the

[3] If the cycle covers only a subspace of all the machine levels, then the populations are determined with respect to the total population of the subspace.

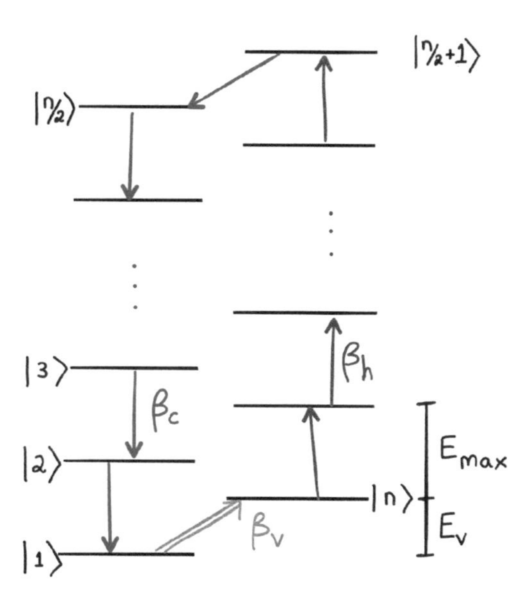

Fig. 11.5 Sketch of the optimal single-cycle refrigerator, for an even number of levels n

case of a refrigerator with an even number of levels n, they read

$$\beta_{\rm v}^{(n)} = \beta_{\rm c} + (\beta_{\rm c} - \beta_{\rm h}) \left(\frac{n}{2} - 1\right) \frac{E_{\max}}{E_{\rm v}} \tag{11.16}$$

$$N_{\rm v}^{(n)} = \frac{1 + e^{-\beta_{\rm v}^{(n)} E_{\rm v}}}{\frac{1-e^{-\frac{n}{2}\beta_{\rm c} E_{\max}}}{1-e^{-\beta_{\rm c} E_{\max}}} + \frac{1-e^{-\frac{n}{2}\beta_{\rm h} E_{\max}}}{1-e^{-\beta_{\rm h} E_{\max}}} e^{-\beta_{\rm v}^{(n)} E_{\rm v}}}, \tag{11.17}$$

while for an odd number of levels n:

$$\beta_{v}^{(n)} = \beta_{\rm c} + (\beta_{\rm c} - \beta_{\rm h}) \left[\left(\frac{n}{2} - \frac{1}{2}\right) \frac{E_{\max}}{E_{\rm v}} - 1\right], \tag{11.18}$$

$$N_{\rm v}^{(n)} = \left(1 + e^{-\beta_{\rm v}^{(n)} E_{\rm v}}\right) \Big[\left(1 - e^{-\beta_{\rm c} E_{\max}}\right)^{-1} \left(1 - e^{-\left(\frac{n+1}{2}\right)\beta_{\rm c} E_{\max}}\right) + e^{-\beta_{\rm v}^{(n)} E_{\rm v}} \left(1 - e^{-\beta_{\rm h} E_{\max}}\right)^{-1} \left(1 - e^{-\left(\frac{n-1}{2}\right)\beta_{\rm h} E_{\max}}\right) \Big]^{-1}. \tag{11.19}$$

The complete results including the case of heat engines are given in Appendix A.2.

Let us now discuss the performance of the optimal machine. As becomes apparent from Eq. (11.16), the number of levels n is clearly a thermodynamical resource, as it allows to reach colder temperatures. Indeed, one finds that the virtual temperature is improved by a fixed amount whenever two extra levels are added, leading to an enhancement

$$\left(\beta_{\rm v}^{(n+2)} - \beta_{\rm v}^{(n)}\right) E_{\rm v} = (\beta_{\rm c} - \beta_{\rm h})\, E_{\max}. \tag{11.20}$$

This relation encapsulates the interplay between the resources involved in constructing a quantum thermal machine - the range of available thermal baths $\{\beta_c, \beta_h\}$, the range of thermal interactions (E_{max}), and the number of levels n. Remarkably, as the inverse virtual temperature β_v increases linearly with n, one can engineer a virtual temperature arbitrarily close to absolute zero. Similarly, for a heat engine, one can obtain a virtual qubit with arbitrarily close to perfect population inversion. This is possible because as n increases, the norm of the virtual qubit does not decrease arbitrarily, but remains bounded below away from zero. Indeed from Eq. (11.17), the norm asymptotically approaches a finite value

$$\lim_{n\to\infty} N_v^{(n)} = \left(1 - e^{-\beta_c E_{max}}\right), \tag{11.21}$$

which is, interestingly, independent of both β_h and E_v.

Finally, we briefly comment on the efficiency [also often referred to as the *coefficient of performance* (COP)] of the optimal single cycle machine. Here we adopt the standard definition of the efficiency of an absorption refrigerator, that is, the ratio between the heat extracted from the object to be cooled and the heat extracted from the hot bath. This can be easily calculated by looking at a single complete cycle of the machine. Imagine that a quantum E_v of heat is extracted from the external qubit in the jump $|1\rangle_M \to |n\rangle_M$ produced by the swap operation. To complete the cycle, the following sequence of jumps must necessarily occur:

$$|n\rangle_M \xrightarrow{\beta_h} \cdots \xrightarrow{\beta_h} |n/2+1\rangle_M \xrightarrow{\beta_c} |n/2\rangle_M \xrightarrow{\beta_c} \cdots \xrightarrow{\beta_c} |1\rangle_M \tag{11.22}$$

where $n/2 - 1$ energy quanta E_{max} of heat are absorbed from the hot bath while releasing $n/2 - 1$ quanta E_{max} and one quantum E_v of heat to the cold bath. The efficiency is hence given by the following expression for the single-cycle COP:

$$\eta_{\text{fridge}}^{(n)} = \frac{E_v}{\left(\frac{n}{2} - 1\right) E_{max}} = \frac{\beta_c - \beta_h}{\beta_v^{(n)} - \beta_c}, \tag{11.23}$$

where the second equality follows by exploiting Eq. (11.16). Crucially, Eq. (11.23) corresponds to Carnot efficiency for an endoreversible absorption refrigerator that is extracting heat from a bath at the (inverse) temperature $\beta_v^{(n)} \geqslant \beta_c \geqslant \beta_h$. That is, if the object to be cooled (now an external bath) is infinitesimally above the temperature of the virtual qubit (such that the virtual qubit cools it down by an infinitesimal amount), then the efficiency (COP) of this process approaches the Carnot limit.

Note that such absorption refrigerators have the property that the COP decreases as the temperature of the cold reservoir drops. In the present case, since $\beta_v^{(n)}$ decreases linearly with n, the same dependence is found in the efficiency of the machine. Intuitively, this makes sense, since the amount of heat drawn from the hot bath (per cycle) increases linearly with n, while the heat extracted from the cold bath remains constant (see Fig. 11.5).

11.5 Multi-cycle Machines

We have seen that the optimal single cycle machine can enhance the virtual temperature by increasing the number of levels n, but this also results in obtaining a norm N_v relatively low. Hence, it is natural to ask if, by adding levels, the norm can be brought back to unity while keeping the same virtual temperature. Below we will see that this is always possible, and in fact, requires only (roughly) twice the number of levels.

For clarity, we illustrate the method starting from the three-level fridge, which has a virtual qubit whose norm is strictly smaller than 1. By adding a fourth level, we can achieve $N_v = 1$, while maintaining the bias, Z_v, as given by its virtual temperature in Eq. (11.7). The fourth level is chosen specifically so that $E_4 = E_v + E_{max}$, and the transition $\mathcal{T}_{2,4}$ is coupled to the cold bath, as schematically depicted in Fig. 11.6a. Hence by design, the new transition $\mathcal{T}_{3,4}$ has the same energy gap E_v as the original virtual qubit $\mathcal{T}_{1,2}$. Furthermore, one can verify that both transitions possess the same virtual temperature $\beta_v^{(3)}$. In fact one can identify two three-level fridge cycles at work in the new system, $\{|2\rangle_M \to |3\rangle_M \to |1\rangle_M\}$ and $\{|4\rangle_M \to |2\rangle_M \to |3\rangle_M\}$. As a consequence, one could also connect $\mathcal{T}_{3,4}$ to the external system that is to be cooled. Since the two transitions can be coupled at the same time to the external system, they will both contribute to the virtual qubit, which is now duplicated. The norm of the (total) virtual qubit is obtained by summing the populations of each transition ($\mathcal{T}_{1,2}$ *and* $\mathcal{T}_{3,4}$). As the two transitions include all four levels, we find that $N_v = p_1 + p_2 + p_3 + p_4 = 1$.

Alternatively, one could view the four level machine as consisting of two real qubits, $\mathcal{H}_M = \mathcal{H}_1 \otimes \mathcal{H}_2$, as in Fig. 11.6b. One of these real qubits corresponds to the virtual qubit, and hence it follows that its norm must be $N_v = 1$. We term this procedure the *virtual qubit amplification* of a single cycle machine. Next, we show explicitly how to perform the above construction starting from any n level single

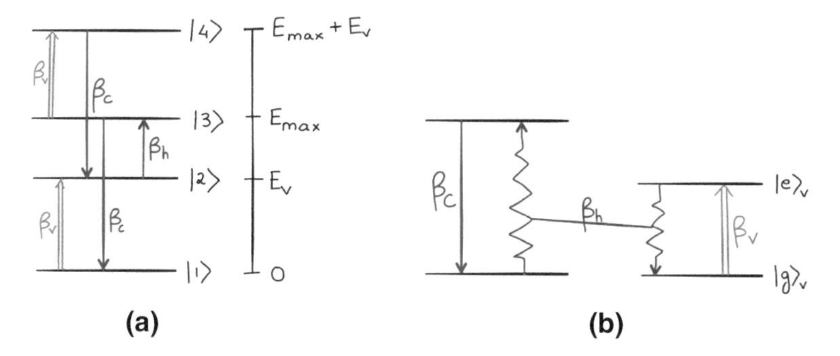

Fig. 11.6 Starting from the three-level fridge, and adding a fourth level $|4\rangle_M$, the norm of the virtual qubit can be increased to $N_v = 1$, while maintaining the same bias Z_v. This four-level fridge thus outperforms the three-level fridge. **b** The four-level fridge viewed as a tensor product of the virtual qubit, now becoming a real qubit since $N_v = 1$, and a simpler thermal cycle. Note the coupling to the hot bath is now nonlocal, between the levels $|0\rangle_M \otimes |e\rangle_v$ and $|1\rangle_M \otimes |g\rangle_v$

cycle machine. This requires the addition of $n-2$ levels. This is the most economical procedure possible, since the original n level cycle contains $n-2$ levels which do not contribute to the virtual qubit.

The general construction works as follows. Consider a single n-level thermal cycle machine as described in Sect. 11.4: a set of n levels with corresponding energies E_j $(1 \leqslant j \leqslant n)$, subsequent $n-1$ transitions coupled to thermal baths at corresponding inverse temperatures $\beta_{j,j+1}$, and virtual qubit $\mathcal{T}_{1,n}$, where $E_n - E_1 = E_{\text{v}}$. To amplify the virtual qubit, one now adds $n-2$ energy levels. Each new level is added in order to form a virtual qubit with each level of the original cycle except for the virtual qubit levels $|1\rangle_M$ and $|n\rangle_M$ [see Fig. 11.7]. The energy of the new levels must be chosen such that

$$E_{j+n-1} = E_j + E_{\text{v}}, \tag{11.24}$$

where j runs from 2 to $n-1$. The corresponding thermal couplings are chosen in such a manner that the structure of the cycle from $j=n$ to $j=2n-2$ is identical to the structure from $j=1$ to $j=n-1$. Specifically, this means choosing

$$\beta_{j+n-2,j+n-1} = \beta_{j-1,j}. \tag{11.25}$$

Following this procedure we finish with a Hilbert space for the machine $\mathcal{H}_M$ with total dimension $n' \equiv \dim\mathcal{H} = 2(n-1)$. One can verify that all the new virtual qubits $(\mathcal{T}_{1+j,n+j})$ have the same virtual temperature β_{v} as the original virtual qubit $\mathcal{T}_{1,n}$. None of these transitions share an energy level, i.e. they are mutually exclusive, and together they comprise all of the $2n-2$ levels present in the system. If every one of these transitions is connected together to the external system, then the effective virtual qubit reaches norm $N_{\text{v}} = 1$ as required. The inverse virtual temperature of the multi-cycle fridge can hence be expressed in terms of the total number of levels n'. For instance in the case of n even, we have:

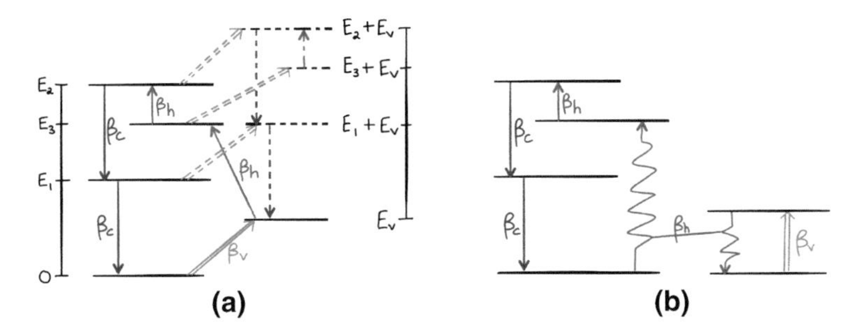

Fig. 11.7 **a** Starting from a 5 level fridge, and adding 3 levels (dashed lines), the norm of the virtual qubit can be boosted to $N_{\text{v}} = 1$ while maintaining the same bias Z_{v}. **b** The resulting 8 level fridge can be viewed as a tensor product of a 4–level cycle and the virtual qubit, which is now a real one since $N_{\text{v}} = 1$

$$\beta_{\mathrm{v}}^{(n')} = \beta_{\mathrm{c}} + (\beta_{\mathrm{c}} - \beta_{\mathrm{h}}) \left(\frac{n'}{4} - \frac{1}{2} \right) \frac{E_{\max}}{E_{\mathrm{v}}}. \tag{11.26}$$

Note that, as in the simple case of amplifying the three-level machine, here too the final machine can be viewed as a tensor product of an $n-1$ level cycle and the virtual qubit (which now becomes a real qubit since $N_{\mathrm{v}} = 1$). In fact, this procedure also allows one to easily convert a fridge into a heat engine and vice versa, as discussed in Appendix A.3. The virtual qubit amplification procedure is schematically depicted for the case of a 5-level fridge cycle in Fig. 11.7.

We also point out that the efficiency of the multi-cycle machine is exactly the same as that of the single cycle it is based upon. This follows from the fact the efficiency is determined by the bias Z_{v}, and does not depend on the norm N_{v}, c.f. Eq. (11.23) (see also Appendix A).

Finally, we note that Ref. [12] presents a different construction for a multi-cycle thermal machine. Compared to a three-level machine, this construction boosts the norm of the virtual qubit to $N_{\mathrm{v}} = 1$, but does not change the bias β_{v}. In comparison, our construction improves both the norm and the bias simultaneously and thus greatly outperforms the former construction, as shown on Fig. 11.4.

11.6 Concatenated Three-Level Machines

As we commented previously, a different possibility for generalizing the simplest three-level machine consists in concatenating several three-level machines. Here we analyze this possibility by characterizing the virtual qubits achievable by concatenating k three-level machines as introduced in Sect. 11.2.

For simplicity we start with case of concatenating $k = 2$ three-level machines in order to obtain a better fridge. The coupling between the two three-level machines can be achieved considering a simple swap Hamiltonian coupling the transitions $\mathcal{T}_{2,3}^{(1)}$ and $\mathcal{T}_{2,3}^{(2)}$:

$$\hat{H}_{\mathrm{int}} = g(|2, 3\rangle\langle 3, 2|_M + \mathrm{h.c.}), \tag{11.27}$$

as shown on Fig. 11.8. Here the first three-level machine represents the actual fridge while the second one works as a heat engine, replacing the hot bath on the transition $\mathcal{T}_{2,3}^{(1)}$. This corresponds to coupling $\mathcal{T}_{2,3}^{(1)}$ to an effective temperature which is hotter than the temperature of the hot bath (or equivalently inverse temperature lower than β_{h}), resulting in a fridge with an improved bias Z_{v}. Indeed the inverse virtual temperature achieved by the concatenated three-level machine is found to be

$$\beta_{\mathrm{v}}^{(2)} = \beta_{\mathrm{c}} + (\beta_{\mathrm{c}} - \beta_{\mathrm{h}}) \frac{E_{\max}}{E_{\mathrm{v}}}, \tag{11.28}$$

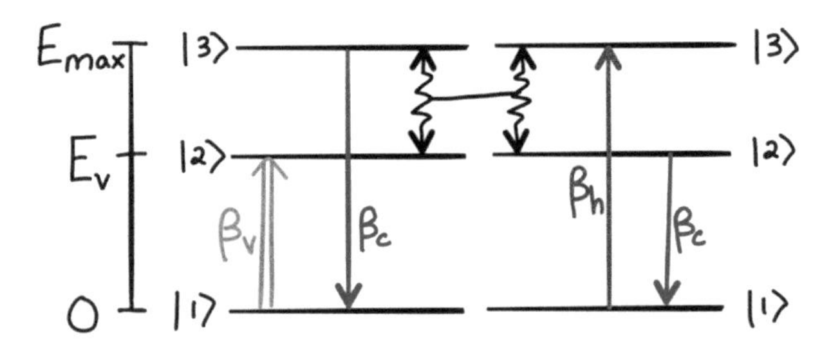

Fig. 11.8 By concatenating two three-level machines, one obtains a better fridge, outperforming the simple three-level fridge. Specifically, the new 6-level machine consists now a three-level fridge (left) which is boosted via the use of a three-level heat engines (right). The role of this heat engine is to create an effectively hotter temperature (hotter than T_h) in order to fuel the fridge

which is colder than the virtual temperature of the simple three-level fridge [see Eq. (11.9)]. Importantly, this enhancement has been achieved without modifying the value of E_{max}, and considering the same temperatures β_c and β_h for the thermal baths. Details about calculations are given in Appendix A.4.

The process may now be iterated, replacing the coupling of $\mathcal{T}^{(2)}_{2,3}$ to the cold bath β_c by a coupling to a third three-level fridge, effectively at a temperature colder than β_c, and so on, as sketched in Fig. 11.9. In this manner one can construct a machine resulting of the concatenation of k three-level machines. Following calculations given in Appendix A.4, we obtain simple expressions for the virtual inverse temperatures

$$\beta_v^{(k)} = \begin{cases} \beta_c + (\beta_c - \beta_h)\frac{k}{2}\frac{E_{max}}{E_v} & \text{if } k \text{ is even,} \\ \beta_c + (\beta_c - \beta_h)\left(\frac{k+1}{2}\frac{E_{max}}{E_v} - 1\right) & \text{if } k \text{ is odd.} \end{cases} \tag{11.29}$$

Again, we see that the virtual temperature approaches absolute zero as k becomes large. Similarly for a concatenated heat engine, one can approach perfect inversion (see details in Appendix).

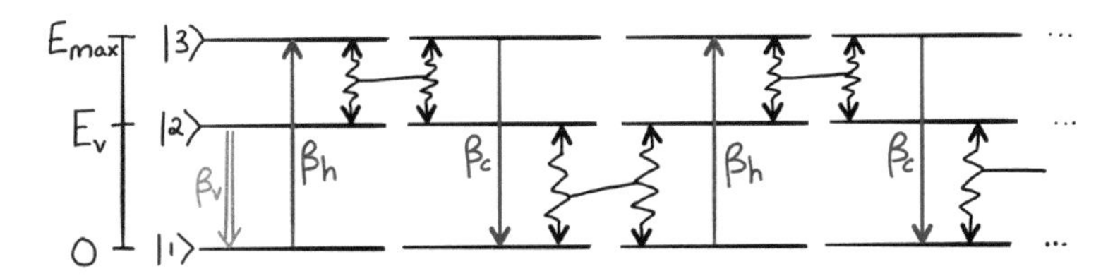

Fig. 11.9 Concatenating many three-level machines

Note that the above expressions are similar to those obtained for the virtual temperature in the case of the single cycle machine. In particular setting $k = n - 2$ we obtain exactly the same result. This correspondence can be intuitively understood via the following observations. First, the single three-level machine is the same as a 3-level cycle. Furthermore, the effect of replacing one of the thermal couplings in a three-level machine by a coupling to an additional three-level system effectively replaces one thermal coupling by two, thus increasing the number of thermal interactions within the working cycle by one. For example, in the two three-level fridge (Fig. 11.8), the effective thermal cycle is

$$|2,2\rangle_M \xrightarrow{\beta_c} |2,1\rangle_M \xrightarrow{\beta_h} |2,3\rangle_M \xrightarrow{H_{int}} |3,2\rangle_M \xrightarrow{\beta_c} |1,2\rangle_M. \tag{11.30}$$

Although this is a cycle of length 5, the virtual temperature is only influenced by the 3 thermal couplings, because the coupling on the degenerate transition $|2,3\rangle_M \leftrightarrow |3,2\rangle_M$ has zero energy gap [see Eq. (11.14)]. Since the thermal couplings are the same as those in the optimal 4-level fridge single cycle, we get the same virtual temperature. By induction, the concatenation of k three-level machines has the same β_v (and indeed the same thermal couplings within its working cycle) as the optimal $(k+2)$-level single cycle.

Finally, it is also important to discuss the behavior of the norm N_v of the virtual qubit in order to characterize the performance of the concatenated machine. Interestingly we find that $N_v \to 1$ in the limit of large k. This can be intuitively understood for the case of the concatenated heat engine, depicted in Fig. 11.9. As k becomes large, the virtual temperature β_v approaches $-\infty$. Thus the population ratio $p_1/p_2 \to 0$, implying that $p_1 \to 0$. However, since $\mathcal{T}_{1,3}^{(1)}$ is coupled to a thermal bath at β_h, the population ratio p_3/p_1 equals $e^{-\beta_h E_{max}}$, implying that $p_3 \to 0$. Thus in the limit $k \to \infty$, the state of the first three-level system approaches the pure state $|2\rangle\langle 2|_1$, and thus $N_v = p_1 + p_2 \to 1$. To understand the case of the fridge, consider in Fig. 11.9 that the machine begins with the second three-level system instead of the first one. This is now a fridge, where the virtual qubit is the transition $\mathcal{T}_{2,3}^{(2)}$. By a similar analysis to the above, we find that the state of the three-level system approaches $|2\rangle\langle 2|_2$ in the limit $k \to \infty$, and thus $N_v \to 1$. It is instructive to observe that in both cases, the concatenation of three-level machines takes the state of the original three-level system closer to the state where all of the population is in the middle level $|2\rangle\langle 2|$, which is both the ideal fridge with respect to $\mathcal{T}_{2,3}$, and the ideal machine with respect to $\mathcal{T}_{1,2}$.

Therefore we can conclude that, again, increasing the number of levels, or equivalently the dimension of the machine Hilbert space, $n \equiv \dim\mathcal{H}_M = 3^k$, the performance is increased. Indeed, as k increase, the virtual qubit bias approaches $Z_v = 1$ (or $Z_v = -1$ for a heat engine), while its norm becomes maximal, i.e. $N_v \to 1$. However notice that in this case the dimension of the machine grows rapidly. Indeed the inverse virtual temperature now grows only logarithmically with the total number of levels, n. For instance when k is even we have:

$$\beta_{\mathrm{v}}^{(n)} = \beta_{\mathrm{c}} + (\beta_{\mathrm{c}} - \beta_{\mathrm{h}}) \left(\frac{\log_3 n}{2} \right) \frac{E_{\max}}{E_{\mathrm{v}}} \tag{11.31}$$

to be compared with the multi-cycle fridge case in Eq. (11.26).

11.7 Third Law

The above results show that when the dimension of the Hilbert space of the thermal machine tends to infinity, the virtual temperature can approach absolute zero even though the maximal energy gap which is coupled to a thermal bath is finite. Nevertheless, an important point is that, in all the constructions given, for any finite n, the lowest possible temperature is always strictly greater than zero. This can be directly seen from the expressions for the inverse virtual temperature of the optimal single-cycle machines, as given in Eq. (11.16). Therefore any single-cycle fridge requires an infinite number of levels in order to cool to absolute zero.

Next, we notice that the lowest temperatures of any other multi-cycle machine with different virtual qubits working in parallel can achieve is bounded by the temperature achieved in any of these cycles. This follows from the fact that the effect of multiple cycles on the virtual qubit can be decomposed as a sum of the effect of each individual cycle. Thus, the bound on the temperature we derive for single-cycle n level machines holds for general machines with n levels.

Therefore we obtain a statement of the third law in terms of Hilbert space dimension. In particular, from (11.5) we see that the bias Z_{v} (and therefore temperature) and norm N_{v} of the virtual qubit determine to what temperature an external object can be cooled down in a single (or multiple) cycles of a thermal machine. The fact that the virtual temperature only approaches zero as the dimension of the thermal machine approaches infinity shows that bringing an external object to absolute zero requires a machine with an infinite number of levels. This is a static version of the third law, complementary to previous statements [3, 13, 14], which is stated in terms of number of steps, time, or energy required in order to reach absolute zero.

Finally, we note that in the case of the multi-cycle machine, since the norm of the virtual qubit is unity, in a single swap operation the external object is brought to exactly the temperature of the virtual qubit, c.f. Eq. (11.5). Thus, using a machine of Hilbert space dimension n, we can cool an external object to the inverse temperature (11.26), which corresponds asymptotically to the scaling

$$T_{\mathrm{S}} \sim \frac{1}{n}, \tag{11.32}$$

that is, the temperature scales inversely with the Hilbert space dimension.

11.8 Statics Versus Dynamics for Single-Cycle Machines

So far, we have discussed improving the *static* configuration of the thermal machine by increasing its dimension. This analysis characterizes the task of cooling (or heating) an external system via a single swap, a so-called *single shot* thermodynamic operation. However, more generally we are interested in continuously cooling the external system, as the latter is unavoidably in contact with its own environment, and thus requires repeated swaps with the virtual qubit in order to maintain the cooling (or heating) effect.

We have seen in Sect. 11.1 that after a single swap between the virtual qubit and the external system, the bias of the virtual qubit Z_v is switched with that of the external system Z_S. Thus the virtual qubit needs to be 'reset' before the next interaction is possible, an operation which should require some time to be performed. This fact hence introduces limitations on the power of the machines. The 'time of reset' will depend in general on the thermalization model, which forces us to go beyond purely static considerations. To illustrate this point we will discuss here the dynamics of the single-cycle refrigerators.

Intuitively one may expect the time of reset of the virtual qubit increases as the number of levels in the cycle increases, i.e. the larger the cycle of the machine, the longer it takes the machine to perform the series of jumps reinitializing it. This introduces the following trade-off. Previously we saw that machines with longer cycles were able to achieve lower temperatures for a single swap. However, they would also take longer to reset. Therefore in order to engineer a good fridge, one could consider (i) a high dimensional fridge (i.e. a long cycle) achieving low temperatures

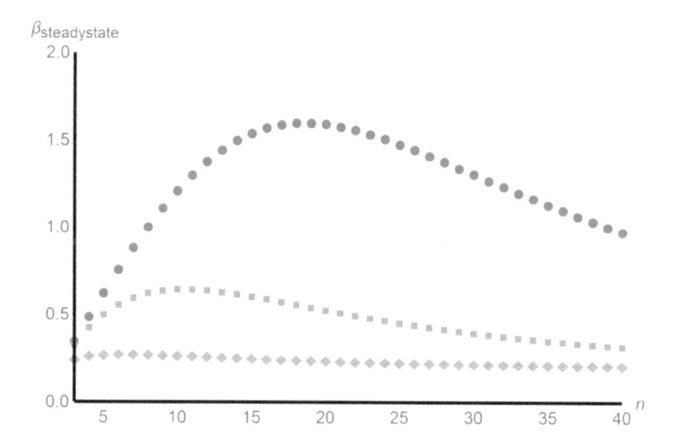

Fig. 11.10 Relationship between the steady-state virtual temperature $\beta_{steadystate}$ and the length of the cycle n. We consider various equilibration timescales, $\tau_S = 1$ (green, diamond), $\tau_S = 10$ (orange, square) and $\tau_S = 100$ (blue, dot). All other parameters are kept fixed: timescale of all thermal couplings of the cycle $\tau_\beta = 1$, bath temperatures $\beta_h = 0.05$, $\beta_c = 0.2$, and energies $E_{max} = 2$, and $E_v = 1$ (as in Fig. 11.4)

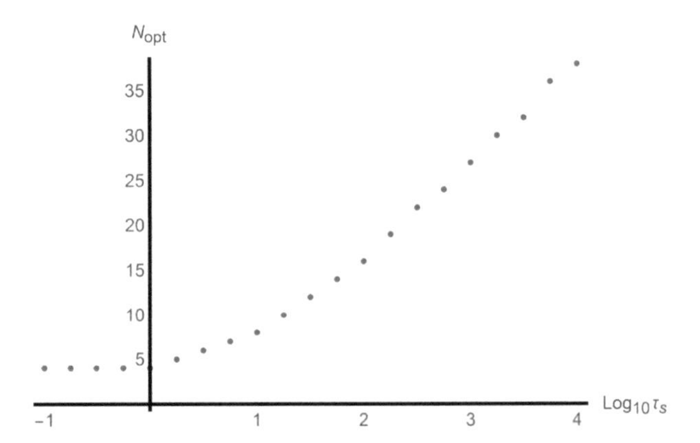

Fig. 11.11 Length of the optimal cycle N_{opt} versus the external system equilibration timescale τ_S in logarithmic units. Other parameters are the same as in Fig. 11.10: $\tau_\beta = 1$, $\beta_h = 0.05$, $\beta_c = 0.2$, $E_{\max} = 2$, and $E_v = 1$

at slower rate, or (ii) a low-dimensional fridge achieving not as low temperatures, but at a faster rate.

In order to find out which regime is better, we consider single-cycle fridges coupled to thermal baths, as modeled by a Markovian master equation in Lindblad form (see Sect. 2.2). Since the thermalization occurs here only on transitions, the specific details of the model are not crucial, and all models (either simple heuristic ones [2, 7–9] or those derived explicitly by microscopic derivations [1, 3–6]) lead to the same qualitative conclusions.

We find that the relevant parameter is the timescale at which the external system interacts with its environment τ_S. If this timescale is short, then the fridge has little time to 'reset' the virtual qubit. Therefore a shorter cycle, that resets quickly, is optimal in this case. If on the contrary the system timescale is long, there is more time available in order to reset the virtual qubit. Thus a longer cycle, providing lower temperatures, is preferable. This trade-off is illustrated in Fig. 11.10.

We also observe from Fig. 11.10 that, for given timescale τ_S, there is an optimal length of the cycle. In Fig. 11.11, we plot the optimal length of the cycle for different timescales. The optimal length appears to grow logarithmically with respect to τ_S for slow timescales (compared with the relaxation time-scale of the machine transitions τ_β), while for fast timescales we observe that the optimal cycle has length 4. This suggests that the simplest three-level machine is always outperformed.

11.9 Conclusions

We discussed the performance of quantum absorption thermal machines, in particular with respect to the size of the machine. Specifically, we considered several designs of machines with n levels and described the static properties of the machine through the range of available virtual qubits, which characterizes the fundamental limit of the machine. Notably, as n increases, a larger range of virtual temperatures becomes available, showing that a machine with $n + 1$ levels can outperform a machine with n levels. In this context, we proposed a design for quantum thermal machines consisting in a single cycle reaching optimal virtual temperatures for a virtual qubit consisting in a single transition of the machine.

Moreover, we also discussed machines with multiple cycles running in parallel. Here performance is increased, as the norm of the virtual qubit can be brought to one, i.e. the virtual qubit becomes a real one. Finally, similar performance is achieved for a design based on the concatenation of the simplest three-level machine. While generally suboptimal in terms of performance, this design gives nevertheless a more intuitive picture and may be more amenable to implementations, as the couplings are simpler.

Furthermore, we have seen that in order to achieve virtual qubits with perfect bias (i.e. achieving a virtual qubit at zero temperature, or with complete population inversion), the required number of levels n of the machine diverges. This can be viewed as a statement of the third law, complementary to previous ones. Usually stated in terms of number of steps, time, or energy required in order to reach absolute zero temperature, we obtain here a statement of the third law in terms of Hilbert space dimension: reaching absolute zero requires infinite dimension.

Finally, we discussed the dynamical performance of autonomous thermal machines, where a trade-off between the achievable virtual temperatures in the machine and the time needed to operate emerges. By studying the particular case of single cycle machines operating on a external qubit coupled to its own environment, we found qualitative results on the scaling of the optimal length of the cycle in terms of the external qubit relaxation timescale.

An outstanding question left open here concerns the performance of machines where multiple single cycle machines or three-level machines run in *parallel*, i.e. are coupled simultaneously to the external system. One may expect that the time necessary to reset the machine is considerably decreased, providing potentially a strong advantage over single-cycle machines. In particular, it would be interesting to understand how to design the most effective machine, given a fixed number of levels or equivalent building blocks (as well as constraints on the energy and temperatures).

Appendix

A.1 Swap Operation Details

This appendix elaborates on the swap as the primitive operation of quantum thermal machines as introduced in Sect. 11.1 (see Fig. 11.1). Consider the setup involving an external qubit system $\{|0\rangle_S, |1\rangle_S\}$ with energy gap E_v, populations p_0 and p_1, and hence bias $Z_S = p_0 - p_1$. In order to modify the bias Z_S (e.g. to cool the system), the external qubit interacts with a virtual qubit (i.e. a transition $\mathcal{T}_{i,j}$ comprising the levels $|i\rangle_M$ and $|j\rangle_M$ within the thermal machine) which has the same energy gap as the system, $E_v = E_j - E_i$. The energy-conserving 'swap' interaction is described by the unitary

$$\hat{U} = \mathbb{1}_{MS} - |i\rangle\langle i|_M \otimes |1\rangle\langle 1|_S - |j\rangle\langle j|_M \otimes |0\rangle\langle 0|_S \\ + |i\rangle\langle j|_M \otimes |1\rangle\langle 0|_S + |j\rangle\langle i|_M \otimes |0\rangle\langle 1|_S. \quad \text{(A.1)}$$

The effect of the swap upon two real qubits would be to swap the states of the qubits for one another (assuming the initial state as diagonal and uncorrelated). However, this is not the case for one real and one virtual qubit, as we show presently.

We assume that the real qubit begins in a diagonal state

$$\rho_S \equiv \frac{1+Z_S}{2}|0\rangle\langle 0|_S + \frac{1-Z_S}{2}|1\rangle\langle 1|_S. \quad \text{(A.2)}$$

For the virtual qubit the sum of the populations is not 1 in general, as the thermal machine comprises many levels in thermal contact with the heat reservoirs, that is $N_v = p_i + p_j < 1$. Assuming that the thermal machine state is block diagonal in the virtual qubit subspace

$$\rho_M \equiv N_v \left(\frac{1+Z_v}{2}|i\rangle\langle i|_M + \frac{1-Z_v}{2}|j\rangle\langle j|_M \right) + (1-N_v)\rho'_M,$$

where ρ'_M is an arbitrary (normalized) state of the remaining levels in the machine. After applying $\hat{U}$, the final state of the external qubit system and the machine containing the virtual qubit is

$$\begin{aligned} \hat{U}\rho_S \otimes \rho_M \hat{U}^\dagger &= \left(\frac{1+Z_S}{2}\right) N_v \left(\frac{1+Z_v}{2}\right) |0\rangle\langle 0|_S \otimes |0\rangle\langle 0|_M \\ &+ \left(\frac{1-Z_S}{2}\right) N_v \left(\frac{1+Z_v}{2}\right) |0\rangle\langle 0|_S \otimes |1\rangle\langle 1|_M \\ &+ \left(\frac{1+Z_S}{2}\right) N_v \left(\frac{1-Z_v}{2}\right) |1\rangle\langle 1|_S \otimes |0\rangle\langle 0|_M \\ &+ \left(\frac{1-Z_S}{2}\right) N_v \left(\frac{1-Z_v}{2}\right) |1\rangle\langle 1|_S \otimes |1\rangle\langle 1|_M \\ &+ (1-N_v)\rho_S \otimes \rho'_M, \end{aligned} \quad \text{(A.3)}$$

from which the final reduced state of the system is

$$\begin{aligned} \rho'_S = &\left[N_v \left(\frac{1+Z_v}{2}\right) + (1-N_v)\left(\frac{1+Z_S}{2}\right) \right] |0\rangle\langle 0|_S \\ + &\left[N_v \left(\frac{1-Z_v}{2}\right) + (1-N_v)\left(\frac{1-Z_S}{2}\right) \right] |1\rangle\langle 1|_S. \end{aligned} \quad \text{(A.4)}$$

Therefore, at the end of the protocol, the bias of the external qubit has been modified to $Z'_{\mathbb{S}} = N_{\mathrm{v}} Z_{\mathrm{v}} + (1 - N_{\mathrm{v}}) Z_{\mathbb{S}}$, which implies a change

$$\Delta Z_{\mathbb{S}} \equiv Z'_{\mathbb{S}} - Z_{\mathbb{S}} = N_{\mathrm{v}} \left(Z_{\mathrm{v}} - Z_{\mathbb{S}} \right). \tag{A.5}$$

A.2 Optimalily Proof for Single Cycle Machines

In this section we prove optimality of the single cycle machine discussed in Sect. 11.4. While there are several ways in which performance could be discussed, we are mainly concerned here with optimality under the swap operation as given by Eq. (A.5). That is, which machine achieves the largest change (A.5) in the bias of the external system acted upon.

Consider an n-level (thermal) cycle machine as introduced in Sect. 11.4, where all transitions are coupled to available temperatures, namely

$$\beta_{\mathrm{h}} \leqslant \beta_{j,j+1} \leqslant \beta_{\mathrm{c}}. \tag{A.6}$$

Notice that intermediate (inverse) temperatures in between β_{c} and β_{h} can be obtained by coupling the corresponding machine transition to both baths at the same time. Furthermore, the energy gaps of the transitions are bounded by

$$-E_{\max} \leqslant \Delta E_{j,j+1} \leqslant E_{\max}. \tag{A.7}$$

We then proceed to determine the unique n-level cycle that *minimizes the ratios of the population of every level j in the cycle with respect to one of the levels of the virtual qubit*. This is then proven to be the optimal cycle. For clarity we detail the proof for the case of the fridge, while the proof for the heat engine follows in a similar way. Let us start by considering the population ratio:

$$\frac{p_j}{p_1} = \prod_{k=1}^{j-1} e^{-\beta_{k,k+1} \Delta E_{k,k+1}} = \exp\left[-\sum_{k=1}^{j-1} \beta_{k,k+1} \Delta E_{k,k+1} \right]. \tag{A.8}$$

To minimize this ratio, one should maximize the summation above. We notice that regardless of the values of any energy gap of the machine transitions $\Delta E_{k,k+1}$, maximizing the sum requires picking the highest possible inverse temperature, β_{c}, if the energy gap is positive, and the smallest possible inverse temperature, β_{h}, if the energy gap is negative. Imposing this condition, one can then collect together the positive and negative energy gaps to simplify the expression. Indeed by labeling the sum of the positive energy gaps as Q_+^j and the sum of the negative ones as Q_-^j, we obtain

$$\frac{p_j}{p_1} = \exp\left[-\left(\beta_{\mathrm{c}} Q_+^j + \beta_{\mathrm{h}} Q_-^j \right) \right]. \tag{A.9}$$

Table A.1 Transition number and size to maximize the heat current Q_+^j associated to an arbitrary level $|j\rangle_M$ with respect to the first energy level $|1\rangle_M$, within a thermal cycle

No. transitions	$+E_{\max}$	$+\delta_j$	$-(E_{\max}-\delta_j)$	$-E_{\max}$
j, m even or odd	$\frac{j+m}{2}-1$	1	0	$\frac{j-m}{2}-1$
j, m opposite parity	$\frac{j+m-1}{2}$	0	1	$\frac{j-m-3}{2}$

In addition, we have the consistency relation

$$Q_+^j + Q_-^j = \Delta E_{1,j} = \sum_{k=1}^{j-1} \Delta E_{k,k+1}. \tag{A.10}$$

This leads to rewrite Eq. (A.9) as

$$\frac{p_j}{p_1} = \exp\left[-\beta_h \Delta E_{1,j} - (\beta_c - \beta_h)\, Q_+^j\right]. \tag{A.11}$$

The ratio (A.9) can hence be minimized in two steps: first we find the optimum Q_+^j for a fixed value of $\Delta E_{1,j}$, and then we optimize over $\Delta E_{1,j}$. For a fixed energy gap $\Delta E_{1,j}$, the minimum ratio is achieved when Q_+^j is as large as possible (since $\beta_c - \beta_h > 0$). Recall that Q_+^j is the sum of positive transitions in the cycle from $|1\rangle_M$ to $|j\rangle_M$, each of which are bounded by $E_{\max}$. Furthermore, also the number of transitions at $E_{\max}$ is limited by the consistency relation (A.10). Optimizing for Q_+^j subject to these constraints results in values for the sizes and number of transition in the cycle in the Table A.1, where we took a fixed $\Delta E_{1,j} = m E_{\max} + \delta_j$, being $m \equiv \Delta E_{1,j} \mod E_{\max}$.

In spite of the dependence on the optimum current Q_+^j upon the relative parities of j and m, it is straightforward to verify that the optimum Q_+^j increases monotonically w.r.t. $\Delta E_{1,j}$. Thus to complete the minimization of (A.11), one has to maximize $\Delta E_{1,j}$. This proceeds in an analogous manner to the optimization of Q_-^j, with the major difference being that $\Delta E_{1,j}$ must be chosen keeping in mind the consistency condition in Eq. (11.11). The result is summarized in Table A.2.

Table A.2 Transition number and size to minimize the population ratio (A.11) of an arbitrary level of the machine $|j\rangle_M$ with respect to the first energy level $|1\rangle_M$, within a thermal cycle

No. transitions	$+E_{\max}$	$+E_v$	$-(E_{\max}-E_v)$	$-E_{\max}$
$j \leqslant \frac{n}{2}$	$j-1$	0	0	0
$j > \frac{n}{2}$, n even	$\frac{n}{2}-1$	1	0	$j-\frac{n}{2}$
$j > \frac{n}{2}$, n odd	$\frac{n-1}{2}$	0	1	$j-\frac{n+1}{2}$

This completes the optimization of the ratio p_j/p_1. From Table A.2 we see that there is a unique construction of the n-level cycle that simultaneously fulfils the optimization criteria for all j: for all $j \leqslant n/2$ fix all of the transitions to be $E_{\max}$, next fix a transition to be $E_{\rm v}$ or $-(E_{\max} - E_{\rm v})$, depending on the parity of n, and continue with all the remaining transitions fixed to be $-E_{\max}$. Finally, connecting all positive transitions to $\beta_{\rm c}$ and negative transitions to $\beta_{\rm h}$, one arrives at the optimal n-level cycle fridge, schematically depicted in Fig. 11.5. If we instead minimize the ratios of populations to the excited state of the virtual qubit (p_j/p_n), we obtain the optimal n-level cycle heat engine, which has the same arrangement of energy levels as the fridge, but swapped temperatures ($\beta_{\rm c} \leftrightarrow \beta_{\rm h}$).

For completeness, we report the virtual temperatures $\beta_{\rm v}^{(n)}$ achieved by the optimal n-level cycle fridge and heat engine, together with its corresponding normalizations $N_{\rm v}^{(n)}$. For the fridge configuration we obtain for n even

$$\beta_v^{(n_{\rm even})} E_{\rm v} = \beta_{\rm c} E_{\rm v} + (\beta_{\rm c} - \beta_{\rm h}) \left(\frac{n}{2} - 1\right) E_{\max}, \tag{A.12}$$

$$N_{\rm v}^{(n_{\rm even})} = \left(1 + e^{-\beta_{\rm v}^{(n)} E_{\rm v}}\right) \Big[\left(1 - e^{-\beta_{\rm c} E_{\max}}\right)^{-1} \left(1 - e^{-\frac{n}{2}\beta_{\rm c} E_{\max}}\right) + e^{-\beta_{\rm v}^{(n)} E_{\rm v}} \left(1 - e^{-\beta_{\rm h} E_{\max}}\right)^{-1} \left(1 - e^{-\frac{n}{2}\beta_{\rm h} E_{\max}}\right) \Big]^{-1}, \tag{A.13}$$

while for n odd we have:

$$\beta_v^{(n_{\rm odd})} E_{\rm v} = \beta_{\rm c} E_{\rm v} + (\beta_{\rm c} - \beta_{\rm h}) \left[\left(\frac{n-1}{2}\right) E_{\max} - E_{\rm v} \right], \tag{A.14}$$

$$N_{\rm v}^{(n_{\rm odd})} = \left(1 + e^{-\beta_{\rm v}^{(n)} E_{\rm v}}\right) \Big[\left(1 - e^{-\beta_{\rm c} E_{\max}}\right)^{-1} \left(1 - e^{-(\frac{n+1}{2})\beta_{\rm c} E_{\max}}\right) + e^{-\beta_{\rm v}^{(n)} E_{\rm v}} \left(1 - e^{-\beta_{\rm h} E_{\max}}\right)^{-1} \left(1 - e^{-(\frac{n-1}{2})\beta_{\rm h} E_{\max}}\right) \Big]^{-1}. \tag{A.15}$$

In the other hand, for the case of the heat engine and n even

$$\beta_v^{(n_{\rm even})} E_{\rm v} = \beta_{\rm h} E_{\rm v} - (\beta_{\rm c} - \beta_{\rm h}) \left(\frac{n}{2} - 1\right) E_{\max}, \tag{A.16}$$

$$N_{\rm v}^{(n_{\rm even})} = \left(1 + e^{+\beta_{\rm v}^{(n)} E_{\rm v}}\right) \Big[\left(1 - e^{-\beta_{\rm c} E_{\max}}\right)^{-1} \left(1 - e^{-\frac{n}{2}\beta_{\rm c} E_{\max}}\right) + e^{+\beta_{\rm v}^{(n)} E_{\rm v}} \left(1 - e^{-\beta_{\rm h} E_{\max}}\right)^{-1} \left(1 - e^{-\frac{n}{2}\beta_{\rm h} E_{\max}}\right) \Big]^{-1}, \tag{A.17}$$

and finally for a heat engine with n odd:

$$\beta_v^{(n_{\rm odd})} E_{\rm v} = \beta_{\rm h} E_{\rm v} - (\beta_{\rm c} - \beta_{\rm h}) \left[\left(\frac{n-1}{2}\right) E_{\max} - E_{\rm v} \right], \tag{A.18}$$

$$N_{\rm v}^{(n_{\rm odd})} = \left(1 + e^{+\beta_{\rm v}^{(n)} E_{\rm v}}\right) \Big[\left(1 - e^{-\beta_{\rm c} E_{\max}}\right)^{-1} \left(1 - e^{-(\frac{n-1}{2})\beta_{\rm c} E_{\max}}\right) + e^{+\beta_{\rm v}^{(n)} E_{\rm v}} \left(1 - e^{-\beta_{\rm h} E_{\max}}\right)^{-1} \left(1 - e^{-(\frac{n+1}{2})\beta_{\rm h} E_{\max}}\right) \Big]^{-1}. \tag{A.19}$$

In the remainder of the section, we demonstrate some useful properties of the optimal n-level cycle. In particular are interested in proving that it achieves the largest change in the bias of an external qubit under the swap operation [see Eq. (A.5)]. We first recall the technical definition of the optimal cycle above as the unique cycle that minimizes the ratios of every single population p_j to the ground state of the virtual qubit p_1 (fridge). In particular, this includes the population ratio of the virtual qubit itself, that is p_n/p_1, which implies that the optimal cycle *maximizes the bias* Z_v. In addition, using $\sum_j p_j = 1$, one can express the norm of the virtual qubit in the useful form

$$N_v = \left(\frac{1 + e^{-\beta_v E_v}}{1 + \sum_{j=2}^{n} p_j / p_n} \right). \tag{A.20}$$

Since the optimal cycle is the unique cycle that minimizes the denominator above, the optimal cycle also *achieves the highest norm* N_v *given the maximum bias* Z_v. Expressing the population of the ground state of the virtual qubit as

$$p_1 = \frac{1}{1 + \sum_{j=2}^{n} p_j / p_n}, \tag{A.21}$$

it is clear that the optimal cycle also *maximizes the population* p_1, which is equivalently the *maximal value of* $N_v(1 + Z_v)$. Since the optimal cycle both maximizes p_1 and minimizes p_n/p_1, we may conclude that it *maximizes the difference between the populations*:

$$p_1 - p_n = N_v Z_v = p_1 \left(1 - \frac{p_n}{p_1} \right). \tag{A.22}$$

Equivalently, in the case of the heat engine, the optimal n-level cycle:

- minimizes Z_v,
- maximizes N_v given the minimum Z_v,
- maximizes $p_n = N_v(1 - Z_v)/2$,
- maximizes $p_n - p_1 = -N_v Z_v$.

We may now prove that the optimal cycle achieves the largest change in the bias of an external qubit via the swap operation. Labeling the norm and bias of the optimal n-level fridge as $\{N_v^+, Z_v^+\}$, and that of an arbitrary n-level cycle as $\{N_v, Z_v\}$, we have

$$Z_v \leqslant Z_v^+, \qquad N_v Z_v \leqslant N_v^+ Z_v^+. \tag{A.23}$$

Thus for the swap using an arbitrary cycle,

$$\begin{aligned} \Delta Z_{\mathbb{S}} &< \frac{N_{\mathrm{v}}^{+} Z_{\mathrm{v}}^{+}}{Z_{\mathrm{v}}} \left(Z_{\mathrm{v}} - Z_{\mathbb{S}}\right) = N_{\mathrm{v}}^{+} Z_{\mathrm{v}}^{+} \left(1 - \frac{Z_{\mathbb{S}}}{Z_{\mathrm{v}}}\right), \\ &< N_{\mathrm{v}}^{+} Z_{\mathrm{v}}^{+} \left(1 - \frac{Z_{\mathbb{S}}}{Z_{\mathrm{v}}^{+}}\right) = N_{\mathrm{v}}^{+} \left(Z_{\mathrm{v}}^{+} - Z_{\mathbb{S}}\right), \end{aligned} \tag{A.24}$$

and the change in the bias is upper bounded by the one achieved by the optimal fridge cycle. One may also prove the analogous result involving the optimal engine cycle, that is

$$Z_{\mathbb{S}} - Z_{\mathbb{S}}' = N_{\mathrm{v}}(Z_{\mathbb{S}} - Z_{\mathrm{v}}) < N_{\mathrm{v}}^{-} \left(Z_{\mathbb{S}} - Z_{\mathrm{v}}^{-}\right), \tag{A.25}$$

where $\{N_{\mathrm{v}}^{-}, Z_{\mathrm{v}}^{-}\}$ are the norm and bias of the optimal engine cycle.

Finally, we discuss on the efficiency of optimal n-level cycle thermal machines. The customary definition for the efficiency of fridges, that is, the so-called *coefficient of performance* (COP), is defined as the ratio between the heat drawn from the object to be cooled to the heat drawn from the hot bath, while for the case of a heat engine, it is defined as the ratio between the work extracted and the heat drawn from the hot bath. We may here apply those definitions for the case of the n-level thermal cycle as we already done in Sect. 11.2 for the three-level thermal machine. Notice that the energy gap of the virtual qubit, E_{v}, represents both the heat drawn from the cold bath in the case of a fridge, and the work extracted in the case of an engine, when a complete cycle is achieved. Furthermore, every time the virtual qubit exchanges E_{v} with an external system, it has to be reset by moving through the entire cycle. Indeed, by applying (A.9) to the ratio of populations of the virtual qubit, one finds that the virtual temperature is determined by the heat dissipated to the cold bath, Q_{c}, and drawn from the hot bath, Q_{h}, in the course of a single cycle. Hence we can identify the terms Q_{+}^{j} and Q_{-}^{j} above with Q_{c} and Q_{h} respectively, for the case of the fridge, and the opposite for the heat engine. One can thus re-express the inverse virtual temperature of the thermal cycle in terms of the heat currents:

$$\text{(fridge)} \quad \beta_{v}^{(n)} E_{\mathrm{v}} = \beta_{\mathrm{c}} \left(Q_{\mathrm{h}} + E_{\mathrm{v}}\right) - \beta_{\mathrm{h}} Q_{\mathrm{h}}, \tag{A.26}$$

$$\text{(engine)} \quad \beta_{v}^{(n)} E_{\mathrm{v}} = \beta_{\mathrm{h}} Q_{\mathrm{h}} - \beta_{\mathrm{c}} \left(Q_{\mathrm{h}} - E_{\mathrm{v}}\right). \tag{A.27}$$

Finally, solving for the efficiency $\eta = E_{\mathrm{v}}/Q_{\mathrm{h}}$, one recovers the efficiencies of the n-level thermal cycle:

$$\eta_{\text{fridge}}^{(n)} = \frac{\beta_{\mathrm{c}} - \beta_{\mathrm{h}}}{\beta_{\mathrm{v}}^{(n)} - \beta_{\mathrm{c}}}, \qquad \eta_{\text{engine}}^{(n)} = \frac{\beta_{\mathrm{c}} - \beta_{\mathrm{h}}}{\beta_{\mathrm{c}} - \beta_{\mathrm{v}}^{(n)}}. \tag{A.28}$$

Notice that in both cases the efficiency falls off with increasing β_{v}. Henceforth for the case of the optimal n-level cycle, as the magnitude of $\beta_{\mathrm{v}}^{(n)}$ increases linearly with n, the efficiency η falls off inversely with n.

A.3 Amplification Methods and Switching Regimes

The amplification of the norm of a virtual qubit (see Sect. 11.5) presents itself as a novel method to amplify the norm $N_{\text{v}}^{(n)}$ of any n-level cycle to one; simply connect its virtual qubit $\mathcal{T}_{1,n}$ to an additional real qubit via a suitable interaction Hamiltonian, and use the latter to interact with the external system.

To be more precise, consider that one has a single n-level cycle thermal machine, whose virtual qubit, labeled by the states $|1\rangle_M$ and $|n\rangle_M$, has an energy gap E_{v} and a inverse virtual temperature $\beta_{\text{v}}^{(n)}$. Then couple this transition to an additional (real) qubit (labeled by $|g\rangle_{\text{v}}$ and $|e\rangle_{\text{v}}$) with the same energy gap E_{v} via a swap-like Hamiltonian

$$\hat{H}_{\text{int}} = g\left(|1\rangle\langle n|_M \otimes |e\rangle\langle g|_{\text{v}} + \text{h.c.}\right). \tag{A.29}$$

This arrangement is depicted in Fig. A.12a. Letting the global system equilibrate in absence of the external object, the populations of the levels must satisfy

$$p(|1\rangle_M \otimes |e\rangle_{\text{v}}) = p(|n\rangle_M \otimes |g\rangle_{\text{v}}). \tag{A.30}$$

But since $p_n/p_1 = e^{-\beta_{\text{v}}^{(n)} E_{\text{v}}}$ for the n-level cycle machine, it follows that the real qubit levels exhibit the same population ratio than the virtual qubit of the machine, that is

$$\frac{p_{e_{\text{v}}}}{p_{g_{\text{v}}}} = e^{-\beta_{\text{v}} E_{\text{v}}}. \tag{A.31}$$

Henceforth taking the additional (real) qubit as the new virtual qubit, we completed the amplification procedure, since now $N_{\text{v}} = 1$.

One can do even more if the states $|1\rangle_M \otimes |e\rangle_{\text{v}}$ and $|n\rangle_M \otimes |g\rangle_{\text{v}}$ are coupled via a thermal bath rather than an energy conserving interaction. In this case the two states need not be degenerate. If the energy gap of the additional qubit is labeled as E'_{v}, and the two states above are coupled to some thermal bath at inverse temperature β_{bath}, as in Fig. A.12b, then the populations now satisfy

$$\frac{p_1 p_{e_{\text{v}}}}{p_n p_{g_{\text{v}}}} = e^{-\beta_{\text{bath}}(E'_{\text{v}} - E_{\text{v}})}. \tag{A.32}$$

Once again the virtual temperature of the n-level cycle obeys $p_n/p_1 = e^{-\beta_{\text{v}} E_{\text{v}}}$, and the virtual temperature β'_{v} of the additional qubit may be determined by

$$\beta_{\text{v}}^{(n)\prime} = \beta_{\text{v}}^{(n)} \frac{E_{\text{v}}}{E'_{\text{v}}} + \beta_{\text{bath}}\left(1 - \frac{E_{\text{v}}}{E'_{\text{v}}}\right), \tag{A.33}$$

which can be made greater than $\beta_{\text{v}}^{(n)}$ by choosing $E'_{\text{v}} \leqslant E_{\text{v}}$ for any inverse temperature of the thermal bath such that $0 \leqslant \beta_{\text{bath}} < \beta_{\text{v}}^{(n)}$.

Finally, consider the case in which rather than coupling the states $|1\rangle_M \otimes |e\rangle_{\text{v}}$ and $|n\rangle_M \otimes |g\rangle_{\text{v}}$, one couples instead $|1\rangle_M \otimes |g\rangle_{\text{v}}$ and $|n\rangle_M \otimes |e\rangle_{\text{v}}$ to a thermal bath,

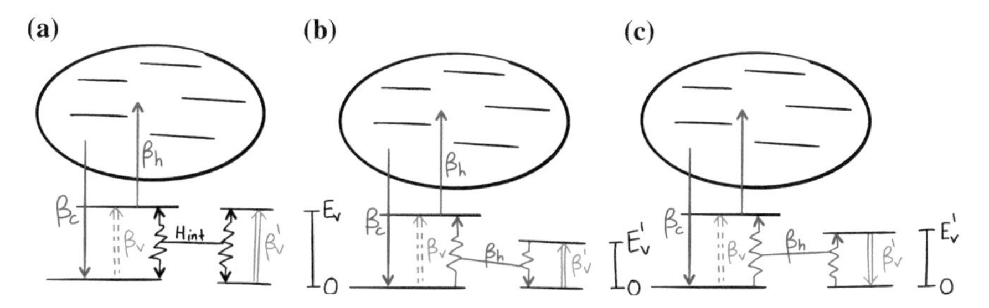

Fig. A.12 Different methods for the amplification of the virtual qubit in an arbitrary cycle. **a** Amplification that maintains the energy and bias of the virtual qubit. **b** Amplification that modifies (possibly amplifies) the bias of the virtual qubit. **c** Amplification that flips the bias of the virtual qubit

see Fig. A.12c. Similarly to the above, one may then determine the additional qubit virtual temperature as

$$\beta_{\text{v}}' = -\beta_{\text{v}} \frac{E_{\text{v}}}{E_{\text{v}}'} + \beta_{\text{bath}} (\frac{E_{\text{v}}}{E_{\text{v}}'} + 1). \tag{A.34}$$

However, in this case the contribution of the original virtual temperature is multiplied by -1. This effectively switches the machine from a fridge to an engine or vice versa! Therefore given a n-level fridge cycle, one may switch to a heat engine and vice-versa, by using the appropriate thermal coupling between the cycle and the additional qubit.

A.4 Concatenated Three-Level Machines Details

In this section we consider the concatenation of three-level machines (see Fig. A.13), and determine the bias Z_{v} and norm N_{v} of the virtual qubit in its steady state (in absence of the external object to be cooled). It is simpler to begin from the end of the concatenation, and derive the state of the machine inductively. Consider the final (rightmost) three-level system in Fig. A.13, ignoring it's interaction with the penultimate three-level system. It is just a single three-level fridge, and it's populations are completely determined by the two thermal couplings to the hot and the cold baths. One now introduces a swap-like interaction between the uncoupled transition of the final three-level system and the corresponding transition of the penultimate one, that is

$$\hat{H}_{\text{int}} = g(|1\rangle\langle 2|_n \otimes |2\rangle\langle 1|_{n-1} + \text{h.c.}). \tag{A.35}$$

This interaction induces the transition of the penultimate qutrit $\mathcal{T}_{1,2}^{(n-1)}$ to have the same population ratio as that of $\mathcal{T}_{1,2}^{(n)}$.

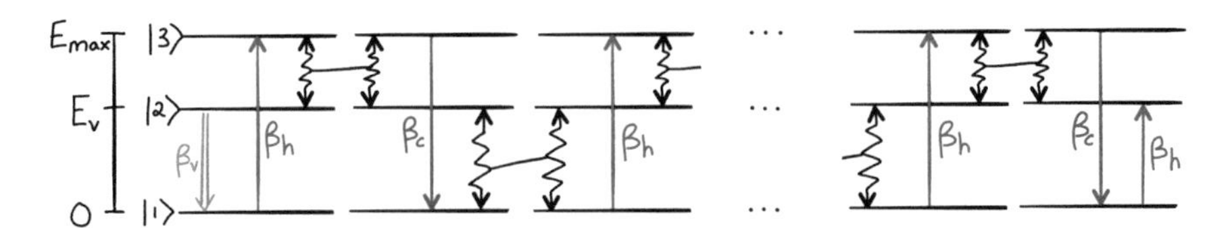

Fig. A.13 Engine formed out of the concatenation of many three-level machines

If one also couples $\mathcal{T}_{1,3}^{(n-1)}$ to the hot bath (at β_{h}), a second population ratio on the penultimate three-level system becomes fixed, which implies that the populations of its three levels are completely determined. The state is still diagonal and a product state of the two three-level systems. Note that the same state of the penultimate three-level system would have been found if one had simply assumed that in place of the final three-level system, there was instead a thermal bath at the virtual temperature of $\mathcal{T}_{1,2}^{(n)}$.

One may repeat this process inductively to determine the state of the first three-level system in the sequence, and in turn the virtual temperature of the transition $\mathcal{T}_{0,1}^{(1)}$. The result is [see Eq. (11.29)]

$$\beta_{\mathrm{v}}^{(k)} = \begin{cases} \beta_{\mathrm{c}} + (\beta_{\mathrm{c}} - \beta_{\mathrm{h}}) \frac{k}{2} \frac{E_{\mathrm{max}}}{E_{\mathrm{v}}} & \text{if } k \text{ is even,} \\ \beta_{\mathrm{c}} + (\beta_{\mathrm{c}} - \beta_{\mathrm{h}}) \left(\frac{k+1}{2} \frac{E_{\mathrm{max}}}{E_{\mathrm{v}}} - 1 \right) & \text{if } k \text{ is odd.} \end{cases} \tag{A.36}$$

We stress that the inverse virtual temperatures for the engine are the same as above with β_{c} and β_{h} switched. Note that the (inverse) virtual temperature of k concatenated three-level systems is identical to that of the optimal $k+2$-levels cycle thermal machine, as given in Eq. (A.12)

We are also interested in calculating the norm N_{v} of the virtual qubit. An interesting freedom in the case of the single three-level machine is the choice of whether to have the virtual qubit as the transition between the lower two levels $\mathcal{T}_{1,2}$ or $\mathcal{T}_{2,3}$ (modifying the energies accordingly so that the energy gap is always E_{v}). We are especially interested in the behavior of the norm as the number of concatenated three-level systems becomes large (and β_{v} approaches $\pm\infty$.) While this choice has no bearing on the bias of the virtual qubit, it does affects its norm. In the fridge configuration, the norm of the virtual qubit $\mathcal{T}_{2,3}$ is:

$$N_{\mathrm{v}}^{(2,3)} = \frac{1 + e^{-\beta_{\mathrm{v}} E_{\mathrm{v}}}}{1 + e^{-\beta_{\mathrm{v}} E_{\mathrm{v}}} + e^{-\beta_{\mathrm{v}} E_{\mathrm{v}}} e^{+\beta_{\mathrm{c}} E_{\mathrm{max}}}}, \tag{A.37}$$

and hence $\lim_{\beta_{\mathrm{v}} \to +\infty} N_{\mathrm{v}}^{(2,3)} = 1$, while if we choose as the virtual qubit the transition $\mathcal{T}_{1,2}$, we have

$$N_{\mathrm{v}}^{(1,2)} = \frac{1 + e^{-\beta_{\mathrm{v}} E_{\mathrm{v}}}}{1 + e^{-\beta_{\mathrm{v}} E_{\mathrm{v}}} + e^{-\beta_{\mathrm{c}} E_{\max}}}, \tag{A.38}$$

and $\lim_{\beta_{\mathrm{v}} \to +\infty} N_{\mathrm{v}}^{(1,2)} = \frac{1}{1+e^{-\beta_{\mathrm{c}} E_{\max}}}$. Comparing the two above equations becomes clear that it is advantageous to place the virtual qubit on the upper two levels. This effect occurs in the opposite sense for the case of the heat engine. We find that the corresponding norms for the case of lower and upper virtual qubits are respectively

$$N_{\mathrm{v}}^{(2,3)} = \frac{1 + e^{+\beta_{\mathrm{v}} E_{\mathrm{v}}}}{1 + e^{+\beta_{\mathrm{v}} E_{\mathrm{v}}} + e^{+\beta_{\mathrm{h}} E_{\max}}}, \tag{A.39}$$

$$N_{\mathrm{v}}^{(1,2)} = \frac{1 + e^{+\beta_{\mathrm{v}} E_{\mathrm{v}}}}{1 + e^{+\beta_{\mathrm{v}} E_{\mathrm{v}}} + e^{+\beta_{\mathrm{v}} E_{\mathrm{v}}} e^{\beta_{\mathrm{h}} E_{\max}}}. \tag{A.40}$$

This motivates the choice of $\mathcal{T}_{2,3}$ as the virtual qubit for the fridge, and $\mathcal{T}_{1,2}$ as the virtual qubit for the engine we performed in Sect. 11.6. It is also worth noticing that via this choice, in the limit $n \to \infty$, both the three-level fridge and heat engine approach the same state, that is a three-level system with all of its population in the middle energy level.

References

1. J.P. Palao, R. Kosloff, Quantum thermodynamic cooling cycle. Phys. Rev. E **64**, 056130 (2001)
2. N. Linden, S. Popescu, P. Skrzypczyk, How small can thermal machines be? The smallest possible refrigerator. Phys. Rev. Lett. **105**, 130401 (2010)
3. A. Levy, R. Alicki, R. Kosloff, Quantum refrigerators and the third law of thermodynamics. Phys. Rev. E **85**, 061126 (2012)
4. L.A. Correa, J.P. Palao, G. Adesso, D. Alonso, Performance bound for quantum absorption refrigerators. Phys. Rev. E 042131 (2013)
5. L.A. Correa, J.P. Palao, D. Alonso, G. Adesso, Quantum enhanced absorption refrigerators. Sci. Rep. 3949 (2014)
6. R. Kosloff, A. Levy, Quantum heat engines and refrigerators: continuous devices. Annu. Rev. Phys. Chem. **65**, 365–393 (2014)
7. P. Skrzypczyk, N. Brunner, N. Linden, S. Popescu, The smallest refrigerators can reach maximal efficiency. J. Phys. A: Math. Theor. **44**, 492002 (2011)
8. N. Brunner, M. Huber, N. Linden, S. Popescu, R. Silva, P. Skrzypczyk, Entanglement enhances cooling in microscopic quantum refrigerators. Phys. Rev. E **89**, 032115 (2014)
9. J.B. Brask, N. Brunner, Small quantum absorption refrigerator in the transient regime: time scales, enhanced cooling, and entanglement. Phys. Rev. E **92**, 062101 (2015)
10. D. Janzing, P. Wocjan, R. Zeier, R. Geiss, T. Beth, Thermodynamic cost of reliability and low temperatures: tightening Landauer's principle and the second law. Int. J. Theor. Phys. **39**, 2717–2753 (2000)
11. R. Silva, G. Manzano, P. Skrzypczyk, N. Brunner, Performance of autonomous quantum thermal machines: Hilbert space dimension as a thermodynamical resource. Phys. Rev. E **94**, 032120 (2016)
12. L.A. Correa, Multistage quantum absorption heat pumps. Phys. Rev. E, 042128 (2014)
13. L. Masanes, J. Oppenheim, A derivation (and quantification) of the third law of thermodynamics (2016), arXiv:1412.3828
14. D. Reeb, M.M. Wolf, An improved Landauer principle with finite-size corrections. New J. Phys. **16**, 103011 (2014)

Part V
Conclusions

Chapter 12
Summary and Outlook

The three main topics in which this thesis has focused can be resumed as:

- (i) The emergence of synchronization phenomena induced by dissipation in harmonic systems and their relation with quantum correlations.
- (ii) The thermodynamical properties of fluctuations in open quantum systems undergoing general irreversible evolution.
- (iii) The characterization of different quantum resources in the performance of small thermal machines operating either in autonomous or non-autonomous ways.

which coincide respectively with parts II, III and IV of the PhD thesis. In the following we provide a summary of our contributions in the three mentioned points, together with the formulation of some open questions pointing possible paths for future research.

12.1 Quantum Synchronization Induced by Dissipation in Many-Body Systems

Large quantum correlations can be an indicator of the presence of quantum phase transitions [1–3], while there have been proposals for revealing or even measuring those correlations from other more accessible quantities acting as witnesses [4, 5]. In part II of this thesis we explored the connection between the emergence of spontaneous synchronization and the dynamical evolution of classical and quantum correlations, as measured respectively by the quantum mutual information and the quantum discord. Pursuing this objective we have demonstrated:

- (a) The possibility to have both transient or asymptotic synchronization in linear systems induced by the proper dissipation.

G. Manzano Paule, *Thermodynamics and Synchronization in Open Quantum Systems*, Springer Theses, https://doi.org/10.1007/978-3-319-93964-3_12

- (b) That the presence of synchronization implies the slow decay (transient synchronization) or even the preservation (asymptotic synchronization) of classical and quantum correlations.
- (c) The possibility of engineering asymptotic synchronization, discord and entanglement in different clusters of a dissipative complex network by tuning one or few local parameters.

The first case study has been presented in Chap. 4 for an open system of two interacting quantum harmonic oscillators with different frequencies. We compared the cases in which the dissipation is modeled by two separate thermal baths with identical properties and at the same temperature, or a common bath which couples to the center of mass of the system. The main conclusion of this first approach is that the common reservoir allows both for the emergence of transient synchronization and for the slow decay of quantum and classical correlations over time characterized by a *plateau* shape. In the long time limit the correlations eventually degrade to those present in the Gibbs thermal state. The underlying mechanism responsible of this effect is the generation of disparate decay rates for the normal modes of the system [Eqs. (4.15), (4.16) and Fig. 4.2] as a result of the symmetry properties of the system-bath coupling. This does not occur in the case of separate baths, displaying a fast decay in all correlations and no synchronization even in the presence of strong coupling between the oscillators (see Fig. 4.3). The same phenomenon has been later reported in the case of spin synchronization through a common dissipative reservoir (but not for a purely dephasing one) [6].

Exploring the role of the different frequencies and the coupling strength between the two oscillators, we observed that both phenomena degrades for high detuning and weak direct coupling strength. In the opposite limit, when the frequencies are exactly equal, one of the two normal modes completely decouples from the environmental action. In such case, a noiseless or decoherence-free subsystem [7–9] is obtained. This may lead to high values for the entanglement between the oscillators, as studied in Refs. [10–13].

In Chap. 5 we have investigated the extension to the case of three interacting harmonic oscillators in the presence of a common environment. We have identified specific symmetry conditions leading to noiseless subsystems (NS), that is, when one or two normal modes of the system effectively decouple from the environmental action. For the case of one normal mode we obtained the condition in Eq. (5.19), representing an hypersurface in the d-dimensional parameter space [$d = (N + 1)N/2 = 6$ for $N = 3$ oscillators]. The condition for two non-dissipative normal modes has been also reported in Eq. (5.21). We then analyzed two specific open-chain configurations with equal frequencies in the external oscillators, for which analytical expressions for the asymptotic entanglement (as given by the logarithmic negativity) have been obtained, and used to construct the phase diagrams of Figs. 5.2 and 5.3. The parameter manifold leading to noiseless subsystems includes several non symmetric configurations: for instance an hyperbolic relation among frequencies can be satisfied for identical couplings in an open chain. This allows both asymptotic entanglement and asymptotic synchronization even if all the oscillators natural frequencies are different, a possibility offered by a chain of three oscillators and absent in the case of two (where only transient synchronization was possible for detuned oscillators). From the analysis of different configurations and general choices of parameters, we conclude that synchronization may not be always a witness of the presence of quantum correlations: when the choice of parameters is very close to a two-mode NS, synchronization degrades while quantum correlations may still be slow decaying. However, the existence of transient synchronization always ensure that the quantum mutual information and the quantum discord will be slow decaying functions of time.

A further implication of the above results is that it is sufficient to tune a single parameter of the system (oscillators frequencies or coupling strength between the oscillators) to obtain a NS. Henceforth, one may engineer the parameters of a system in order to obtain a desired NS in order to use it in quantum information or quantum computational tasks. This option is fully explored in Chap. 6, in which the case of arbitrary complex networks of dissipative harmonic oscillators is analyzed. Extended systems can dissipate in different ways across the structure and we have considered the paradigmatic cases of independent losses, of a common bath to which all the elements in the network equally couple, and the case of a local bath acting on a single element of the network. Global or partial synchronization have been shown in a random network through local tuning in one node (synchronizer) [see Eqs. (6.13), (6.14) and Fig. 6.3, and Eqs. (6.19), (6.20) with Fig. 6.6], as well as the possibility of entangle and synchronize two nodes (not linked between them) through a network, even when starting from separable states [Eq. (6.21) and Fig. 6.7]. Our analysis can be extended to more complex settings such as the production of independently synchronized parts of the system beating at different frequencies, or in the presence of more complicated dissipation situations (for example several local baths of different strengths).

In some sense, tuning part of a network so that the rest reaches a synchronous, highly correlated state can be seen as a kind of reservoir engineering, where here the tuned part of the network play the role of an extension of the reservoir. Indeed, this perspective has been recently adopted in Ref. [14] in which probing of the spectral density, structure, and topology of harmonic networks are also considered. This is to be compared with recent proposals of dissipative engineering for quantum information, where special actions are performed to target a desired non-classical state [15–18]. In the context of quantum communications and considering recent results on quantum Internet [19, 20], our studies can offer some insight in designing a network with coherent information transport properties. Furthermore, implications of our approach can be explored in the context of efficient transport in biological systems [21, 22]. Our analysis, when restricted to the classical limit, also gives some insight about vibrations in an engineering context, providing the conditions for undamped normal modes and their effect [23–25].

12.2 Quantum Fluctuation Theorems and Entropy Production

The research presented in this thesis contributes to the extension of fluctuation theorems to the quantum regime by:

- (a) Deriving general detailed and integral fluctuation theorems for nonequilibrium systems in arbitrary initial states which evolve under the action of a broad class of CPTP maps beyond the unital case.
- (b) Identifying a suitable form for the total (von Neumann) entropy production in open quantum systems from an inclusive approach considering the interaction between a quantum system and a general reservoir.
- (c) Developing a split of this total entropy production in adiabatic and non-adiabatic contributions (fulfilling independent fluctuation theorems) which, importantly, only applies under specific symmetry conditions of the maps governing the dynamical evolution.
- (d) Illustrating the applicability of our framework to understand relevant situations in quantum thermodynamic setups, including those in which quantum effects play a prominent role.

Contribution (a) has been developed in Chap. 7, where we have shown how a general fluctuation theorem both in detailed and integral forms [Eqs. (7.17) and (7.18) respectively] can be derived for systems which evolve according to very arbitrary CPTP maps. The maps fulfilling our theorem verify a detailed balance condition in operator version [Eq. (7.15)] linking the Kraus operators of the map with the ones of the map governing a suitable dual-reverse dynamics [26]. This includes general classical stochastic dynamics and quantum operations inducing jumps between eigenstates of the invariant state of the dynamics, as well as some specific superpositions of them.

The most important feature of our theorem is that it can be applied to arbitrary situations without caring about the specific characteristics of the environment. When specialized to maps induced by thermal reservoirs, our results reproduce known quantum fluctuation theorems for work and different versions of the entropy production. However, this can be extended as well to the case of reservoirs in generalized Gibbs ensembles inducing the corresponding generalized Gibbs-preserving maps. This includes as particular cases: heat and particle reservoirs, angular momentum reservoirs, coherent thermal reservoirs, squeezed thermal reservoirs, or information reservoirs.

The key point to derive our theorem has been the introduction of the nonequilibrium potential as a fluctuating quantity [Eq. (7.11)], which overcomes the efficacy reductions previously pointed in the literature [27–30]. The resulting entropy production for single trajectories, Σ, can then be interpreted a physical entropy production in most situations of interest.

The characterization of entropy production in situations going beyond the assumption of ideal equilibrium reservoirs constitutes an open challenge [31, 32], in which different approaches using quantum trajectories [33, 34], a phase-space perspective [35] or feedback protocols [36] have been considered. In Chap. 8 we trace the induced irreversibility in an otherwise general evolution, deriving an expression for the total (von Neumann) entropy produced in the process [Eq. (8.13)]. This expression, called the *inclusive* entropy production, measures the quantum correlations lost in the measurement process plus the local measurement-induced disturbance on system and environment. When the remaining classical correlations are further inaccessible, the entropy production increases to give Eq. (8.15), which is just the sum of the entropy changes in system and environment, and has been instead labeled *non-inclusive*. With those definitions at hand we identified trajectory versions, which fulfill a universal fluctuation theorem [Eqs. 8.23 and 8.27]. We notice that this fluctuation theorem is a particular case of the one derived in Refs. [32, 37] for isolated systems when introducing the proper partition and the local measurements giving a precise meaning to the entropy production, which is indeed the important point.

Once identified the total entropy production per trajectory we have addressed point (c), that is, we develop its decomposition into *adiabatic* and *non-adiabatic* contributions, accounting for different sources of irreversibility in processes with a steady state [38–40]. This can be done for the non-inclusive version of the entropy production, which allows relating the TMP approach to the CPTP maps formalism developed in Chap. 7. The split of the entropy production requires the introduction of three different thermodynamic processes, namely, the backward (or time-reverse) process, the dual process, and the dual-reverse process, which are described by three different CPTP maps with their corresponding Kraus operators exploiting the symmetries of the setup, c.f. Eqs. (8.34), (8.42) and (8.46). Then fluctuation theorems for the adiabatic [Eq. (8.48)] and non-adiabatic [Eq. (8.43)] entropy productions follow (8.49).

The above results also extend to the case of concatenations of CPTP maps, where the maps act in sequence also with different invariant states. Taking the continuous limit leads to quantum trajectories generated by unraveling driven Lindblad master equations like Eq. (8.84). In the latter case we developed a general method to identify the environmental entropy changes during the trajectories induced by the quantum jumps, allowing us to recover the fluctuation theorems [Eqs. (8.100)–(8.102)]. The meaning of the terms adiabatic and non-adiabatic become clear in this situation, as the non-adiabatic contribution becomes zero for quasi-static drivings following the instantaneous steady state of the dynamics. Importantly, the fluctuation theorems for the adiabatic and non-adiabatic entropy productions do not need to be always fulfilled in the quantum case. Two conditions are needed: the maps must fulfill Eq. (8.41), and the backward (or time-reverse) dynamics must preserve the (inverted) invariant state of the original dynamics, Eq. (8.44). Those requirements are always fulfilled for classical Markov dynamics, but may fail in the quantum case.

In order to clarify this and other issues concerning the abstract quantities introduced in the derivations, we considered in Chap. 9 three relevant examples which constitute our contribution (d) above: an autonomous three-level thermal machine, a dissipative cavity mode resonantly driven by a classical field, and a Maxwell's demon toy model. The first example is used to clarify the meaning of the adiabatic and non-adiabatic entropy productions as well as the backward, dual, and dual-reverse maps in a simple but important setup. We indeed obtain expressions for the entropy production [Eqs. (9.28)–(9.30)], reminiscent of phenomenological thermodynamics using the local equilibrium approach at the average level [Eqs. (9.34)–(9.36)]. Our results are compatible with the average thermodynamic description usually employed to describe autonomous thermal machines operating at steady state conditions [41–43]. Moreover, we gain insight into the transient regime which may be explored e.g. in relation to recent proposals of single-shot refrigeration in autonomous fridges [44, 45].

The second example, a periodically driven cavity mode at resonance in contact with a single thermal reservoir, has been selected to illustrate the lack of entropy production split into (positive) adiabatic and non-adiabatic terms. In this case, only the fluctuation theorem for the total entropy production is valid, while the adiabatic/non-adiabatic entropy production split breaks down at the trajectory level. This has interesting consequences: the non-adiabatic entropy production rate, measuring the rate of convergence of the system to its steady state, may be boosted in comparison with the total entropy production rate, which is proportional to the input power dissipated into the thermal reservoir. As a consequence, a negative adiabatic entropy production rate can emerge in the initial transient dynamics, and the cavity mode experiences an accelerated energy gain. Moreover, this analysis predicts similar breaking of the split whenever the different dynamical contributions in the dynamics promote jumps between eigenstates of different system observables. The consequences of this quantum effect are an open question for future research.

Our third example allows to relate information and thermodynamic effects. We studied a Maxwell demon toy model consisting of a semi-infinite array of degenerated energy levels (called the external memory) monitoring the heat exchange of quanta between two thermal reservoirs at different temperatures. Our framework provides a meaningful interpretation of the thermodynamics in the setup in which the direction of the flux of heat can be controlled by means of the entropy of the initial state introduced in the memory. In this case the non-adiabatic entropy production and the total entropy production are equal and provide a particular formulation of Landauer's principle [46–48]. Furthermore we show how this setup can be generalized by replacing the thermal reservoirs by squeezed thermal ones. In such case, the flow of heat and coherence between the reservoirs induce squeezing in the memory, which in turn enhances the performance in different regimes surpassing the Landauer's limit in the thermal case. This example indicates the feasibility of our framework to deal with information thermodynamics in quantum devices in more real situations, while opening interesting questions about the use of genuine quantum resources to enhance its performance.

12.3 Quantum Thermal Machines

Different works in the literature have pointed that nonequilibrium quantum reservoirs may be used to increase both power and efficiency of quantum thermal machines, including coherent [49–52], correlated [53], or squeezed thermal reservoirs [54–57]. In this respect, equipped with the general findings for quantum fluctuation theorems and entropy production developed in the previous chapters, we were in position to address the thermodynamical consequences of nonequilibrium thermal reservoirs in work extraction. This analysis has been carried out along Chap. 10. In particular, we focused on the case of the squeezed thermal reservoir. In contrast to part III of the thesis, here we consider only the average thermodynamical behavior. We found that the total average entropy production of a bosonic mode relaxing in the presence of a squeezed thermal reservoir is due to a single non-adiabatic contribution, given by Eq. (10.8). The entropy exchange term, as characterized by the nonequilibrium thermodynamic potential, between system and reservoir just equals the entropy decrease in the reservoir, but differs from the purely thermal configuration, $\Delta\Phi = \beta Q$. Instead, it includes both energetic and coherence contributions, weighted according to the value of the modulus of the squeezing parameter [Eq. (10.9)]. Analyzing the genuine quantum entropy exchange term, Eq. (10.10), we found that it is proportional to the asymmetry in orthogonal quadratures of the bosonic mode induced by the squeezing in reservoir. Remarkably, this modifies the second-law inequality in such a way that processes reducing the entropy of the system without exchanging heat are now allowed. It also opens the possibility of work extraction from a single squeezed thermal reservoir. In this respect, we designed a simple cyclic two-stroke protocol consisting of unitary and relaxation steps which is optimized to irreversibly extract a maximum amount of work. That is, work can be extracted from a single squeezed thermal reservoir using the entropic contribution, which acts as a thermodynamical resource.

These first results have been then applied to a more elaborated model of a four-stroke quantum heat engine: a quantum Otto engine. We have analyzed in detail the cycle and optimized it to profit from the squeezing effects. This has been done by combining unitary unesqueezing of the mode with adiabatic frequency modulation in the isentropic compression stroke. This possibility has also been independently pointed in the work of Niedenzu et al. [58]. The consequences are important when comparing to previous approaches [55, 59]. We found that new regimes of operation emerge, as simultaneous refrigeration and work extraction, as well as perfect heat-to-work conversion from both reservoirs. Using the entropy production approach we are able to obtain general bounds on the energetic efficiency in all regimes, where previous expressions fail to bound the actual machine efficiency. Furthermore, the engine power is no longer constrained to low-frequency modulation and, consequently, efficiency at maximum power is just 1. An experimental realization of our findings is proposed by building on the single-ion heat engine recently realized in the laboratory [55, 60].

Finally, we proposed an interpretation of the squeezed thermal reservoir as an extra source of nonequilibirum free energy. We derived an exact expression for this input free energy rate [Eq. (10.30)] with thermal and squeezing contributions, and suggested the consideration of a thermodynamical efficiency for work extraction always bounded by one, as follows from the expression of the entropy production [Eq. (10.29)]. An interesting question raised from this fact is whether it is better to use energetic or thermodynamic efficiency to characterize the performance of the thermal machine. If one is just interested in work extraction and heat flows, the energetic efficiency seems to be the correct quantity to use, but if one wants to keep trace of all the resources invested in work extraction, the thermodynamic efficiency should be used.

Most of the above results focused on the heat engine operation of the modified Otto cycle. They may be complemented with an analysis of the refrigeration efficiency, to be compared with the regular Otto cycle fridge reported in Ref. [57], the simple power-driven refrigeration, and autonomous fridges. In addition, a comparison with other kind of cycles such as Carnot-like cycles are appealing. In order to do that, one must first address the implementation of reversible transformations for the bosonic mode coupled to the squeezed thermal reservoir, a question we left for future research. Finally, it should be also desirable a comparison of the performance enhancements induced by the squeezed thermal reservoir and other nonequilibrium reservoirs, both classical and quantum, as for instance the coherent thermal reservoir.

In the last Chap. 11, we addressed the question of whether the performance of small autonomous thermal machines coupled to two thermal reservoirs at different temperatures can be enhanced by increasing the number of energy levels, or in other words, the dependence of the performance of the machine on its Hilbert space dimension. We developed a systematic way to evaluate the performance of this kind of machines by focusing on its static properties once a small set of physically motivated assumptions are considered. Several designs of thermal machines can be compared on this basis. The key property we detected is the range of available virtual qubits in the machine as characterized by its bias and normalization [Eq. (11.1)]. These properties delimit e.g. the minimum temperature achievable for fridges.

For the considered designs, we obtained that the performance of the machine can be always increased by using additional levels, that is, the Hilbert space dimension of the machine constitutes a thermodynamical resource. Among different designs, we introduced quantum thermal machines consisting in a single cycle reaching optimal bias (virtual temperatures) for a virtual qubit consisting in a single transition of the machine. As a second step we then showed that those machines can be outperformed by machines with multiple cycles running in parallel, which improve the normalization of the virtual qubit. Finally, as an alternative model we considered the concatenation of the simplest three-level machines. In the latter case, unless the performance turns out to be suboptimal, they are more suitable to practical implementations.

Another important finding is that in all the considered models, the number of levels diverge when trying to obtain a perfect bias. In the case of fridges, this is the same as saying that a perfect zero temperature is achievable only by using a machine with infinite dimension. We notice that this can be seen as a statement of the third law of quantum thermodynamics, alternative to previous ones using the number of steps in the refrigeration process, the time needed to reach absolute zero, or the energy invested in the cooling process (see e.g. Refs. [46, 61, 62]). This statement can be made precise in the case of multi-cycle machines in which we obtained that the minimum achievable temperature scales as $T \sim N^{-1}$, where N is the Hilbert space dimension of the machine.

Finally, we explored the interplay between optimizing the performance of the multi-level machines at the static level and dynamical features. As improved machines will have an increasing number of levels, they will also need an increasing time to be reset (thermalized with the baths) in order to perform continuous operation. We considered a particular modeling to give us some qualitative understanding of this tradeoff, in which the machine operates over an external two-level system coupled to its own thermal reservoir. Results concerning the scaling of the length of the cycle in terms of the relaxation time-scale of the external system have been obtained for the single-cycle machines, showing that for fast relaxation time-scales the optimal length of the cycle converges to the case of fourth levels. It would be also interesting to consider the case of multiple single-cycle machines operating on the external system in parallel, as they may potentially achieve a faster reseting of the machine.

Here we have reported the optimal design of a machine consisting in a single cycle. However, an important question left open is the design of the optimal multi-level thermal machine, that is, the one achieving better bias and normalization in the virtual qubit by using the minimum possible number of levels. Our results indicate that this machine would lie in between the single-cycle and multi-cycle cases.

References

1. A. Osterloh, L. Amico, G. Falci, R. Fazio, Scaling of entanglement close to a quantum phase transition. Nature **416**, 608–610 (2002)
2. L.-A. Wu, M.S. Sarandy, D.A. Lidar, Quantum phase transitions and bipartite entanglement. Phys. Rev. Lett. **93**, 250404 (2004)
3. T. Werlang, C. Trippe, G.A.P. Ribeiro, G. Rigolin, Quantum correlations in spin chains at finite temperatures and quantum phase transitions. Phys. Rev. Lett. **105**, 095702 (2010)
4. G. Tóth, Entanglement witnesses in spin models. Phys. Rev. A **71**, 010301(R) (2005)
5. M. Wieśniak, V. Vedral, V. Brukner, Heat capacity as an indicator of entanglement. Phys. Rev. B **78**, 064108 (2008)
6. G.L. Giorgi, F. Plastina, G. Francica, R. Zambrini, Spontaneous synchronization and quantum correlation dynamics of open spin systems. Phys. Rev. A **88**, 042115 (2013)
7. D.A. Lidar, I.L. Chuang, K.B. Whaley, Decoherence-free subspaces for quantum computation. Phys. Rev. Lett. **81**, 2594 (1998)
8. E. Knill, R. Laflamme, L. Viola, Theory of quantum error correction for general noise. Phys. Rev. Lett. **84**, 2525–2528 (2000)
9. D.A. Lidar, K.B. Whaley, Decoherence-free subspaces and subsystems, in *Irreversible Quantum Dynamics*, vol. 622, ed. by F. Benatti, R. Floreanini (Springer, Berlin Heidelberg, Germany, 2003), pp. 83–120
10. J.S. Prauzner-Bechcicki, Two-mode squeezed vacuum state coupled to the common thermal reservoir. J. Phys. A: Math. Gen. **37**, L173 (2004)

11. K.-L. Liu, H.-S. Goan, Non-Markovian entanglement dynamics of quantum continuous variable systems in thermal environments. Phys. Rev. A **76**, 022312 (2007)
12. J.P. Paz, A.J. Roncaglia, Dynamics of the Entanglement between two oscillators in the same environment. Phys. Rev. Lett. **100**, 220401 (2008)
13. F. Galve, G.L. Giorgi, R. Zambrini, Entanglement dynamics of nonidentical oscillators under decohering environments. Phys. Rev. A **81**, 062117 (2010)
14. J. Nokkala, F. Galve, R. Zambrini, S. Maniscalco, J. Piilo, Complex quantum networks as structured environments: engineering and probing. Sci. Rep. **6**, 26861 (2016)
15. S. Diehl, A. Micheli, A. Kantian, B. Kraus, H.P. Büchler, P. Zoller, Quantum states and phases in driven open quantum systems with cold atoms. Nat. Phys. **4**, 878–883 (2008)
16. J.T. Barreiro, P. Schindler, O. Gühne, T. Monz, M. Chwalla, C.F. Roos, M. Hennrich, R. Blatt, Experimental multiparticle entanglement dynamics induced by decoherence. Nat. Phys. **6**, 943–946 (2010)
17. F. Verstraete, M.M. Wolf, J.I. Cirac, Quantum computation and quantum-state engineering driven by dissipation. Nat. Phys. **5**, 633–636 (2009)
18. J.T. Barreiro, M. Müller, P. Schindler, D. Nigg, T. Monz, M. Chwalla, M. Hennrich, C.F. Roos, P. Zoller, R. Blatt, An open-system quantum simulator with trapped ions. Nature **470**, 486–491 (2011)
19. H.J. Kimble, The quantum internet. Nature **453**, 1023–1030 (2008)
20. S. Ritter, C. Nölleke, C. Hahn, A. Reiserer, A. Neuzner, M. Uphoff, M. Mücke, E. Figueroa, J. Bochmann, G. Rempe, An elementary quantum network of single atoms in optical cavities. Nature **484**, 195–200 (2012)
21. G. Panitchayangkoon, D.V. Voronine, D. Abramavicius, J.R. Caram, N.H.C. Lewis, S. Mukamel, G.S. Engel, Direct evidence of quantum transport in photosynthetic light-harvesting complexes. Proc. Natl. Acad. Sci. **108**, 20908–20912 (2011)
22. G.S. Engel, T.R. Calhoun, E.L. Read, T.-K. Ahn, T. Manĉal, Y.-C. Cheng, R.E. Blankenship, G.R. Fleming, Evidence for wavelike energy transfer through quantum coherence in photosynthetic systems. Nature **446**, 782–786 (2007)
23. J.W.S. Rayleigh, *The Theory of Sound* (Dover Publishers, USA, 1945)
24. S. Adhikari, Damping Models for Structural Vibration, Ph.D. thesis University of Cambridge, UK, 2000
25. W.J. Bottega, *Engineering Vibrations* (CRC Taylor and Francis, USA, 2006)
26. G.E. Crooks, Quantum operation time reversal. Phys. Rev. A **77**, 034101 (2008)
27. D. Kafri, S. Deffner, Holevo's bound from a gernal quantum fluctuation theorem. Phys. Rev. A **86**, 044302 (2012)
28. T. Albash, D.A. Lidar, M. Marvian, P. Zanardi, Fluctuation theorems for quantum processes. Phys. Rev. E **88**, 032146 (2013)
29. A.E. Rastegin, K. Życzkowski, Jarzynski equality for quantum stochastic maps. Phys. Rev. E **89**, 012127 (2014)
30. J. Goold, M. Paternostro, K. Modi, Nonequilibrium quantum landauer principle. Phys. Rev. Lett. **114**, 060602 (2015)
31. M. Esposito, K. Lindenberg, C. Van den Broeck, Entropy production as correlation between system and reservoir. New J. Phys. **12**, 013013 (2010)
32. T. Sagawa, Second law-like inequalitites with quantum relative entropy: An introduction, in *Lectures on quantum computing, thermodynamics and statistical physics*, vol. 8, ed. by M. Nakahara (Kinki University Series on Quantum Computing (World Scientific), USA, 2013)
33. J.M. Horowitz, J.M.R. Parrondo, Entropy production along nonequilibrium quantum jump trajectories. New. J. Phys. **15**, 085028 (2013)
34. B. Leggio, A. Napoli, A. Messina, H.P. Breuer, Entropy production and information fluctuations along quantum trajectories. Phys. Rev. A **88**, 042111 (2013)
35. S. Deffner, Quantum entropy production in phase space. Europhys. Lett. **103**, 30001 (2013)
36. K. Funo, Y. Watanabe, M. Ueda, Integral quantum fluctuation theorems under measurement and feedback control. Phys. Rev. E **88**, 052121 (2013)

37. M. Esposito, U. Harbola, S. Mukamel, Nonequilibrium fluctuations, fluctuation theorems, and counting statistics in quantum systems. Rev. Mod. Phys. **81**, 1665–1702 (2009)
38. M. Esposito, C. Van den Broeck, Three detailed fluctuation theorems. Phys. Rev. Lett. **104**, 090601 (2010)
39. M. Esposito, C. Van den Broeck, Three faces of the second law. I. Master equation formulation. Phys. Rev. E **82**, 011143 (2010)
40. C. Van den Broeck, M. Esposito, Three faces of the second law. II. Fokker-Planck formulation. Phys. Rev. E **82**, 011144 (2010)
41. H.E.D. Scovil, E.O. Schulz-DuBois, Three-level masers as heat engines. Phys. Rev. Lett. **2**, 262 (1959)
42. P. Skrzypczyk, N. Brunner, N. Linden, S. Popescu, The smallest refrigerators can reach maximal efficiency. J. Phys. A: Math. Theor. **44**, 492002 (2011)
43. L.A. Correa, J.P. Palao, G. Adesso, D. Alonso, Performance bound for quantum absorption refrigerators. Phys. Rev. E, 042131 (2013)
44. M.T. Mitchison, M.P. Woods, J. Prior, M. Huber, Coherenceassisted single-shot cooling by quantum absorption refrigerators. New J. Phys. **17**, 115013 (2015)
45. J.B. Brask, N. Brunner, Small quantum absorption refrigerator in the transient regime: time scales, enhanced cooling, and entanglement. Phys. Rev. E **92**, 062101 (2015)
46. D. Reeb, M.M. Wolf, An improved Landauer principle with finite-size corrections. New J. Phys. **16**, 103011 (2014)
47. S. Deffner, C. Jarzynski, Information processing and the second law of thermodynamics: an inclusive, hamiltonian approach. Phys. Rev. X **3**, 041003 (2013)
48. D. Mandal, H.T. Quan, C. Jarzynski, Maxwell's refrigerator: an exactly solvable model. Phys. Rev. Lett. **111**, 030602 (2013)
49. M.O. Scully, M.S. Zubairy, G.S. Agarwal, H. Walther, Extracting work from a single heat bath via vanishing quantum coherence. Science **299**, 862–864 (2003)
50. H.T. Quan, P. Zhang, C.P. Sun, Quantum-classical transition of photon-Carnot engine induced by quantum decoherence. Phys. Rev. E **73**, 036122 (2006)
51. H. Li, J. Zou, W.-L. Yu, B.-M. Xu, J.-G. Li, B. Shao, Quantum coherence rather than quantum correlations reflect the effects of a reservoir on a system's work capability. Phys. Rev. E **89**, 052132 (2014)
52. A.Ü.C. Hardal, Ö.E. Müstecaphoglu, Superradiant Quantum Heat Engine. Sci. Rep. **5**, 12953 (2015)
53. R. Dillenschneider, E. Lutz, Energetics of quantum correlations. Europhys. Lett. **88**, 50003 (2009)
54. X.L. Huan, T. Wang, X.X. Yi, Effects of reservoir squeezing on quantum systems and work extraction. Phys. Rev. E **86**, 051105 (2012)
55. J. Roßnagel, O. Abah, F. Schmidt-Kaler, K. Singer, E. Lutz, Nanoscale heat engine beyond the carnot limit. Phys. Rev. Lett. **112**, 030602 (2014)
56. L.A. Correa, J.P. Palao, D. Alonso, G. Adesso, Quantumenhanced absorption refrigerators. Sci. Rep., 3949 (2014)
57. R. Long, W. Liu, Performance of quantum Otto refrigerators with squeezing. Phys. Rev. E **91**, 062137 (2015)
58. W. Niedenzu, D. Gelbwaser-Klimovsky, A.G. Kofman, G. Kurizki, On the operation of machines powered by quantum nonthermal baths. New J. Phys. **18**, 083012 (2016)
59. O. Abah, E. Lutz, Efficiency of heat engines coupled to nonequilibrium reservoirs. Europhys. Lett. **106**, 20001 (2014)
60. J. Roßnagel, S.T. Dawkins, K.N. Tolazzi, O. Abah, E. Lutz, F. Schmidt-Kaler, K. Singer, A single-atom heat engine. Science **352**, 325–329 (2016)
61. A. Levy, R. Alicki, R. Kosloff, Quantum refrigerators and the third law of thermodynamics. Phys. Rev. E **85**, 061126 (2012)
62. L. Masanes, J. Oppenheim, A derivation (and quantification) of the third law of thermodynamics (2016). arXiv:1412.3828

Printed in the United States
By Bookmasters